Advances in Computer Vision and Pattern Recognition

Titles in this series now included in the Thomson Reuters Book Citation Index!

Advances in Computer Vision and Pattern Recognition is a series of books which brings together current developments in this multi-disciplinary area. It covers both theoretical and applied aspects of computer vision, and provides texts for students and senior researchers in topics including, but not limited to:

- Deep learning for vision applications
- Computational photography
- Biological vision
- Image and video processing
- Document analysis and character recognition
- Biometrics
- Multimedia
- Virtual and augmented reality
- Vision for graphics
- Vision and language
- Robotics

Lu Fang

Plenoptic Imaging and Processing

Lu Fang
Department of Electronic Engineering
Tsinghua University
Beijing, China

ISSN 2191-6586 ISSN 2191-6594 (electronic)
Advances in Computer Vision and Pattern Recognition
ISBN 978-981-97-6914-8 ISBN 978-981-97-6915-5 (eBook)
https://doi.org/10.1007/978-981-97-6915-5

This work was supported by Lu Fang.

This Springer imprint is published by the registered company Springer Nature Singapore Pte Ltd.
The registered company address is: 152 Beach Road, #21-01/04 Gateway East, Singapore 189721, Singapore

If disposing of this product, please recycle the paper.

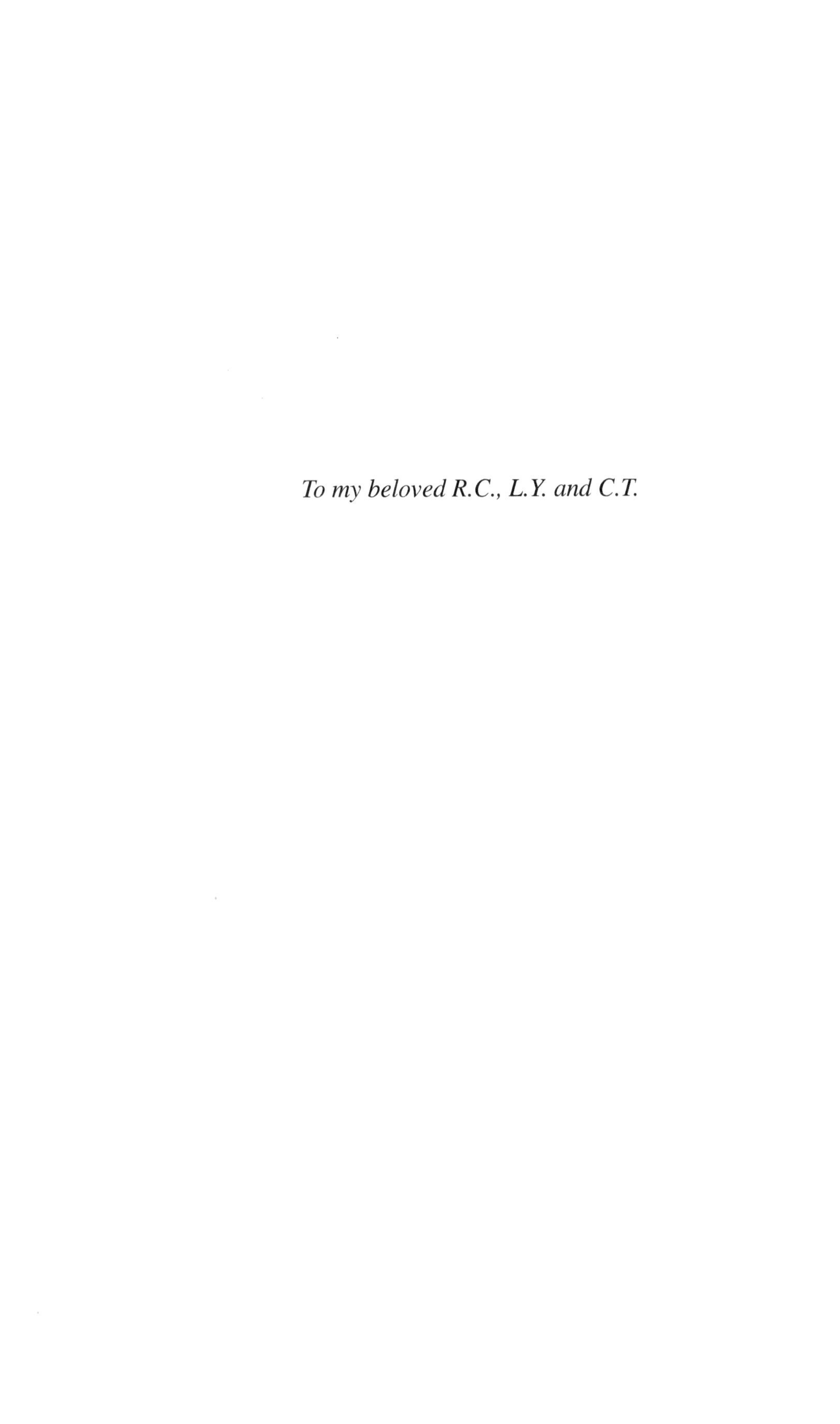

To my beloved R.C., L.Y. and C.T.

Preface

Plenoptic imaging, derived from the Latin words "plenus" (meaning "full") and "optic," introduces a revolutionary approach to imaging. Unlike conventional imaging systems that usually record the intensity of light rays merely, plenoptic imaging uses an array of microlenses or a specialized camera setup to capture additional information about the light field, including its direction, color, etc. Computational plenoptic imaging utilizes various hardware and software co-optimized techniques such as multiplexing, coding, and computational algorithms to extract valuable information from the captured raw data. By leveraging advanced algorithms and computational power, it enables new capabilities in image processing, analysis, and manipulation, including depth sensing, image enhancement, super-resolution, and even post-capture refocusing that traditional imaging systems cannot achieve. Such a technology has various applications in photography, computer vision, machine vision, unmanned systems, robot intelligence, augmented reality, virtual reality, and other fields where the ability to extract three-dimensional information from a scene is important.

In this book, we start by introducing the fundamental principle of the plenoptic function and tracing the historical development of plenoptic imaging. We then delve into various representative plenoptic sensing systems, including single-sensor devices equipped with lenslet arrays, coded-aperture masks, structured camera arrays, and unstructured camera arrays. After discussing different hardware components, we introduce advanced gigapixel plenoptic sensing techniques that excel in capturing large-scale dynamic scenes with exceptional resolution. Subsequently, we examine typical plenoptic reconstruction methods, such as light-field image reconstruction, image-based geometry reconstruction, and RGBD-based geometry reconstruction. Moving forward, we tackle the challenges of large-scale plenoptic reconstruction by incorporating sparse-view priors, high-resolution observations, and semantic information. Lastly, we showcase the gigapixel-level dataset PANDA, obtained through plenoptic imaging systems, and explore the intricacies of processing plenoptic images.

Beijing, China Lu Fang

Acknowledgments

Undertaking the endeavor of writing a book is a significant achievement, and I owe its completion to the invaluable contributions and support of numerous individuals. I would like to express my deepest gratitude to my collaborators, colleagues, and students, particularly Haozhe Lin, Xiaoyun Yuan, Jun Zhang, Guangyu Wang, Jinzhi Zhang, and Haiyang Ying, for generously sharing their expertise, insights, and time throughout this book project.

I offer my heartfelt thanks to Tsinghua University, the Hong Kong University of Science and Technology, the University of Science and Technology of China, and the National Natural Science Foundation of China. Their steadfast support has been instrumental in the successful execution of the research projects featured in this book. My gratitude extends to the publishing team for their commitment and efforts in bringing this book to life.

Lastly, I am deeply indebted to my cherished family for their unwavering support and encouragement throughout this journey. Their boundless love and profound understanding have been a perpetual source of inspiration and strength for me, propelling me forward with renewed vigor and determination.

Contents

Chapter 1
Introduction to Plenoptic Imaging

Plenoptic is derived from the Latin words plenus ("full") + optic and was proposed by Edward Adelson in 1991. The plenoptic function is a seven-dimensional function describing the three-dimensional viewing position, two-dimensional visual angle, one-dimensional wavelength, and one-dimensional time of a light ray in space. Conventional imaging model sums all the light rays emitted from a spatial location regardless of their angle, such that the imaging systems can determine the spatial information in the scene merely. Plenoptic imaging aims to detect and reconstruct the multidimensional and multiscale information of light rays in space, demonstrating a novel approach to optical imaging.

In this chapter of this book, we briefly introduce the basic principle of the plenoptic function and the historical development of plenoptic imaging. Next, in Chap. 2, we describe representative plenoptic sensing systems, including single-sensor devices with lenslet arrays, coded-aperture masks, structured camera arrays, and unstructured camera arrays. After discussing the various hardware components, in Chap. 3, we introduce gigapixel plenoptic sensing techniques, which can be used to capture large-scale dynamic scenes with extremely high resolution. In Chap. 4, we introduce representative plenoptic reconstruction approaches, including light-field image reconstruction, image-based geometry reconstruction, and RGBD-based geometry reconstruction. In Chap. 5, we discuss large-scale plenoptic reconstruction, aiming to address this challenge by introducing sparse-view priors, high-resolution observations, semantic information, etc. Finally, in Chap. 6, we present the gigapixel-level dataset PANDA, which was captured by the high-performance cameras, and discuss the corresponding opportunities and challenges raised for large-scale visual understanding.

L. Fang, *Plenoptic Imaging and Processing*, Advances in Computer Vision and Pattern Recognition, https://doi.org/10.1007/978-981-97-6915-5_1

1.1 What Is Plenoptic Imaging?

The plenoptic function parameterizes a light ray in space using a seven-dimensional function considering the position (x, y, z), angle (θ, Φ), wavelength (λ), and time (t):

$$P(x, y, z, \theta, \Phi, \lambda, t). \tag{1.1}$$

Conventional imaging techniques based on the pinhole camera model only capture the two-dimensional position, wavelength, and time information $(x/z, y/z, \lambda, t)$. The restricted perception model cannot meet the needs of various visual applications. For example, due to the missing depth information, unmanned systems cannot recover the 3D environment based on a single image, limiting navigation and interaction applications. Humans and most animals have evolved binocular vision systems to obtain 3D perspectives of reality, and plenoptic imaging extends this concept.

Plenoptic imaging: Detecting and reconstructing the multidimensional and multiscale information of light rays in space.

In this chapter, the basic principles and development of plenoptic imaging techniques are briefly discussed.

1.2 Principle of Plenoptic Function

Before discussing the plenoptic function, we first introduce the basic pinhole camera model. Have you ever wondered why cameras require lenses? Fig. 1.1a answers this question. Without a lens or pinhole, the light rays that originate from points A, B, and C reach point A' in the sensor, which means that the pixel intensity at A' represents the average light intensity of the three light rays. As a result, the sensor receives a very blurry image. Conventional cameras are constructed by placing a lens or pinhole in the center (Fig. 1.1b). With this approach, only the light rays that originate from point A reach A', and the sensor receives a sharp image. However, as discussed above, this type of imaging model directly maps the 3D physical world to the 2D sensor plane, which results in the loss of the visual angle information (θ, Φ), and the full 3D position (x, y, z) cannot be recovered. As all the light rays that originate from A are received at A', only the direction of A relative to A' is known, while the distance is unknown.

Figure 1.1c, d describe two widely used models of the plenoptic function: the light field and the reflection field. In the light-field model shown in (c), (x, y, z) denotes the 3D position information of the sensor and (θ, Φ) denotes the direction of the light ray hitting the sensor. λ denotes the wavelength and t denotes the time. In the reflectance field model, (x, y, z) denotes the 3D position information of the source

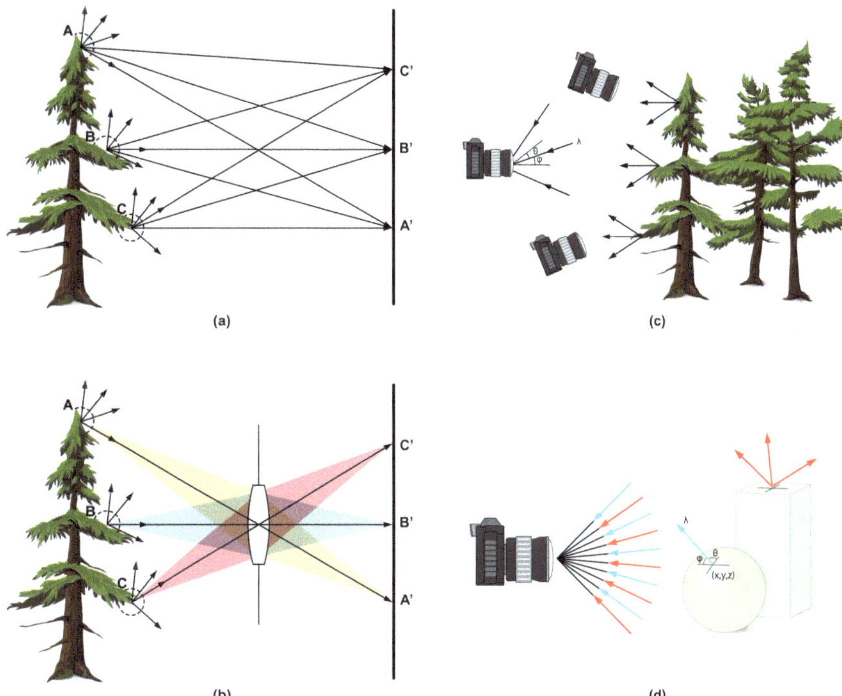

Fig. 1.1 Basic principle of plenoptic function

that reflects or emits the light ray, and (θ, Φ) denotes the direction of the light ray originating at that position.

Theoretically, the plenoptic function uses geometric rays to completely represent the light in space, which is similar to the holographic representation based on wave optics. The photographs or videos used in our daily lives can be viewed as a portion of the plenoptic function representation.[1]

1.3 Development of Plenoptic Imaging

As mentioned above, the aim of plenoptic imaging is to detect and reconstruct the multidimensional and multiscale information of light rays in space, as shown in Fig. 1.2. In the past two decades, a variety of plenoptic imaging techniques have been proposed. In general, these approaches can be divided into two categories: plenoptic sensing and plenoptic reconstruction.

[1] Only general natural scenes are considered here, and the instantaneous phase and polarization of light are ignored.

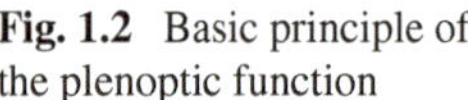

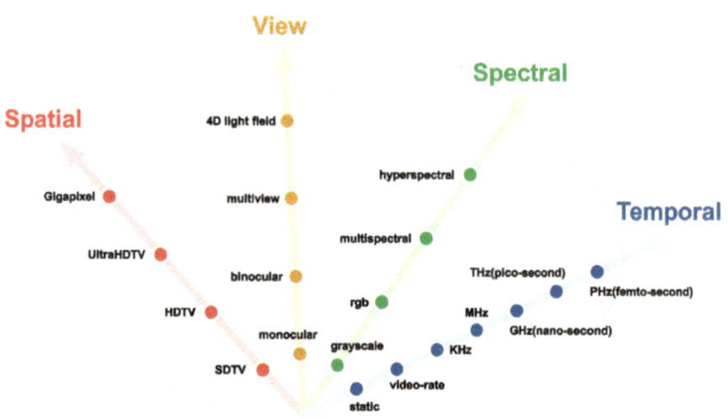

Fig. 1.2 Basic principle of the plenoptic function

Plenoptic sensing aims to capture multidimensional or large-scale light-field information using 2D sensors. In 2005, Ng et al. proposed a hand-held plenoptic camera with a microlens array that can sample both the spatial (x, y) and angular information (θ, Φ) based on its sensor in a single photographic exposure [6]. In this camera, multiple sensor pixels are required to record incoming light angles, which reduces the spatial resolution. Marwah et al. proposed replacing the microlens array with an intensity mask to compressively encode the spatial-angular information and reconstructed this information using dictionaries [4]. Camera arrays are another popular approach for plenoptic sensing. Wilburn et al. first proposed a large-scale camera array [7], which can capture light fields with high spatial and angular resolution. This camera array can also be extended to other dimensions, e.g., multispectral imaging [11].

In addition, plenoptic sensing addresses the scale restriction of conventional imaging. Plenoptic cameras perform digital refocusing by backprojecting the light rays in the plenoptic camera, leading to a large depth-of-field [6]. The multiple microcameras in the camera array can be configured for high-resolution imaging, high-frame-rate imaging, and large-aperture imaging [7]. In 2012, Brady et al. presented the AWARE2 plenoptic camera, which increased the number of pixels in the camera to the Giga level for the first time [1]. Yuan et al. proposed a multiscale unstructured camera array, enabling gigapixel imaging an arbitrary field-of-view (FoV) and nonuniform spatial resolution [8, 9].

Plenoptic reconstruction aims to reconstruct the reflectance field using the captured light field. Early reconstruction approaches utilized pure light-ray geometry information to estimate the reflectance field and adopted discrete explicit representations, such as 4D images, point clouds, and meshes. Recently, continuous neural network representations have been proposed, significantly improving the angular resolution of the reconstructed reflectance field [5]. Neural networks have also been widely used to infer the 3D structures of scenes [2, 3]. These methods utilize the powerful learning ability of neural networks to address the challenging feature matching problem, outperforming conventional pure geometry-based methods. In terms of content, plenoptic reconstruction methods have been extended from small objects to large-scale outdoor scenes, controlled to natural lighting conditions, and low-resolution to high-resolution images. Zhang et al. proposed GigaMVS, the first benchmark for ultralarge-scale gigapixel-level plenoptic reconstruction [10].

1.4 Roadmap of This Book

This book consists of three main parts, as illustrated in Fig. 1.3. The first part of this book introduces the fundamental knowledge of plenoptic sensing. In addition, we briefly review representative plenoptic sensing systems (Chap. 2) and gigapixel plenoptic sensing techniques (Chap. 3). The second part of this book briefly reviews plenoptic reconstruction techniques, including conventional small-scale reconstruction (Chap. 4) and large-scale reconstruction methods (Chap. 5). The third part of this book discusses the effect of plenoptic imaging technology on computer vision applications (Chap. 6).

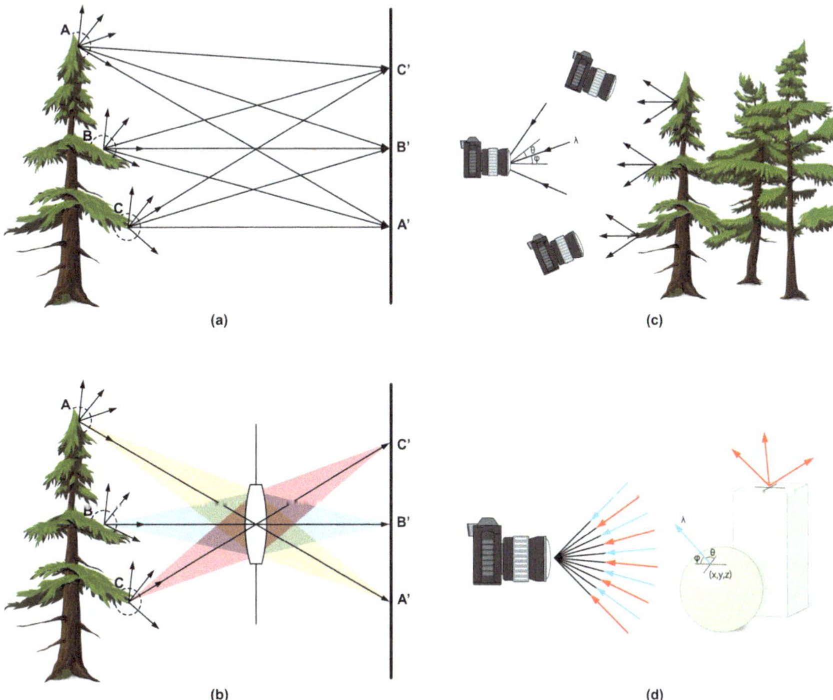

Fig. 1.3 Roadmap of this book

References

1. David J Brady, Michael E Gehm, Ronald A Stack, Daniel L Marks, David S Kittle, Dathon R Golish, EM Vera, and Steven D Feller. Multiscale gigapixel photography. *Nature*, 486(7403):386–389, 2012.
2. Mengqi Ji, Juergen Gall, Haitian Zheng, Yebin Liu, and Lu Fang. Surfacenet: An end-to-end 3d neural network for multiview stereopsis. In *Proceedings of the IEEE International Conference on Computer Vision*, pages 2307–2315, 2017.
3. Mengqi Ji, Jinzhi Zhang, Qionghai Dai, and Lu Fang. Surfacenet+: An end-to-end 3d neural network for very sparse multi-view stereopsis. *IEEE Transactions on Pattern Analysis and Machine Intelligence*, 43(11):4078–4093, 2020.
4. Kshitij Marwah, Gordon Wetzstein, Yosuke Bando, and Ramesh Raskar. Compressive light field photography using overcomplete dictionaries and optimized projections. *ACM Transactions on Graphics (TOG)*, 32(4):1–12, 2013.
5. Ben Mildenhall, Pratul P. Srinivasan, Matthew Tancik, Jonathan T. Barron, Ravi Ramamoorthi, and Ren Ng. Nerf: Representing scenes as neural radiance fields for view synthesis. In *ECCV*, pages 405–421. Springer, 2020.
6. Ren Ng, Marc Levoy, Mathieu Brédif, Gene Duval, Mark Horowitz, and Pat Hanrahan. *Light field photography with a hand-held plenoptic camera*. PhD thesis, Stanford University, 2005.
7. Bennett Wilburn, Neel Joshi, Vaibhav Vaish, Eino-Ville Talvala, Emilio Antunez, Adam Barth, Andrew Adams, Mark Horowitz, and Marc Levoy. High performance imaging using large camera arrays. In *ACM SIGGRAPH 2005 Papers*, volume 24, pages 765–776. ACM, 2005.
8. Xiaoyun Yuan, Lu Fang, Qionghai Dai, David J Brady, and Yebin Liu. Multiscale gigapixel video: A cross resolution image matching and warping approach. In *2017 IEEE International Conference on Computational Photography (ICCP)*, pages 1–9. IEEE, 2017.
9. Xiaoyun Yuan, Mengqi Ji, Jiamin Wu, David J Brady, Qionghai Dai, and Lu Fang. A modular hierarchical array camera. *Light: Science & Applications*, 10(1):1–9, 2021.
10. Jianing Zhang, Jinzhi Zhang, Shi Mao, Mengqi Ji, Guangyu Wang, Zequn Chen, Tian Zhang, Xiaoyun Yuan, Qionghai Dai, and Lu Fang. Gigamvs: A benchmark for ultra-large-scale gigapixel-level 3d reconstruction. *IEEE Transactions on Pattern Analysis & Machine Intelligence*, (01):1–1, 2021.
11. Yang Zhao, Tao Yue, Linsen Chen, Hongyuan Wang, Zhan Ma, David J Brady, and Xun Cao. Heterogeneous camera array for multispectral light field imaging. *Optics Express*, 25(13):14008–14022, 2017.

Chapter 2
Plenoptic Sensing Systems

Plenoptic sensing systems are a type of camera that can capture the multidimensional information of the light field, e.g., capture both the light rays' intensities and directions. This allows for reconstructing the high-dimensional light fields and supporting various post-capture features such as depth perception, refocusing, synthetic aperture, etc. As the widely used CMOS-based image sensor can only record the intensity of light rays, conventional camera cannot directly capture the multidimensional light field. To address this challenge, a variety of plenoptic sensing systems with specialized hardware and algorithms have been proposed.

In Sect. 2.1 of this chapter, we introduce the single-camera-based plenoptic sensing systems. These systems place a lenslet array or a coded mask in front of the image sensor to simultaneously capture the intensity and direction of the incoming light rays. The lenslet array approaches face the trade-off between spatial resolution and angular resolution. The coded aperture approaches utilize compressive sensing technology to overcome this limitation, but their imaging throughput are still restricted by the pixel count of a single sensor. Then, in Sect. 2.2, we move to the structured camera array-based plenoptic sensing systems, including large-scale camera array system, hybrid camera array system, and the portable camera array systems. These systems break the throughput limitation of the single sensor, but dramatically increase the hardware cost and require careful camera array calibration. Finally, in Sect. 2.3, we describe unstructured camera arrays that can capture and reconstruct light fields without the initial camera array calibration. Further, we also discuss the novel multiscale unstructured camera array using heterogeneous design, which can significantly reduce the hardware and computational costs.

L. Fang, *Plenoptic Imaging and Processing*, Advances in Computer Vision and Pattern Recognition, https://doi.org/10.1007/978-981-97-6915-5_2

2.1 Single Camera Sensing

As mentioned in the previous chapter, the five-dimensional model $L(x, y, z, \theta, \Phi)$ is widely used to describe the light ray as a function of the position and angle, where (x, y, z) denotes the 3D position in space, and (θ, Φ) denotes the angle of the light ray direction. As the plenoptic function of all points in a light ray can represent the whole light ray, one dimension can be eliminated [12]. Therefore, the four-dimensional plenoptic function $L(u, v, s, t)$ can also be used to describe the radiance of light, where (u, v) and (s, t) represent two points on two predefined parallel planes.

In previous works, light-field images were typically captured by an array consisting of multiple cameras, with the images captured at different viewpoints referred to as subaperture images (SAIs). The location of the camera in the array is denoted by (s, t) and is called the angular dimension. In contrast, the point (u, v) on the other plane is called the spatial dimension. These notations are used in the following sections.

2.1.1 Light-Field Camera with a Lenslet Array

Dense camera arrays can obtain high-quality light-field information with dense viewpoints for real scenes. However, synchronization and control issues limit the use of camera arrays. To obtain light-field information more conveniently and with lower costs. Ren Ng et al. [18] designed and implemented a hand-held plenoptic camera by placing a lenslet array between the main lens and the sensor in a typical camera to resolve the trade-off problem between the angle and spatial domains.

Figure 2.1 shows the structure of the plenoptic camera. The main lens, lenslet array and sensor are placed on parallel planes, the lenslet array is placed at the focal plane of the main lens, and the distance between the lenslet array and sensor is the focal length of the lenslet.

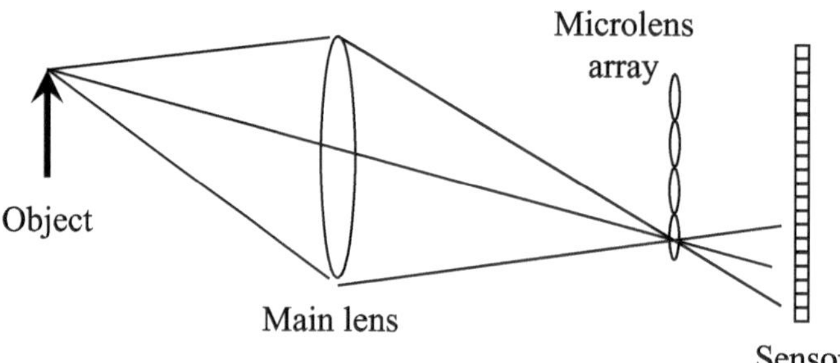

Fig. 2.1 Plenoptic camera structure designed by Ren Ng et al.

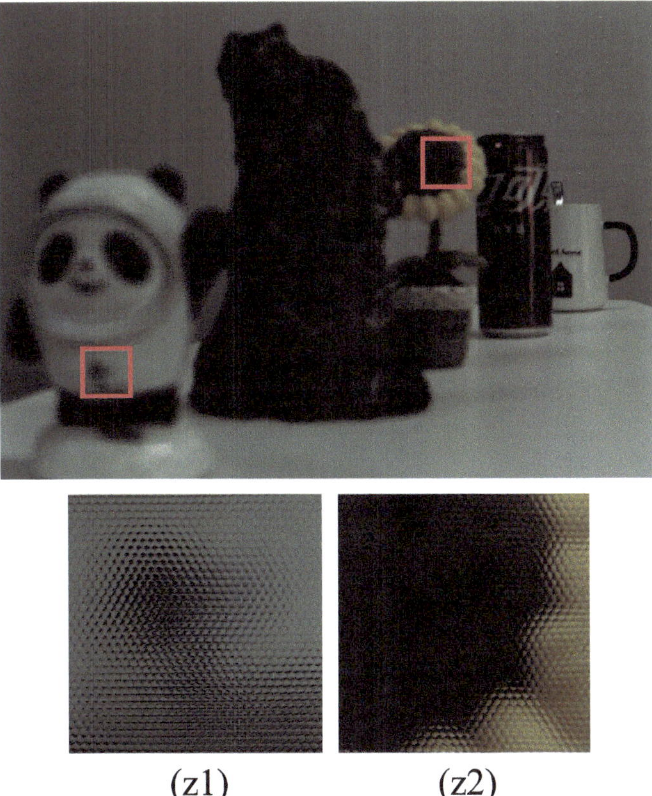

Fig. 2.2 A raw image captured by the plenoptic camera

Figure 2.2 shows a sample raw image acquired by a plenoptic camera. The raw image consists of an array of patches, and each patch represents the image acquired from the sensor area under one lenslet. Similar to the above notation, the raw image is an (N_x, N_y) grid of image patches, where (x, y) represents the corresponding lenslet location in the lenslet array. Each patch shows the light information through different (u, v) positions at the main lens. In other words, the spatial domain resolution is indicated by the number of lenslets in the array, and the angle domain resolution is determined by the ratio of the sensor resolution to the spatial resolution.

The detailed image blocks are shown at the bottom of Fig. 2.2, where image z1 shows an area of a Bing Dwen Dwen ceramic figurine which is out of focus, and image z2 shows an area of a sunflower figurine who is just in focus.

To reconstruct the SAI of a plenoptic camera, all the pixels are passed through the same (u, v) angle in each lenslet patch to produce a 2D array. As shown in Fig. 2.3, all the pixels in Fig. 2.2 are transposed into different subaperture images, and the bottom detailed images show the "real" observation result at two different viewpoints.

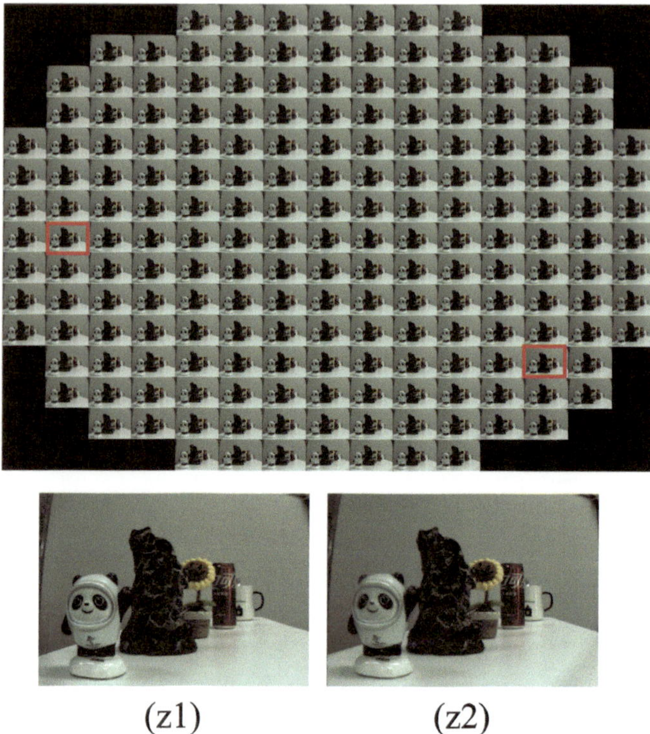

Fig. 2.3 Subaperture images of the light-field image in Fig. 2.2

In addition to capturing the 4D (x, y, u, v) light-field information in a single image and producing $N_u \times N_v$ different perspective images by transposing the pixels in the raw image, the plenoptic camera can generate new images focused on different areas in the scene. By applying a physically based simulation of a synthetic conventional camera, the refocused synthesized image can be generated by summing the corresponding cone of rays. Images A1–A5 in Fig. 2.4 show the refocused images computed based on the raw image, and image B is an extended depth-of-field image computed by combining the sharpest areas in the previous 5 images [4].

Based on a previous camera prototype, a light-field camera based on a lenslet array has been applied in industrial production. The Lytro [2] and Raytrix [1] light-field cameras, which focus on shooting first and then focusing, have considerable market potential for applications such as volumetric velocimetry, plant phenotyping, and automated optical inspection (Fig. 2.5).

Fig. 2.4 Refocused and extended depth-of-field images

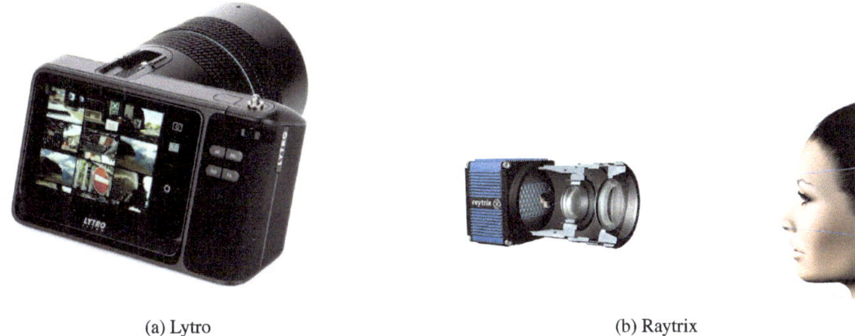

(a) Lytro (b) Raytrix

Fig. 2.5 Commercial plenoptic cameras

2.1.2 Lenslet Array Light-Field Microscope

Lenslet arrays can also be implemented in traditional microscopes to capture the 4D light field of microscopic objects. One disadvantage of traditional microscopy is that microscopes are orthographic devices, and the specimen can only be observed from a fixed vertical perspective. Another limitation is the shallow depth of field, especially when the magnification and numerical apertures are high. Moreover, it is difficult to adjust the stage position to observe specimens clearly when the specimens are living or light-sensitive.

Light-field microscopy (LFM) [19] is a scanless 3D computational imaging approach that records both the 2D spatial and 2D angular distributions of light passing through a specimen. This kind of spatio-angular data can be used to computationally synthesize focal stacks, flexibly control the depth-of-field, and achieve full volumetric reconstruction; thus, LFM has important applications in optical bioimaging [22]. The current optical schematic of LFM, which was first proposed by Lippmann in 1908 [15], is implemented by inserting a microlens array at the intermediate image plane of an optical microscope, with the sensor pixels capturing the rays of the 4D light field during a single exposure. However, microlens array-based light-field microscopy (MALM) suffers from inherent trade-offs between sensor spatial resolution and angular resolution measurements [14], which degrades the final achievable image resolution by orders of magnitude compared with the raw sensor resolution.

Figure 2.6 shows the optical layout of a typical microscope and a light-field microscope from [13]. The light rays are focused by condenser lens A onto specimen B, and objective lens C magnifies the specimen in both microscopes. Then, the ocular lens at E in the typical microscope is replaced by a sensor array in the LFM, and a lenslet array is placed in the imaging plane between C and E. Similar to the macroscopic light-field capturing system, (s, t) denotes the spatial resolution, and (u, v) denotes the angle resolution (Fig. 2.7).

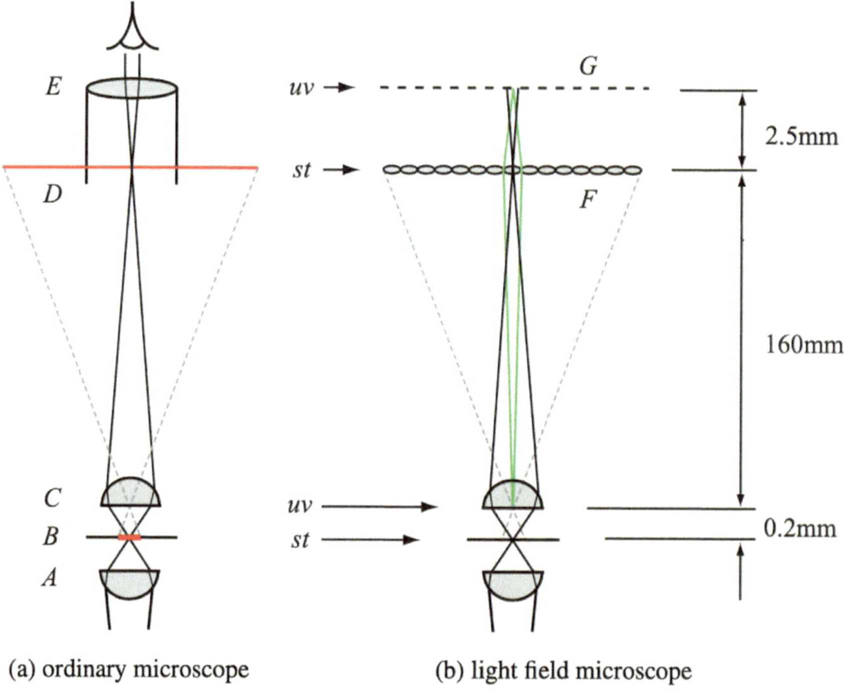

Fig. 2.6 Optical layouts of a typical microscope and a light-field microscope

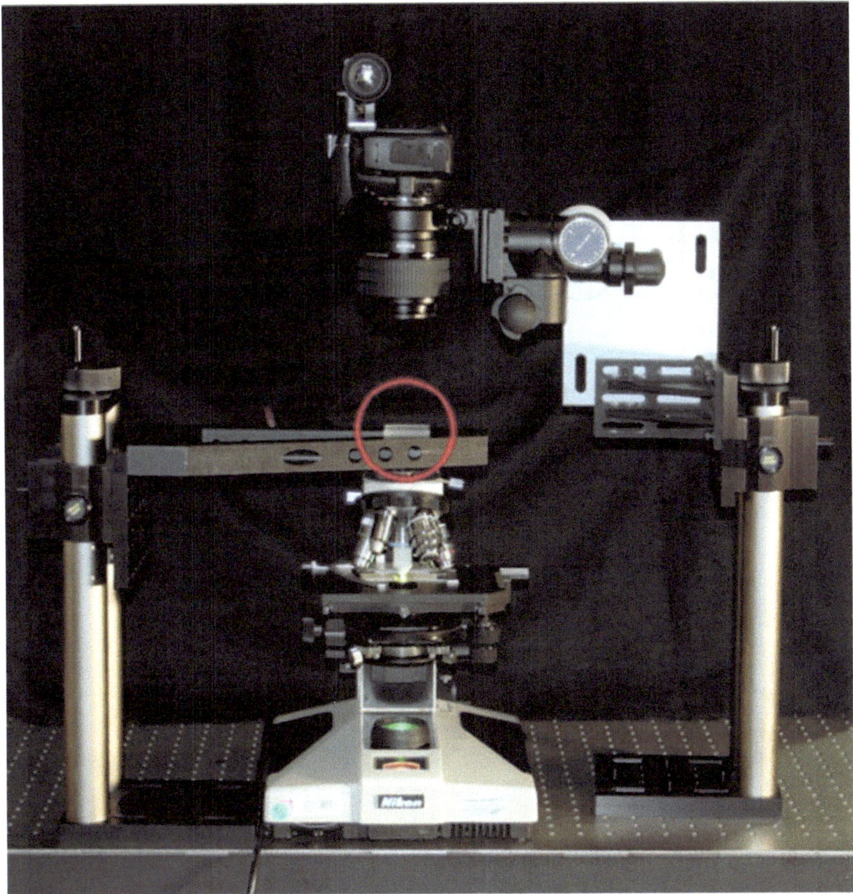

Fig. 2.7 Prototype of the microscope

2.1.3 Coded Mask Sensing

Light-field cameras based on lenslet arrays have the advantages of a compact size and low hardware costs. However, they also have several disadvantages: (1) The spatial resolution loss of the light-field viewpoint is large. As the number of viewpoints increases, the subaperture image resolution rapidly decreases. (2) The observation range in the light field is limited by the aperture of the camera. These disadvantages directly impact the increasing resolution demands of users. To address these problems, compressive sensing techniques have been used to reduce the redundant information in the light field; thus, complete light fields can be reconstructed with less data.

As shown in Fig. 2.8, a translucent coded mask is placed in front of the sensor of a traditional camera, and the light transmittance of each pixel through the mask

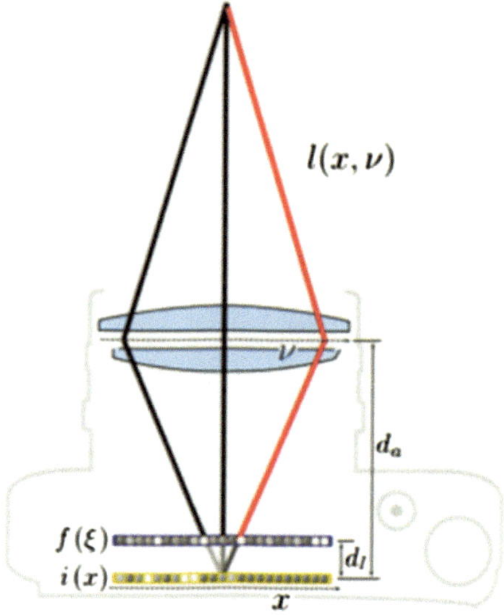

Fig. 2.8 The optical setup with a coded attenuation mask

is different (also called the mask pattern). The light entering the aperture is modulated by the mask and reaches the imaging sensor. Using the pretrained light-field dictionary, the complete light field can be reconstructed based on a single modulated image. Marwah et al. [16] proposed a compressive light-field camera architecture for high-resolution light-field imaging, as shown in Fig. 2.9. The proposed architecture captures coded 2D light-field projections and recovers a 4D light field by using robust sparse reconstruction methods. Figure 2.10 demonstrates that a well-designed coded mask pattern can lead to a higher quality light field than a random mask pattern, which is a popular choice in other compressive computational photography applications.

The drawback of the compressive sensing light-field camera is the reduction in the signal-to-noise ratio, which is mainly due to two reasons: (1) The coded masks reduce the light efficiency, and the optical signal intensity decreases because the transmittance of the mask cannot reach 100%; and (2) the reconstructed light field is not directly collected by the imaging sensor, and the final light field is distorted due to the modulation and demodulation procedure. The main advantage of compressive sensing light-field cameras is that they improve the angular resolution without reducing the spatial resolution; thus, these cameras have attracted much interest.

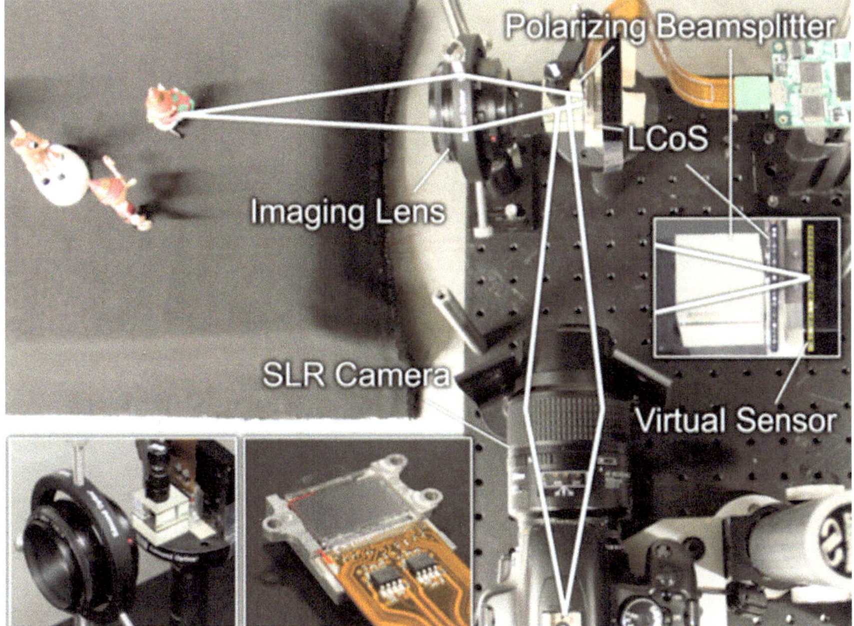

Fig. 2.9 Prototype light-field camera proposed in [16]. An optical relay system is implemented to emulate a spatial light modulator (SLM), which is mounted at a slight offset in front of the sensor (right inset). The lower left inset shows a liquid-crystal-on-silicon (LCoS) display, which aims to easily change the mask pattern

2.2 Structured Camera Array Sensing

2.2.1 Multicamera Array

As mentioned by Mehmood [17], the development of multicamera system techniques is motivated not only by the demand for high-performance and low-cost imaging systems but also by advances in multiple areas of research, such as image processing, signal processing, and machine learning. The first multicamera system was proposed by Triboulet [26] in 1884, where multiple cameras were attached to a balloon for aerial imaging. Moreover, multicamera systems have been used in other industrial applications [7]. Research on multicamera systems has rapidly advanced.

A camera array is a collection of camera sensors used to capture images or videos of a scene. A camera array system can be homogeneous, meaning that it consists of camera sensors with the same configuration, or heterogeneous, meaning that it consists of different types of camera sensors. By integrating multiple camera sensors into a single camera array, high-performance imaging results can be achieved, i.e., larger FoVs and increased spatial or angular resolution. Because of the high performance of these systems, Wilburn et al. [30], Yang et al. [31], and other researchers

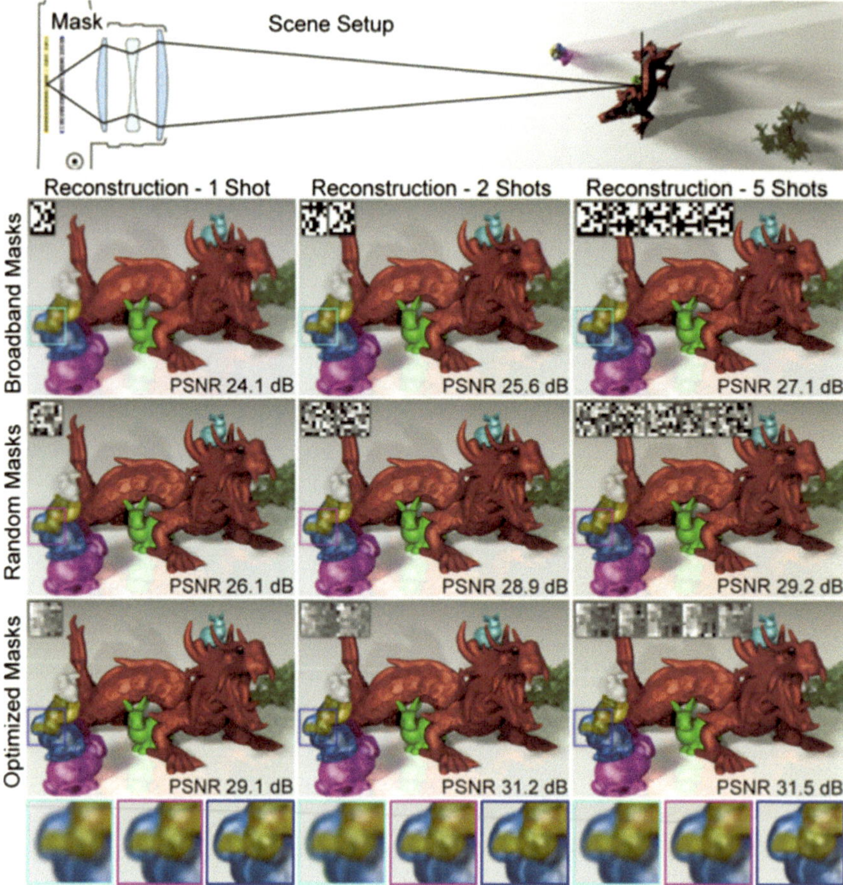

Fig. 2.10 Light-field reconstructions are evaluated based on different coded projections for one, two, and five captured camera images. For all optical codes, the reconstruction quality improves as the number of images increases. However, the optimized mask patterns proposed by Marwah et al. [16] achieve single-shot reconstruction quality that the other patterns can achieve only with multiple shots

used structured camera arrays to acquire multiview images of single scenes. In 2006, researchers at Stanford University invented the first large-scale camera array, which consisted of 96 cameras. Since the structure of the array camera is fixed, these camera arrays are called structured camera arrays. Based on reasonable configurations of the image sensor array, virtual light-field cameras can be established to achieve computational light-field imaging.

The camera arrays use different configurations depending on the application and the scenery, as shown in Fig. 2.11. When the distance between all cameras with telephoto lenses is relatively small, the camera array can be regarded as a monocular camera that can capture ultrahigh resolution, high signal-to-noise ratio, and high

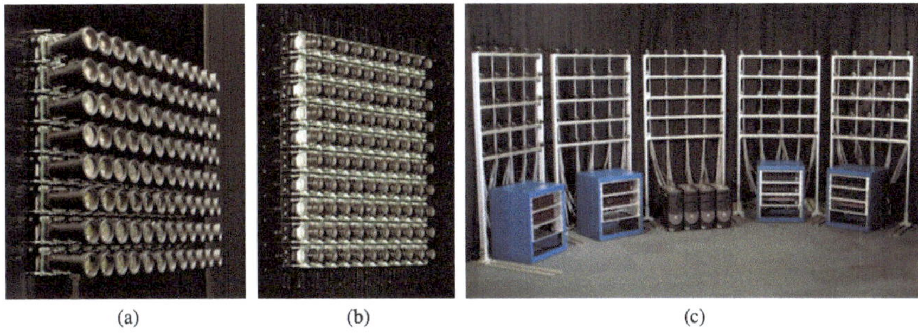

Fig. 2.11 Prototype of the camera array developed by Wilburn et al.

Fig. 2.12 Ultrahigh-resolution images captured by the system proposed by Wilburn et al.

dynamic range images, as shown in Fig. 2.12. When cameras with wide-angle lenses are tightly packed, the entire camera array can be regarded as a camera with a synthetic aperture for temporal superresolution and hybrid aperture imaging; thus, the array can capture high-speed videos and detect occluded objects through tree branches or crowds. Figure 2.13 shows the synthetic aperture image computed by combining the recorded light rays that pass through the occluders in all the cameras. When the distance between the cameras is large, the array becomes a multiview camera system that can obtain multiview information to construct panoramic images. However, the parallax between different cameras causes artifacts such as patch misalignment and discontinuities.

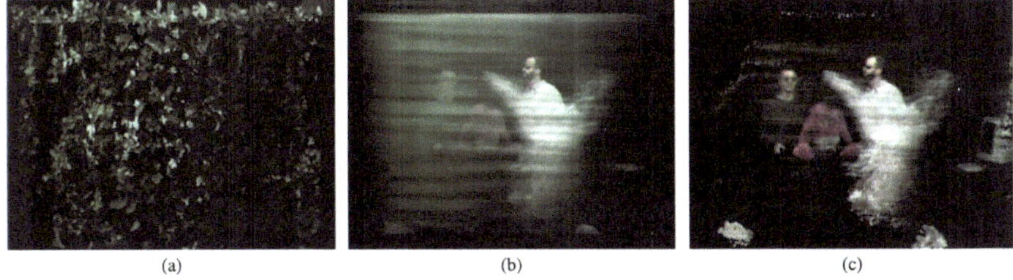

Fig. 2.13 Synthetic aperture image computed by the system proposed by Wilburn et al.

2.2.2 *Hybrid Array with a DSLR and Camera Ring*

Wang et al. proposed a hybrid structured camera array, denoted as the "light field attachment", which can transform a DSLR into a plenoptic camera [29]. As shown in Fig. 2.14a, a high-resolution DSLR camera is placed at the center (center view), and the surrounding rings are low-cost low-resolution microcameras (side view). In contrast to most existing light-field camera architectures, this design simultaneously maintains the high-resolution imaging ability while enabling a new light-field imaging mode with little added weight and cost. The high-quality and high-resolution light field can be reconstructed using a new light-field superresolution algorithm known as the iterative patch- and depth-based synthesis (iPADS) method. This approach combines patch-based and depth-based synthesis in a novel manner, as demonstrated in Fig. 2.14b.

Specifically, the iPADS method consists of five steps: (1) A patch feature dictionary is constructed based on the high-spatial-resolution image; Then, patch-based superresolution (PaSR) is performed [6] based on the low-resolution side-view images using this dictionary. During this step, some of the high-frequency texture components are lost due to the parallax caused by the scene depth changes. (2) Multiview depth estimation is performed based on the center high-resolution image and

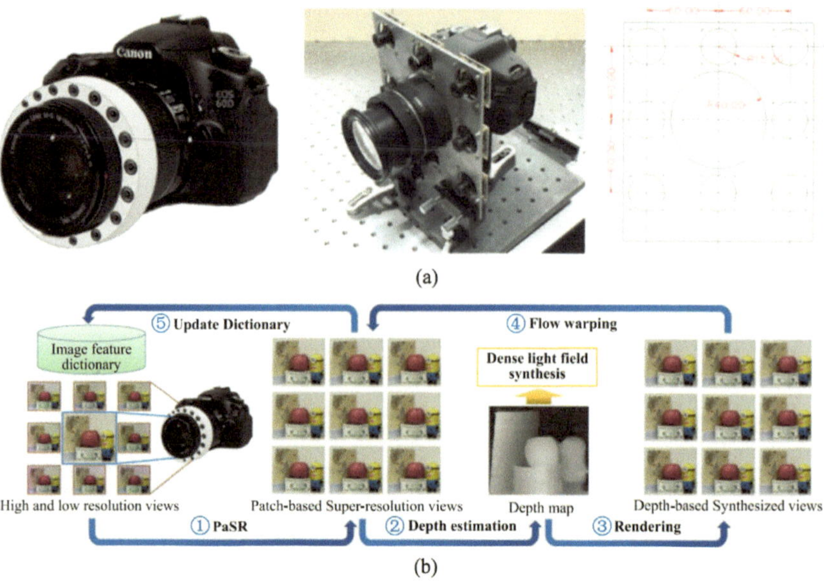

Fig. 2.14 Light-field attachment using a high-resolution DSLR camera and a light-field ring. **a** Left: Conceptual design of our light-field lens attachment. Right: Prototype and design of the camera layout. Units: millimeters. **b** Pipeline of the iterative patch- and depth-based synthesis (iPADS) method. Reprinted from Wang (2016)

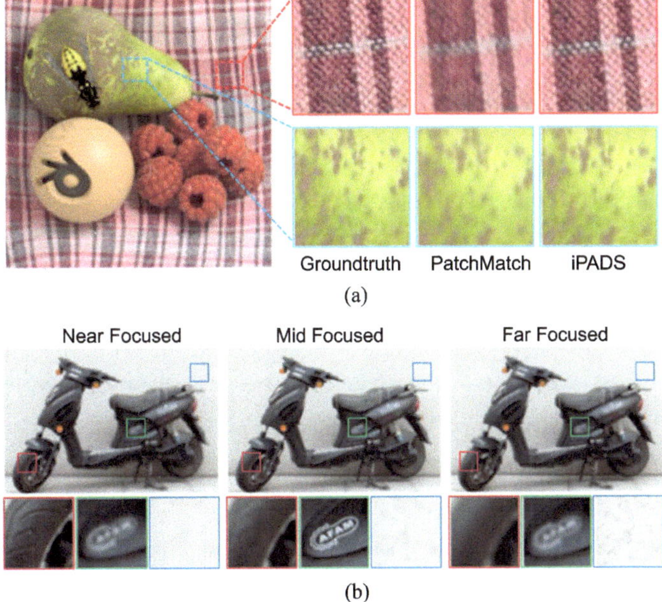

Fig. 2.15 Imaging results of the light-field attachment system. **a** The iPAD results based on a simulated light field. **b** The refocusing results of a real captured light field. Reprinted from Wang (2016)

the superresolution side-view images. (3) A depth-aided phase-based synthesis algorithm [34] is used to render new side-view images. This step restores most of the high-frequency texture information. (4) An optical flow-based warping algorithm is applied to rectify the synthesis errors caused by the depth estimation errors. (5) The warped views are added to the dictionary to act as candidates for the next iteration of the superresolution process.

Figure 2.15a demonstrates the results based on simulation data. The iPADS method generates better high-frequency details than the original PatchMatch method. Figure 2.15b shows the results of refocusing a real captured light field. As shown by the zoomed-in patches, the light field is focused at three different depths.

2.2.3 Hybrid Array with DSLR and Lenslet Cameras

As demonstrated in Fig. 2.16a, Wang et al. developed a dual-camera system [27] consisting of a DSLR camera for high-frame-rate monocular video capturing (30 frames per second (fps)) and a light-field camera (Lytro ILLUM) for slow-frame-rate light-field video capturing (3 fps). The proposed system can generate full light-field videos at 30 fps by combining these two data sources. Specifically, Wang et al. proposed a

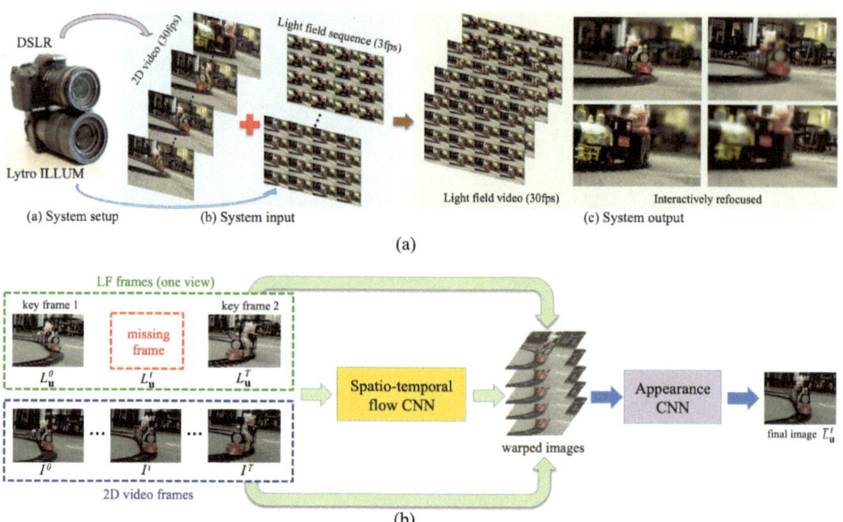

Fig. 2.16 The hybrid structured camera array system. **a** The camera array is composed of a DSLR camera and a Lytro ILLUM light-field camera. **b** The learning-based approach for generating light-field videos at 30 fps. Reprinted from Wang (2017)

learning-based approach consisting of two convolutional neural networks (CNNs): a spatiotemporal flow estimation-based network and an appearance estimation-based network (Fig. 2.16b). The first CNN warps the input video frames and light-field images based on the target angular view. The second CNN combines the warped views to generate the final image.

Figure 2.17 presents the comparison results with two optical flow approaches: EpicFlow [23] and FlowNet [10]. The left scene shows a toy train moving toward the camera. EpicFlow and FlowNet generate ghosting around the train head, especially

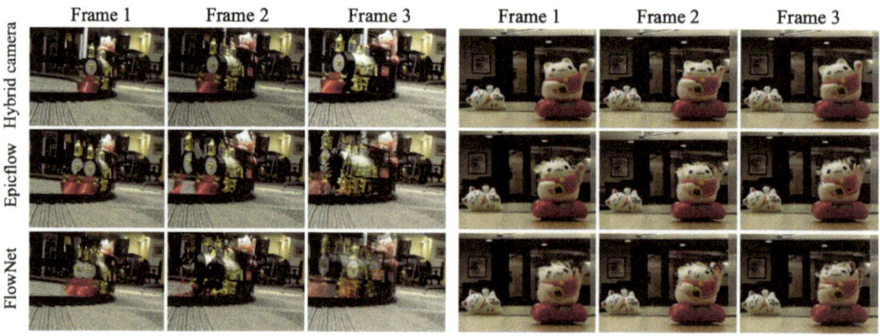

Fig. 2.17 Comparison results. EpicFlow [23] and FlowNet [10] produce ghosting or visible artifacts around the moving objects. Reprinted from Wang (2017)

as the train moves closer to the camera, indicating that the train moves more quickly. The right scene shows a cat with a swinging body. EpicFlow and FlowNet cannot address the fast-moving head and arm of the cat.

Cross-Baseline Hybrid Array

The lenslet-based plenoptic camera shows great potential for passive/general depth estimation due to its high angular resolution; however, this system has limited sensing distance due to its small baseline. A stereo camera with a large baseline can better handle distant scenes but may fail for closer objects due to its limited angular resolution. Aiming at developing an all-in-depth solution, Ding et al. proposed a hybrid camera array with a cross-baseline monocular and lenslet camera [11]: a commercial lenslet-based plenoptic camera (Lytro ILLUM) and a high-resolution monocular camera based on a stereo camera. The idea is simple yet nontrivial due to the significant difference in the angular resolution and baseline between the two cameras.

As shown in Fig. 2.18a, the cross-baseline camera array consists of a commercial lenslet-based plenoptic camera (denoted as LF camera) and a high-resolution monocular camera mounted side-by-side with a fixed baseline. Before depth estimation, the camera array is calibrated as follows: (1) The central view of the LF camera is used as a reference to calibrate and rectify the monocular camera. (2) The other views of the LF camera are adjusted using the calibrated lenslet parameters.

Depth Sensing Pipeline

Cross-baseline Matching Confidence. As shown in Fig. 2.18b, as large-baseline stereo depth estimation performs well for distant objects and the LF camera performs

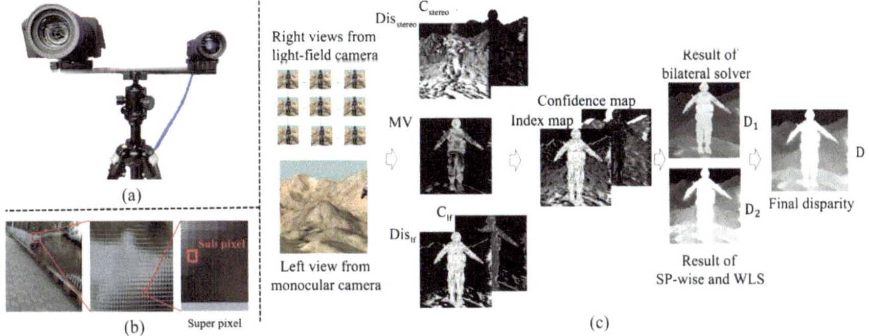

Fig. 2.18 **a** The proposed system. **b** Raw image acquired by the lenslet camera. **c** All-in-focus depth estimation pipeline

well for nearby objects, a confidence map is proposed to fuse these two depth maps. For a depth range of $0 \sim Z_{res}$, the variance $V_i(x, y)$ of the echo planar imaging (EPI) line at different shearing degrees α_i is

$$\begin{aligned} V_i(x, y) &= \frac{1}{N}\sum_{u,v}(L_{\alpha_i}(x, y, u, v) - \overline{L_{\alpha_i}})^2, \\ \alpha_i &= \alpha_{min} + \frac{(\alpha_{max} - \alpha_{min})}{Z_{res}} i, i = 0, 2, \cdots, Z_{res}, \end{aligned} \tag{2.1}$$

where α_{max} and α_{min} are the maximum and minimum shearing degrees based on the EPI results. The shearing degree range can be determined based on the sensing distance range. $L_{\alpha_i}(x, y, u, v)$ represents the intensity value of the pixel at (x, y, u, v) on an EPI line with a shearing degree of α_i. The bar above L_{α_i} denotes the mean operation. N is the number of pixels along the line in the EPI image. The MV map, disparity map Dis_{stereo}, and confidence map C_{stereo} can be computed based on $V_i(x, y)$:

$$\begin{aligned} MV(x, y) &= \min_i V_i(x, y), \\ Dis_{stereo}(x, y) &= \arg\min_i V_i(x, y), \\ C_{stereo}(x, y) &= \frac{Var_{mean}(x, y)}{MV(x, y)}, \\ Var_{mean}(x, y) &= \frac{1}{Z_{res}}\sum_i V_i(x, y). \end{aligned} \tag{2.2}$$

The values in the Dis_{stereo} map represent the relative disparity. $MV(x, y)$ denotes the quality; the higher the value is, the less likely it is to find a matching point. C_{stereo} denotes the confidence of the matching results. Hence, pixels with MV values greater than the threshold are removed. For these pixels, a disparity map Dis_{lf} and a confidence map C_{lf} are obtained with a similar approach using only the images acquired by the LF camera.

Cross-baseline Disparity Fusion. Based on the computed disparity map and confidence map, the coarse disparity d can be calculated as follows:

$$d(x, y) = \begin{cases} \left[\alpha_{min1} + \frac{(\alpha_{max1} - \alpha_{min1})}{Z_{res}} Dis_{stereo}(x, y)\right] B, \text{for } MV(x, y) \leqslant Var_{th} \\ \left[\alpha_{min2} + \frac{(\alpha_{max2} - \alpha_{min2})}{Z_{res}} Dis_{lf}(x, y)\right] b, \text{for } MV(x, y) > Var_{th} \end{cases}$$

where α_{max1} and α_{max2} represent the maximum shearing ranges of the large (stereo system with the monocular and LF cameras) and small baseline (LF camera only) cases, while α_{min1} and α_{min2} are the minimum shearing ranges. When MV is greater than the threshold $Varth$, the final confidence map can be obtained by replacing

C_{stereo} with C_{lf} after calculating the disparity map. To combine the final confidence map and the coarse disparity map, two methods, the bilateral solver and superpixel-wise (SP-wise) regularization, are used to generate a refined disparity map D. The bilateral solver [5] is effective for constructing the edge of disparity map D_1, and the SP-wise method is useful for revealing the fuzzy entity structure [8] in disparity map D_2. The final disparity map D is obtained as

$$D = D_1^{\alpha} D_2^{1-\alpha}, \tag{2.3}$$

where α ($\alpha \in (0, 1)$) is a parameter for adjusting the proportion of the contributions of the two methods. The final depth map Z can be calculated based on the disparity map:

$$Z = \begin{cases} \dfrac{fB}{D}, & \text{for} \quad MV \leqslant Var_{th} \\ \dfrac{fb}{D}, & \text{for} \quad MV > Var_{th}. \end{cases} \tag{2.4}$$

Experimental Results

The present state-of-the-art depth estimation methods, including Tao et al. [25] and Chen et al. [8] for LF and Olsson et al. [20] and Sun et al. [24] for stereo matching, are used to evaluate the superiority of the proposed system. For both the synthetic dataset (Fig. 2.19) and the real captured dataset (Fig. 2.20), the proposed system outperforms existing LF camera-based methods and stereo-based methods in terms of both the depth sensing range and accuracy.

2.2.4 *Portable Camera Devices*

Structured camera arrays have been widely used in portable cameras and personal mobile phones. Light L16 [3] is a DSLR camera with 16 camera modules, including five 28 mm, five 70 mm, and six 150 mm lenses and their associated image sensors, as shown in Fig. 2.21. Each module captures images at different focal lengths, and all the images can be combined into a large 52-megapixel photo. The L16 camera can also refocus images after they are acquired, similar to Lytro and Ratrix.

Modern smartphones also include multiple camera modules with different functions. For example, the Samsung Galaxy S22 Ultra, as shown in Fig. 2.22, uses one 23 mm wild lens, one 230 mm periscope telephoto lens, one 70 mm telephoto lens and one 13 mm ultrawide lens to enhance the depth-of-field and bokeh effect performance for daily photography applications.

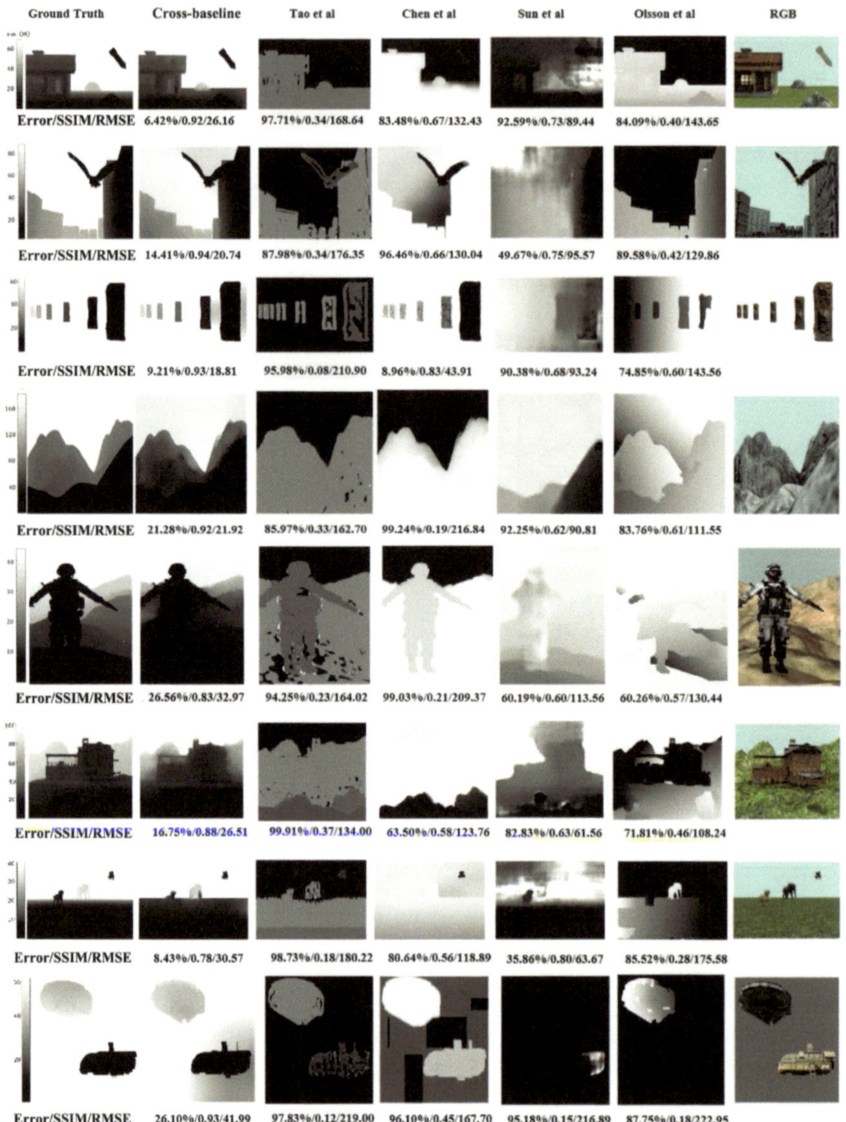

Fig. 2.19 Comparison of the depth maps estimated based on synthetic data among the ground-truth results, our method, and the methods developed by Tao et al. [25], Chen et al. [8], Sun et al. [24], and Olsson et al. [20]

Fig. 2.20 Comparisons of the results obtained based on real-world data. The comparison methods include those developed by Tao et al. [25], Chen et al. [8], Sun et al. [24], and Olsson et al. [20]

Fig. 2.21 Light L16 camera with 16 camera modules

2.3 Unstructured Camera Array

2.3.1 Unstructured Light Field

Davis et al. first proposed capturing and rendering light fields using a hand-held unstructured camera in 2012 [9] (Fig. 2.23). A laptop was used to provide feedback to the user to control the camera. The user first initialized the pose estimation process and

Fig. 2.22 Samsung Galaxy S22 Ultra

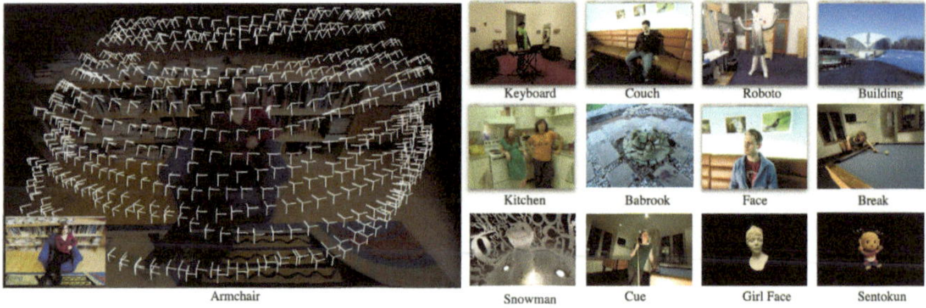

Fig. 2.23 Light fields captured using a simple hand-held camera. Left: The poses of the captured images during the light-field capturing process. Right: The captured light fields

then selected a subject of interest before the images were captured. The acquisition process takes 1–11 minutes, depending on the size of the object, the view density, and the user's expertise. When the light field is captured, the laptop displays a coverage map overlaid on the current view, as shown in Fig. 2.24(a). Previously captured images of the scene are also projected onto the coverage map. The user needs to control the camera to "paint" the surface of the sphere. When the sphere is well painted, enough data have been captured to generate a high-quality light field. The coverage is defined by the bounding reprojection errors between different viewpoints, accounting for all four dimensions of the light field. The camera pose is computed using the simultaneous localization and mapping (SLAM) technique. The system can provide users with real-time feedback and guide them toward undersampled parts of the light field.

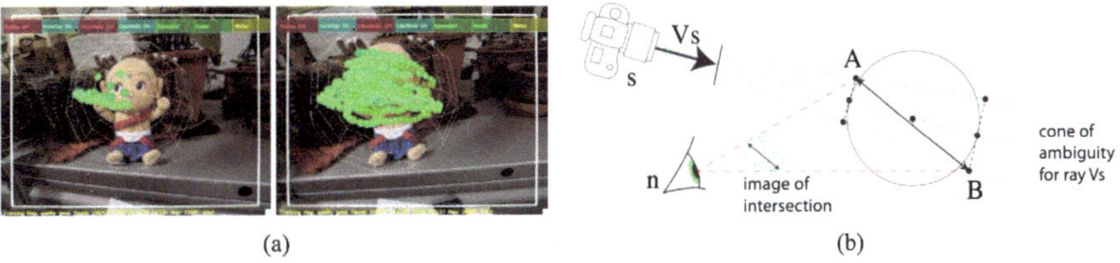

Fig. 2.24 Unstructured light field. **a** Screen capture of our visualization. The virtual meshed sphere serves as a bound on the scene to be captured and a coverage map showing the range of viewpoints that have already been covered. Initially (left), only a limited range is covered, and the user moves the camera to "paint" the scene. **b** The coverage criterion is used to compute a bound on the reprojection error from view s to view n associated with pixel V_s. This bound is sensitive to both parallax error and the ambiguity resulting from resolution differences, covering all four degrees of freedom in the light field

More specifically, to compute the worst-case reprojection error, the ambiguity cone associated with a pixel V_s in s (Fig. 2.24b) is considered. This cone covers all the points in the scene that might contribute to V_s. The cone can be viewed as the projection of a circle centered at V_s onto the scene. The radius of the circle is equal to the distance between adjacent pixels. The intersection of this cone with the subject sphere contains all the points that might contribute to V_s. The image of this intersection in n represents the range of possible reprojections for these points. The reprojection error is then bounded by the longest distance in the imaged intersection region. This distance is the projection of AB in Fig. 2.24b, where AB lies on the epipolar plane defined by s, n, and the center of the subject sphere. The reprojection error is sensitive to both parallax error and resolution changes and thus accounts for all the dimensions of the 4D light field.

2.3.2 Unstructured Camera Array for Panoramic Videos

By combining structured camera arrays, high resolution photography and panoramic images can be achieved. The main problem with existing methods is that the position and orientation of the array cameras need to be manually adjusted. As shown in Fig. 2.25, Perazzi et al. [21] introduced two unstructured camera arrays to capture panoramic videos: the first array consists of 14 machine vision cameras, and the second array consists of 5 GoPro or RED cameras. To determine the warp order of the cameras in the array, a patch-based error metric is defined based on image gradients. The patch size in this configuration is 25^2 pixels. As this error metric is more sensitive to parallax errors in highly structured image regions than the typical per-pixel error metric, Perazzi et al. warped appropriate input views and panoramic fragments using the selection strategy demonstrated in Fig. 2.26 for the structure shown in Fig. 2.25a to minimize the parallax error.

Fig. 2.25 The two unstructured camera arrays used in [21]. Left: 14 Lumenera LT425 machine vision cameras, each recording at a resolution of 2048 × 2048. Right: 5 GoPro Hero3 cameras, each recording at a resolution of 1920 × 1080

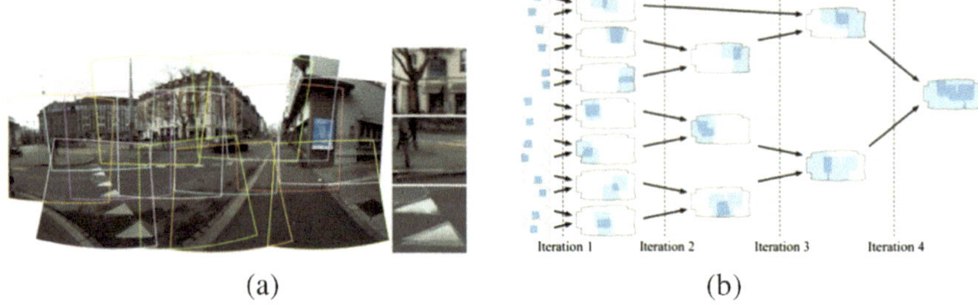

(a) (b)

Fig. 2.26 **a** Example of a reference projection generated based on 14 unstructured input views. The fields of view of 14 individual cameras are highlighted. The zoomed-in views show the average images to illustrate parallax. **b** Illustration of the selection strategy for computing an optimal warping sequence

Local warp refinement in overlap regions usually causes spatial discontinuities in the overall panoramic images. The panoramic images are calculated independently for each frame, which leads to interframe jitter. Considering these two issues, Perazzi et al. proposed a globally coherent warping method to eliminate visual artifacts. Figure 2.27 shows the panoramas generated by the system with two camera arrays. When unstructured camera array systems are used, the camera layout does not need to be manually refined and calibrated; however, a large overlapping area is required for sparse feature matching, and considerable computational resources are needed for spatiotemporal optimization to reduce the artifacts caused by parallax.

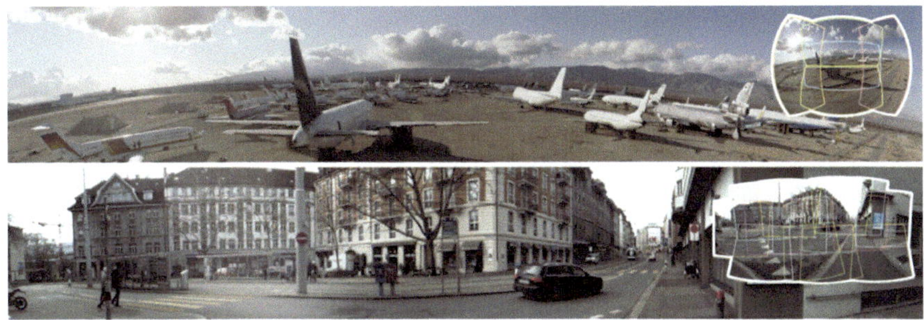

Fig. 2.27 Two panoramas generated by Perazzi's camera arrays. Top: A cropped frame from a 160 megapixel panoramic video generated based on five input videos. Bottom: A cropped frame from a 20 megapixel panoramic created based on a 14 camera array system

2.3.3 *UnstructuredCam: Multiscale Unstructured Camera Array*

Size, weight, power, and cost are critical challenges for structured camera array-based plenoptic imaging systems. Although the curved focal plane design of the AWARE2 objective lens effectively reduced the size and weight of gigapixel-level optical systems [7], the volume and weight of the camera electronics required to collect the large amount of data were more than 10× larger than those of the optical components. The information distribution in natural scenes is usually nonuniform, and objects of interest appear in only small parts of the scene. In conventional uniform sensing systems, a large number of pixels are included in unimportant regions, such as the sky, static buildings, and trees. To address this issue, Yuan et al. presented a multiscale unstructured camera array (denoted as UnstructuredCam) [32, 33], as shown in Fig. 2.28. The global microcamera captures a large field-of-view (FoV) of the scene, and the local microcameras are used to adaptively sample the scene. Since the information in natural scenes is usually sparsely and nonuniformly distributed, unstructured sampling methods significantly reduce the hardware, bandwidth, and computational costs.

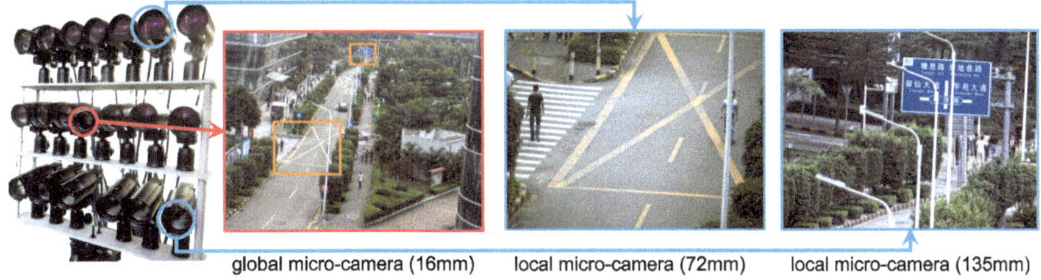

Fig. 2.28 UnstructuredCam and the captured raw videos

UnstructuredCam Hardware

Figure 2.28 illustrates a prototype of UnstructuredCam. All the microcameras are FLIR FL3-U3-120S3C-C industrial machine vision cameras with 4000 × 3000 resolution. The central global microcamera (red circle) is equipped with a 16-mm lens, and the surrounding local microcameras are equipped with varifocal 25–135 mm lenses. The global microcamera captures a wide FoV, while the local microcameras capture the local high-resolution details. The poses of all the local microcameras can be arbitrarily set, as long as the local microcameras are within the view of the global microcamera. This design addresses the overlapping requirements of the local microcameras, enabling content-adaptive unstructured sampling. For example, one possible set of local microcameras may attempt to cover scenes at different distances with approximately the same spatial resolution. As shown in Fig. 2.29, the car maintains a consistent resolution in the videos captured by the local microcameras (top), and its resolution degrades with increasing distance in the video captured by a conventional camera (bottom).

Scene-Adaptive Unstructured Sensing

As shown in Fig. 2.30a, the image sequences captured by the global microcamera are used to compute the temporal entropy map, and the poses of all the local microcameras can be optimized by maximizing the covered entropy:

$$\max_{x_i, y_i, w_i, h_i} \sum \bigcup_{i=1}^{n} E(x_i, y_i, w_i, h_i), \tag{2.5}$$

Fig. 2.29 Illustration of a captured car at 70, 150, and 250 m. Top: Captured by the local microcameras in UnstructuredCam. Bottom: Captured by a conventional camera

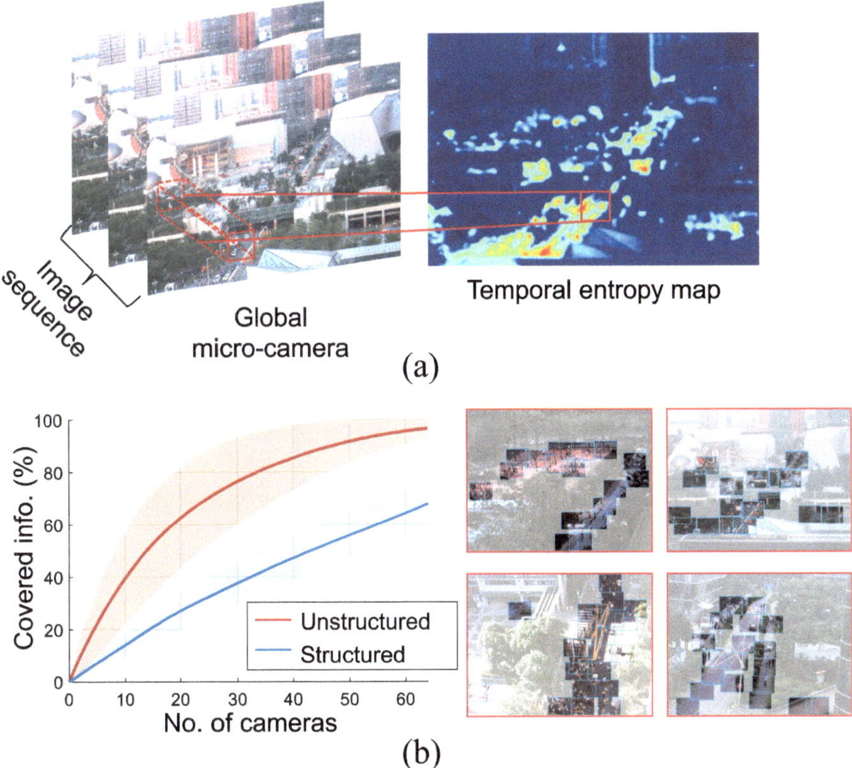

Fig. 2.30 Schematic of our unstructured sampling strategy. **a** Temporal entropy map computed based on the image sequences captured by the global camera node. Unstructured sampling can be applied based on the guidance of the temporal entropy map. **b** Illustration of the unstructured sampling results. Left: Plots of the covered information (%) versus the number of camera nodes. The red color represents our UnstructuredCam, and the blue color represents the structured camera array. The shadow regions denote the standard deviation across the whole dataset. Right: The positions assigned to the local cameras by the unstructured sampling algorithm. Four representative scenes are shown

where i denotes the index of the local microcamera, and E is the computed entropy map. For simplicity, the FoV of the local cameras is represented using rectangles, and the width w_i and height h_i are determined based on the CMOS sensor size and lens. x_i and y_i are the center positions of the ith local microcamera. $E(x_i, y_i, w_i, h_i)$ represents the entropy of the ith local microcamera. The union operator $\bigcup$ merges all the rectangles, and the sum operator $\sum$ computes the total covered entropy. The objective is to maximize the covered entropy. An acceptable solution can be found using a greedy searching algorithm. The entropy map is computed based on the specific visual computing task. One possible approach is to view the image sequences as multiple 1D temporal signals and compute the entropy of each pixel.

The advantage of the unstructured sampling strategy is shown in Fig. 2.30b. The blue curve represents the structured camera array, while the red curve denotes the UnstructuredCam module. The curves were computed based on the video sequences in the PANDA dataset [28]. The performance improvement is due to two main reasons: (1) Conventional camera arrays usually need large overlapping regions between neighboring cameras (approximately 30%) for calibration, while our system reduces this requirement. The sensor in AWARE2 has a pixel utilization of only approximately 58%. However, because our UnstructuredCam uses only the global camera node for calibration, the sensor utilization approaches 100% as the number of local camera nodes is increased. (2) The information in most natural scenes is distributed unevenly and sparsely, similar to the activated neurons in the biological/human vision system. The conventional structured design inherently assumes a uniform distribution, which means that many pixels are wasted in unimportant regions. The right side of Fig. 2.30b demonstrates the positions of the local camera nodes assigned by our unstructured sampling strategy for 4 representative scenes. The local microcameras are mainly located on roads, in crowds, and on other moving objects, which contain valid information.

Scene-Adaptive Unstructured Light Field

Figure 2.31 demonstrates a scene-adaptive unstructured light field captured in downtown Shenzhen, approximately 400 m to 1000 m away from our camera array. The pixels with a translucent white mask are captured by the global microcamera, while the remaining pixels are captured by the local microcameras. The positions of the global and local microcameras are presented in the top-left corner of the figure. Three regions at 400, 450, and 700 m are shown on the right side of the figure. The system

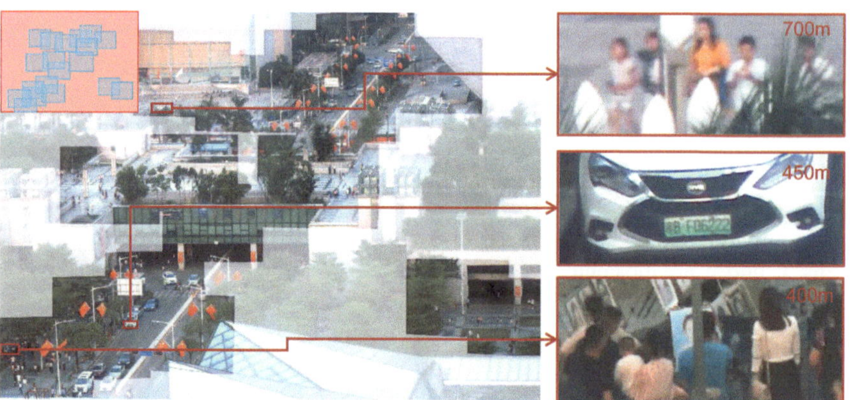

Fig. 2.31 Gigapixel-level videography captured using our UnstructuredCam. The red and blue rectangles on the top left represent the poses of the global and local cameras. The zoomed-in details of three regions at 400, 450, and 700 m are shown in the right column

could recognize a car license plate at 450 m. Moreover, at 700 m, the poses of the pedestrians were still very clear. Compared with existing structured camera arrays and unstructured camera arrays, UnstructuredCam can capture both wide field-of-view and high-spatial-resolution videos of natural scenes with only approximately 10–30% of the microcameras required by previous systems.

2.3.4 UnstructuredCam3D: Panoramic 3D Unstructured Camera Array

The UnstructuredCam proposed in Sect. 2.3.3 achieved content-adaptive nonuniform sampling. However, as the local microcameras have to be linked to the center global microcamera, the field-of-view (FoV) is still restricted. Zhang et al. extended the UnstructuredCam to UnstructuredCam3D, which has panoramic FoV and 3D perception capabilities. As illustrated in Fig. 2.32, UnstructuredCam3D uses a novel hybrid cylindrical distributed camera array to achieve multiscale, gigapixel-level, and 3D panoramic light-field sensing. This system has the following unique features.

- Multiscale: The hybrid camera column is the basic component of UnstructureCam3D, which is composed of two global stereo microcameras and several local microcameras with telephoto lenses.
- Unstructured: To achieve content-adaptive sensing in different regions, the camera column and the local microcameras are assembled in an unstructured and adjustable manner to capture various scenes.
- Scalable: A flexible global gimbals is used to connect the camera columns. The numbers of camera columns and local microcameras are both scalable.

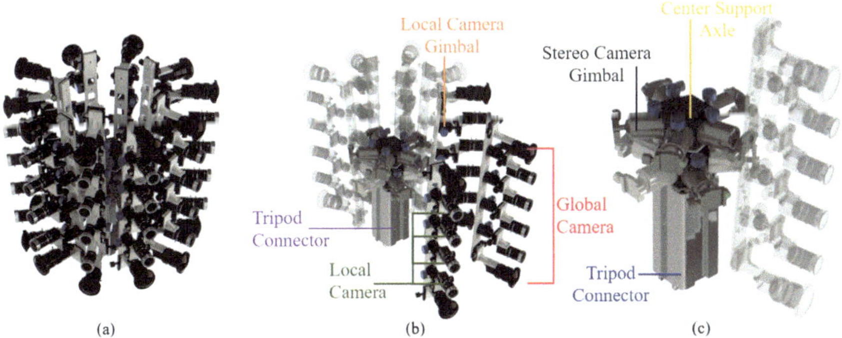

Fig. 2.32 Illustration of our hardware design. **a** UnstructuredCam3D consists of multiple camera columns. **b** Each camera column consists of a pair of global microcameras and multiple local microcameras. **c** The pluggable middle supporting structure enables a scalable setup over a 360° FoV

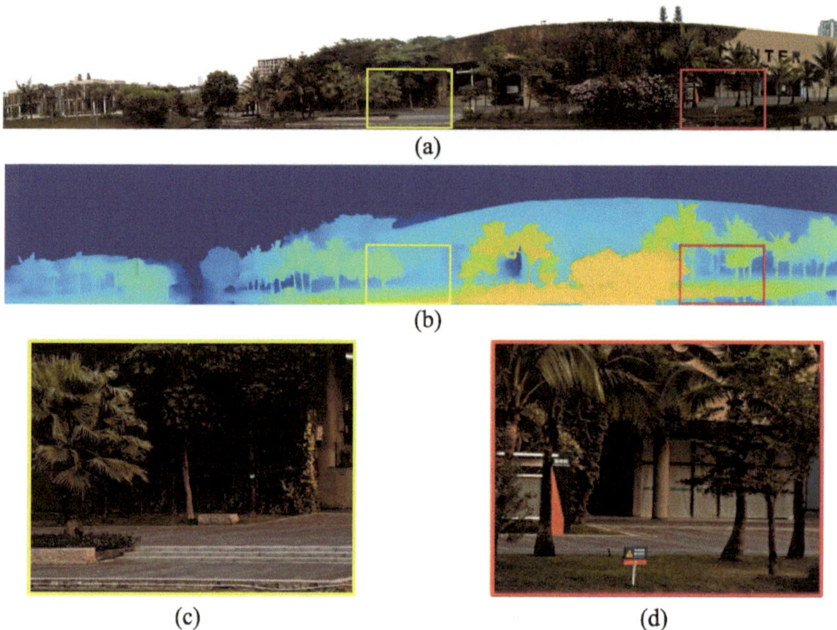

Fig. 2.33 A representative scene captured by UnstructuredCam3D. **a** A wide FoV light field is captured by the global microcameras. **b** The wide FoV depth map estimated based on the global microcamera pairs. **c**, **d** Two magnified regions of interest

Global microcameras employ 12-mm lenses with $2/3''$ CMOS sensors (pixel resolution: 2464×2056) to capture wide FOV scenes. The baseline of each global stereo microcamera pair is set to 450 mm to estimate a depth map over a large distance ($\sim$100 m). In addition, the local microcameras employ 12-36-mm varifocal telephoto lenses with $1/1.8''$ CMOS sensors (pixel resolution: 2064×1544) to capture local details with high resolution. The focal lengths of the local microcameras can be adjusted to adapt to various scenes with different distances.

Figure 2.33 demonstrates a representative scene captured by UnstructuredCam3D. A wide FoV is achieved by combining the global microcameras of multiple camera columns, and the depth information can be estimated based on the global microcamera pairs. Two regions of interest are magnified and presented in Figs. 2.33c, d.

References

1. RayTrix. Avaliable: http://www.raytrix.de/, 2010. 2018. [Online].
2. Lytro. Avaliable: https://www.lytro.com/, 2011. 2018. [Online].
3. Light. Avaliable: http://www.light.co/, 2016. 2018. [Online].

4. Aseem Agarwala, Mira Dontcheva, Maneesh Agrawala, Steven Drucker, Alex Colburn, Brian Curless, David Salesin, and Michael Cohen. Interactive digital photomontage. In *ACM SIGGRAPH 2004 Papers*, pages 294–302. 2004.
5. Jonathan T Barron and Ben Poole. The fast bilateral solver. In *European Conference on Computer Vision*, pages 617–632. Springer, 2016.
6. Vivek Boominathan, Kaushik Mitra, and Ashok Veeraraghavan. Improving resolution and depth-of-field of light field cameras using a hybrid imaging system. In *2014 IEEE International Conference on Computational Photography (ICCP)*, pages 1–10. IEEE, 2014.
7. David J Brady, Michael E Gehm, Ronald A Stack, Daniel L Marks, David S Kittle, Dathon R Golish, EM Vera, and Steven D Feller. Multiscale gigapixel photography. *Nature*, 486(7403):386–389, 2012.
8. Jie Chen, Junhui Hou, Yun Ni, and Lap-Pui Chau. Accurate light field depth estimation with superpixel regularization over partially occluded regions. *IEEE Transactions on Image Processing*, 27(10):4889–4900, 2018.
9. Abe Davis, Marc Levoy, and Fredo Durand. Unstructured light fields. In *Computer Graphics Forum*, volume 31, pages 305–314. Wiley Online Library, 2012.
10. Alexey Dosovitskiy, Philipp Fischer, Eddy Ilg, Philip Hausser, Caner Hazirbas, Vladimir Golkov, Patrick Van Der Smagt, Daniel Cremers, and Thomas Brox. Flownet: Learning optical flow with convolutional networks. In *Proceedings of the IEEE international conference on computer vision*, pages 2758–2766, 2015.
11. Dingjian Jin, Anke Zhang, Jiamin Wu, Gaochang Wu, Haoqian Wang, and Lu Fang. All-in-depth via cross-baseline light field camera. In *Proceedings of the 28th ACM International Conference on Multimedia*, pages 3559–3567, 2020.
12. Marc Levoy and Pat Hanrahan. Light field rendering. In *Proceedings of the 23rd annual conference on Computer graphics and interactive techniques*, pages 31–42, 1996.
13. Marc Levoy, Ren Ng, Andrew Adams, Matthew Footer, and Mark Horowitz. Light field microscopy. In *ACM SIGGRAPH 2006 Papers*, volume 39, pages 924–934. IEEE, 2006.
14. Marc Levoy, Zhengyun Zhang, and Ian McDowall. Recording and controlling the 4d light field in a microscope using microlens arrays. *Journal of microscopy*, 235(2):144–162, 2009.
15. Maurice Gabriel Lippmann. La photographies integrals. *Compt. rend.*, 146(9):446–451, 1908.
16. Kshitij Marwah, Gordon Wetzstein, Yosuke Bando, and Ramesh Raskar. Compressive light field photography using overcomplete dictionaries and optimized projections. *ACM Transactions on Graphics (TOG)*, 32(4):1–12, 2013.
17. Muhammad Owais Mehmood. *People detection methods for intelligent multi-Camera surveillance systems*. PhD thesis, Ecole Centrale de Lille, 2015.
18. Ren Ng. *Digital light field photography*. stanford university, 2006.
19. Ren Ng, Marc Levoy, Mathieu Brédif, Gene Duval, Mark Horowitz, and Pat Hanrahan. *Light field photography with a hand-held plenoptic camera*. PhD thesis, Stanford University, 2005.
20. Carl Olsson, Johannes Ulén, and Yuri Boykov. In defense of 3d-label stereo. In *Proceedings of the IEEE Conference on Computer Vision and Pattern Recognition*, pages 1730–1737, 2013.
21. Federico Perazzi, Alexander Sorkine-Hornung, Henning Zimmer, Peter Kaufmann, Oliver Wang, Scott Watson, and Markus Gross. Panoramic video from unstructured camera arrays. In *Computer Graphics Forum*, volume 34, pages 57–68. Wiley Online Library, 2015.
22. Robert Prevedel, Young-Gyu Yoon, Maximilian Hoffmann, Nikita Pak, Gordon Wetzstein, Saul Kato, Tina Schrödel, Ramesh Raskar, Manuel Zimmer, Edward S Boyden, et al. Simultaneous whole-animal 3d imaging of neuronal activity using light-field microscopy. *Nature methods*, 11(7):727–730, 2014.
23. Jerome Revaud, Philippe Weinzaepfel, Zaid Harchaoui, and Cordelia Schmid. Epicflow: Edge-preserving interpolation of correspondences for optical flow. In *Proceedings of the IEEE conference on computer vision and pattern recognition*, pages 1164–1172, 2015.
24. Deqing Sun, Xiaodong Yang, Ming-Yu Liu, and Jan Kautz. Pwc-net: Cnns for optical flow using pyramid, warping, and cost volume. In *Proceedings of the IEEE Conference on Computer Vision and Pattern Recognition*, pages 8934–8943, 2018.

25. Michael W Tao, Sunil Hadap, Jitendra Malik, and Ravi Ramamoorthi. Depth from combining defocus and correspondence using light-field cameras. In *Proceedings of the IEEE International Conference on Computer Vision*, pages 673–680, 2013.
26. Gaston Tissandier. *La photographie en ballon*. Gauthier-Villars, 1886.
27. Ting-Chun Wang, Jun-Yan Zhu, Nima Khademi Kalantari, Alexei A Efros, and Ravi Ramamoorthi. Light field video capture using a learning-based hybrid imaging system. *ACM Transactions on Graphics (TOG)*, 36(4):1–13, 2017.
28. Xueyang Wang, Xiya Zhang, Yinheng Zhu, Yuchen Guo, Xiaoyun Yuan, Liuyu Xiang, Zerun Wang, Guiguang Ding, David Brady, Qionghai Dai, et al. Panda: A gigapixel-level human-centric video dataset. In *Proceedings of the IEEE/CVF conference on computer vision and pattern recognition*, pages 3268–3278, 2020.
29. Yuwang Wang, Yebin Liu, Wolfgang Heidrich, and Qionghai Dai. The light field attachment: Turning a dslr into a light field camera using a low budget camera ring. *IEEE transactions on visualization and computer graphics*, 23(10):2357–2364, 2016.
30. Bennett Wilburn, Neel Joshi, Vaibhav Vaish, Eino-Ville Talvala, Emilio Antunez, Adam Barth, Andrew Adams, Mark Horowitz, and Marc Levoy. High performance imaging using large camera arrays. In *ACM SIGGRAPH 2005 Papers*, volume 24, pages 765–776. ACM, 2005.
31. Jason C Yang, Matthew Everett, Chris Buehler, and Leonard McMillan. A real-time distributed light field camera. *Rendering Techniques*, 2002:77–86, 2002.
32. Xiaoyun Yuan, Lu Fang, Qionghai Dai, David J Brady, and Yebin Liu. Multiscale gigapixel video: A cross resolution image matching and warping approach. In *2017 IEEE International Conference on Computational Photography (ICCP)*, pages 1–9. IEEE, 2017.
33. Xiaoyun Yuan, Mengqi Ji, Jiamin Wu, David J Brady, Qionghai Dai, and Lu Fang. A modular hierarchical array camera. *Light: Science & Applications*, 10(1):1–9, 2021.
34. Zhoutong Zhang, Yebin Liu, and Qionghai Dai. Light field from micro-baseline image pair. In *Proceedings of the IEEE conference on computer vision and pattern recognition*, pages 3800–3809, 2015.

Chapter 3
High-Resolution Plenoptic Sensing

The high-resolution plenoptic sensing system can capture light field with a high level of detail and clarity. Resolution refers to the amount of detail that can be discerned in the reconstructed image or light field. The cutting edge of high-resolution plenoptic sensing is gigapixel plenoptic sensing, specifically refers to a system capable of capturing images with billions of pixels, resulting in extremely high-resolution light field.

In Sect. 3.1 of this Chapter, we briefly introduce the optical imaging resolution model and analyze the challenges associated with high-resolution plenoptic sensing. Then, in Sect. 3.2, we describe plenoptic sensing systems for high-resolution photography, including a motor-controlled camera mount system for gigapixel image stitching, together with an image superresolution algorithm with a reference-based neural network and an implicit multi-image prior. Finally, in Sect. 3.3, we discuss plenoptic sensing systems for gigapixel videography capturing. We first present two gigapixel imaging systems with two-scale optical lens design. After that, we introduce content-adaptive gigapixel videography with the multiscale unstructured camera array. The system can be further extended to capture panoramic 3D gigapixel videography.

3.1 Challenges of High-Resolution Plenoptic Sensing

Gigapixel imaging aims to capture large-scale dynamic scenes with high resolution and a wide FoV, which is indispensable to many research areas. However, gigapixel imaging is associated with a large space-bandwidth product, and it is almost impossible to achieve this with conventional optical imaging systems, which are restricted by optical diffraction, geometric/chromatic aberration, and sensor-pixel limitations.

An imaging system is composed of optical components (lenses) and electronic components (sensors). The resolution of an optical lens is restricted by optical diffraction and aberration issues, and plane wavefronts focus at circular spots on the sensor

L. Fang, *Plenoptic Imaging and Processing*, Advances in Computer Vision and Pattern Recognition, https://doi.org/10.1007/978-981-97-6915-5_3

instead of at the ideal point. The resolution of a lens is defined as the maximum number of resolvable spots in its FoV, which is also referred to as the space-bandwidth product (SBP).

Based on the scaling law of the lens system [27], the SBPs of a lens are

$$\begin{aligned} R_{diff}(M) &= \frac{M^2 \Delta x \Delta y}{(\lambda F/\#)^2}, \\ R_{aber}(M) &= \frac{M^2 \Delta x \Delta y}{M^2 \sigma_g^2}, \\ R_{conv}(M) &= \frac{M^2 \Delta x \Delta y}{(\lambda F/\#)^2 + M^2 \sigma_g^2}, \end{aligned} \tag{3.1}$$

where $R_{diff}(M)$, $R_{aber}(M)$ and $R_{conv}(M)$ represent the SBP of a lens with diffraction only, aberration only and both, respectively. M is the scaling factor, Δx and Δy denote the CMOS sensor size, λ represents the light wavelength, $F/\#$ represents the F-number of the lens and $M\sigma_g$ is the blur spot size caused by aberrations (both geometric and chromatic). As $M\sigma_g$ increases linearly with the scaling factor M, directly increasing the scaling factor M does not improve the SBP when aberrations are dominant in the system ($M\sigma_g \gg \lambda F/\#$).

Conventional optical lens designs must balance optical diffraction and aberration to guarantee a sharp focus on the sensor plane. A general rule is to increase $F/\#$ when scaling up, which leads to a focal length that is approximately equal to $(F/\#)^3$. Based on this rule, the aberration blur spot size σ_g is a constant, and the final SBP becomes

$$\begin{aligned} R_{conv}(M) &= \frac{M^2 \Delta x \Delta y}{\lambda^2 M^{2/3} + \sigma_g^2} \\ &\propto M^{4/3}. \end{aligned} \tag{3.2}$$

Hence, the volume of a conventional lens increases dramatically with the designed SBP, increasing the difficultly of manufacturing a practical gigapixel SBP lens. Cossairt et al. presented a gigapixel lens with a flat focal plane designed using conventional rules [6]; however, this system needs a 75 × 75-mm sensor (larger than most commercial CMOS/CCD sensors) and requires 11 elements to address the diffraction limitations, leading to an extremely high cost.

3.2 High-Resolution Plenoptic Photography

3.2.1 Gigapixel Images from Stitching

As shown in Fig. 3.1, Kopf et al. first proposed capturing and stitching gigapixel images using a long-focus DSLR camera with a motor-controlled camera mount [20], which works similarly to the human eye: the system examines the scene and generates

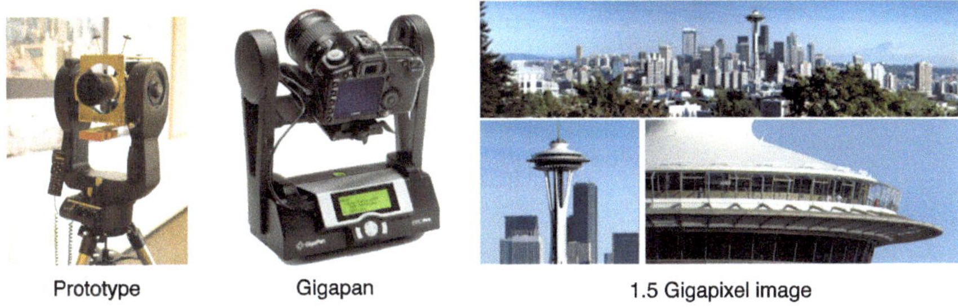

Fig. 3.1 Motor-controlled camera mount for capturing gigapixel images. From left to right, the prototype, the commercial product Gigapan, and the captured gigapixel images of the skyline of Seattle, WA, USA. Reprinted from Kopf (2007)

a large image by stitching all the captured images. To capture and merge hundreds of images, the design of the camera mount system must consider the following factors:

- A long telephoto lens is needed to capture local details with high resolution.
- The images must be captured as quickly as possible to minimize the artifacts caused by the moving elements in the scene and the changing light conditions.
- The images must be captured on as regular a grid as possible to improve the stitching performance.
- The stitched wide FoV image should support a high dynamic range (HDR).

For the discussed imaging system (Fig. 3.1), a 16 Megapixel Canon 1DS Mark II or an 8 Megapixel Canon 20D with a 100-400 mm zoom lens was employed. The wide dynamic range of most scenes was captured by setting a fixed lens aperture and automatic exposure times. An f-11 aperture was used for the lens to balance the image resolution/quality and the depth of field. The camera recorded the images in raw format. The user selects the area of the panoramic sphere to be captured, and the rest of the procedure is automatic. Based on the user input and a desired 16% overlap between images, the system generates a sequence of pan and pitch angles for the motors. A total of 250 to 800 high-resolution images are captured and combined to generate the final gigapixel image, which takes approximately 30 to 90 minutes.

Based on the good initial camera poses provided by the motor, the captured images can be combined to generate a seamless gigapixel panorama. However, several technical challenges must be addressed to create an HDR Gigapixel image based on the captured images with varying exposures. Kopf et al. proposed a geometric and radiometric alignment pipeline, as shown in Fig. 3.2. The first step in the radiometric alignment pipeline is to estimate the radiance values of all the input images. Although the camera does not capture the full dynamic range in a small FOV, the stitched wide FoV gigapixel image covers a high dynamic range. The first step in the radiometric processing pipeline is demosaicing, which is used to convert the raw images to RGB images. A vignette adjustment map is then estimated to compensate for the lens vignetting. White balance and exposure normalization are then performed to convert the pixel values to radiance values. Here, exposure normalization is performed by

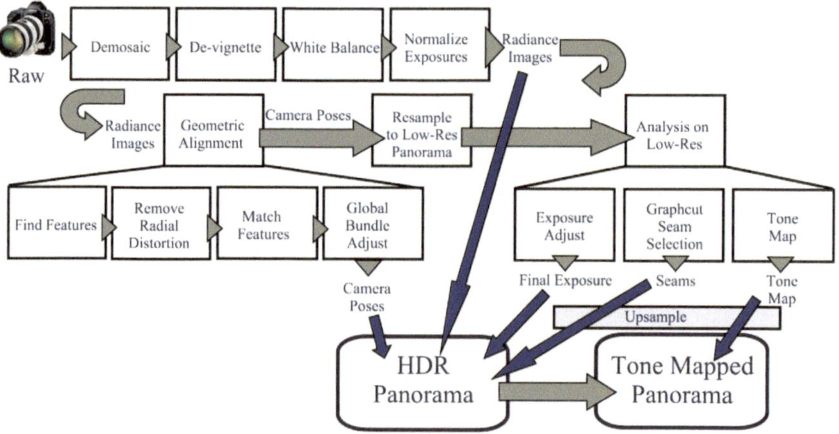

Fig. 3.2 The processing pipeline of GigaPan. Reprinted from Kopf (2007)

dividing the pixel value by the exposure time. For the geometric alignment pipeline, a feature-based image alignment and stitching method with radial distortion removal and global bundle adjustment is used to estimate the camera poses of all the images. A graph-cut-based seam generator is used to minimize the artifacts in the images. All the images are then mapped to a plane (perspective for a narrow FOV) or a cylinder or sphere (wide FOV) to generate the composite gigapixel image. Finally, a tone mapping step is used to convert the stitched HDR image to a low dynamic range image for visualization. This approach is robust and efficient for gigapixel image generation and has been extended to the commercial product Gigapan.

3.2.2 Cross-Resolution Reference-Based Superresolution

One of the largest challenges in image stitching is addressing the parallax among the images caused by the center shift of the cameras. To address the large parallax caused by short-distance objects, Tan et al. presented CrossNet++, an end-to-end network for cross-resolution reference-based superresolution (RefSR) [37]. RefSR aims to superresolve low-resolution (LR) images based on reference high-resolution (HR) images, which suffer from resolution gaps and large parallax. As shown in Fig. 3.3, CrossNet++ includes two-stage cross-scale warping modules, an image encoder, and a fusion decoder. The first-stage warping module learns to narrow the parallax based on sparse landmarks and intensity distribution consensus. The second-stage warping module employs finer-grained alignment and aggregation in the feature domain to generate the final superresolved HR image. To further address the large parallax, new hybrid loss functions consisting of the warping loss, landmark loss, and superresolution loss are proposed to regularize the training process and improve convergence.

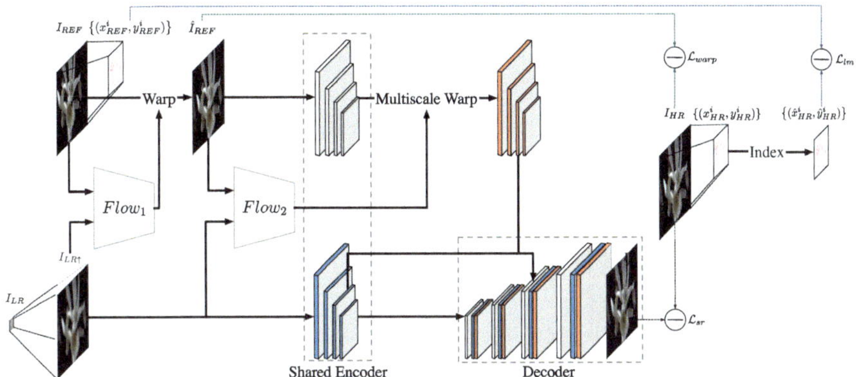

Fig. 3.3 The pipeline of CrossNet++, which consists of two-stage cross-scale warping modules, an image encoder, and a fusion decoder. The first warping module roughly aligns the HR reference image to the LR image. The encoder extracts multiscale features from the LR image and the prealigned reference image. The second warping module aligns the reference feature map with the LR feature map at a finer spatial resolution. Finally, the decoder merges the feature maps from both domains to generate the HR output

Network Structure of CrossNet++

In the first stage, a flow estimator is used to predict the flow based on the input images. Then, warping is used to align the reference image to the LR image. The proposed warping loss and landmark loss supervise the generation of the warped reference image. In the second stage, the cross-scale flow estimator is utilized to align the feature maps at multiple scales and synthesize the superresolution image via an encoder–decoder module.

Alignment Module. To align the reference image to the LR image, a modified FlowNet [10] is utilized as the flow estimator to generate cross-scale feature map correspondences at four different scales, which is denoted as

$$\{V^3, V^2, V^1, V^0\} = Flow(I_1, I_2), \tag{3.3}$$

where I_1 and I_2 are two input images and $V^i, i = 0, 1, 2, 3$ are the flow maps at scale i. The size of the flow map at scale $i - 1$ is half the size of the flow map at scale i, and the size of the flow map at scale 0 is the same as the size of the input images.

Figure 3.3 shows the network architecture, including the two-stage alignment modules, image encoder, and fusion decoder. In the first stage, $Flow_1$ estimates the warping field from the reference image to the LR image. Mathematically, this can be represented as

$$V_1^0 = Flow_1(I_{LR\uparrow}, I_{REF}), \tag{3.4}$$

where I_{REF} is the reference image, and $I_{LR\uparrow}$ denotes the upsampled LR image (I_{LR}) obtained using a single-image superresolution (SISR) approach: $I_{LR\uparrow} = SISR(I_{LR})$. V_1^0 denotes the warping field from I_{REF} to $I_{LR\uparrow}$, which is used at scale 0 only to align the original input images and is not used for the multiscale feature maps. The prealigned reference image $\hat{I}_{REF}$ is then generated by warping I_{REF} using V_1^0, $\hat{I}_{REF} = Warp(I_{REF}, V_1^0)$.

In the second stage, $\hat{I}_{REF}$ and $I_{LR\uparrow}$ are fed into the flow estimator $Flow_2$ to estimate the multiscale warping fields, which are used by the encoder and decoder to generate the final superresolved image.

$$\{V_2^3, V_2^2, V_2^1, V_2^0\} = Flow_2(I_{LR\uparrow}, \hat{I}_{REF}). \tag{3.5}$$

Encoder. The encoder extracts 4-scale feature maps from $\hat{I}_{REF}$ and $I_{LR\uparrow}$, and the second-stage alignment module produces 4-scale warping fields for feature map registration. The encoder has 5 convolutional layers with 64 5×5 filters. The first two layers, which have strides of 1, are used to extract the feature map at scale 0. The remaining three layers, which have strides of 2, extract the feature maps at scales 1, 2, and 3 as follows:

$$\begin{aligned} F^0 &= \sigma(\mathbf{W}^0 I), \\ F^i &= \sigma(\mathbf{W}^i F^{i-1}),\ i = 1, 2, 3, \end{aligned} \tag{3.6}$$

where F^i denotes the feature map at scale i and σ represents the rectified linear unit (ReLU) activation function. The same encoder is used for both $\hat{I}_{REF}$ and $I_{LR\uparrow}$ to reduce the redundancy of the model parameters. After the feature maps are extracted, warping is performed based on the reference image features F^i_{REF} using the multi-scale flow fields V_2^i derived from Eq. (3.5) to align the feature maps:

$$\hat{F}^i_{REF} = Warp(F^i_{REF}, V_2^i),\ i = 0, 1, 2, 3. \tag{3.7}$$

Decoder. A U-Net-like decoder is proposed to merge the aligned feature maps and estimate the final HR image.

The warped features, the LR image features at scale i, and the decoder features at scale $i-1$ (if any) are concatenated and input to a deconvolution layer with 64 filters (4 × 4, stride 2) to generate the decoded features at scale i:

$$\begin{aligned} F_D^3 &= \sigma(\mathbf{W}_D^3(F^3_{LR}, \hat{F}^3_{REF})), \\ F_D^i &= \sigma(\mathbf{W}^i_D(F^{i+1}_{LR}, \hat{F}^{i+1}_{REF}, F_D^{i+1})), i = 2, 1, 0, \end{aligned} \tag{3.8}$$

In addition, three convolution layers with a filter size of 5 × 5 and filter numbers of {64, 64, 3} are added to F_D^0 (decoded feature map at scale 0) to generate the final HR image I_p.

Loss Function. A new hybrid loss function consisting of the warping loss, landmark loss, and superresolution loss is proposed to achieve better convergence.

(1) Warping loss: The warping loss is defined based on the pixel intensity between the warped reference image and the ground truth image:

$$\mathcal{L}_{warp} = \sum_i \|\hat{I}^i_{REF} - I^i_{HR}\|^2_2, \tag{3.9}$$

where $\hat{I}_{REF}$ is obtained from Eq. 3.7.

(2) Landmark loss: The landmark loss is introduced to achieve accurate directional guidance. The widely used SIFT feature points are first extracted from pairs of the ground truth HR and reference images. For each point $\mathbf{p} = (x_{HR}, y_{HR})$ in the ground truth HD image, its corresponding landmark $\mathbf{q} = (x_{REF}, y_{REF})$ can be found in the reference image. After removing the outliers, the generated flow field V^0_1 is used to warp the landmarks $\mathbf{p}^j$ to $\hat{\mathbf{p}}^j$:

$$\hat{\mathbf{p}}^j = \mathbf{p}^j + V^0_1[\mathbf{p}], \tag{3.10}$$

and the landmark loss is formulated as

$$\mathcal{L}_{lm} = \sum_{j=1} \left\|\hat{\mathbf{p}}^j - \mathbf{q}^j\right\|^2_2, \tag{3.11}$$

where $\hat{\mathbf{p}}^j$ and $\mathbf{q}^j$ represent the jth landmark pair (Fig. 3.4).

(3) Superresolution loss: Given the network output I_p and the ground truth HR image I_{HR}, the superresolution loss is defined as

$$\mathcal{L}_{sr} = \rho(I^i_{HR} - I^i_p), \tag{3.12}$$

where $\rho(x) = \sqrt{x^2 + 0.001^2}$ is the Charbonnier penalty function (Tables 3.1 and 3.2).

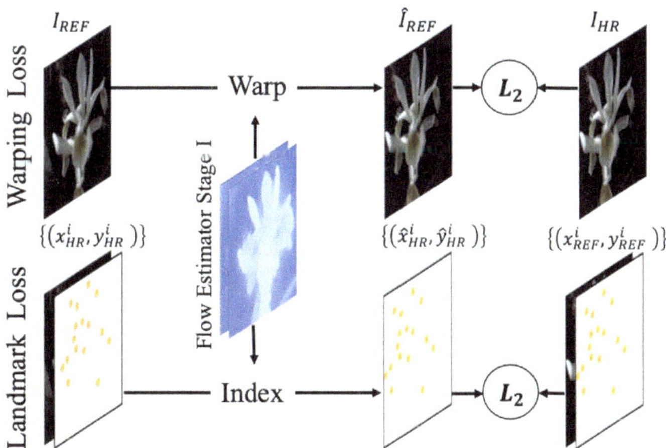

Fig. 3.4 Interpretations of the warping loss and landmark loss

Table 3.1 Quantitative evaluations of the state-of-the-art SISR and RefSR algorithms based on light-field datasets, showing the PSNR/SSIM/IFC for scale factors of $4\times$ and $8\times$

	Methods	Scale	Flower (1, 1)			Flower (7, 7)			LFVideo (1, 1)			LFVideo (7, 7)		
			PSNR	SSIM	IFC	PSNR	SSIM	IFC	PSNR	SSIM	IFC	PSNR	SSIM	IFC
SISR	SRCNN [9]	4×	32.76	0.8910	2.46	32.96	0.8957	2.49	32.98	0.8592	2.07	33.27	0.8649	2.08
	VDSR [19]	4×	33.34	0.9027	2.73	33.58	0.9070	2.76	33.58	0.8733	2.29	33.87	0.8787	2.30
	MDSR [23]	4×	34.40	0.9158	3.04	34.65	0.9195	3.07	34.62	0.8912	2.62	34.91	0.8959	2.63
	LapSR [21]	4×	32.51	0.9063	2.87	31.97	0.8672	2.89	31.97	0.8672	2.24	32.27	0.8727	2.26
	DBPN [12]	4×	33.52	0.9042	2.65	33.69	0.9079	2.67	33.51	0.8694	2.44	33.78	0.8744	2.45
	RCAN [42]	4×	34.75	0.9152	3.24	34.96	0.9186	3.26	34.27	0.8820	2.60	34.57	0.8868	2.61
RefSR	PatchMatch [4]	4×	38.03	0.9655	5.11	35.23	0.9373	3.85	38.22	0.9536	4.60	37.08	0.9404	3.99
	SRNTT [43]	4×	34.76	0.9440	4.20	32.23	0.8848	2.66	37.47	0.9518	4.48	31.95	0.8485	2.13
	CrossNet [45]	4×	42.09	0.9830	6.70	38.49	0.9662	5.02	42.21	0.9756	5.96	39.03	0.9565	4.61
	CrossNet++	4×	42.51	0.9831	6.87	39.23	0.9708	5.37	42.33	0.9748	6.02	39.52	0.9613	4.88
SISR	SRCNN [9]	8×	28.17	0.7653	0.98	28.25	0.7713	1.00	29.43	0.7540	0.82	29.63	0.7601	0.82
	VDSR [19]	8×	28.58	0.7792	1.04	28.68	0.7849	1.06	29.83	0.7672	0.89	30.04	0.7732	0.89
	MDSR [23]	8×	29.15	0.7946	1.17	29.26	0.8001	1.19	30.43	0.7850	1.04	30.65	0.7906	1.05
	LapSR [21]	8×	28.23	0.7919	0.97	28.20	0.7970	0.98	28.35	0.7693	0.73	28.47	0.7750	0.73
	DBPN [12]	8×	27.17	0.7732	0.78	27.14	0.7772	0.78	29.39	0.7673	0.81	29.58	0.7727	0.81
	RCAN [42]	8×	29.27	0.8036	1.17	29.33	0.8085	1.18	29.95	0.7844	0.99	30.15	0.7900	1.00
RefSR	PatchMatch [4]	8×	35.26	0.9471	4.00	30.41	0.8485	2.07	36.72	0.9419	3.81	34.48	0.9066	2.84
	SS-Net [44]	8×	37.46	0.9677	4.72	32.42	0.9086	2.95	37.93	0.9510	4.06	35.81	0.9335	3.30
	SRNTT [43]	8×	27.26	0.7817	1.20	26.37	0.7488	0.86	28.76	0.7689	0.94	28.22	0.7425	0.69
	CrossNet++	8×	40.94	0.9792	5.96	35.49	0.9445	3.93	41.34	0.9719	5.25	37.04	0.9394	3.69

Table 3.2 No-reference metrics PI [3], IS [31] and FID [14] are applied to quantitatively evaluate the results based on the dual-camera dataset (scale 4×)

Methods	Dual-camera dataset		
	PI ↓	IS ↑	FID ↓
MDSR [23]	6.6044	3.296 ± 0.367	134.22
PatchMatch [4]	6.6251	3.298 ± 0.346	118.34
SRNTT [43]	5.8895	3.417 ± 0.563	132.35
CrossNet++	6.7810	3.445 ± 0.443	85.35

Reference-Based Superresolution Experimental Results

To evaluate the reference-based superresolution performance of CrossNet++, the light-field datasets Flower [35] and LFVideo [40] are used to generate LR and reference image pairs with both parallax and a resolution gap. Moreover, the system is evaluated based on a real-world dataset captured by a dual-camera system.

Figure 3.5 qualitatively compares CrossNet++ with SISR approaches, including SRCNN, VDSR, MDSR, LapSR, DBPN, and RCAN, and RefSR approaches, including PatchMatch and SRNTT, for an 8× resolution gap. The RefSR approaches outperform the SISR methods by a large margin because more information is included in the reference image, and the high-frequency details are better preserved. The proposed CrossNet++ approach obtains more fine-grained details and achieves better performance based on quantitative measures, including the peak signal-to-noise ratio (PSNR), structural similarity index measure (SSIM), and information fidelity criterion (IFC).

Figure 3.6 demonstrates the results based on real-world data captured by a dual-camera system. CrossNet++ generates much better results than PatchMatch [4] and SRNTT [43], and the high-frequency details are visually pleasing without disturbing artifacts. As there are no ground truth HR images for the real-world dual-camera dataset, no-reference metrics are adopted as quantitative measures, including the perception index (PI) [3], inception score (IS) [31] and FID [14] (widely used evaluation metrics for generated images), where lower PI and FID scores and higher IS scores indicate better performance. CrossNet++ also achieves the best performance among the considered approaches, with the results exactly matching the visual comparisons shown in Fig. 3.6.

3.2.3 Improving the Superresolution Performance with Multi-Image Priors

The goal of single-image superresolution (SISR) is to recover high-resolution (HR) images based on their low-resolution (LR) counterparts. As a basic low-level vision problem, SISR has been an actively researched subject for several decades; however,

Fig. 3.5 Visual comparisons for large parallax cases, i.e., 8× RefSR based on the Flower dataset viewpoint (7, 7) and the LFVideo dataset viewpoint (7, 7). The CrossNet and CrossNet++ approaches outperform several SISR methods, including SRCNN [9], VDSR [19], MDSR [23], LapSR [21], DBPN [12], and RCAN [42], and several RefSR methods, including PatchMatch [4] and SRNTT [43]. The details can be viewed by zooming in

how multiple light-field views can be used in superresolution applications has not been well studied. Jin et al. proposed a novel learning framework that implicitly utilizes multiple image priors of a light field for SISR problems [18].

Implicit Multi-Image Prior

Jin et al. discovered that by using implicit multi-image priors obtained during the training stage, the effectiveness and efficiency of SISR methods can be improved without increasing the computational load of the inference stage. The architecture is depicted in Fig. 3.7. In the training stage, given a 4D LR LF image $\mathbf{l} \in \mathbb{R}^{U\times V\times H_l\times W_l\times C}$, the LR subaperture images (SAIs) $\{l^i\} \in \mathbb{R}^{H_l\times W_l\times C}$ can be obtained by decomposing the first two dimensions of $\mathbf{l}$ (angular dimension). A single-image

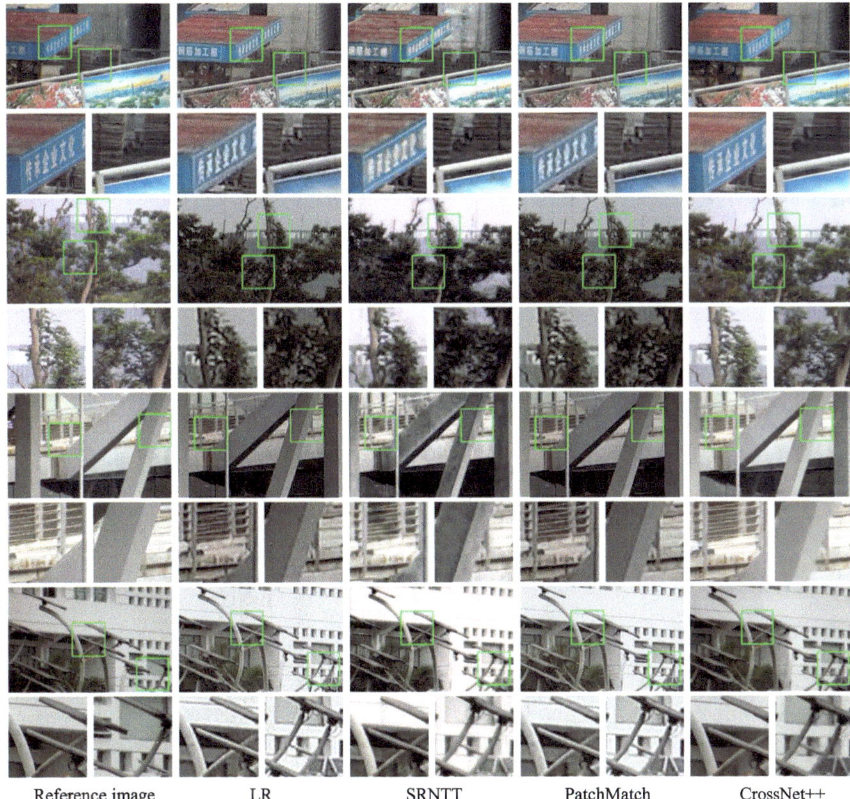

Fig. 3.6 Superresolution comparisons among different algorithms based on our real dual-camera dataset. Our proposed CrossNet++ achieves significant visual improvements compared with the LR image and the RefSR methods PatchMatch [4] and SRNTT [43]. The details can be viewed by zooming in

superresolution (SISR) network is then used to generate high-resolution (HD) SAIs $\{\hat{h}^i \in \mathbb{R}^{H_h \times W_h \times C}\}$. These HD SAIs are further composed into 4D SR LF images $\hat{\mathbf{h}} \in \mathbb{R}^{U \times V \times H_h \times W_h \times C}$.

The multi-image prior is learned using a hybrid loss term $\mathcal{L}$ in the training stage and improves the superresolution performance in the inference stage.

$$\mathcal{L} = \lambda_1 \mathcal{L}_{cs} + \lambda_2 \mathcal{L}_v + \lambda_3 \mathcal{L}_d, \tag{3.13}$$

where the content and structure loss $\mathcal{L}_{cs}$ regularizes the pixel intensity of the restored LF images, the variance loss $\mathcal{L}_v$ regularizes the restored object edges, and the disparity loss $\mathcal{L}_d$ regularizes the multiview consistency in the disparity domain.

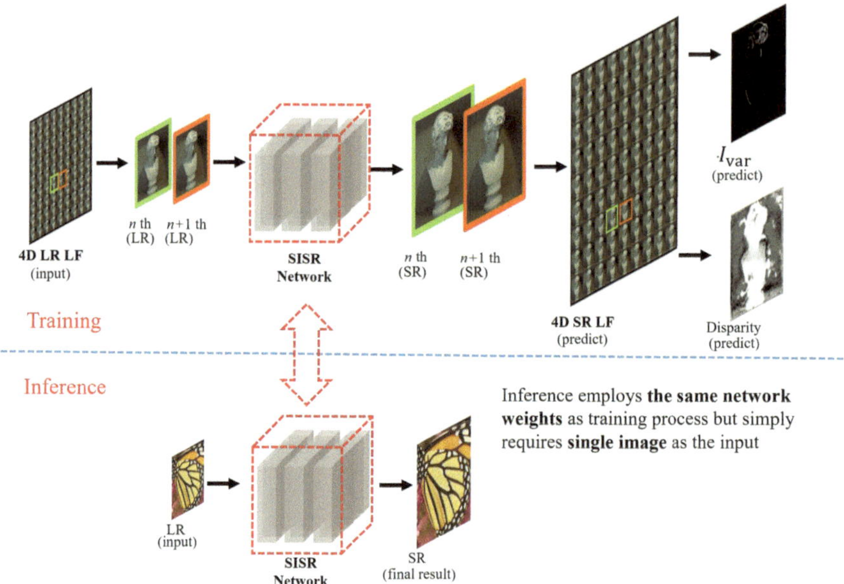

Fig. 3.7 Architecture of our learning framework

Content and Structure Loss. The 4D L2 loss and 4D SSIM loss are combined as the content and structure loss. In contrast to the 2D case, the 4D loss terms consider both the spatial and angular information.

$$\begin{aligned} \mathcal{L}_{cs} &= \mathcal{L}_c + \mathcal{L}_s, \\ \mathcal{L}_c &= (1-\alpha)||\hat{\mathbf{h}} - \mathbf{h}||_2^2, \\ \mathcal{L}_s &= \alpha \frac{1 - SSIM(\hat{\mathbf{h}}, \mathbf{h})}{2}, \end{aligned} \tag{3.14}$$

where $\mathcal{L}_c$ denotes the L2 loss and $\mathcal{L}_s$ denotes the SSIM loss. In addition, α represents the SSIM loss ratio, $\mathbf{h}$ represents the ground truth 4D light field, and $\hat{\mathbf{h}}$ represents the prediction results.

Variance Loss. The commercial LF camera records a lenslet format LF image, as shown in Fig. 3.8, which represents a 4D LF in a 2D sensor plane. One superpixel is formed by one microlens. Here, a superpixel consists of 9×9 subpixels, as shown in Fig. 3.8. Each subpixel represents a single view of a world point with angular details. The pixel values of the subpixels within one superpixel may differ because of occlusion. The variance increases in the edge area but remains low in the smooth area. The variance of each superpixel is calculated as

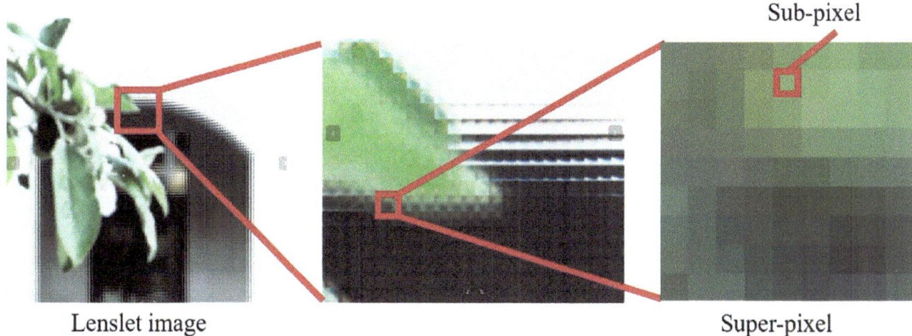

Fig. 3.8 In the zoomed-in view of the lenslet image (generated from the sensor of the LF camera), each superpixel contains 81 subpixels, which represent 81 different angular views

$$
\begin{aligned}
I_{var}^{p}(x, y) &= \frac{\sum_{u'=1}^{N_u} \sum_{v'=1}^{N_v} (\hat{\mathbf{h}}[u', v', x, y] - M(\hat{\mathbf{h}}[u', v', x, y]))^2}{N_u N_v}, \\
I_{var}^{g}(x, y) &= \frac{\sum_{u'=1}^{N_u} \sum_{v'=1}^{N_v} (\mathbf{h}[u', v', x, y] - M(\mathbf{h}[u', v', x, y]))^2}{N_u N_v},
\end{aligned}
\tag{3.15}
$$

where I_{var}^{p}, I_{var}^{g} are the variance maps of the predicted 4D LF and the ground truth 4D HR LF, respectively. N_u and N_v denote the size of a superpixel. $M(\cdot)$ calculates the average pixel intensity inside the superpixel. The variance loss is formulated as

$$
\mathcal{L}_v = ||I_{var}^{p} - I_{var}^{g}||_2^2. \tag{3.16}
$$

Disparity Loss. The disparity map is first estimated in an end-to-end manner based on the 4D SR LF and is then refined by simultaneously minimizing the disparity loss and the other two losses.

$$
\mathcal{L}_d = ||Dis(\hat{\mathbf{h}}) - Dis_{GT}||_2^2, \tag{3.17}
$$

where $Dis(\cdot)$ denotes the disparity estimation network [22], and Dis_{GT} is the ground truth disparity based on the LF dataset.

Improving the Performance of Existing SISR Methods

Superresolution Public Datasets. The comparison results of the proposed methods (ESPCN-LF, VDSR-LF, and RCAN-LF) and the original methods (ESPCN [34], VDSR [19], and RCAN [42]) are presented in Fig. 3.9. Five benchmarks, Set5 [2],

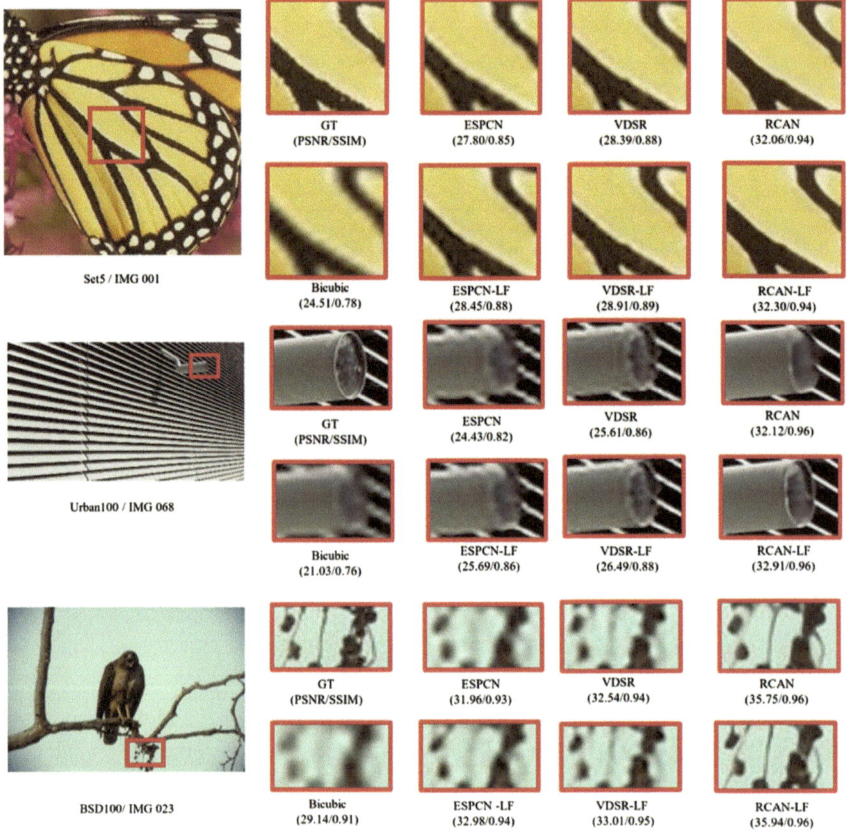

Fig. 3.9 Visual comparison for the 3× scale based on Set5, Urban100, and BSD100. These datasets are widely used in SISR comparison tasks

Set14 [41], Urban100 [16], BSD100 [29], and Campus 17, are used to evaluate the performance of the various methods.

For a quantitative comparison, Table 3.3 shows the PSNR evaluation based on the Set5 [2], Set14 [41], Urban100 [16], BSD100 [29]and Campus17 datasets with scales of 2×, 3× and 4×.

For the sake of fairness, when training the original methods of ESPCN, VDSR and RCAN, the methods are extended based on the results with each single image training dataset [32, 38, 39], with all the SAIs transformed from the LF dataset [7, 15]; these methods are denoted as ESPCN+, VDSR+ and RCAN+.

As shown in Fig. 3.9, with the implicit multi-image prior learned from the LF structure in the training phase, the proposed methods generate visually pleasing results, especially for the edge regions. For example, in the Urban100 scene, ESPCN-LF generates the tube's edge and the line on the wall much sharper than the original methods.

Table 3.3 PSNR evaluation results. Model indicates the original methods. Model+ indicates that the original methods were trained with an additional light-field training dataset. Model-LF indicates the proposed method

Method	Scale	Set5	Set14	Urban100	BSD100	Campus17
		PSNR/SSIM	PSNR/SSIM	PSNR/SSIM	PSNR/SSIM	PSNR/SSIM
Bicubic	×2	33.66/0.93	30.24/0.87	26.88/0.84	29.56/0.84	31.29/0.90
ESPCN+	×2	36.53/0.94	32.22/0.89	29.09/0.87	31.21/0.87	34.26/0.93
ESPCN-LF	×2	36.95/0.94	32.80/0.89	29.73/0.88	31.66/0.87	34.79/0.94
VDSR [19]	×2	37.53/0.96	33.03/0.91	30.76/0.91	31.90/0.90	None
VDSR+	×2	37.71/0.96	33.16/0.91	30.84/0.91	31.99/0.90	34.84/0.94
VDSR-LF	×2	37.99/0.96	33.60/0.92	31.21/0.92	32.27/0.90	35.06/0.94
RCAN [42]	×2	38.33/0.96	34.23/0.92	33.54/0.94	32.46/0.90	None
RCAN+	×2	38.39/0.96	34.25/0.92	33.63/0.94	32.52/0.90	35.54/0.94
RCAN-LF	×2	38.41/0.96	34.50/0.92	33.95/0.94	32.77/0.91	35.87/0.94
Bicubic	×3	30.39/0.87	27.55/0.77	24.46/0.73	27.21/0.74	27.88/0.82
ESPCN [34]	×3	32.55/None	29.08/None	None	28.26/None	None
ESPCN+	×3	32.84/0.90	29.27/0.82	26.63/0.79	28.64/0.79	29.80/0.86
ESPCN-LF	×3	33.21/0.91	29.73/0.83	27.23/0.82	28.90/0.80	30.34/0.87
VDSR [19]	×3	33.66/0.92	29.77/0.83	27.14/0.83	28.82/0.80	None
VDSR+	×3	33.77/0.92	29.84/0.83	27.21/0.83	28.97/0.80	30.51/0.87
VDSR-LF	×3	33.94/0.93	30.31/0.84	27.74/0.84	29.24/0.80	30.73/0.87
RCAN [42]	×3	34.85/0.93	30.76/0.85	29.31/0.87	29.39/0.81	None
RCAN+	×3	34.89/0.93	30.83/0.85	29.49/0.87	29.51/0.90	31.82/0.89
RCAN-LF	×3	34.90/0.93	31.03/0.85	29.65/0.88	29.64/0.90	32.06/0.90
Bicubic	×4	28.42/0.81	26.00/0.70	23.14/0.66	25.96/0.67	25.34/0.72
ESPCN+	×4	30.12/0.84	27.04/0.74	24.13/0.70	26.77/0.71	26.77/0.76
ESPCN-LF	×4	30.28/0.85	27.35/0.75	24.50/0.72	26.92/0.72	27.18/0.77
VDSR [19]	×4	31.35/0.88	28.01/0.77	25.18/0.75	27.29/0.73	None
VDSR+	×4	31.46/0.88	28.14/0.77	25.29/0.75	27.42/0.73	27.63/0.78
VDSR-LF	×4	31.51/0.88	28.25/0.77	25.53/0.76	27.53/0.73	27.74/0.78
RCAN [42]	×4	32.73/0.90	28.98/0.79	27.10/0.81	27.85/0.75	None
RCAN+	×4	32.74/0.90	29.06/0.79	27.19/0.81	27.88/0.75	28.93/0.80
RCAN-LF	×4	32.74/0.90	29.16/0.80	27.26/0.81	27.94/0.75	28.99/0.80

The experimental results verify that existing SISR networks can be improved by the proposed implicit multi-image prior without any modifications to the network.

Ablation Study. The effectiveness of the proposed loss terms is studied. The scale is set to 3×, and ESPCN is used as the SISR network. As shown in Fig. 3.10, the Stanford LF dataset [7](split) was chosen as the training/testing dataset. The loss terms $\mathcal{L}_c$, $\mathcal{L}_s$, $\mathcal{L}_v$, and $\mathcal{L}_d$ are incrementally calculated, e.g., $\mathcal{L}_{cs}$ and $\mathcal{L}_{cs} + \mathcal{L}_v$.

As shown in Fig. 3.10, the bicubic and ESPCN methods generate blurry EPIs. ESPCN(A) retains more structural details and achieves a higher PSNR value (+0.65

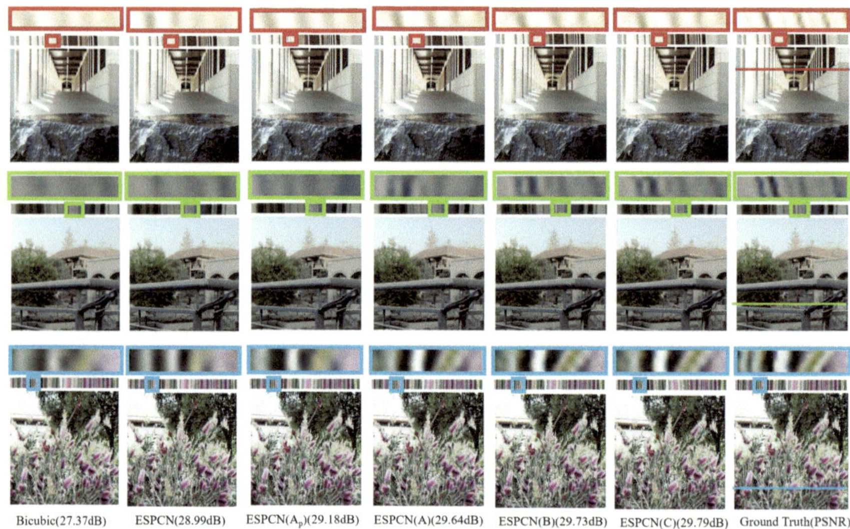

Fig. 3.10 Results of the ablation study. The epipolar images (EPIs) are shown in the red boxes. Clearer EPIs indicate a better LF structure. ESPCN(A_p) considers the content loss term $\mathcal{L}_c$ only. ESPCN(A) considers both the content loss and structure loss. ESPCN(B) adds the variance loss, and ESPCN(C) adds the disparity loss

Table 3.4 Ablation study based on the Set5, Set14, Urban100, BSD100, and Campus17 datasets with 3× scale. The settings are the same as those used in Fig. 3.10

Method	Scale	Set5	Set14	Urban100	BSD100	Campus17
		PSNR/SSIM	PSNR/SSIM	PSNR/SSIM	PSNR/SSIM	PSNR/SSIM
Bicubic	×3	30.39/0.87	27.55/0.77	24.46/0.73	27.21/0.74	27.88/0.82
ESPCN	×3	32.84/0.90	29.27/0.82	26.63/0.79	28.64/0.79	29.80/0.86
ESPCN(A_p)	x3	32.92/0.90	29.36/0.82	26.74/0.79	28.68/0.79	29.89/0.86
ESPCN(A)	×3	33.11/0.91	29.52/0.83	26.95/0.82	28.80/0.80	30.21/0.87
ESPCN(B)	×3	33.17/0.91	29.63/0.83	27.16/0.82	28.86/0.80	30.29/0.87
ESPCN(C)	×3	**33.21/0.91**	**29.73/0.83**	**27.23/0.83**	**28.90/0.80**	**30.34/0.87**

dB) and a better visual result than the other methods. The variance loss and disparity loss are considered in ESPCN(B) and ESPCN(C), respectively, leading to 0.09 dB and 0.06 dB improvements in the PSNR. Table 3.4 shows the ablation study results based on other datasets. The largest PSNR improvements achieved with $\mathcal{L}_c$, $\mathcal{L}_s$, $\mathcal{L}_v$, and $\mathcal{L}_d$ are 0.11 dB, 0.32 dB, 0.21 dB, and 0.1 dB, respectively.

3.3 Gigapixel Plenoptic Videography

3.3.1 Gigapixel Videography with a Spherical Optical Lens

In 2012, Brady et al. built the world's first gigapixel video camera AWARE2 [5], which adopts a two-scale optical imaging system: a spherical objective lens is used to capture a large FOV image of the whole scene, and 98 microcameras relay parts of the image onto the sensor (Fig. 3.11a).

The novel spherical objective lens has a curved focal plane, simplifying the lens design and supporting a very wide FoV. The objective lens used in AWARE2 is called Gigagon [28], which only has 9 layers but supports 40 gigapixels within a 120-degree FoV (far better than the conventional lens). Cossairt et al. also proposed a computational gigapixel imaging method using a spherical objective lens and derived a scaled raw image to analyze the system performance [6]. In contrast, they only used a simple one-layer spherical lens and utilized a deconvolution-based postprocessing step to obtain sharp images. As no curved sensor supports such a high resolution, a two-scale imaging architecture was proposed. Several microcameras relay small parts of the wide FoV image onto the sensor, correct the objective aberrations, and generate diffraction-limited small-portion images. In AWARE2, 98 microcameras are used to cover a 120 × 40-degree FOV (Fig. 3.11b).

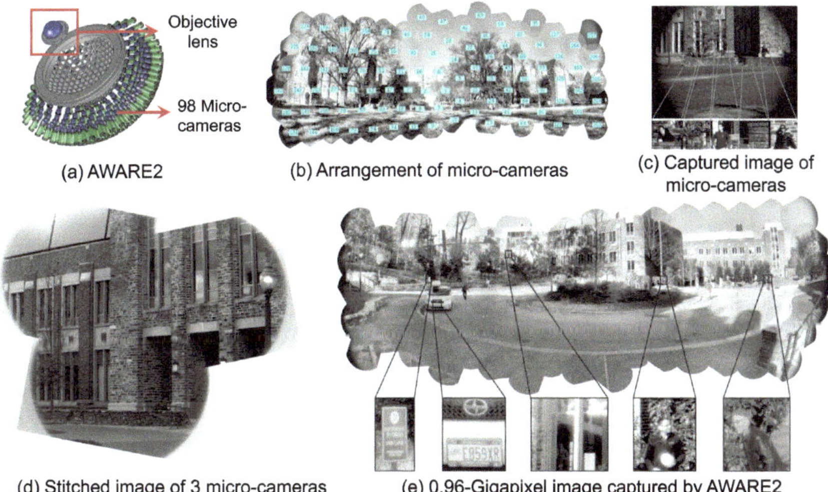

Fig. 3.11 AWARE2 gigapixel camera. **a** The two-scale imaging architecture with a spherical objective lens and 98 microcameras. **b** The spatial arrangement of the 98 microcameras. **c** An image captured by a single microcamera. Due to the limited size of the microcamera lenses, there are severe vignetting effects in the captured image, and only the center portions are useful. (**c**) A stitched image from 3 microcameras. **d** A 0.96 Gigapixel image is captured by the AWARE2 camera

Before stitching these small images, calibration is essential to determine the most accurate position and pose for the microcameras. Therefore, a specific testbed is designed and built to calibrate and characterize the AWARE2 system. Similar to existing imaging stitching methods, overlapping FoVs between adjacent microcameras are required for calibration. As shown in Fig. 3.11c, d, the microcamera is designed to image up to 4.6 degrees off axis, and the field beyond 3.5 degrees is used to provide the overlap, which means that only approximately 58% ($3.5^2/4.6^2$) of the sensor pixels contribute to the final gigapixel imaging results. Figure 3.11e demonstrates a gigapixel-level image captured by the AWARE2 system. Due to the large data bandwidth, the image processing speed of AWARE2 is only 3 frames per minute.

3.3.2 Gigapixel Macroscopy with Flat-Curved-Flat Optical Lens

Fan et al. extended the structured camera array to a gigapixel macroscope [11]. Imaging biological dynamics at a large scale with a high spatiotemporal resolution is essential for systems biology research. However, due to the space-bandwidth product theorem, conventional microscopes have an inherent trade-off between the FoV and the spatial resolution. In addition, handling the enormous amount of data generated by large-scale imaging platforms is challenging.

These bottlenecks were addressed by designing a flat-curved-flat imaging system, in which the sample plane is magnified onto a large curved surface and seamlessly relayed to multiple planar sensors. As shown in Fig. 3.12, the real-time, ultralarge-scale, high-resolution (RUSH) imaging platform operates with a $10 \times 12\,\text{mm}^2$ FoV, a uniform resolution of $\sim 1.20\,\mu\text{m}$, and a data throughput of 5.1 gigapixels per second.

The RUSH platform can capture video-rate, gigapixel-level biological dynamics at the centimeter scale with micrometer resolution, supporting in vivo fluorescent imaging in awake, behaving animals. As shown in Fig. 3.13, in vivo long-trace tracking of single leukocyte trafficking along vasculatures was performed to demonstrate the imaging capability of the RUSH macroscope. The RUSH system can be used to observe the migration of leukocytes in the vasculature across the superficial cortex, and the average speed of the cells was measured as $\sim 65\,\mu\text{ms}^{-1}$.

3.3.3 Unstructured Gigapixel Videography

Although the feasibility of the gigapixel camera was verified with AWARE2, existing techniques are still far from practical use because of the low imaging frame rate and low sensor-pixel utilization. Yuan et al. proposed UnstructuredCam (Sect. 2.3.3), which does not require overlapping FoVs between adjacent microcameras, improving the sensor-pixel utilization. In contrast to AWARE2, UnstructuredCam generates

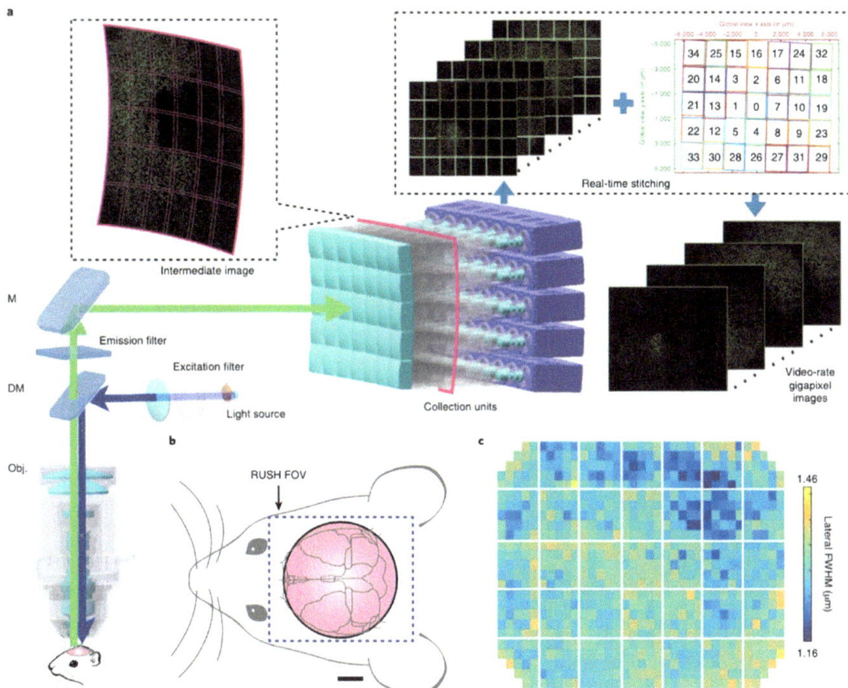

Fig. 3.12 Schematic and characterization of the gigapixel microscope system. **a** Diagram of the RUSH system. In fluorescence imaging mode, the excitation beam from the light source is filtered with an excitation filter and reflected by a dichroic mirror (DM) before passing through a customized objective and being projected onto biological samples. The fluorescence is collected by the same objective, filtered by the DM and an emission filter, and reflected by a mirror (M) to generate a spherical intermediate image. The spherical field is divided into 5 × 7 sub-FOVs and imaged with corresponding collection units composed of a relay lens array and 35 sCMOS cameras. The customized objective lens is designed with 0.35 NA and 10 × 12 mm^2 FOV; the collection units provide a data throughput of up to 5.1 gigapixels per second. **b** Illustration of the FOV of the RUSH system. Scale bar, 2,000 μm. **c** Average FWHM of the 500-nm-diameter fluorescence beads across the whole FOV

gigapixel images/videos by embedding the local microcameras into the global microcamera array using cross-resolution matching and warping techniques (Unstructured Embedding). Figure 3.14 demonstrates this pipeline, which consists of three steps: coarse localization, local video embedding and color correction. The first two steps utilize a coarse-to-fine strategy to identify feature matches and estimate a mesh-based embedding field. The third step corrects the color of the embedded local microcameras.

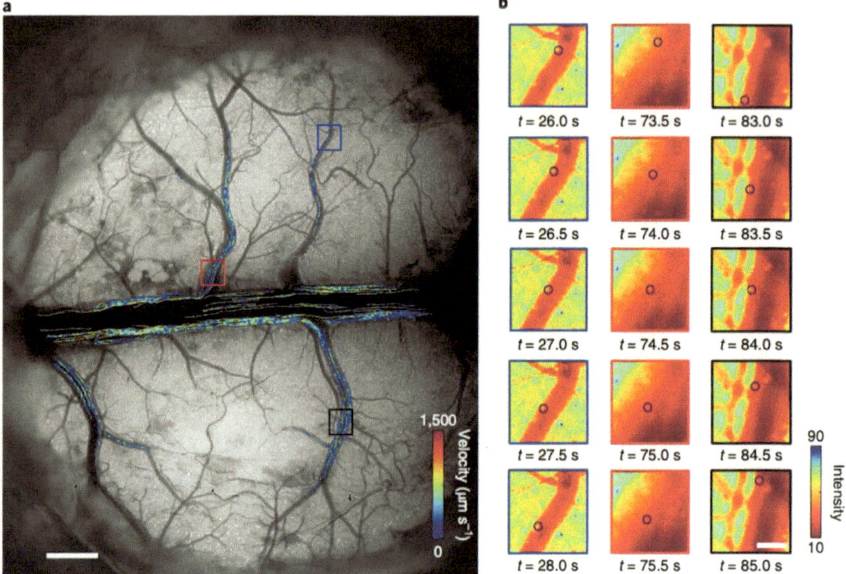

Fig. 3.13 In vivo brain-wide imaging of a Cx3Cr1-GFP mouse. **a** Tracking traces of GFP-labeled immune cells, with the different colors representing different velocities. **b** Enlarged views of the areas indicated in a. Each column corresponds to one area (color coded) at different time stamps, denoting the movement of the immune cell. Scale bars, 1,000 μm (**a**), 200 μm (**b**)

Coarse Localization

This step aims to identify a small block in the global video (denoted as the "cropped global block"), as shown in Fig. 3.14a. The structured edge map [8] is used to increase the robustness of the results. More specifically, the response map is computed as

$$\mathbf{R} = ZNCC\{\mathbf{I}^g, \mathbf{I}^l_{\downarrow}\} \odot ZNCC\{E(\mathbf{I}^g), E(\mathbf{I}^l_{\downarrow})\}, \tag{3.18}$$

where $\mathbf{I}^g$ and $\mathbf{I}^l$ denote the global video and local video, respectively. $\odot$ denotes the elementwise product. $E(\cdot)$ is the structured edge operator [8], and $\downarrow$ indicates the downsampling operator. The sampling scale can be calculated based on the ratio of the focal lengths of the global and local microcameras f_g/f_l. $ZNCC\{A, T\}$ denotes the zero-mean normalized cross correlation (ZNCC) metric, where T is the template and A is the input image containing the template. Then, a global pre-warping step is applied: the cropped global image is divided into 2×2 blocks, and the center of each block is extracted as a template. ZNCC block matching is applied again to identify 4 matching point pairs, and a global homography matrix H_g is estimated using the 4 point pairs to pre-warp the local microcamera to the global microcamera.

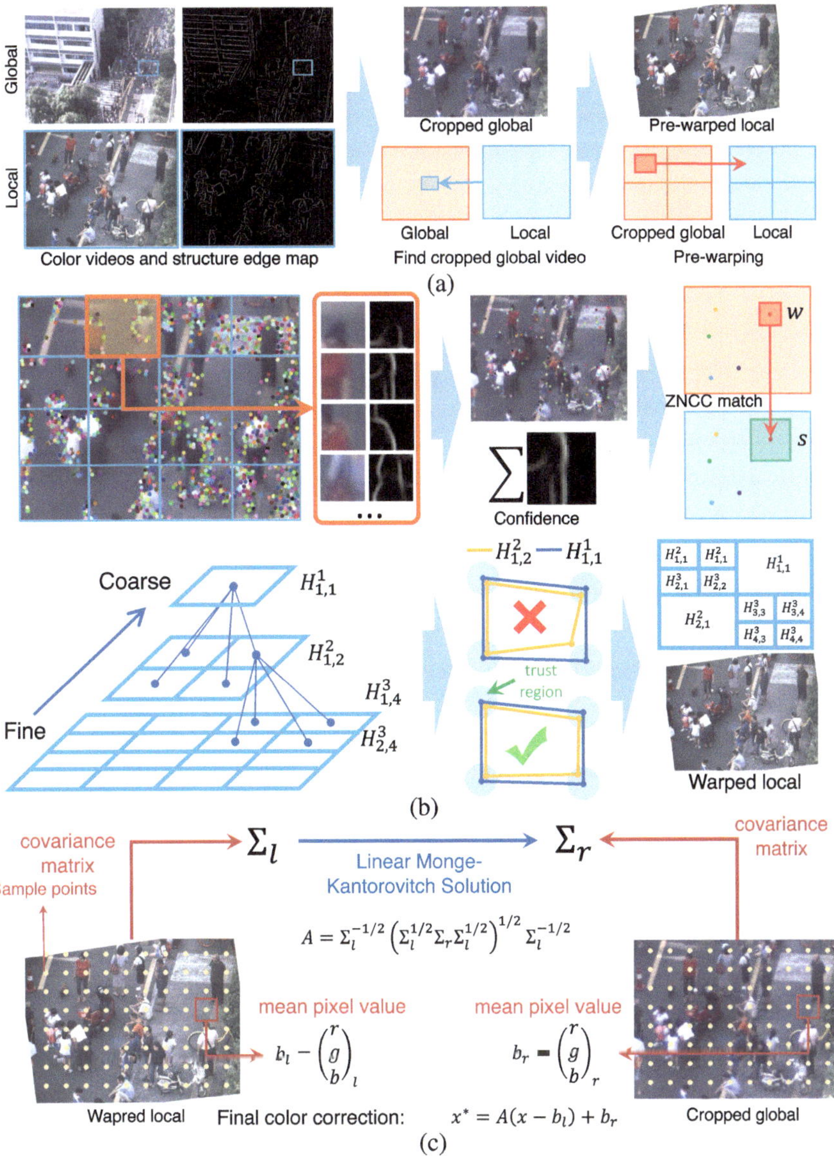

Fig. 3.14 Unstructured embedding with cross-resolution matching and warping. **a** Coarse localization. **b** Local video embedding. The top row shows the feature matching, and the bottom row shows the multiscale mesh fusion. **c** Color correction

Local Embedding

Because the optical centers of the global and local microcameras are different, global homography mapping cannot accurately map the pixels. Hence, local embedding is proposed to establish a mesh-based multiple homography model [17, 25] to embed local videos into global videos. As demonstrated in Fig. 3.14b, the cropped global block is first divided into $N \times N$ subblocks. For each subblock, keypoints (denoted as $\{p_i^g\}$ are detected using the GoodFeatureToTrack [33] algorithm, and only the first k keypoints with the highest confidence scores are preserved. The confidence score is defined as the sum of a $w \times w$ patch centered at p_i^g,

$$\mathbf{C}(p_i^g) = \sum \mathbf{E}(\mathbf{G})(p_i^g, w), \tag{3.19}$$

where $E(G)$ is the structured edge map of the cropped global block. ZNCC matching is then used to find matching points:

$$\begin{aligned} p_i^l &= \arg\max_{p_i^l} \mathrm{ZNCC}(\mathbf{I}^g(p_i^g, w), \mathbf{I}^l(p_i^l, w)) \\ \text{s.t.} \quad & \|p_i^l - p_i^r\|_2 < \epsilon, \end{aligned} \tag{3.20}$$

where p_i^g is the detected keypoint in the cropped global block, p_i^l is its corresponding matching point in the pre-warped local image, and w is the patch size used for the ZNCC computation. With this approach, many matching points can be found to estimate the warping field. For a resized 1500×1000 image, five-layer meshes $(1 \times 1, 2 \times 2, 4 \times 4, 8 \times 8, and 16 \times 16)$ are used, with $N = 16$, $k = 3$, $w = 256$, and $\epsilon = 128$ or 256.

Next, a multiscale fusion method is proposed to estimate the mesh-based multiple tomography model, as shown at the bottom of Fig. 3.14b. More specifically, a mesh is used to represent the mesh-based homography model of each layer. The coarsest layer has only one quad, similar to the global homography model. In the next finest layer, the one quad is divided into 4 quads. In each layer, each quad is used to estimate a homography model based on the matching point pairs inside the quad. As a result, the mapping model in the finest layer is more accurate, while the mapping model in the coarsest layer is more robust. Hence, an accurate and robust mesh-based multiple homography model can be estimated by merging the coarse and fine layers. A trust region-based fusion method is proposed to combine the layers. If the 4 vertices of a quad in the finer mesh are inside the trust region of the 4 coarser mesh vertices, the quad in finer scale mesh is accepted; otherwise, it is rejected. The trust region is defined as a circle with radius $r = \alpha w$, where w is the width of the mesh quad and $\alpha = 0.1 \sim 0.2$ is the parameter used to adjust the rigidity of the mapping field (Fig. 3.15).

Color Correction

The purpose of this step is to correct the color of each local microcamera based on its corresponding global microcamera. First, uniform anchor points in the warped local video and the cropped global block are sampled. For each point, two patches are cropped based on these two images, and the affine matrix A for color correction is estimated using the linear Monge-Kantorovitch solution [30]. The bias is calculated based on the mean pixel values of the two patches. Finally, for each pixel in the warped local video, bilinear interpolation is used to obtain the color correction parameter $\{A, b_l, b_r\}$. Figure 3.16 demonstrates two gigapixel images before and after color correction. Without color correction, the color differences between the background global microcamera and the embedded local microcameras are quite obvious. After color correction, the differences are essentially neglible.

Results

Figure 3.15 demonstrates one generated gigapixel videography result. Four regions are magnified twice to visualize the details in the scene, including the traffic board (more than 1 km away), car license plate (approximately 500 m away), and human bodies and faces (approximately 200 m away). In addition, UnstructuredCam supports movable local microcameras during imaging, as shown in Fig. 3.17. The pixels in the red rectangle are captured by a movable microcamera, and the image content inside the rectangle has significantly higher resolution than the background

Fig. 3.15 Light field captured by UnstructuredCam

Fig. 3.16 Images before and after color correction

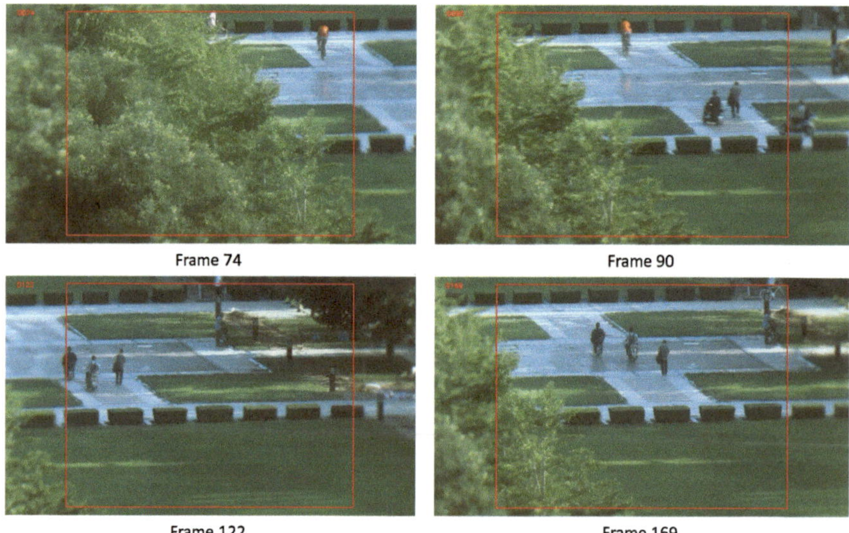

Fig. 3.17 Gigapixel-level videography with a movable local microcamera

region outside the rectangle. Figure 3.18 demonstrates that UnstructuredCam supports online calibration during real-time streaming. After disturbing a local microcamera ($t = 13$s), the unstructured embedding strategy can recalibrate the local microcamera within 1 s ($t = 14$s).

Figure 3.19 shows a comparison between the results of the unstructured embedding algorithm and the Microsoft Image Composite Editor (ICE). ICE is a state-of-the-art stitching software using the conventional overlapping-based method. Figure 3.19a shows the results generated using ICE, and the red crosses denote the images that could not be stitched. ICE is limited by the nonconvex global optimization process and becomes unstable as the number of local cameras increases. Adding one local camera affects all the other cameras. Figure 3.19b shows the unstructured

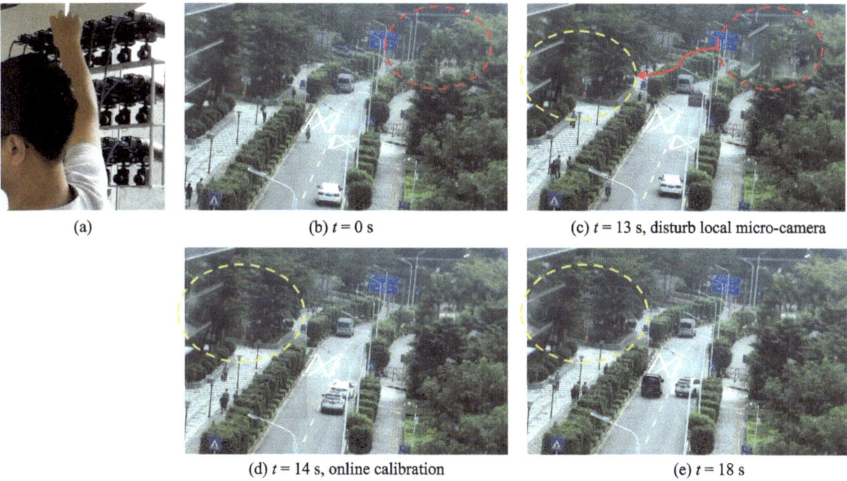

Fig. 3.18 Gigapixel-level videography with online calibration. **a** A local microcamera is manually disturbed. **b** Gigapixel videography at t = 0; the local microcamera capturing the red dashed circle region is disturbed. **c** At t = 13 s, the capturing region of the local microcamera moves from the red circle to the yellow circle. **d** At t = 14 s, the local microcamera is recalibrated online based on the new region. **e** At t = 18 s, after online calibration, the resolution of the yellow region is higher than that at t = 0

embedding results, which were always consistent, regardless of the number of input local cameras. All the local microcameras were independent of each other, leading to more stable and robust results. The two curves in Fig. 3.19c present the running times of the two strategies for the same number of microcameras. The computational complexity of the conventional method increases dramatically as the number of cameras increases, while the unstructured embedding approach has approximately linear complexity with the number of cameras, showing a higher degree of parallelism. All the running times were measured on a PC with an Intel Core i7-7700 CPU.

3.3.4 Panoramic 3D Unstructured Gigapixel Videography

Zhang et al. proposed UnstructuredCam3D, enabling 3D gigapixel videography. The hardware used by UnstructuredCam3D is discussed in Sect. 2.3.4. However, UnstructuredCam3D also poses significant challenges to the reconstruction of 3D panoramic gigapixel videography results. Figure 3 illustrates the 3D gigapixel videography generation pipeline. Our system uses the video streams captured by all the microcameras as input and generates and renders multiscale immersive content as output. Auto white balance and stereo rectification are used to maintain color consistency and reduce distortion. The 3 main components of our pipeline are described as follows (Fig. 3.20):

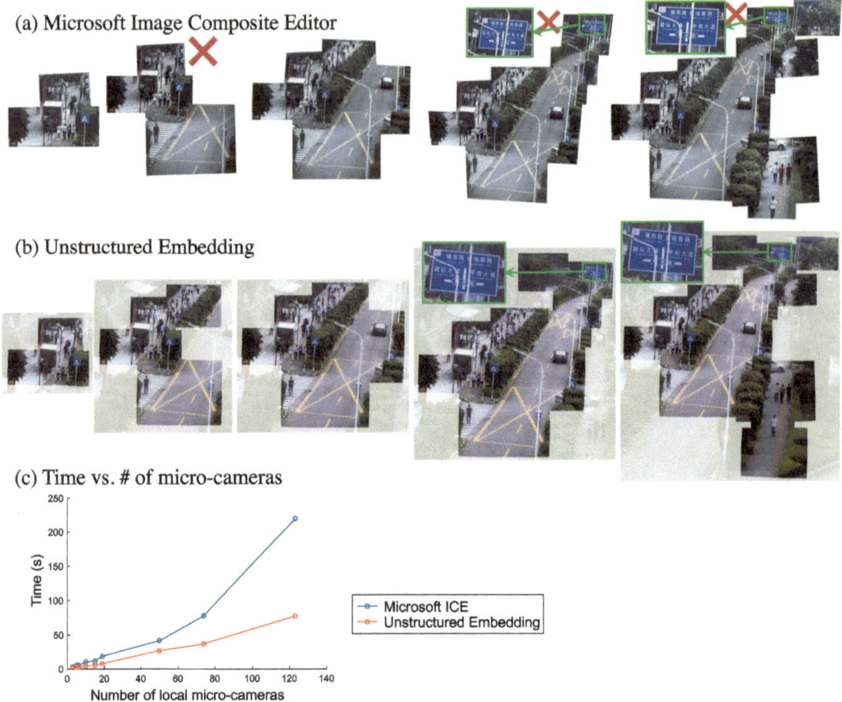

Fig. 3.19 Comparison between the unstructured embedding algorithm and the Microsoft Image Composite Editor (ICE). **a** Results generated using ICE. From left to right, stitching results using 3, 5, 6, 10, and 15 local cameras. **b** Stitched results using the unstructured embedding algorithm. **c** Running time versus the number of local microcameras

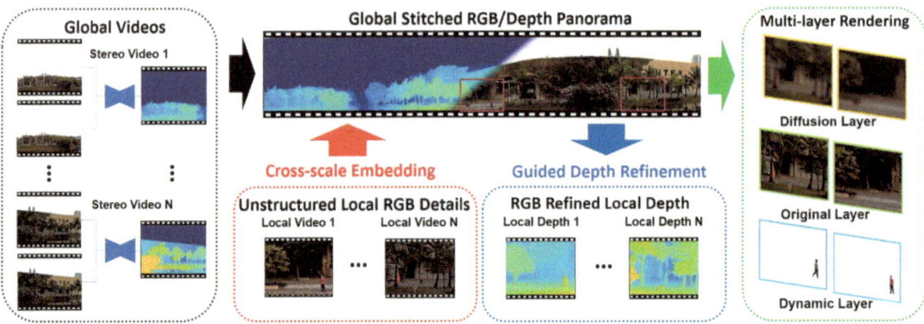

Fig. 3.20 3D gigapixel videography generation pipeline

1. Gigapixel videography The videos captured by the global microcameras are stitched to generate a large-scale panorama scene, while the videos captured by the local microcameras are seamlessly embedded into the panorama to enhance the local details.
2. Depth estimation A learning-based depth estimation network is proposed to generate a panoramic depth map based on the global microcameras, and the local depth map is estimated using the high-resolution RGB information captured by the local microcameras.
3. Layer-based rendering A three-layer-based rendering pipeline is proposed, in which the dynamic foreground and static background are processed separately to achieve realistic and fluent VR content rendering.

2D Gigapixel Videography Generation

In UnstructuredCam3D, none of the microcameras need to be calibrated, except for the two stereo global microcameras in the same column, which are calibrated for depth perception. A novel two-stage stitching scheme is proposed for unstructured gigapixel videography generation, including a global stage for global microcamera video stitching and a local stage for embedding the local microcamera videos into the global panorama. In the global stage, SIFT feature matching, RANSAC, and bundle adjustment are used to estimate the pose of each global microcamera pair. Then, graph cutting and multiband blending are used to eliminate the visual artifacts. In the local stage, an unstructured embedding module embeds all the local microcamera videos into the stitched global panorama to enhance the high-resolution details. Specifically, a cross-resolution matching algorithm is applied to identify matching points between global-local pairs, and a mesh-based multiple homography model is estimated to warp the local microcamera videos. Finally, a linear Monge-Kantorovitch algorithm is utilized to perform color correction.

3D Gigapixel Depth Estimation

Depth estimation technology plays a pivotal role in the proposed system. First, a global depth map is generated based on the semantic information. Next, a color image-guided local depth refinement step is applied to the regions covered by the local microcameras.

As shown in Fig. 3.21, the depth estimation pipeline consists of two levels, a global level for robust depth map estimation and a local level for local depth map refinement.

Global Level Depth Estimation. A good depth map should maintain the hierarchy of the scene and the semantic information, as the human vision system is sensitive to visual artifacts but insensitive to accurate absolute depth information. For instance, the rendering result may be seriously affected if the foreground-background depth

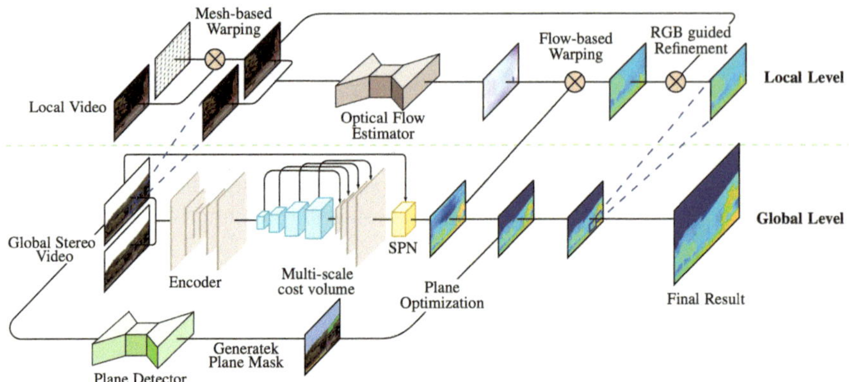

Fig. 3.21 Depth estimation pipeline

relationship is incorrect or the building plane is distorted. On the basis of this insight, a learning-based global-level depth estimation algorithm is developed. More specifically, spatial propagation layers, a plane-based correction module, and a unique hierarchical supervision loss are proposed to consider the semantic information and generate a high-quality panoramic depth map.

Feature Encoder. A feature pyramid network is used to extract the feature maps from two stereo global microcameras. A coarse-to-fine strategy is proposed to extract 4-scale feature maps to reduce the difficulty of estimating large-scale depth maps.

Multiscale 4D Cost Volume. A 4D (channel number C, height H, width W and disparity D) disparity cost volume is constructed using the extracted multiscale feature maps. Here, 4 different scales are used, corresponding to the layers of the feature pyramid. Then, 3D convolution layers are introduced to extract the semantic information and aggregate the matching cost based on the 4D cost volume. The differentiable soft-argmin operation is utilized to estimate the disparity/depth information. The probability of each candidate disparity in the cost volume is computed based on the cost c_d using the $softmax$ operation $\sigma()$. The predicted disparity value is computed by determining the weighted average of all the disparity values:

$$\hat{d} = \sum_{d=0}^{D_{max}} d \times \sigma(-c_d). \tag{3.21}$$

Spatial Propagation Network (SPN). Spatial propagation is used to improve the predicted disparity maps based on the RGB images, which are generated based on the SPNetwork [26]. Specifically, the spatial propagation layers extract the affinity matrices based on the RGB images, and the predicted disparity, which has high accuracy, can be propagated to the neighboring areas based on the affinity matrices. Hence, the noise in the estimated disparity map can be reduced.

Loss Function. First, the smooth $L1$ loss is adopted for better convergence:

$$
\begin{aligned}
L_{s1}(d, \hat{d}) &= \frac{1}{N}\sum_{i=1}^{N} l(d_i - \hat{d}_i),\\
l(x) &= \begin{cases} 0.5x^2 & |x| < 1 \\ |x| - 0.5 & |x| \geqslant 1, \end{cases}
\end{aligned}
\tag{3.22}
$$

where N is the number of disparities in the cost volume, d is the ground truth disparity, and $\hat{d}$ is the predicted disparity. A hierarchical loss function is designed to further improve the disparity map. Candidate disparity maps can be regarded as the decomposition of the final disparity map, which describe the hierarchy of the foreground and background information. Compared with supervising the final disparity maps, directly supervising the candidate disparity maps can better preserve the hierarchy of the scene. Correspondingly, the ground truth disparity map can also be decomposed into subdisparity maps, and the candidate disparity maps can be supervised by

$$
L_{submap}(D, \hat{D}) = \frac{1}{D_{max}}\sum_{i=0}^{D_{max}} l(D - \hat{D}). \tag{3.23}
$$

where D is the sliced ground truth disparity map and $\hat{D}$ is the predicted candidate disparity map. The final loss function is defined as the weighted sum of the losses computed based on the 4-scale disparity maps:

$$
L = L_1 + 0.5L_2 + 0.25L_3 + 0.125L_4. \tag{3.24}
$$

Plane Based Correction. To enhance the smoothness of the disparity map and eliminate abrupt errors, a plane detection and segmentation method is used [24] to regularize the disparity estimation process. The plane equations of the detected planes can be fitted by minimizing

$$
\begin{aligned}
&\min \sum_i (ax_i + by_i + cz_i + d)^2\\
&\quad \text{s.t.} \quad a^2 + b^2 + c^2 = 1,
\end{aligned}
\tag{3.25}
$$

where (x_i, y_i) and z_i are the coordinate and depth value of pixel i, respectively. With the plane equation, the disparity values of the pixels inside the plane can be refined.

Local Depth Refinement. The resolution of the estimated depth/disparity maps in the final step is limited by the resolution of the global microcameras. However, the local microcameras can capture high-resolution RGB images, which can be used to obtain superresolved high-resolution depth maps in the local regions. The embedding scheme embeds the local images into the global panorama. To reduce visual artifacts in the scene, the embedding scheme does not align the global and local videos pixel-by-pixel. Hence, before superresolving the depth map using the high-resolution RGB information, the estimated disparity maps and the high-resolution local RGB images must be aligned at the pixel level. Here, an optical flow approach [36] is adopted to estimate the warping field to warp the disparity maps based on high-resolution local images.

After the alignment process is performed, a bilateral solver is utilized [1] to superresolve the warped disparity maps based on the structural information of the high-resolution local RGB images. We assume that the target disparity map is t, the per-pixel confidence map is c, and the refined disparity map x can be generated by optimizing

$$\min \sum_{i,j} \hat{W}_{i,j}(x_i - x_j)^2 + \sum_i c_i(x_i - t_i)^2, \tag{3.26}$$

where $\hat{W}_{i,j}$ is the affinity matrix, which is computed based on the high-resolution local RGB images. Here, we assume that all the pixels in the disparity map have the same confidence value.

Layer-Based Rendering

A novel three-layer rendering pipeline is designed for rendering multiscale gigapixel 3D videography. As shown in Fig. 3.22, the pipeline consists of an original layer to render the background with small-range motion, a diffusion layer to handle the occluded area, and a dynamic layer to render the foreground objects with large-range motions.

Original Layer. The original layer aims to render the high-resolution background. The estimated and stitched panoramic disparity maps are backprojected to the 3D world coordinates to generate a background mesh.

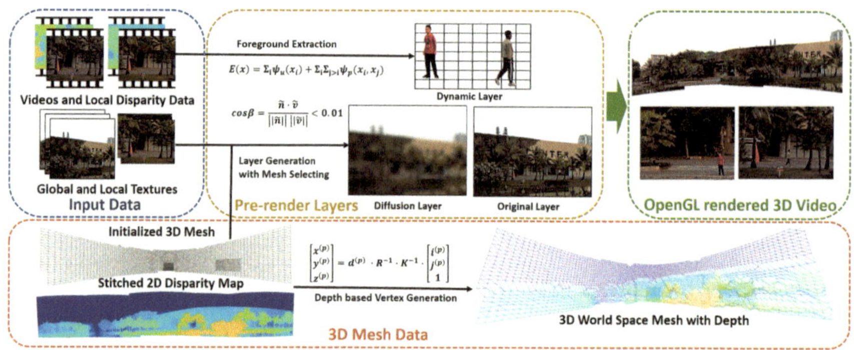

Fig. 3.22 The multilayer-based rendering pipeline

$$\begin{bmatrix} x^{(p)} \\ y^{(p)} \\ z^{(p)} \end{bmatrix} = d^{(p)} \cdot R^{-1} \cdot K^{-1} \cdot \begin{bmatrix} i^{(p)} \\ j^{(p)} \\ 1 \end{bmatrix}, \tag{3.27}$$

where $\{K, R\}$ represent the intrinsic and extrinsic camera parameters. K is precalibrated for all cameras, and R is estimated during the 2D gigapixel videography generation process. $j^{(p)}$ and $i^{(p)}$ are the pixel positions of point p in the image plane, $d^{(p)}$ is the estimated depth value, and $x^{(p)}$, $y^{(p)}$ and $z^{(p)}$ denote the 3D backprojected position of pixel p. The videos captured by the global microcameras are stitched and drawn on the background mesh using OpenGL. For regions covered by local microcameras, the mesh vertex density is increased to obtain better depth quality when zooming in.

Diffusion Layer. A simple single-layer mesh causes stretched triangle artifacts at the depth edges when moving the viewpoints, as shown in Fig. 3.26a. To eliminate these artifacts, the triangles with normal directions with large angles with respect to the view direction are removed in the original layer mesh:

$$\cos\beta = \frac{\mathbf{n} \cdot \mathbf{v}}{\|\mathbf{n}\| \|\mathbf{v}\|} < 0.01, \tag{3.28}$$

where $\boldsymbol{n}$ represents the normal direction, $\boldsymbol{v}$ represents the view direction from the face center to the original point (optical center of the imaging system), and β denotes the angle between the two directions. After tearing, holes appear when moving the viewpoint. Inspired by the two-layer rendering scheme [13], a diffusion layer is designed to for inpainting. The holes are filled by rendering a diffusion layer on the back of the original layer with a blurred texture.

Dynamic Layer. To efficiently render gigapixel-level 3D videography, only the mesh of the dynamic foreground objects is updated on a frame-by-frame basis. The foreground objects are first segmented using a Gaussian mixture model (GMM)-based background subtraction approach [46, 47] and refined by a dense conditional random field (CRF) algorithm. Then, the 3D mesh vertices belonging to the dynamic foreground objects are recomputed and rendered. The proposed rendering pipeline generates gigapixel 3D panoramic content and supports the novel zoom function for users to observe high-resolution details in local areas.

Experimental Results

Figure 3.23b shows the stitched panoramic disparity map of the global scene. The approach guarantees the integrity of the depth map by introducing semantic information, even for regions with repeated color textures (e.g., meadow, tarmac). By merging the depth information obtained by the stereo global microcameras and the local microcameras, the approach can be used to estimate high-resolution and high-quality depth maps for long-distance locations (Figs. 3.23c, d, approximately 80 m).

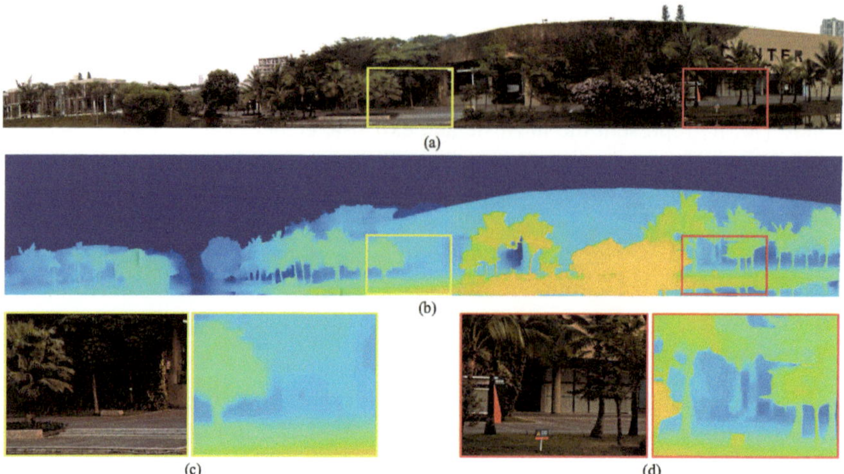

Fig. 3.23 Experimental result based on the "Campus" scene. **a** The Campus scene is captured by the UnstructuredCam3D system at the given video frame rate. The global videos are stitched and merged to generate a multiscale gigapixel panoramic video. **b** The panoramic depth video is stitched and estimated based on the stereo global microcameras. **c**, **d** The high-resolution RGB and depth images of two selected local areas

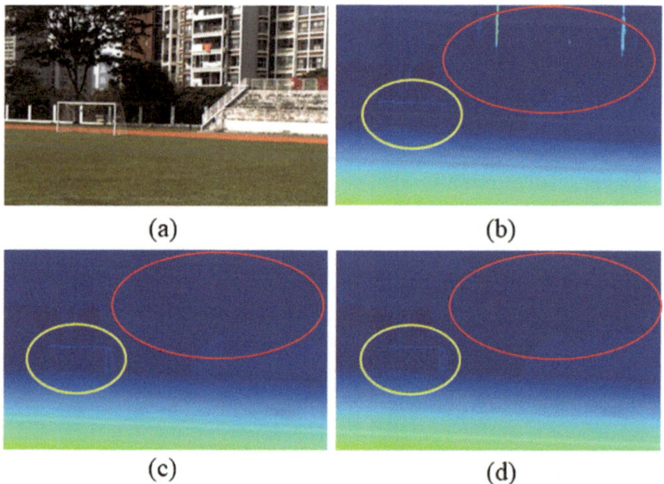

Fig. 3.24 Evaluation of the global depth estimation results. **a** Color images. **b** With the SPN layer but without hierarchical supervision. **c** Without the SPN layer but with hierarchical supervision. **d** With both the SPN layer and hierarchical supervision

In Fig. 3.24, the hierarchical supervision strategy and the spatial propagation layers used for global depth estimation are evaluated. With the hierarchical supervision strategy, abrupt disparity errors in the building areas are effectively suppressed. The

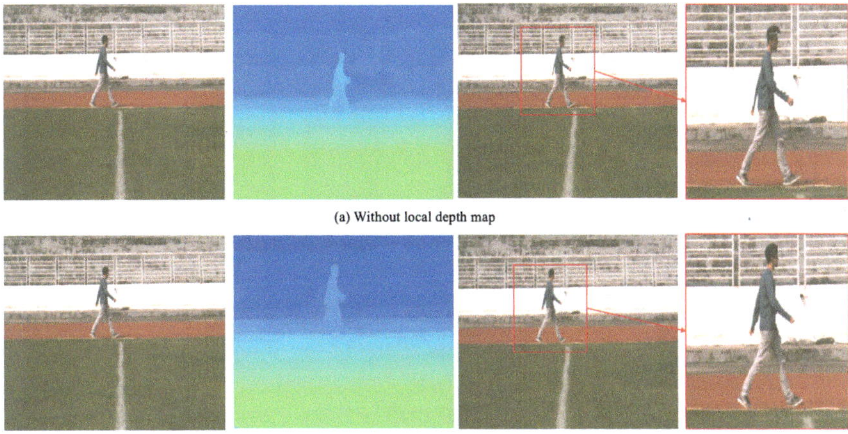

(a) Without local depth map
(b) With RGB refined local depth map

Fig. 3.25 Evaluating the performance of the local depth refinement scheme

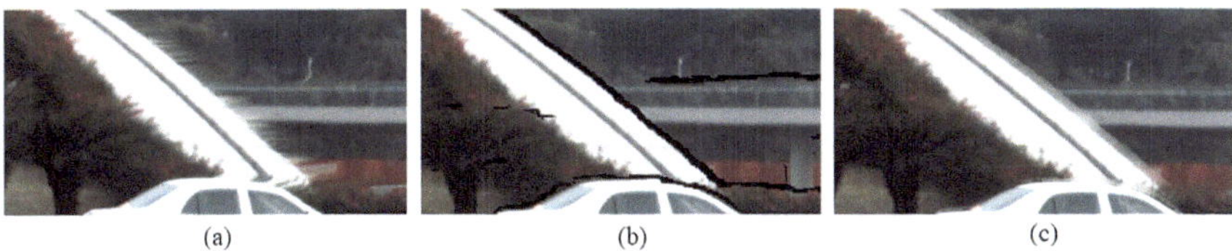

Fig. 3.26 Rendering the discontinuous depth regions. **a** Without removing the triangles at the depth edges. The mesh at the edges shows stretching artifacts. **b** Simply removing the triangles at the depth edges introduces holes in the mesh. **c** The rendering result after adding the diffusion layer. The stretching artifacts are suppressed, and the holes are filled

spatial propagation strategy further improves the depth estimation results, leading to more intact building areas.

The local depth refinement scheme not only enhances the structural details of the depth map but also aligns the depth map to the RGB image, leading to better and more accurate rendering results. As demonstrated in Fig. 3.25, the results without local depth refinement suffer from severe artifacts. The arms and the head are not well rendered because of the poor depth quality and the misalignment between the depth and RGB images. However, the proposed local depth refinement scheme addresses the aforementioned issues, leading to better rendering results that contain more semantic details.

As shown in Fig. 3.26, the proposed rendering pipeline can suppress the stretching artifacts caused by depth discontinuities, and the holes are filled by the diffusion layer. As shown in Fig. 3.27b, the GMM-based approach is used to generate an initial rough mask, and the dense CRF algorithm is applied to remove the outliers, fill the holes, and refine the outline of the mask (Fig. 3.27c). Figure 3.27f demonstrates that the

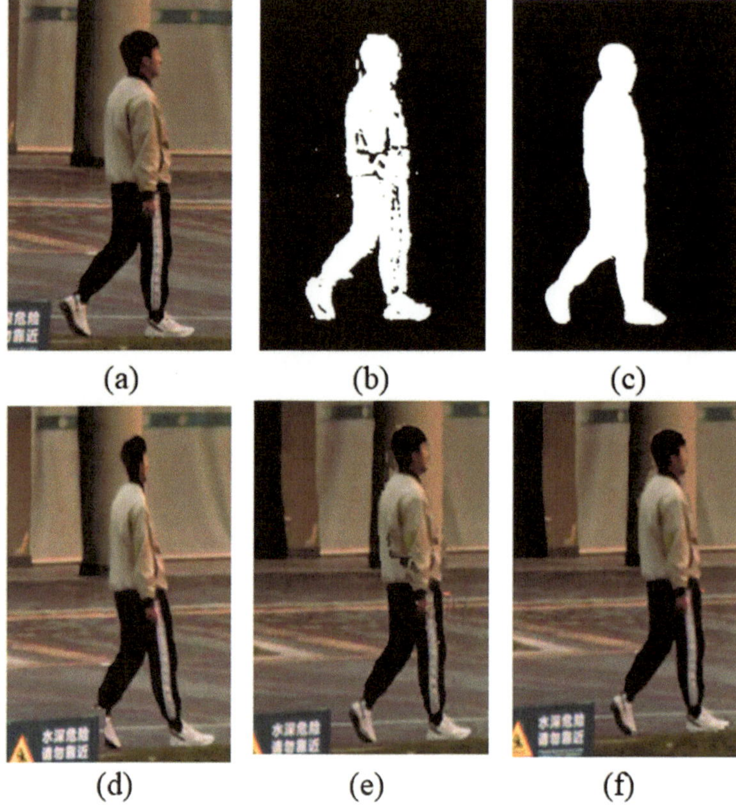

Fig. 3.27 Evaluation of the dynamic layer updating strategy. **a** The input RGB image. **b** Foreground object mask generated by the GMM. **c** Dense CRF refined mask. **d** Rendering result without a dynamic mask. **e** Rendering result with a dynamic mask generated by the GMM. **f** Rendering result with a dynamic mask refined by the dense CRF

refined foreground-object mask rejects the artifacts and leads to more immersive rendering results in the local regions.

In summary, the proposed efficient rendering pipeline uses the advantages of the high-resolution RGB panorama and the high-quality depth maps to render vivid panoramic 3D gigapixel videography.

References

1. Jonathan T Barron and Ben Poole. The fast bilateral solver. In *European Conference on Computer Vision*, pages 617–632. Springer, 2016.
2. Marco Bevilacqua, Aline Roumy, Christine Guillemot, and Marie Line Alberi-Morel. Low-complexity single-image super-resolution based on nonnegative neighbor embedding. *BMVA*

press, 2012.

3. Yochai Blau, Roey Mechrez, Radu Timofte, Tomer Michaeli, and Lihi Zelnik-Manor. The 2018 pirm challenge on perceptual image super-resolution. In *Proceedings of the European Conference on Computer Vision (ECCV)*, pages 0–0, 2018.
4. Vivek Boominathan, Kaushik Mitra, and Ashok Veeraraghavan. Improving resolution and depth-of-field of light field cameras using a hybrid imaging system. In *2014 IEEE International Conference on Computational Photography (ICCP)*, pages 1–10. IEEE, 2014.
5. David J Brady, Michael E Gehm, Ronald A Stack, Daniel L Marks, David S Kittle, Dathon R Golish, EM Vera, and Steven D Feller. Multiscale gigapixel photography. *Nature*, 486(7403):386–389, 2012.
6. Oliver S Cossairt, Daniel Miau, and Shree K Nayar. Scaling law for computational imaging using spherical optics. *JOSA A*, 28(12):2540–2553, 2011.
7. Donald G Dansereau, Bernd Girod, and Gordon Wetzstein. Liff: Light field features in scale and depth. In *Proceedings of the IEEE/CVF Conference on Computer Vision and Pattern Recognition*, pages 8042–8051, 2019.
8. Piotr Dollár and C Lawrence Zitnick. Structured forests for fast edge detection. In *Proceedings of the IEEE international conference on computer vision*, pages 1841–1848, 2013.
9. Chao Dong, Chen Change Loy, Kaiming He, and Xiaoou Tang. Image super-resolution using deep convolutional networks. *IEEE transactions on pattern analysis and machine intelligence*, 38(2):295–307, 2015.
10. Alexey Dosovitskiy, Philipp Fischer, Eddy Ilg, Philip Hausser, Caner Hazirbas, Vladimir Golkov, Patrick Van Der Smagt, Daniel Cremers, and Thomas Brox. Flownet: Learning optical flow with convolutional networks. In *Proceedings of the IEEE international conference on computer vision*, pages 2758–2766, 2015.
11. Jingtao Fan, Jinli Suo, Jiamin Wu, Hao Xie, Yibing Shen, Feng Chen, Guijin Wang, Liangcai Cao, Guofan Jin, Quansheng He, et al. Video-rate imaging of biological dynamics at centimetre scale and micrometre resolution. *Nature Photonics*, 13(11):809–816, 2019.
12. Muhammad Haris, Gregory Shakhnarovich, and Norimichi Ukita. Deep back-projection networks for super-resolution. In *Proceedings of the IEEE conference on computer vision and pattern recognition*, pages 1664–1673, 2018.
13. Peter Hedman, Suhib Alsisan, Richard Szeliski, and Johannes Kopf. Casual 3d photography. *ACM Transactions on Graphics (TOG)*, 36(6):1–15, 2017.
14. Martin Heusel, Hubert Ramsauer, Thomas Unterthiner, Bernhard Nessler, and Sepp Hochreiter. Gans trained by a two time-scale update rule converge to a local nash equilibrium. *Advances in neural information processing systems*, 30, 2017.
15. Katrin Honauer, Ole Johannsen, Daniel Kondermann, and Bastian Goldluecke. A dataset and evaluation methodology for depth estimation on 4d light fields. In *Asian Conference on Computer Vision*, pages 19–34. Springer, 2016.
16. Jia-Bin Huang, Abhishek Singh, and Narendra Ahuja. Single image super-resolution from transformed self-exemplars. In *Proceedings of the IEEE Conference on Computer Vision and Pattern Recognition*, pages 5197–5206, 2015.
17. Takeo Igarashi, Tomer Moscovich, and John F Hughes. As-rigid-as-possible shape manipulation. *ACM transactions on Graphics (TOG)*, 24(3):1134–1141, 2005.
18. Dingjian Jin, Mengqi Ji, Lan Xu, Gaochang Wu, Liejun Wang, and Lu Fang. Boosting single image super-resolution learnt from implicit multi-image prior. *IEEE Transactions on Image Processing*, 30:3240–3251, 2021.
19. Jiwon Kim, Jung Kwon Lee, and Kyoung Mu Lee. Accurate image super-resolution using very deep convolutional networks. In *Proceedings of the IEEE conference on computer vision and pattern recognition*, pages 1646–1654, 2016.
20. Johannes Kopf, Matt Uyttendaele, Oliver Deussen, and Michael F Cohen. Capturing and viewing gigapixel images. *aCm Transactions on Graphics (TOG)*, 26(3):93–es, 2007.
21. Wei-Sheng Lai, Jia-Bin Huang, Narendra Ahuja, and Ming-Hsuan Yang. Deep laplacian pyramid networks for fast and accurate super-resolution. In *Proceedings of the IEEE conference on computer vision and pattern recognition*, pages 624–632, 2017.

22. Titus Leistner, Hendrik Schilling, Radek Mackowiak, Stefan Gumhold, and Carsten Rother. Learning to think outside the box: Wide-baseline light field depth estimation with epi-shift. *arXiv preprint* arXiv:1909.09059, 2019.
23. Bee Lim, Sanghyun Son, Heewon Kim, Seungjun Nah, and Kyoung Mu Lee. Enhanced deep residual networks for single image super-resolution. In *Proceedings of the IEEE conference on computer vision and pattern recognition workshops*, pages 136–144, 2017.
24. Chen Liu, Kihwan Kim, Jinwei Gu, Yasutaka Furukawa, and Jan Kautz. Planercnn: 3d plane detection and reconstruction from a single image. In *Proceedings of the IEEE/CVF Conference on Computer Vision and Pattern Recognition*, pages 4450–4459, 2019.
25. Shuaicheng Liu, Lu Yuan, Ping Tan, and Jian Sun. Bundled camera paths for video stabilization. *ACM Transactions on Graphics (TOG)*, 32(4):1–10, 2013.
26. Sifei Liu, Shalini De Mello, Jinwei Gu, Guangyu Zhong, Ming-Hsuan Yang, and Jan Kautz. Learning affinity via spatial propagation networks. In *Proceedings of the Advances in Neural Information Processing Systems*, pages 1520–1530, 2017.
27. Adolf W Lohmann. Scaling laws for lens systems. *Applied optics*, 28(23):4996–4998, 1989.
28. Daniel L Marks and David J Brady. Gigagon: a monocentric lens design imaging 40 gigapixels. In *Imaging Systems*, page ITuC2. Optica Publishing Group, 2010.
29. David Martin, Charless Fowlkes, Doron Tal, and Jitendra Malik. A database of human segmented natural images and its application to evaluating segmentation algorithms and measuring ecological statistics. In *Proceedings Eighth IEEE International Conference on Computer Vision. ICCV 2001*, volume 2, pages 416–423. IEEE, 2001.
30. François Pitié and Anil Kokaram. The linear monge-kantorovitch linear colour mapping for example-based colour transfer. *IET Conference Publications*, pages 1–9, 12 2007. 10.1049/cp:20070055.
31. Tim Salimans, Ian Goodfellow, Wojciech Zaremba, Vicki Cheung, Alec Radford, and Xi Chen. Improved techniques for training gans. *Advances in neural information processing systems*, 29, 2016.
32. Samuel Schulter, Christian Leistner, and Horst Bischof. Fast and accurate image upscaling with super-resolution forests. In *Proceedings of the IEEE Conference on Computer Vision and Pattern Recognition*, pages 3791–3799, 2015.
33. Jianbo Shi et al. Good features to track. In *1994 Proceedings of IEEE conference on computer vision and pattern recognition*, pages 593–600. IEEE, 1994.
34. Wenzhe Shi, Jose Caballero, Ferenc Huszár, Johannes Totz, Andrew P Aitken, Rob Bishop, Daniel Rueckert, and Zehan Wang. Real-time single image and video super-resolution using an efficient sub-pixel convolutional neural network. In *Proceedings of the IEEE conference on computer vision and pattern recognition*, pages 1874–1883, 2016.
35. Pratul P. Srinivasan, Tongzhou Wang, Ashwin Sreelal, Ravi Ramamoorthi, and Ren Ng. Learning to synthesize a 4D RGBD light field from a single image. In *IEEE International Conference on Computer Vision*, pages 2262–2270, 2017.
36. Deqing Sun, Xiaodong Yang, Ming-Yu Liu, and Jan Kautz. PWC-Net: CNNs for optical flow using pyramid, warping, and cost volume. In *CVPR*, 2018.
37. Yang Tan, Haitian Zheng, Yinheng Zhu, Xiaoyun Yuan, Xing Lin, David Brady, and Lu Fang. Crossnet++: Cross-scale large-parallax warping for reference-based super-resolution. *IEEE Transactions on Pattern Analysis and Machine Intelligence*, 43(12):4291–4305, 2020.
38. Radu Timofte, Vincent De Smet, and Luc Van Gool. A+: Adjusted anchored neighborhood regression for fast super-resolution. In *Asian conference on computer vision*, pages 111–126. Springer, 2014.
39. Radu Timofte, Eirikur Agustsson, Luc Van Gool, Ming-Hsuan Yang, and Lei Zhang. Ntire 2017 challenge on single image super-resolution: Methods and results. In *Proceedings of the IEEE conference on computer vision and pattern recognition workshops*, pages 114–125, 2017.
40. Ting-Chun Wang, Jun-Yan Zhu, Nima Khademi Kalantari, Alexei A Efros, and Ravi Ramamoorthi. Light field video capture using a learning-based hybrid imaging system. *ACM Transactions on Graphics (TOG)*, 36(4):1–13, 2017.

41. Roman Zeyde, Michael Elad, and Matan Protter. On single image scale-up using sparse-representations. In *International conference on curves and surfaces*, pages 711–730. Springer, 2010.
42. Yulun Zhang, Kunpeng Li, Kai Li, Lichen Wang, Bineng Zhong, and Yun Fu. Image super-resolution using very deep residual channel attention networks. In *Proceedings of the European conference on computer vision (ECCV)*, pages 286–301, 2018.
43. Zhifei Zhang, Zhaowen Wang, Zhe Lin, and Hairong Qi. Image super-resolution by neural texture transfer. In *Proceedings of the IEEE/CVF conference on computer vision and pattern recognition*, pages 7982–7991, 2019.
44. Haitian Zheng, Mengqi Ji, Lei Han, Ziwei Xu, Haoqian Wang, Yebin Liu, and Lu Fang. Learning cross-scale correspondence and patch-based synthesis for reference-based super-resolution. In *BMVC*, volume 1, page 2, 2017.
45. Haitian Zheng, Mengqi Ji, Haoqian Wang, Yebin Liu, and Lu Fang. Crossnet: An end-to-end reference-based super resolution network using cross-scale warping. In *Proceedings of the European conference on computer vision (ECCV)*, pages 88–104. Springer, 2018.
46. Zoran Zivkovic and Ferdinand Van Der Heijden. Efficient adaptive density estimation per image pixel for the task of background subtraction. *Pattern Recognition Letters*, 27(7):773–780, 2006.
47. Zoran Zivkovic et al. Improved adaptive Gaussian mixture model for background subtraction. In *Proceedings of the International Conference on Pattern Recognition*, pages 28–31. Citeseer, 2004.

Chapter 4
Plenoptic Reconstruction

Empowered by advanced plenoptic sensing systems, light-field imaging becomes one of the most extensively used methods for capturing 3D views of a scene. In contrast to the traditional input to a 3D graphics system, namely, scenes consisting of predefined geometric primitives with different materials and sets of lights, the input to a light field is only a set of 2D images which are informative and cost effective. Unfortunately, due to the limited sensor resolution, existing systems must balance the spatial and angular resolution, i.e., one can obtain dense sampling images in the spatial dimension but only sparse sampling images in the angular (viewing angle) dimension or vice versa. This necessitates more advanced light-field reconstruction algorithms to recover high-fidelity, large-scale scenes from sparser input views with wider baselines.

This chapter covers the key aspects of light-field reconstruction. In Sect. 4.1, we introduce the theoretical foundation of light-field imaging and briefly describe the EPI-based representation and more advanced learning-based optimization. Then, in Sect. 4.2, we cast light on geometric reconstruction techniques based on multiview RGB images, including the depth map-based representation featuring scalability, the volumetric representation with global structural awareness, and the implicit neural representation with superior physical fidelity. Lastly, in Sect. 4.3, we introduce the RGB-D-based algorithms, highlighting the unprecedented robustness and efficiency for high-quality 3D or 4D geometric reconstruction and material perception.

4.1 Light-Field Image Reconstruction

Light-field imaging [65] is a widely employed technique for depicting the 3D appearance of a scene. Light-field imaging provides detailed representations of real-world environments, enabling exciting capabilities such as refocusing and altering the point of view. In contrast to conventional inputs to 3D graphics systems, namely, scenes composed of various materials and lights [65], a light-field camera can capture only a

L. Fang, *Plenoptic Imaging and Processing*, Advances in Computer Vision and Pattern Recognition, https://doi.org/10.1007/978-981-97-6915-5_4

set of 2D images that include not only the accumulated intensity at each pixel but also the light rays in various directions. Typically, these data are obtained by acquiring multiple images from different perspectives or using a microlens array.

In previous works, light-field cameras required specialized and expensive equipment, making them inaccessible to the general public. However, with the development of commercial light-field cameras such as Lytro [2] and RayTrix [1], interest in light-field imaging has increased, ushering in a new era in the field. These plenoptic (light field) cameras are composed of microlens arrays and can capture multiple images simultaneously [115]. However, in practical scenarios, issues such as dynamic scenes or limited acquisition times lead to insufficient sampling in the angular dimension. To achieve high-quality rendering, the disparities between adjacent views must be less than one pixel. This is also known as a densely sampled light field (LF). As a result, the quality of the novel rendered views is inevitably affected by the large disparity (range) in the sampled LF. Furthermore, non-Lambertian effects in the scene, such as those found in jewelry, fur, glass, and faces, exacerbate this issue [131].

To address the difficulties of large disparities and non-Lambertian effects, traditional LF rendering methods [131] often approach plenoptic sampling and reconstruction using signal processing techniques. These methods use powerful antialiasing filters to reconstruct the LF in the Fourier domain. However, these approaches rely on precisely designed filters that depend on depth information and the degree of aliasing, which can result in blurry images due to the loss of high-frequency components [58].

With the advancement of deep learning techniques, existing methods have achieved promising results, such as depth estimation followed by view synthesis [58] or direct LF reconstruction [125]. However, existing view synthesis and LF reconstruction methods have difficulty simultaneously addressing the aforementioned challenges.

In this work, we present a theoretical analysis (Sect. 4.1.1) in the Fourier domain, revealing that the aliasing problem is the underlying issue leading to these challenges [131]. Based on this theoretical framework, we propose an improvement to existing learning-based methods in Sect. 4.1.2. Our proposed method decomposes the conventional antialiasing filtering in the Fourier domain into a sequence of operations, including shearing, downscaling, and prefiltering in the image domain. We extensively tested our method with various light fields containing complex occlusion regions, non-Lambertian surfaces, and challenging microscope scenes, and the results demonstrate the effectiveness of our approach.

4.1.1 EPI-Based Optimization

In 4D light fields $L(x, y, s, t)$, where x and y represent the spatial dimensions and s and t represent the angular dimensions, an epipolar plane image (EPI) can be acquired by gathering horizontal lines with a fixed y^* value along a constant camera coordinate t^*, represented by $E_{y^*,t^*}(x, s)$. This 2D slice can be used to obtain a 2D representation of the light field. The low-angular-resolution EPI, E_L, is obtained by downsampling the high-angular-resolution EPI, E_H, as follows:

$$\mathbf{E}_L = \mathbf{E}_{H\downarrow} \downarrow$$

Here, $\downarrow$ represents the downsampling operation. Our objective is to find an inverse operation, F, that can minimize the error between the reconstructed EPI and the original high-angular-resolution EPI:

$$\hat{F} = \min_{F} \|\mathbf{E}_H - F(\mathbf{E}_L)\| .$$

When working with a densely sampled light field, where the difference between neighboring views is less than 1 pixel, the angular sampling rate meets the Nyquist sampling criterion. The specific calculations can be found in [67]. The plenoptic function can be used to reconstruct the light field. However, for light fields sampled under the Nyquist sampling rate in the angular dimension, the difference between neighboring views is always more than 1 pixel. This results in undersampling of the light field, which leads to the loss of high-frequency components in the angular dimension while preserving the spatial information. This uneven distribution of information between the angular and spatial dimensions causes aliasing effects in the EPI, which worsen as the angular resolution increases. The ghosting effects in the reconstructed light field are due to the aliasing in the EPI. This uneven information distribution always occurs when the difference between the neighboring views is more than 1 pixel. A more detailed analysis of angular aliasing effects in the spatial domain can be found in [116].

To balance the information in the spatial and angular dimensions in the EPI, the spatial resolution of the light field can be decreased to an appropriate level. However, with this approach, novel views with the original spatial quality become difficult to recover, especially if a large downsampling rate is needed. Thus, instead of reducing the spatial resolution of the light field, we extract the low-frequency information by convolving the EPI with a 1D blur kernel in the spatial dimension. Due to the relationship between the spatial and angular dimensions [67], the blur step is equivalent to an antialiasing operation in the angular dimension. Additionally, since the kernel is predesigned, the spatial details can be easily obtained using a nonblind deblur operation after the angular restoration process.

We reformulated the reconstruction of EPI_L as follows:

$$f = \arg\min_{f} \|\mathbf{E}_H - D_{\kappa} f\left((\mathbf{E}_L * \kappa) \uparrow\right)\| ,$$

where $*$ is the convolution operator, κ is the blur kernel, $\uparrow$ is a bicubic interpolation operation that upsamples the EPI to the desired angular resolution, f represents an operation to recover the high-frequency details in the angular dimension, and D_K is a nonblind deblur operator that uses the kernel κ to recover the spatial details of the EPI suppressed by the EPI blur operation.

We next discuss a new framework to address the issues of large disparities and non-Lambertian surfaces in plenoptic theory [67]. The results of a Fourier analysis indicate that aliasing is the main problem that must be addressed. The level of aliasing is influenced by the non-Lambertian effect's highest frequency and the maximum disparity. To address these issues, we propose a comprehensive approach involving shearing, downscaling, and prefiltering operations to ensure that the LF reconstruction results are antialiased. The execution of each of these operations in the image domain enables the development of an effective learning-based pipeline.

Large Disparity and Non-Lambertian Challenges

Consider a 4D light field (LF) consisting of two spatial dimensions (x, y) and two angular dimensions (θ, ϕ), represented by $LF(x, y, \theta, \phi)$. To extract an epipolar plane image (EPI), we fix the angular coordinate ϕ and capture a set of vertical lines with a constant camera coordinate $\theta = \theta_0$, denoted as $E_{\theta_0}(x, y)$. Similarly, we can extract a horizontal EPI $E_{\phi_0}(x, y)$ by fixing the angular coordinate θ and capturing a set of horizontal lines with a constant camera coordinate $\theta = \theta_0$. For convenience, we denote the high-angular-resolution EPI as E_{HAR}, and the low-angular-resolution EPI can be viewed as an undersampled version of E_{HAR}, i.e., $E_{LAR} = E_{HAR} \downarrow$, where $\downarrow$ represents the downsampling operation in the angular dimension (θ or ϕ).

Consider a complex image consisting of multiple textured and nontextured regions at different depths. Figure 4.1a (top) shows an EPI extracted from an LF image, where the disparities in the textured regions are within 1 pixel, while the disparities in the nontextured region are larger. The Fourier transform and its simplified diagram are shown in the middle and bottom of Fig. 4.1a, respectively. The spectral support of the line with disparity d is defined by $d\Omega_u + \Omega_s = 0$, where Ω_u and Ω_s denote the frequencies along the u and s dimensions. According to previous work [131], downsampling in the angular dimension causes spectrum replicas, leading to aliasing effects, as shown in Fig. 4.1b. The overlap between the original spectrum and its replicas leads to aliasing effects, even when the corresponding disparity is less

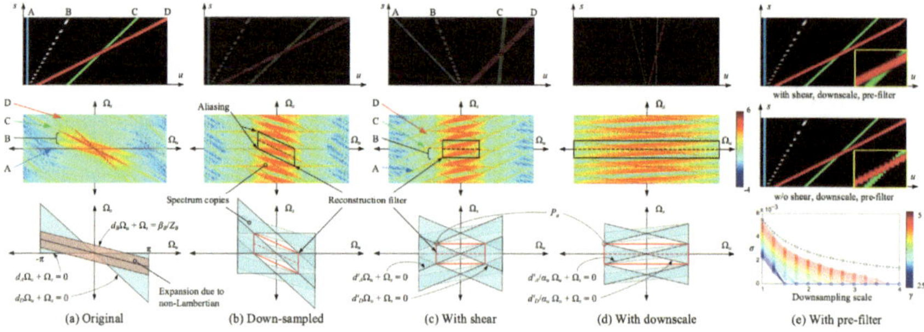

Fig. 4.1 Fourier analysis of shear, downscaling (spatial) and prefilter operations

than 1 pixel. The disparity and the highest frequency in the nontextured regions are independent factors that contribute to the aliasing problem. As the disparity or bandwidth in the nontextured region increases, the aliasing problem worsens. Therefore, addressing the aliasing problem is essential for solving challenges related to nontextured regions and large disparities in a unified framework.

An effective approach to address the aliasing problem is to apply a reconstruction filter for high-resolution EPI reconstruction. The parallelogram in Fig. 4.1b illustrates this process. The optimal rendering disparity, d_{opt}, is defined as the midpoint between the minimum and maximum disparities, d_{min} and d_{max}. The cutoff frequency can be determined based on the degree of aliasing. While it is possible to design a network to determine the optimal disparity and the cutoff frequency by combining deep learning techniques with reconstruction filters in the Fourier domain, this approach requires a global understanding of the entire spectrogram, resulting in a large number of required parameters. This issue can be addressed by using local receptive fields in the network architecture, allowing for more efficient parameterization while still achieving accurate reconstruction results.

Framework for Antialiasing Reconstruction

Spatial Transformation Pipeline. Our proposed antialiasing framework is designed to address the large disparity and non-Lambertian effects in light-field imaging. Unlike previous methods that operate in the Fourier domain, we propose a spatial transformation pipeline composed of three sequential operations. The first operation is a shearing operation that translates the disparity range of the input EPI so that the center is at zero, thus minimizing the maximum (absolute) disparity value [101]. The shearing operation shifts each subaperture view according to the following equation:

$$f_h\left(E(u,s);\alpha_h\right) = E\left(u + s \times \alpha_h, s\right),$$

where f_h is the shearing operation and α_h is the shear amount. By shearing the EPI with the optimal rendering disparity d_{opt}, the minimum and maximum disparities become $d'_{min} = \frac{d_{min}-d_{max}}{2}$ and $d'_{max} = \frac{d_{max}-d_{min}}{2}$, while the optimal rendering disparity becomes $d'_{opt} = 0$. Figure 4.1c illustrates the EPI, the Fourier transform, and the simplified diagram after the shearing operation.

Shearing is an essential step in our method and has been found to be highly efficient in addressing small disparity ranges within EPI patches with large disparities. This step allows the reconstruction filter to be conveniently implemented in the image domain, for example, by using a Gaussian function. Nonetheless, it is important to note that the shearing operation only reduces the maximum disparity to some extent and keeps the size of the disparity range constant, which may cause aliasing problems.

Downsampling Operation. Instead of applying a straightforward Gaussian filter for antialiasing, we propose a downsampling operation that enhances the sampling interval in the spatial dimension [131]. Here, an antialiasing interpolation method is used to decrease the spatial resolution. The spatial downsampling operation also reduces the disparity range. Therefore, by rescaling a sheared EPI by a factor α_u, the spectral support of the EPI is limited to the following frequency range:

$$d'/\alpha_u \Omega_u + \Omega_s = \pm\beta/Z.$$

Here, $\beta = 0$ and $\beta > 0$ for the Lambertian and non-Lambertian cases, respectively. The minimum and maximum disparities are $d''_{\text{min}} = \frac{d_{\text{min}} - d_{\text{max}}}{2\alpha_u}$ and $d''_{\text{max}} = \frac{d_{\text{max}} - d_{\text{min}}}{2\alpha_u}$.

Downsampling Technique. A downsampling technique is applied to mitigate the effects of high-frequency components in the original signal that may cause aliasing problems. By decreasing the sampling rate, the original signal is compressed toward the desired axis, as illustrated in Fig. 4.1d. The width of the reconstruction filter is then increased by a factor of α_u to further reduce aliasing effects, as shown in Fig. 4.1d (middle and bottom). However, depending on the downsampling factor, the downsampling operation may lead to a loss of high-frequency information. To address this issue, a multiscale approach is adopted by exploiting deep learning techniques to balance the trade-off between reducing aliasing effects and maintaining high-frequency information.

Prefiltering Approach. The prefiltering technique is used to suppress high-frequency components in the EPI, thereby reducing the aliasing effects introduced by the reduced disparity range and non-Lambertian effects [8]. In this study, we demonstrate that a combination of downsampling and prefiltering operations is more effective than simply applying a prefilter or conventional reconstruction filter [131]. The prefilter is designed using a Gaussian kernel κ in the image domain, which can be transformed into a Gaussian function $\mathcal{F}_\kappa$ in the Fourier domain as follows:

$$\kappa(u;\sigma) = \frac{1}{\sqrt{2\pi}\sigma} e^{-\frac{u^2}{2\sigma^2}} \Leftrightarrow \mathcal{F}_\kappa\left(\Omega_u;\sigma\right) = e^{-\frac{\Omega_u^2}{1/\left(2\pi^2\sigma^2\right)}}.$$

Here, σ is a shape parameter. The prefilter in the Fourier domain is equivalent to a convolution with the Gaussian kernel in the image domain.

The prefilter $\mathcal{F}n$ is designed such that the amplitude of a reference aliasing point Pa does not exceed the baseline spectrum amplitude γ after the filtering operation:

$$\mathcal{F}_{E_{LRN}}\left(P_\alpha\right)\mathcal{F}_*\left(\Omega_{u,P_w};\sigma\right) \leq \gamma,$$

where $F_{E_{Ln}}$ is the Fourier transform of the EPI E_{LR}, the reference aliasing point P_a has the lowest spatial frequency in the given range, and Ω_{u,P_a} is the Ω_u coordinate of point P_a, as shown in Fig. 4.1c, d. Different γ values can be set for various degrees of antialiasing. The shape parameter σ is determined as follows:

$$\sigma \geq \sqrt{\frac{1}{2\pi^2 \times \Omega_{u,P_a}^2} \ln\left(\frac{\mathcal{F}_{E_{LR}}\left(P_a\right)}{\gamma}\right)},$$

where $\Omega_{u,P_a} = \alpha_u \left[\left(\Omega_{s,P_a} \pm \beta_{P_a} / Z_{P_a} \right) / \left(-d'_{P_a} \right) \right]$. As the depth is not influenced by the shearing or downscaling operations, we consider β_{P_a} / Z_{P_a} a constant. Additionally, the vertical coordinate Ω_{s,P_0} and the disparity after shearing d'_{P_a} are fixed. Thus, we have

$$\sigma \propto \frac{1}{\alpha_u} \sqrt{\ln \left(\frac{\mathcal{F}_{E_{L\pi}} (P_a)}{\gamma} \right)}.$$

.

4.1.2 Learning-Based Optimization

The fundamental unit in a neural network is the neuron, which processes both spatial and temporal information. We exploit this property to develop a deep learning model for image inpainting by treating each pixel as a neuron and leveraging temporal information to reconstruct the missing pixels. Specifically, we formulate the problem as an optimization task using the following equation:

$$f = \arg\min_{f} \left\| \mathbf{E}_H - D_\kappa f \left(\left(\mathbf{E}_L * \kappa \right) \uparrow \right) \right\| . \quad (4.1)$$

To prevent information asymmetry, we propose a blur-restoration-deblur framework, as shown in Fig. 4.2.

The initial step in our approach involves extracting the high-frequency spatial information from the EPI using an EPI sharpening technique (see Fig. 4.1a). This is followed by an upsampling operation that increases the angular resolution of the EPI to the desired level using bicubic interpolation. In the second step, a convolutional

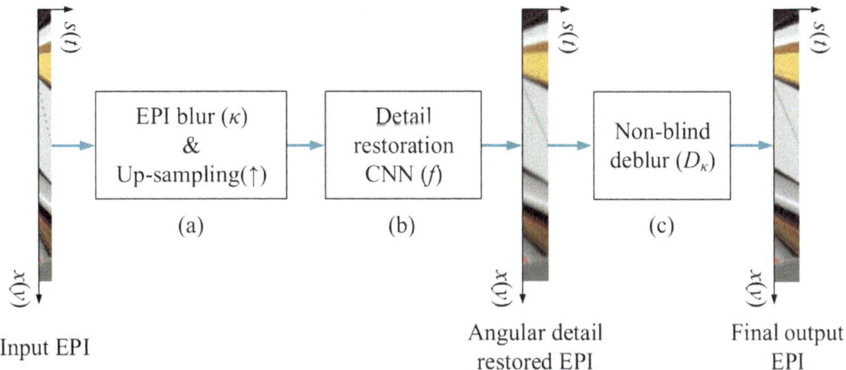

Fig. 4.2 The proposed blur-restoration-deblur framework for light-field reconstruction based on an EPIThe proposed blur-restoration-deblur framework for light-field reconstruction based on an EPI

neural network (CNN)-based restoration algorithm is employed to restore the angular details of the EPI (see Fig. 4.2b). The CNN architecture is inspired by the network proposed in [18], with the main difference being the use of residual learning to predict only the angular details. Finally, a nonblind deconvolution operation is performed to recover the high-frequency spatial details in the EPI, as shown in Fig. 4.2c. The resulting EPIs are then used to reconstruct the final high-angular-resolution light field. It is worth noting that the CNN is designed to restore the angular details lost due to undersampling of the light field, rather than the spatial details that are suppressed by EPI sharpening. While the deconvolution operation can be incorporated into the CNN, this would increase the depth of the network, slow down convergence, and complicate optimization. Therefore, we find that the nonblind deconvolution method is more appropriate for this task, as the kernel is known.

Our approach to generate high-angular-resolution light fields from sparsely sampled light fields involves a hierarchical reconstruction process, as illustrated in Fig. 4.3. In Step 1, which is depicted in Fig. 4.3b, we create a set of novel views using the input light field, indicated by the red views in Fig. 4.3a. Specifically, we use EPIs from the horizontal views to generate the green views and EPIs from the vertical views to generate the blue views. In Step 2, as shown in Fig. 4.3c, we use the images produced in Step 1 (indicated by the blue views) to generate the remaining views (indicated by the yellow views) in the final high-angular-resolution light field. We attempted to use the images indicated by the green views to generate the remaining views in Step 2, but we found that the choice of input images has minimal impact on the final results.

To reconstruct the EPIs in each step with the hierarchical process, we utilize a "blur-restoration-deblur" framework. To extract only the spatial low-frequency information from the EPI, we use a 1D blur kernel instead of a 2D image blur kernel. We consider three candidate kernels for this purpose: the sinc function, the spatial

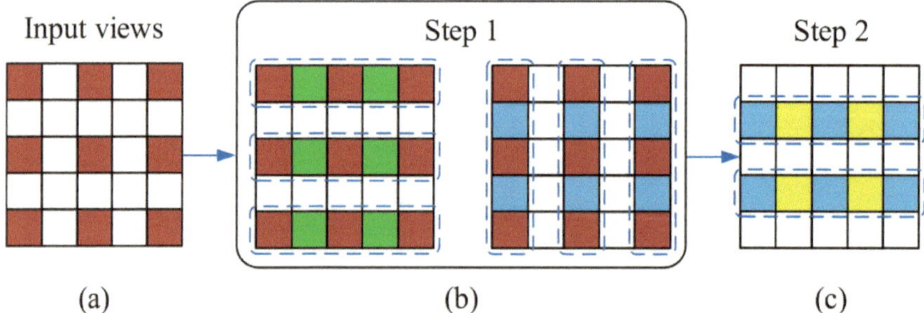

Fig. 4.3 Hierarchical reconstruction of the full light field. **a** The input light field is composed of the images marked in red. **b** In Step 1, the EPIs from the horizontal views (in the left dashed boxes) are used to generate the novel views marked in green, and the EPIs from the vertical views (in the right dashed boxes) are used to generate the novel views marked in blue. **c** In Step 2, the views generated in Step 1 (in the dashed boxes) are used to produce the images of the remaining views (marked in yellow)

representation of a Butterworth low-pass filter of order 2, and the Gaussian function. These kernels are defined in Eq. 4.2, where c_1, c_2, and c_3 are normalization constants, and σ is a shape parameter. The kernel size is selected based on the largest disparity. We evaluate these kernels based on the visual coherence of the deblurred result and the mean squared error (MSE) between the blurred low-angular-resolution EPI and the blurred ground truth EPI, as defined in Eq. 4.3. Our experiments show that the Gaussian function performs the best among the considered kernels, as the Gaussian function does not produce ringing artifacts and has the lowest MSE.

To extract only the low-frequency components of the EPI in the spatial dimension, we define the blur kernel in 1D space rather than defining a 2D image blur kernel. The following candidates were considered to extract the low-frequency components of the EPIs: the sinc function, the spatial representation of a Butterworth low-pass filter of order 2 and the Gaussian function. The spatial representations of the filters are defined as follows:

$$\begin{aligned} \kappa_s(x) &= c_1 \operatorname{sinc}(x/(2|\sigma|)) \\ \kappa_b(x) &= c_2 e^{-|x/\sigma|}(\cos(|x/\sigma|) + \sin(|x/\sigma|)) \\ \kappa_g(x) &= c_3 e^{-x^2/(2\sigma^2)}, \end{aligned} \tag{4.2}$$

where c_1, c_2 and c_3 are scale parameters, and σ is a shape parameter. The kernel size is determined by the largest disparity. The scale parameters are used to normalize the kernels.

We evaluate these three kernels based on the following two principles: the final deblurred result must show visual coherency with the ground truth EPI, and the mean squared error (MSE) between the blurred low-angular-resolution EPI and the blurred ground truth EPI must be as small as possible:

$$\hat{\kappa} = \min_{\kappa} \frac{1}{n} \sum_{i=1}^{n} \left\| \left(\mathbf{E}_L^{(i)} * \kappa \right) \uparrow - \mathbf{E}^{(i)} * \kappa \right\|^2, \tag{4.3}$$

where i is the index of the EPIs, n is the number of EPIs, E_L represents the low-angular-resolution EPI, and E represents the ground truth high-angular-resolution EPI. We evaluate the kernels based on the Stanford Light-Field Archive, and the errors between the processed (blurred and upsampled) EPIs and the blurred ground truth EPIs are 0.153, 0.089, and 0.061 for the sinc, Butterworth, and Gaussian kernels, respectively. The sinc function represents an ideal low-pass filter in the spatial dimension, and the low-frequency components can pass through the filter without distortion. However, this ideal low-pass filter causes ringing artifacts in the EPIs. The Butterworth kernel generates imperceptible ringing artifacts, whereas the Gaussian kernel does not produce ringing artifacts. Based on this observation and the numerical evaluation results, the Gaussian kernel was selected for the EPI blur operation.

Detail Restoration Based on a CNN

We developed a novel method for restoring the fine image details using a convolutional neural network (CNN) based on a residual network with three convolutional layers. This method aims to recover the fine image features that are lost during the downsampling process. Our approach is inspired by methods developed in previous works, including the SRCNN method proposed by Dong et al. [18] and the deeper network structure proposed by Kim et al. [60]. Our restoration network is composed of three convolution layers. The first layer has 64 filters of size $1 \times 9 \times 9$ for feature extraction, the second layer has 32 filters of size $64 \times 5 \times 5$ for nonlinear mapping, and the final layer has 1 filter of size $32 \times 5 \times 5$ for detail reconstruction. Each layer is followed by a rectified linear unit (ReLU) activation function. We padded the data with zeros before each convolution operation to preserve the input and output size.

Our method uses a residual learning approach to focus on restoring the high-frequency components of images that were lost during the downsampling operation while ignoring the low-frequency components. This approach better uses resources while achieving higher accuracy. The loss function is formulated as follows:

$$L = \frac{1}{n} \sum_{i=1}^{n} \left| \mathbf{R}^{(i)} - \mathcal{R}\left(\mathbf{E}_L^{\prime(i)}\right) \right|^2, \tag{4.4}$$

where n is the number of training images. We use the CIFAR-10 dataset as the training data. To improve the performance of the network and prevent overfitting, we utilize several data augmentation techniques to increase the amount of training data, including flipping the images, reducing the spatial resolution of the images, and adding Gaussian noise.

To handle different upsampling factors, we employed scale augmentation by downsampling some of the images by a factor of 4 and the desired output images by a factor of 2 and then upsampled the resulting images to their original resolution. This approach allows the network to be trained with various upsampling factors. To further improve the network convergence, we adjusted the learning rate as the training iterations increased. The network was trained in 8×10^5 iterations, and the initial learning rate was set to 0.01 and decreased by a factor of 0.001.

Evaluation

We tested the proposed algorithm based on various datasets, including real-world scenes, microscope light-field data, and synthetic scenes, and compared the performance of our algorithm with that of traditional depth-based methods and the approach proposed by Kalantari et al. [58]. The individual steps in the framework, including the blur-deblur steps and the restoration step, were evaluated separately, and the quality of the synthesized views was measured using the peak signal-to-noise ratio

Table 4.1 Quantitative results (PSNR/MS-SSIM) of reconstructed light fields based on real-world scenes

	30 scenes	Reflective 29	Occlusion 16
Wang et al. [112]	33.03/0.9766	28.97/0.9613	25.94/0.9244
Jeon et al. [52]	34.42/0.9841	40.27/0.9946	32.10/0.9830
Kalantari et al. [58]	37.78/0.9912	37.70/0.9798	32.24/0.9842
Bicubic only	34.97/0.9861	40.28/0.9952	32.97/0.9815
FSRCNN only	37.23/0.9901	43.68/0.9961	35.04/0.9848
Our CNN only	37.15/0.9889	44.84/0.9962	35.89/0.9835
Our proposed method	41.02/0.9968	46.10/0.9981	38.86/0.9970

(PSNR) and multiscale structural similarity index measure (MS-SSIM) metrics with the ground truth image. Additionally, we provide a video with the comparison results.

To assess the algorithm's performance for real-world scenes, we used 30 test scenes captured using a Lytro Illum camera by Kalantari et al. [58], as well as two representative scenes, Reflective 29 and Occlusion 18, from the Stanford Lytro Light-Field Archive. We reconstructed 7x7 light fields using 3x3 views.

We present the quantitative results of our proposed method based on real-world datasets in Table 4.1. We evaluate the performance of different view synthesis methods using the PSNR metric based on the average results of 30 scenes. We compare our proposed method with the approach by Kalantari et al. [58], which uses CNNs to minimize the error between the synthesized views and the ground truth views. However, their CNNs are designed specifically for Lambertian regions, which limits their performance for reflective surfaces, as seen in the Reflective 29 case. Our proposed method shows better performance for real-world scenes than the other approaches.

To further analyze the effectiveness of our proposed method, we compare the results of using a single CNN without the blur-deblur framework, including our network (denoted as "Our CNN only") and FSRCNN (fine-tuned based on EPIs, denoted as "FSRCNN only"). The quantitative results show that a single CNN produces lower quality light fields than the complete framework.

In summary, our proposed method outperforms existing approaches in terms of quantitative metrics, especially for reflective surfaces. The complete framework, including the blur-deblur and restoration steps, obtains better results than the single CNN.

Figure 4.4 displays several visual results, including the Leaves scenario from 30 scenes and the Reflective 29 and Occlusion 16 scenarios from the Stanford Lytro Light-Field Archive. The Leaves scene is a challenging situation due to overexposure of the sky and occlusions near the leaves, as shown in the blue box. The methods of Wang et al. [52] and Jeon et al. [52] suffer from blurring artifacts around the leaves, while the method by Kalantari et al. [58] exhibits ghosting artifacts. The Reflective 29 scene is also challenging due to the presence of reflective surfaces on the pot and kettle. The approach of Wang et al. [52] produces blurring artifacts around the pot and

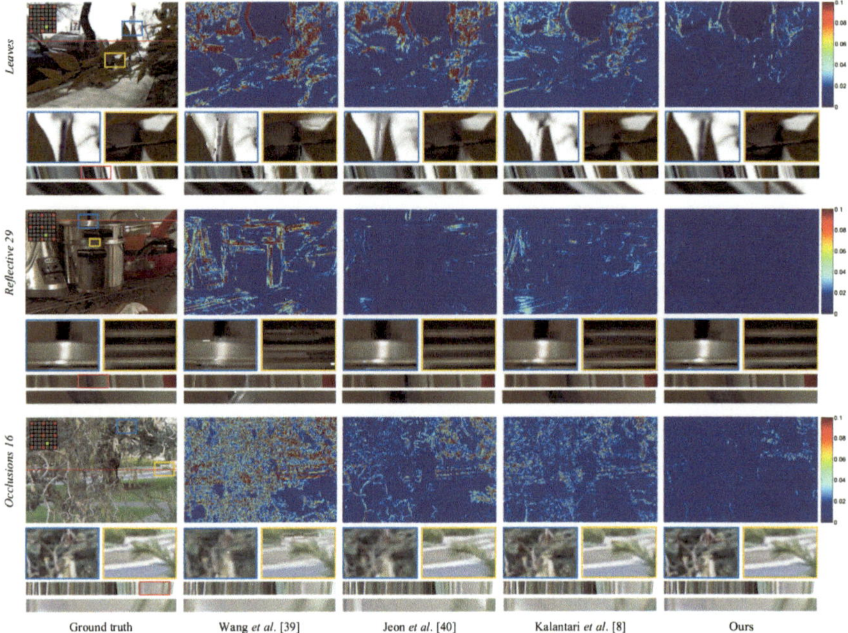

Fig. 4.4 Comparison of the proposed approach and other methods for real-world scenes. The results show the ground truth images, error maps of the synthesized results in the Y channel, zoomed-in views of the images in the blue and yellow boxes, and the EPIs located at the red line shown in the ground truth view. The EPIs are upsampled to an appropriate scale in the angular dimension for better viewing. The bottom image in each block shows a zoomed-in view of the EPIs in the red box

the kettle, while the approaches of Jeon et al. [52] and Kalantari et al. [58] exhibit discontinuities in the EPIs. The Occlusion 16 scene contains complex occlusions that make view synthesis challenging, and as a result, previous methods suffer from blurring artifacts near the occluded regions such as the branches and leaves. As depicted by the error maps and zoomed-in images, the proposed approach generates highly coherent visual results for both the synthesized views and the EPIs.

Microscope Light-Field Dataset. In this section, the methods are tested with the Stanford Light-Field microscope datasets [113] and the camera array-based light-field microscope datasets by Lin et al. [66]. These datasets include challenging light fields containing complicated occlusions and translucency. The numerical results are presented in Table 4.2, and the reconstructed views are shown in Fig. 4.5.

We used 3×3 views to reconstruct the 7×7 light fields in the Neurons 40 case and 5×5 light fields in the Neurons 40 case. Wang et al. [112] produced blurry results due to errors in the estimated depth. Jeon et al. [52] achieved higher PSNR values but failed to estimate the scene depth, which can be determined based on the EPI. Kalantari et al. [58] produced high-quality visually coherent results; however, the results were blurred and contained tearing artifacts in the occluded regions. The Worm and Cells cases have simpler structures but contain transparent objects such as

Table 4.2 Quantitative results (PSNR/MS-SSIM) of reconstructed light fields based on real-world scenes

	Neurons 20×	Neurons 40×
Wang et al. [52]	17.45/0.7368	13.21/0.7206
Jeon et al. [52]	23.02/0.9338	23.07/0.9092
Kalantari et al. [58]	20.94/0.9169	19.02/0.8847
Our proposed method	29.34/0.9741	32.47/0.9901

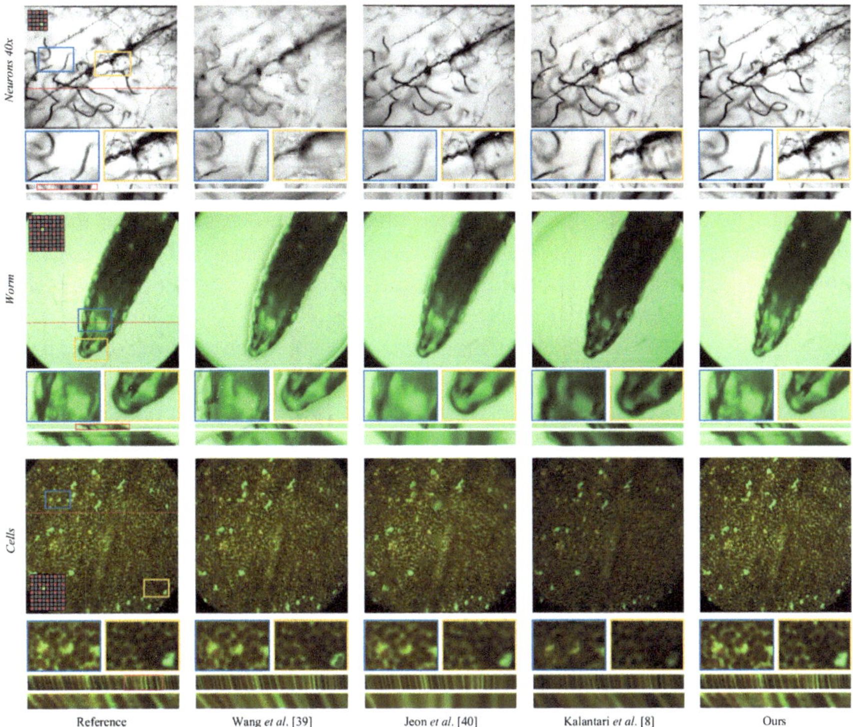

Fig. 4.5 Comparison of the proposed approach with other methods based on microscope light-field datasets. The results show the ground truth views, synthesized results, zoomed-in results, and EPIs on the red line shown in the ground truth view

the head of the worm. Depth-based approaches fail to estimate accurate depth maps in the translucent regions, resulting in tearing and ghosting artifacts. Our approach produces plausible results in both the occluded and translucent regions, even for these challenging cases.

4.2 Image-Based Geometric Reconstruction

Reconstructing 3D structures based on light fields, usually sets of input images, has been a classic topic in computer vision for more than 40 years. This is still a challenging task since the problem is generally ill posed. For example, one can sample exactly the same images from a light field with different geometries, materials, and lighting conditions. However, without further assumptions, no single algorithm can correctly reconstruct the 3D geometry based on the images alone [32]. Therefore, a set of assumptions, such as rigid Lambertian textured surfaces, are required to recover highly detailed reconstruction results. In this section, we introduce several methods to address this ambiguity with different assumptions.

All the methods described in this section have the same pipeline: recover a dense 3D model based on a set of 2D images with known camera parameters. Roughly speaking, light-field reconstruction methods can be roughly categorized into 1) depth map fusion algorithms, 2) direct point cloud reconstruction approaches, and 3) volumetric methods. Depth map fusion algorithms, which densely sample thousands of images in a scene, are the conventional approach. These methods have attracted considerable attention in light-field imaging and rendering. The advantages of these depth map fusion methods, such as robustness to occlusions [129] and reducing image noise [9], have been well studied. Unfortunately, it is impractical to densely sample a scene for high-resolution 3D reconstruction, especially for large-scale scenes. Point cloud-based methods directly use 3D points and usually apply propagation strategies to gradually densify the reconstruction results [54]. However, as the point cloud propagation proceeds sequentially, these methods are difficult to fully parallelize and usually require long processing times. Volumetric-based methods [53] divide the 3D space into regular grids and estimate if each voxel is adhered to the surface. These methods are easy to parallelize for multiview processes and handle the problem in global coordinates. However, this method has several disadvantages, including space discretization errors and high memory consumption.

In the following section, we first analyze the advantages and disadvantages of depth map-based methods. Then, an end-to-end point cloud-based algorithm with faster speed and higher accuracy is proposed. Finally, we discuss two popular volumetric-based methods to address the issue of high memory consumption.

4.2.1 ElasticMVS: Depth Map Reconstruction

Given a set of images with known camera parameters, depth map reconstruction methods aim to densely and accurately reconstruct the geometry of the scene.

Most works in this area focus on view-wise depth map estimation [33, 106] and fuse the results to generate the point cloud. Among these methods, patch match-based depth optimization methods [35, 90], which are initialized based on randomly sampled depth hypotheses, have great run times and memory efficiency for depth map estimation. Some learning-based methods [111] introduce iterative coarse-to-fine optimization methods in end-to-end trainable architectures to improve the core steps in the patch match algorithm. However, these algorithms obtain incomplete and noisy depth predictions in areas with no texture and high reflection.

Currently, learning methods based on 3D cost volume regularization have achieved state-of-the-art results according to several benchmarks. These methods optimize the 3D geometry with 3D CNN by projecting 2D images or features into 3D volumes [123]. However, these supervised methods rely on ground truth 3D data, which are difficult to generalize in real-world environments. To realize self-supervised MVS without ground truth 3D training data, [16, 80, 133] replaced the supervision signals with an unsupervised rendering loss or a cross-view feature consistency loss. However, reprojections in multiple views are highly sensitive to the environment illumination conditions, which are difficult to generalize in highly variable scenarios.

Thus, although existing learning-based approaches [123], which typically formulate the problem as an end-to-end depth regression task, outperform traditional geometry-based approaches [90], their generalizability in open environments is severely hindered by the availability of laser-scanned 3D training data. Self-supervised MVS [59] methods address this limitation by leveraging the multiview photometric consistency instead of using supervisory signals. Unfortunately, pixel-level photometric regularization methods are susceptible to textureless patterns and illumination variations, leading to incomplete and inaccurate reconstruction results, especially in outdoor environments.

We address these challenges with a novel image parsing paradigm. Instead of treating the images as projections of 3D scenes, we show that the inherent geometric correlations, such as surface connectedness, smoothness, and boundaries, can be implicitly inferred from the images and serve as reliable guidance for the pixelwise multiview correspondence estimation. The final goal is to determine the correlations between the latent space of the representation and the scene space of the surface in a self-supervised manner. In the following section, we present a detailed introduction to the novel elastic part representation framework, which encodes physically connected segments with elastically varying scales, shapes, and boundaries.

Matching in Light-Field Reconstruction

We estimate the depth maps of a given reference image x and the corresponding M unstructured source images with known camera parameters. The final point cloud reconstruction result is obtained by fusing all of the depth maps. Since the geometric information obtained from the photometric loss contains missing data and artifacts in some areas of the image (especially in textureless regions), heuristic geometry regularization methods for depth prediction cannot be used in these regions. Therefore, the key innovation in our framework is a novel *elastic part representation*, which encodes sufficient geometric details to obtain piecewise-smooth depth map predictions. We first formally define our part-aware representation. Based on the learned representation, we introduce two improvements based on the hypothesized depth propagation and evaluate the part-aware patch match algorithm. We present a self-supervised training scheme to robustly obtain the representation, even with a noisy initialization.

Elastic Part Representation Suppose that $(\tilde{d}, \tilde{n}) \in \mathbb{R}^{H\times W\times 4}$ are the depth and normal map of image x. We first define the geometry-aware part segmentation S_p related to each pixel location $p \in \mathbb{R}^2$, which is geometrically concentrated in the region $S_p \triangleq \{q \in \mathbb{R}^2 | \Gamma\{(\tilde{d}_p, \tilde{n}_p); (\tilde{d}_q, \tilde{n}_q)\} \leq \epsilon\}$, where $\Gamma\{(\tilde{d}_p, \tilde{n}_p); (\tilde{d}_q, \tilde{n}_q)\} \triangleq |([q, \tilde{d}_q] - [p, \tilde{d}_p]) \cdot \tilde{n}_p|$ is the 3D point-to-plane distance between two pixels.

Based on the part segmentation S_p of each pixel p, we propose learning a per-pixel elastic part representation $z_p \in \mathbb{R}^D$ for each $p \in \mathbb{R}^2$ such that z_p is compact only in the geometrically concentrated region S_p.

Specifically, we use the soft-nearest neighbor loss [88] to formulate the following loss function:

$$L_p = -\sum_p \log \frac{\sum_{p^+ \in S_p} \exp\left(\langle z_p, z_{p^+}\rangle/\tau\right)}{\sum_{q \neq p} \exp\left(\langle z_p, z_q\rangle/\tau\right)}, \tag{4.5}$$

where $z_{p^*} \in z$ is the per-pixel elastic part representation based on the pixel location p^*, $\langle\cdot,\cdot\rangle$ is the dot product operation, and τ is the temperature.

Finally, we use a ConvNet, $F_\Theta(\cdot)$, to model the generation of the elastic part representation, i.e., for the reference image $x \in \mathbb{R}^{H\times W\times 3}$, $F_\Theta(x) \in \mathbb{R}^{H\times W\times D}$ denotes the collection of the per-pixel elastic part representations.

Part-Aware Matching After training the part-aware representation, we build our part-aware matching module based on the seminal work of the patch match algorithm [35]. The final depth prediction is iteratively refined based on the random initialization. In particular, we incorporate our elastic part representation and propose the following two novel improvements. 1. Propagation: We propagate the depth hypotheses to neighbors based on our learned elastic part representation. 2. Evaluation: We evaluate and choose the best depth hypotheses based on the photometric consistency, feature-level correspondence, and geometric smoothness. The resulting module is differentiable, and the architecture of the part-aware patch match module is illustrated in Fig. 4.6.

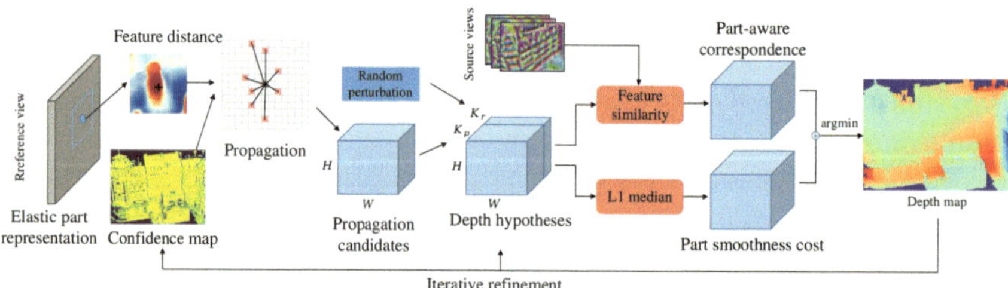

Fig. 4.6 Detailed architecture of part-aware patch match algorithm. Given the pixelwise elastic part representation obtained by ConvNet, depth hypotheses are sampled from high-confidence regions with similar features. The optimal depth values obtained during each iteration are chosen based on a combination of the part-aware representation and the part smoothness cost

Propagation

The key idea of part-aware propagation is to gather hypotheses based on nearby pixels in the same physical surface component. As illustrated in Fig. 4.7, instead of naively propagating the depth hypotheses from a set of static neighbors [35], our depth hypotheses are sampled from nearby pixels that have similar features to the learned elastic part representation.

Since the depth value is unreliable in textureless and highly reflecting areas, a confidence map $c \in [0, 1]^{H \times W}$ is generated to distinguish the reliable candidates from noise, which considers the photometric and geometric consistency [90]. A more detailed explanation is provided in the Supplementary Material. For each pixel p in the depth map to be updated, K_p patch hypotheses are propagated from the propagation candidates $\mathcal{T}_p$, which is a set of nearby pixels q with features z_q that are similar to feature z_p with high confidence c_q:

$$\mathcal{T}_p = \left\{ q \in R^2 \middle| \|z_p - z_q\| \leq \eta, c_q \geq \xi \right\}. \tag{4.6}$$

According to the property of the elastic part representation in Eq. 4.5, the depth hypotheses generated from $\mathcal{T}_p$ remain close to the underlying surface.

In addition to the photometric consistency loss, we propose two novel losses to evaluate the hypotheses: the part-aware correspondence loss and the part smoothness loss.

Part-Aware Correspondence Loss. This loss evaluates the feature similarity of the elastic part representations between the reference view and the warped source views. Given homogeneous coordinates and the depth hypothesis d, we obtain the warped source feature $z^{[i]}_{p_i(d)}$ of the i-th source view via differentiable bilinear interpolation. Therefore, given the features $z^{[0]}_p, z^{[i]}_{p_i(d)}$ of the elastic part representation in the reference and source views, the part-aware correspondence at depth d_p in the normal direction n_p is given by

$$M_s(d_p, n_p|x, z) - \left[\sum_i \langle z^{[0]}_p, z^{[i]}_{p_i(d)} \rangle + \alpha_s \cdot \rho(d_p, n_p|x, x^{[i]}) \right], \tag{4.7}$$

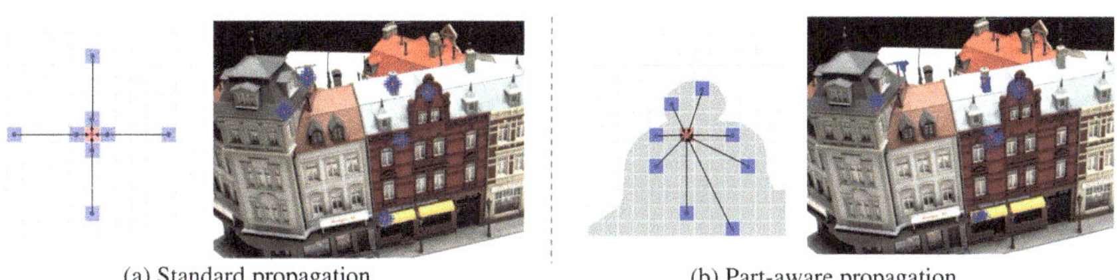
(a) Standard propagation
(b) Part-aware propagation

Fig. 4.7 Visualization of all sampled locations for propagation (blue points) during 3 iterations of the patch match algorithm given six initial pixels (red points)

where $\rho(\cdot)$ is the bilateral weighted normalized cross correlation [90] implemented in the traditional patch match algorithm.

Part Smoothness Loss. This loss aims to augment the local patch match evaluation with a nonlocal piecewise smoothness term corresponding to the same physical surface component. Inspired by the work on locally optimal projections (LOPs) for point cloud-based surface reconstruction, the part smoothness loss is formulated using an L1 median loss, which is robust to outliers and produces a piecewise second-order surface approximation for each surface component.

More specifically, the depth map $d \in \mathbb{R}^{H\times W\times 1}$ is first transferred to the point set $e \in \mathbb{R}^{H\times W\times 3}$ in the scene space based on the camera matrix $\mathbf{M}$: $e = \{e_{p^*} \in \mathbb{R}^3 | [e_{p^*}^\top, 1]^\top = \mathbf{M}^{-1}[p^{*\top}, d_{p^*}, 1]^\top\}$. Then, the part smoothness loss is defined as the sum of the Euclidean distance to the points from the propagation candidates in $\mathcal{T}_p$ (Eq. 4.6):

$$M_g\left(d_p, n_p \mid z\right) = \sum_{q\in T_p} \omega_q \left\|e_p - e_q\right\|, \tag{4.8}$$

where $\omega_q = \exp\left\{-c_p - \alpha_n\left\langle n_p, n_q\right\rangle\right\}$ is a weight correlated with the confidence c_q and the normal similarity.

Based on these components, as illustrated in Fig. 4.6, the depth map initialization consists of K_p random values within the depth range of interest. For subsequent iterations, K_p hypotheses are propagated based on the previous depth values. To obtain a more diverse set of hypotheses than can be obtained using only propagation, K_r hypotheses are sampled by adding a small random perturbation to the previous estimation.

This function is not directly solved using first-order optimization mechanisms such as gradient descent; instead, the function is approximated using a discrete sampling strategy, which has long been applied in the field of stereo matching [90]. Empirically, 5 iterations are sufficient to obtain an accurate piecewise smoothness depth map.

Network Training

The elastic part representation z is compact in areas with smooth surface components. We next describe how to train ConvNet $F_\Theta(x)$ using the self-supervised contrastive learning strategy without knowing the ground truth depth and normal information.

The core of the contrastive learning approach is constructing a set of positive and negative samples, with the representations of the positive samples being close and the representation of the negative samples being far apart. Before training, a depth map d and a normal map n of each reference image x are propagated and generated by the initialized ElasticMVS network without the part representation, which is similar to traditional patch match-based algorithms, e.g., Gipuma [35].

When selecting the positive and negative samples during the training stage, we eliminate the noise and errors in the initial depth map using the confidence map. For each pixel p with high confidence ($c_p \geq \xi$), we construct a set of pixel candidates $\hat{\mathcal{S}}_p$ for positive samples, where each sample is geometrically close to p with high confidence:

$$\hat{\mathcal{S}}_p = \{q \in \mathbb{R}^2 | \Gamma\{(d_p, n_p); (d_q, n_q)\} \leq \epsilon, c_q \geq \xi\}. \tag{4.9}$$

Therefore, the dense contrastive loss for self-supervised training is defined as

$$L_c = -\sum_p \mathbf{1}_{[c_p \geq \xi]} \log \frac{\sum_{p^+ \in \hat{\mathcal{S}}_p} \exp\left(\langle z_p, z_{p^+}\rangle / \tau\right)}{\sum_{q \neq p} \exp\left(\langle z_p, z_q\rangle / \tau\right)}. \tag{4.10}$$

This loss function encourages compact representations in regions with small surface distances and separation otherwise, i.e., the distance in the representation space naturally reflects the distance in the 3D scene space. Since this loss function only acts on pixels with confidence values greater than the set threshold, initial reconstruction artifacts do not impact the contrastive learning process. Due to memory limitations, it is impossible to densely sample all positive and negative samples in the image with Eq. 4.10. Therefore, we randomly select N_c points from all samples in $\hat{S}_p$ using a probability distribution that is inversely proportional to the distance from p.

To obtain spatially concentrated representations, we utilize a spatial concentration loss [121] that encourages isotropically isolated pixel embeddings:

$$L_s = \sum_p \left\| \frac{\sum_{q \in \hat{\mathcal{S}}_p} \exp\left(\beta \cdot \langle z_p, z_q\rangle\right) \cdot q}{\sum_{q \in \hat{\mathcal{S}}_p} \exp\left(\beta \cdot \langle z_p, z_q\rangle\right)} - p \right\|, \tag{4.11}$$

where β is a constant parameter that controls the weight of the feature similarity between p and q, which encourages the weighted average of the sampled neighboring points q to be close to point p. Therefore, the pixels far away from the sampled point have low weights, leading to a spatially concentrated representation. The concentration loss enables a reasonable distance between the representations in high-confidence regions and low-confidence regions, leading to robust training of the representations toward noisy and incomplete depth predictions.

Our final loss function is a weighted sum of the contrastive loss and concentration loss. The final loss function is defined as

$$L_{total} = L_c + \gamma_s \cdot L_s, \tag{4.12}$$

where γ_s denotes the weight of the concentration loss in the total loss function.

Experiment

A qualitative comparison of state-of-the-art MVS methods based on all scans in the Tanks and Temples dataset is shown in Fig. 4.8. Our method achieves state-of-the-art

Fig. 4.8 Point cloud visualization of the T&T intermediate and advanced benchmarks with three types of reconstruction methods: supervised methods (**a**) (**b**), self-supervised methods (**d**) (**e**), and a traditional geometry-based method (**c**)

performance among all supervised and self-supervised MVS methods and obtains a substantial performance improvement based on the *T&T advanced* benchmark, demonstrating the effectiveness and robustness of the proposed method for complex categories and geometries. The qualitative comparison of the methods for all scenes in the *T&T* dataset demonstrates that our method precisely reconstructs more complete scenes with less noise.

4.2.2 SurfaceNet: Volumetric Reconstruction

Point cloud reconstruction methods directly utilize global 3D points, with sparse features detected and propagated to obtain gradually denser reconstruction results [33, 37].

Recent advances in deep learning have revolutionized multiview stereopsis, with neural networks utilized as basic building blocks to replace the key components in traditional pipelines. Replacing the entire pipeline with an end-to-end learning framework that uses images with known camera parameters as inputs to infer the surface of a 3D object can realize the full potential of deep learning for multiview stereopsis. This end-to-end approach allows the image consistency and geometric context for dense reconstructions to be simply inferred based on data without the need to manually design separate processing stages.

In the following sections, several multiview point cloud reconstruction frameworks are introduced. First, we present technical details about SurfaceNet [53], which is the first end-to-end learning framework for multiview stereopsis. An extension of SurfaceNet, which addresses very sparse multiview stereopsis problems, is also briefly discussed. Finally, we discuss recent unsupervised point cloud reconstruction methods and neural representations.

Overall Framework

SurfaceNet was the first successful end-to-end learning framework for multiview stereopsis that did not focus on enhancing individual pipeline steps. SurfaceNet directly infers 3D models based on image sequences and the related camera settings. The main benefit of the framework is that end-to-end learning of image consistency and geometric relationships related to the surface structure can be determined for multiview stereopsis.

SurfaceNet, a 3D CNN, can process two or more views, and all accessible views of the predicted surface are used to directly compute the loss function. The colored voxel cube (CVC), a novel representation of accessible viewpoints, was proposed to implicitly encode the camera parameters via a simple perspective projection operation outside the network to obtain a fully convolutional network.

Colored Voxel Cube Representation

As shown in Fig. 4.9, for a given view v, the image I_v is converted into a 3D colored cube I_v^C by projecting each voxel $x \in C$ onto the image I_v and storing the RGB values i_x for each voxel. The mean color is determined based on the color values. The voxels that are on the same projection ray have the same color i_x because the same

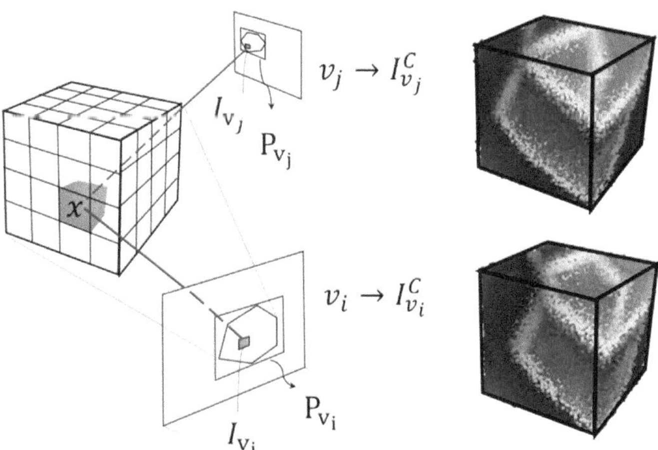

Fig. 4.9 Illustration of two colored voxel cubes (CVCs)

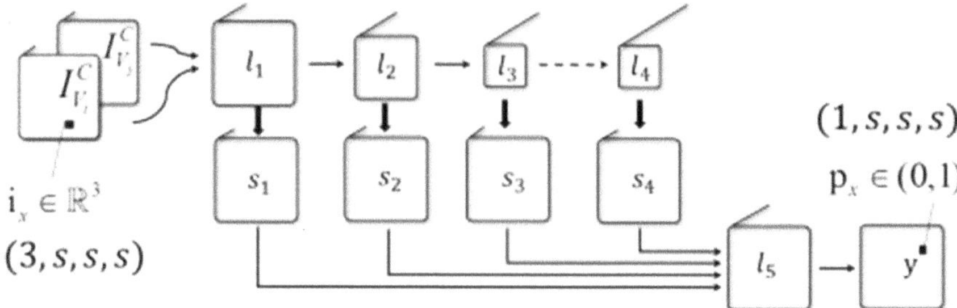

Fig. 4.10 SurfaceNet uses two CVCs from different viewpoints as input. There are four groups of convolutional layers in the forward direction. Two dilated convolution layers make up the l_4 layers. The output layer y predicts the surface probability for each voxel position by combining the multiscale information extracted by the side layers s_i

representation is generated for each voxel. In other words, the camera parameters can be encoded with the CVCs, resulting in projection-specific stripe patterns.

The network architecture is shown in Fig. 4.10. The network uses two CVCs with distinct angles as input and predicts the confidence $p_x \in (0, 1)$ for each voxel $x \in C$, which indicates if the voxel is on the surface. The basic building blocks of the network are 3D convolutional layers $l(\cdot)$, 3D pooling layers $p(\cdot)$ and 3D up-convolutional layers $s(\cdot)$. A rectified linear unit (ReLU) is appended to each convolutional layer l_i, and the sigmoid function is added after layers s_i and y. To decrease the training time and increase the robustness of training, batch normalization [50] is utilized before each layer. The layers in l_4 are dilated convolutions [126] with a dilation factor of 2. These layers are designed to exponentially increase the receptive field without reducing the feature map resolution. Layer l_5 improves network performance by aggregating multiscale contextual information from the side layers s_i, allowing the network to consider multiscale geometric features. Because the network is completely convolutional, the CVC cube size can be changed as needed. The sizes of the output and the CVC cubes are always the same.

Inference

A point cloud can be reconstructed by an additional binarization step based on the surface probabilities predicted by the neural network. This simple thresholding operation converts all voxels with $p_x > \tau$ into surface voxels and sets all other voxels to zero. Additionally, since the surface may be several voxels thick after the binarization operation, a thinning operation known as ray pooling is an optional procedure that can be used to reduce the surface thickness. Specifically, for each view, a voxel is determined to be on the surface $s_x = 1$ if p_x is the maximum probability along the projection ray.

In other words, during ray pooling, a voxel x is set as a surface voxel based on both perspectives, and if $p_x > \tau$. Other than the binarization and thinning operations, SurfaceNet does not need additional postprocessing or depth fusion operations to obtain accurate and comprehensive reconstruction results.

Training

SurfaceNet was trained based on the DTU dataset [4], which includes images, camera parameters, and reference reconstructions obtained by a structured light system. A single training sample consists of a cube $\hat{S}^C$ cropped from a 3D model and two CVC cubes $I_{v_i}^C$ and $I_{v_j}^C$ obtained from two randomly selected views v_i and v_j. Since most of the voxels do not represent the surface, i.e., $\hat{s}_x = 0$, the surface voxels are weighted by

$$\alpha = \frac{1}{|\mathbf{C}|} \sum_{C \in \mathbf{C}} \frac{\sum_{x \in C} (1 - \hat{s}_x)}{|C|}, \tag{4.13}$$

where $\mathbf{C}$ denotes the set of training samples. A class-balanced cross-entropy function is used as the loss for training, i.e., for each training sample C, we have

$$L(I_{v_i}^C, I_{v_j}^C, \hat{S}^C) = -\sum_{x \in C} \{\alpha \hat{s}_x \log p_x + (1 - \alpha)(1 - \hat{s}_x) \log (1 - p_x)\}. \tag{4.14}$$

The Nesterov momentum updating operation is combined with stochastic gradient descent to update the model's weights. Data augmentation is used to improve model generalizability and prevent overfitting because the dataset contains relatively few 3D models. Each cube C is randomly translated and rotated, and the color is changed by varying the illumination conditions and adding Gaussian noise.

View Selection

Thus far, we have discussed SurfaceNet training and inference using only two views. We next discuss model training and applications for multiview scenarios.

If multiple views $v_1, ..., v_V$ are available, a subset of view pairs (v_i, v_j) is selected, and CVC cubes I_v^C are computed for each cube C and selected view v. For each view pair (v_i, v_j), SurfaceNet predicts $p_x^{(v_i, v_j)}$, i.e., the confidence that a voxel x is on the surface. The predictions of all view pairs can be combined by taking the average of the predictions $p_x^{(v_i, v_j)}$ for each voxel. However, the view pairs should not be considered the same, as each view pair has a different reconstruction accuracy. The accuracy generally depends on the two perspectives' divergent vantage points and the presence of occlusions. To identify occlusions between views v_i and v_j, a 64×64 patch around the projected center voxel of C is cropped in images I_{v_i} and I_{v_j}. A triplet network is trained to learn the mapping $e(\cdot)$ between the images and a condensed 128D Euclidean space, with the distance directly indicating the image similarity to compare the similarity of the two patches. The difference between the two patches is calculated as

$$d_C^{(v_i,v_j)} = \left\| e(C, I_{v_i}) - e(C, I_{v_j}) \right\|_2, \tag{4.15}$$

where $e(C, I_{v_i})$ denotes the feature embedding provided by the triplet network for the patch based on image I_{v_i}. This measurement can be combined with the relation between the two viewpoints v_i and v_j, which is measured by the angle between the projection rays of the center voxel of C, which is denoted by $\theta_C^{(v_i,v_j)}$.

Another 2-layer fully connected neural network $r(\cdot)$ is used to predict the relative weights for each view pair as follows:

$$w_C^{(v_i,v_j)} = r\left(\theta_C^{(v_i,v_j)}, d_C^{(v_i,v_j)}, e(C, I_{v_i})^T, e(C, I_{v_j})^T\right). \tag{4.16}$$

The weighted average of the predicted surface probabilities $p_x^{(v_i,v_j)}$ is given by

$$p_x = \frac{\sum_{(v_i,v_j)\in \mathbf{V}_c} w_C^{(v_i,v_j)} p_x^{(v_i,v_j)}}{\sum_{(v_i,v_j)\in \mathbf{V}_c} w_C^{(v_i,v_j)}}, \tag{4.17}$$

where $\mathbf{V}_c$ denotes the set of selected view pairs. Since not all view pairs need to be considered, only the N_v view pairs with the highest weights $w_C^{(v_i,v_j)}$ are selected.

SurfaceNet can be trained by averaging the results of multiple view pairs (see Eq. 4.17). For each cube C, N_v^{train} random view pairs are selected, and the loss is computed after averaging all the view pairs. $N_v^{train} = 6$ is used as a trade-off since larger values increase the memory required for each sampled cube C; thus, the batch size for training must be reduced due to GPU memory limitations. For the inference process, N_v can be larger or smaller than N_v^{train}. To train the triplet network for the dissimilarity measurement $d_C^{(v_i,v_j)}$ in Eq. 4.15, the cubes C and three random views for which the surface is not occluded are sampled. A positive pair is formed by the patches that are produced when the cube's center is projected onto the first two viewpoints. The negative patch is created by randomly displacing the third patch by at least a quarter of its size. The third view may be the same as one of the other two viewpoints; however, the first two views must be distinct. Data augmentation operations are also employed, including changing the lighting or introducing noise, rotation, scaling, or translation. After SurfaceNet and the triplet network converge, the shallow network $r(\cdot)$ in Eq. 4.16 is finally learned (Fig. 4.3).

Table 4.3 Comparison with other methods. The results are reported based on a test set consisting of 22 models

Method	Med. Acc.	Med. Compl.
Camp [11]	0.335	**0.108**
Furu [33]	0.215	0.246
Tola [106]	0.190	0.268
Gipuma [35]	0.184	0.165
SurfaceNet [53]	**0.183**	0.342

4.2.3 Residual Learning for Implicit Reconstruction

Novel view synthesis, a fundamental technique for virtual reality applications, aims to create new views based on given observation samples of scenes. Recent works such as GoogleJump [6] and DeepView [26] have achieved significant progress by employing synchronized structured camera arrays as capture devices. However, high-quality novel view synthesis based on sparse-view inputs remains a challenging task. Existing methods attempt to solve this problem by either reconstructing an explicit geometric model of the scene [31] or employing probabilistic depth representations [83, 139]. Typically, model-based methods need fewer input views but require high-resolution and precise 3D models. Moreover, these methods cannot reflect changes in the lighting conditions for different views. On the other hand, probabilistic depth-based methods model the scene geometry as a probabilistic distribution instead of as an explicit depth surface. For instance, StereoMagnify [139] employs multiplane images to generate scene representations and renders novel views based on alpha composition. NeRF [75] parametrizes the scene as a radiance field using an implicit scene representation network and applies volumetric rendering to generate novel views.

The NeRF method [75] achieves superior performance by employing a fully connected network to represent the underlying continuous volumetric radiance fields of complex scenes. The network can be trained based on a sparse set of input 2D images without additional 3D supervision. Benefiting from the volumetric scene representation, the NeRF method generates continuous novel views for freely moving cameras. Unfortunately, because neural networks inherently overfit low-frequency information [86], high-frequency texture details are lost in the synthesized images, even when a positional encoding scheme is applied, which leads to disturbing blurry effects.

Existing implicit scene representation networks simply encode the spatial coordinates as the representations of each point and do not consider that the points may have different characteristics when they are backprojected onto the input views. Specifically, the backprojected observations (denoted as spatial color priors) at different view angles are consistent for points on Lambertian surfaces but vary significantly for nonsurface points. As a result, spatial color priors and the actual radiance color of each point are strongly correlated.

Based on this observation, a residual color learning framework for novel view synthesis is proposed. Specifically, this framework uses the spatial color priors of each point as the reference color and employs a scene representation network (e.g., NeRF [75]) to regress residuals between the surface color and reference color. Note that the residuals are small or close to zero for most spatial points. Thus, these values are easier to learn than previous methods that required the network to learn intricate texture details. This scheme preserves clearer details for novel view synthesis, leading to more pleasing visual results than the present state-of-the-art methods. Notably, for complex scenes, previous methods such as NeRF [75] obtain results with blurry artifacts, while the proposed method achieves significantly improved results due to the residual learning scheme.

The proposed approach uses a sparse set of views as input and aims to render novel views from a given viewpoint. The overall framework is illustrated in Fig. 4.13. Spatial color priors are proposed based on the input multiview observations, while occluded pixels are removed by the proposed patch feature filter. The reference color is obtained based on the spatial color priors through a voting strategy. In addition, a residual color learning scheme is introduced in the implicit scene representation network to reduce the network capacity requirements for high-frequency information.

Implicit Scene Representation

Reference [93] employed a fully connected network to implicitly describe scenes. This network learns a function that maps the continuous 3D coordinates to a feature representation of the scene at those feature coordinates. The feature representation may be used to obtain properties such as the density [75] or signed distance function [80] for different targets.

The representative SRN method NeRF [75] models the scene as a neural radiance field and applies volumetric rendering [57] for novel view synthesis. Each spatial point is represented by its 3D coordinates $p = (x, y, z)$ and view direction $d_r = (\theta, \phi)$, which are mapped to the density (opacity) σ and radiance color c using a fully connected network. The expected color $C(r)$ of a camera ray r can be rendered with conventional volumetric rendering techniques, as shown in Eq. 4.18.

$$\bar{C}(r) = \int_{t_n}^{t_f} \exp\big(-\int_{t_n}^{t} \sigma(s)ds\big)\sigma(t)c(t, d_r)dt, \tag{4.18}$$

where t_n and t_f are the near and far bounds of r, respectively, and dt is the distance between the camera rays. d_r indicates the view direction of r, and exp is the exponential function. Based on the volumetric rendering results [57], the continuous integration of Eq. 4.18 can be replaced by the following numerical quadrature:

$$
\begin{aligned}
T_i &= \exp(-\sum_{j=1}^{i-1} \sigma_j \delta_j), \\
w_i &= T_i \big(1 - \exp(-\sigma_i \delta_i)\big), \\
\bar{C}(r) &= \sum_{i=1}^{N} w_i c_i.
\end{aligned}
\tag{4.19}
$$

σ_i and c_i, which denote the density and color of the i-th sampled point, respectively, are represented by the fully connected network $F_\theta(p_i, d_r)$. δ_i represents the distance between two sampling points. $\bar{C}(r)$ is calculated by summing all sampling points for a ray based on weight w_i. Here, $F_\theta(p_i, d_r)$ can be learned based on the given sparse input views by minimizing the difference between the rendered views $\bar{C}(r)$ and the observed views $C(r)$ as follows:

$$
L = \sum_{r \in R} \|\bar{C}(r) - C(r)\|, \tag{4.20}
$$

where R is the set of all camera rays. The size of this set is determined by the number of image pixels.

Spatial Color Prior

Recall that the scene geometry and texture information can be determined based on the color consistency of multiview observations. Thus, we propose spatial color priors and a residual color learning scheme to reduce the network capacity requirements for high-frequency information.

The spatial points are first projected onto the observation images to obtain their projection histogram. The training images are denoted as $\boldsymbol{I} = \{I_i, i \in N\}$, and the corresponding camera poses are denoted as $\boldsymbol{H} = \{H_i, i \in N\}$. The distance between the current camera pose H_c and $\boldsymbol{H}$ can be calculated, and the M closest training images $\boldsymbol{I}$ are selected. The local images are denoted as $\boldsymbol{I}_{local} = \{I^i_{local}, i \in M\}$. Then, the backprojection pixels are calculated based on the multiview geometry as follows [45]:

$$
u_i = K H_i H_c^{-1} p, i \in M, \tag{4.21}
$$

where K represents the intrinsic camera parameters. $\boldsymbol{u} = \{u_i, i \in M\}$ represents the projection pixels in the local images $\boldsymbol{I_{local}}$ near point p. The projection histogram of point p is defined as the statistical histogram of $\boldsymbol{u}$.

The projection histogram has different characteristics when the sampling point is near or far from the object surface. Figure 4.11 illustrates the projection histograms for nonsurface and surface points. For the nonsurface point, the observations from different views are irrelevant, as indicated by its scattered projection histogram. For the point on the object surface, the observations from different views are consistent,

Fig. 4.11 Spatial color priors (the histogram of the projected pixels, which is determined based on 45 projection views) for **a** a nonsurface point and **b** a surface point for the scene shown in Fig. 4.13. From left to right: observed color histogram in red, green, and blue of the point when backprojected to the input views. Note that the histogram for the nonsurface point (**a**) is distributed, while the histogram for the surface point (**b**) is centralized; thus, the reference color can be robustly estimated based on our proposed spatial color priors. The other nonsurface and surface points show similar trends

and its projection histogram is centralized. As the color consistency of the projection histogram can be used to determine the scene geometry and texture information, for each spatial point, we propose spatial color priors based on the information in the corresponding projection histogram.

If the point is on a Lambertian surface, the projection pixels are similar except the occluded pixels. As the occluded pixels are irrelevant to the other projection pixels, they are meaningless noise for the spatial color priors. To address this issue, a patch feature filter is adopted to remove occluded pixels from the projection histogram. The local image patches of the same 3D point in different perspectives are expected to be similar except for occluded points; this information can be used to remove occluded pixels. The 3×3 patches of the half-size image are used as pixel features since the downsampled image has a larger receptive field and the same local patch size. The patch features of the projected pixels are compared with the pixels in the current view. The l_2 norm is calculated, and pixels whose differences with the current view are larger than the threshold are removed. The proposed patch filter is a simple but effective method to handle multiple occlusions in surrounding scenes. This filter does not need to be extremely accurate since the residual color prediction compensates for the small bias.

During the training process, the patch feature of the current view is extracted based on the training images. During the inference process, the patch feature of the current view is extracted based on the predicted radiance color C (see **Residual Color Learning.**). The patch feature filter can remove the occluded pixels. Next, the reference color c_{ref} is calculated based on the remaining projected pixels $\mathbf{u}'$

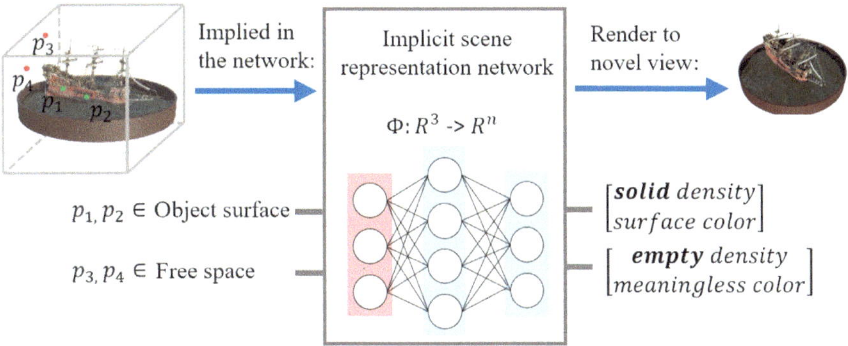

Fig. 4.12 The network architecture. The input is the position (x, y, z) and viewing direction (θ, ϕ). The positional encoding of the input location p is passed through 8 fully connected ReLU layers, each with 256 channels (F_{a1} and F_{a2}). Then, the output 256-dimensional feature is combined with the input positional encoding and viewing direction r and passed through 4 fully connected ReLU layers, each with 128 channels (F_{b1} and F_{b2}). The output includes the density σ, radiance color c and residual color c_r

through a voting strategy. Although the occluded pixels are removed by the feature filter, some projection pixels with strong reflectance may still influence the reference color calculation. Thus, the mean value of $\mathbf{u}'$ is calculated, and the pixel values that are larger than the threshold based on the mean are removed. The pixels with strong reflectance are removed by the voting strategy. Then, the reference color is calculated based on the mean value of the remaining pixels. Note that even without the feature filter, the residual learning approach for the spatial color priors clearly improves the performance in most areas; however, small artifacts are introduced in occluded regions. The feature filter is introduced to handle these occlusions. However, directly using the feature filter may lead to poor results (Fig. 4.18) because although the feature filter provides a more accurate reference color, it causes the projection histograms of some nonsurface points to be more centralized, which decreases the accuracy of the density prediction results. The feature filter must be combined with the joint training shown in Eq. 4.25 to enhance the robustness of the density prediction. Then, the artifacts in the occluded regions can be successfully removed (Fig. 4.12).

Residual Color Learning

According to the reference color calculated based on the spatial points, a residual color learning scheme is proposed to apply the spatial color priors for novel view synthesis. The reference color c_{ref} of each spatial point is calculated based on the spatial color priors, as discussed in **Spatial Color Priors.**, and the residual color c_r is predicted by the SRN F_θ. The reference and residual colors are combined as the predicted color c for volumetric rendering of color $\bar{C}_R$ at ray r, as shown below:

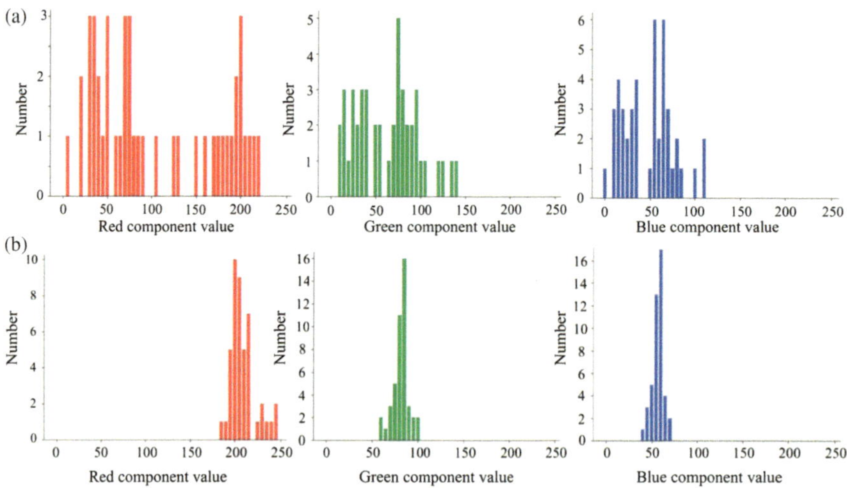

Fig. 4.13 An overview of the proposed residual color learning scheme. For each spatial point, we calculate its reference color c_{ref} based on the input multiview observations and predict the density σ, radiance color c and residual color c_r using a scene representation network. The spatial color priors are the projected pixels (e.g., C_1, C_2, C_3 and C_4 for P_1, which are the center pixels of the corresponding image patches). The image patches are used to remove the occluded pixels. The reference color c_{ref} of point P_1 is estimated with a voting scheme based on backprojected pixels of the point. For novel view synthesis at a given viewpoint, volumetric rendering is applied by integrating $c_r + c_{ref}$ for the spatial points along all pixel rays based on the predicted density σ. The radiance color c is integrated to predict a coarse image for occlusion detection (removing $C4$ in this case) to obtain better reference color prediction results. During the training stage, the input views are sampled as ground truth data, and $F(\theta)$ is trained using the rendering loss of the radiance color and residual color

$$
\begin{aligned}
c^i_{com} &= c^i_{ref} + c^i_r, \\
\bar{C}_R(r) &= \sum_{i=1}^{N} T_i \big(1 - \exp(-\sigma_i \delta_i)\big) c^i_{com}.
\end{aligned}
\tag{4.22}
$$

The pixel colors of different views are similar for the points on the Lambertian surface. With the robust reference color calculation, the residual color predictions of different views are more similar to the original radiance color prediction. The task of learning complex high-frequency texture details is simplified to learning residual colors that are close to 0 for most spatial points, significantly reducing the network costs.

$$
L = \sum_{r \in R} \|\bar{C}_R(r) - C(r)\|. \tag{4.23}
$$

However, learning the network based on only the residual color, as shown in Eq. 4.23, may lead to overfitting, as nonsurface points may be assigned nonzero density values if their reference colors are similar to the target color. The radiance and residual colors can lead to different density predictions. However, to enhance the robustness of the density prediction results after introducing the feature filter, a joint training scheme is proposed, which leverages the radiance color loss for density prediction by learning residual and radiance colors with the same density as follows:

$$\begin{aligned}\bar{C}_W(r) &= \sum_{i=1}^{N} T_i\big(1-\exp(-\sigma_i\delta_i)\big)c_i,\\ \bar{C}_R(r) &= \sum_{i=1}^{N} T_i\big(1-\exp(-\sigma_i\delta_i)\big)(c_{ref}^i + c_r^i),\end{aligned} \tag{4.24}$$

where δ_i, c_i and c_r^i are the outputs of the fully connected network $F_\theta(p_i, d_r)$, δ_i is the density prediction, and c_i and c_r^i are the radiance and residual color outputs, respectively. The network is trained based on the rendering losses of the radiance image $\bar{C}_W(r)$ and residual image $\bar{C}_R(r)$ as follows:

$$L = \sum_{r\in R} \|\bar{C}_W(r) - C(r)\| + \sum_{r\in R} \|\bar{C}_R(r) - C(r)\|. \tag{4.25}$$

The proposed residual color learning scheme greatly reduces the network costs. As a result, our proposed method achieves better performance and converges with fewer iterations than the NeRF method.

Experiment

For a fair comparison with previous methods, the proposed method is evaluated based on various datasets: forward-facing data from the LLFF dataset [74], synthetic data from the NeRF dataset [75], indoor surrounding data from the casual 3D [3] dataset, self-collected large-scale outdoor data ("Auditorium" and "Theater" in Table 4.4), and data from the Tanks and Temples [62] dataset. In the following sections, quantitative and qualitative evaluations are performed to verify the performance of the proposed method.

Quantitative Evaluation

The quantitative evaluation was performed using the PSNR, SSIM, and learned perceptual image patch similarity (LPIPS) metrics [134]. Smaller PSNR and SSIM values indicate higher accuracy, while higher LPIPS values indicate better visual quality. The proposed method was compared with previous state-of-the-art methods, including the SRN [93], NV [71], LLFF [74] and NeRF [75] methods, as shown in Table 4.4.

Table 4.4 Quantitative evaluations based on public datasets in terms of three metrics (PSNR (↑), SSIM (↑) and LPIPS (↓)). The scores are the mean value of all testing images

	Room [74]	Fortress [74]	Drums [75]
	PSNR SSIM LPIPS	PSNR SSIM LPIPS	PSNR SSIM LPIPS
SRN [93]	27.29 0.883 0.240	26.63 0.641 0.453	17.18 0.766 0.267
NV [71]	–	–	22.58 0.873 0.214
LLFF [74]	28.42 0.932 0.155	29.40 0.872 0.173	21.13 0.890 0.126
NeRF [75]	32.70 0.948 0.178	**31.16** 0.881 0.171	25.01 0.925 **0.091**
Ours	**32.89 0.955 0.151**	31.15 **0.905 0.144**	**26.06 0.934** 0.099
	Library [3]	Attic [3]	Kitchen [3]
	PSNR SSIM LPIPS	PSNR SSIM LPIPS	PSNR SSIM LPIPS
NeRF [75]	29.02 0.784 0.481	23.64 0.744 0.535	26.13 0.826 0.334
Ours	**33.08 0.926 0.183**	**25.25 0.780 0.424**	**27.70 0.878 0.229**
	Auditorium	Theater	Family [62]
	PSNR SSIM LPIPS	PSNR SSIM LPIPS	PSNR SSIM LPIPS
NeRF [75]	21.81 0.766 0.334	21.50 0.666 0.425	31.07 0.924 0.126
Ours	**23.58 0.834 0.210**	**23.38 0.691 0.323**	**32.71 0.953 0.069**

For simple scenes with a small point of view, e.g., "Room" and "Fortress" in the LLFF dataset, the NeRF method achieves good performance with sufficient memory capacity. Spatial color priors help to reveal high-frequency details, but the performance improvement is relatively small. For complex scenes with large-scale surrounding views, e.g., "Library" and "Attic" in the casual 3D dataset [3], the NeRF method performs poorly due to network size limitations. The proposed method achieves much better performance since the proposed spatial color priors reduce the network capacity requirements for large-scale scenes. The NeRF method achieves better performance with increasing network size, suggesting that the rendering quality of the NeRF approach is limited by the network capacity. However, increased memory increases the complexity of the network, which limits the network size. Additionally, increasing the network size leads to limited improvement. On the other hand, with the proposed spatial color priors, the network capacity requirements are greatly reduced, and the proposed residual learning scheme achieves much better quality, even with a smaller network.

The performance of the NeRF approach and our proposed method was also compared in terms of rendering novel views at different resolutions, as shown in Table 4.5. As the resolution increases, the gap between the proposed method and NeRF approach increases, demonstrating that the proposed method can generate more realistic high-resolution rendering results.

Table 4.5 PSNR comparison between the NeRF approach and our method at different resolutions. "Library" and "Attic" are categories with surrounding indoor data from the casual 3D [3] dataset

	Library	Library	Attic	Attic
	960×720	1200×900	592×880	886×1330
NeRF	29.02	28.09	23.44	23.64
Ours	33.08	33.21	24.93	25.25

Qualitative Evaluation

The reference color is calculated based on the spatial color priors. It is close to the real rendering result in most areas and may suffer from distortion in the corner regions due to incorrect pixel projections. Residual color prediction has the potential to partially correct these issues. Additionally, this process can add different light shadows from different perspectives (as shown in Fig. 4.14). The following qualitative evaluations show that the proposed method achieves robust reference color calculation and high-quality rendering performance.

Overall Performance The proposed method applies a residual-based framework to utilize the spatial color priors. This idea was investigated by comparing the proposed and NeRF methods. Figure 4.15 shows qualitative comparisons of our method and the NeRF approach [75] for different kinds of scenes. The NeRF method has difficulty learning high-frequency information, such as texture, and detailed information is lost. The proposed method transforms the high-frequency learning task into a low-frequency learning task. The residual color scheme needs only the low-frequency information since the calculated reference color captures the high-frequency texture of the scene. As a result, the high-frequency information is better preserved, and the proposed method obtains clearer details than the NeRF approach. In particular,

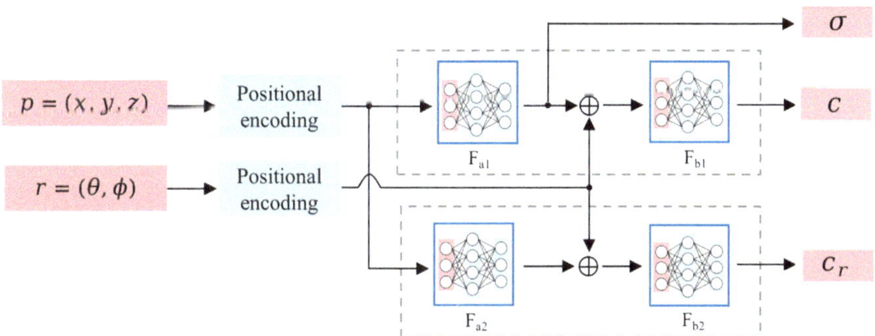

Fig. 4.14 Illustration of the decomposed rendering results. In the residual image, the distortion in the reference image caused by incorrect projection pixels (red block) is corrected, and view-dependent light shadows (green block) are added. After adjusting the residual image, the PSNR of the rendering result is increased from 30.08 to 33.80

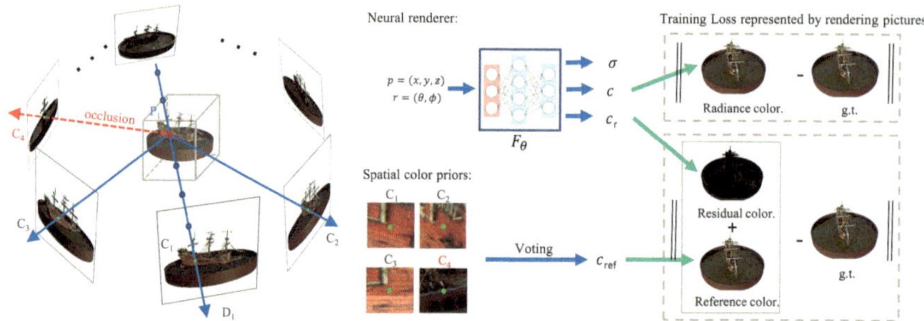

Fig. 4.15 Qualitative evaluations comparing our method with the NeRF method [75] based on public datasets: **a**, **b**, **c** show the results based on the casual 3D dataset [3], and **d**, **e** show the results based on the Tanks and Temples dataset [62]. The experimental results show that our residual learning scheme based on the proposed spatial color priors produces clearer details than previous state-of-the-art methods

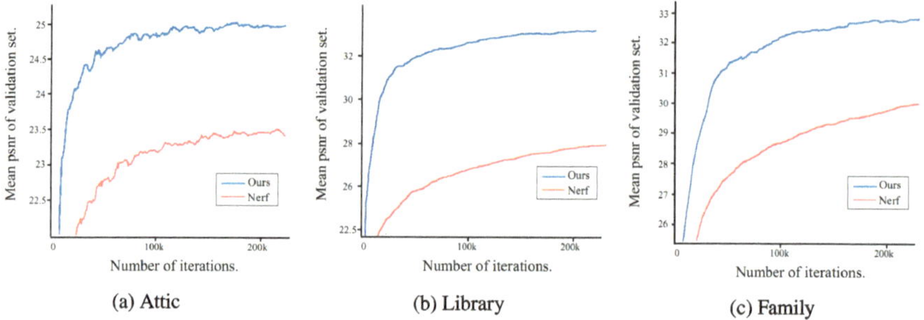

Fig. 4.16 Qualitative evaluations of the SRN [93], NV [71], NeRF [75] and newly published NSVF [69] methods based on the "Jade" data in the BlendedMVS dataset [124] (**a**) and the "Family" data in the Tanks and Temples dataset [62] (**b**). The experiments show that the proposed method achieves considerably better performance than previous methods (SRN, NV, and NeRF) and comparable performance with NSVF

the NeRF method can only recover limited scenes with low quality. For large-scale scenes, the NeRF method tends to perform poorly due to network size limitations. The proposed method can effectively handle large-scale scenes because the proposed spatial color priors reduce the network capacity requirements. For example, for the complex indoor scenes in Figs. 4.15b, c and large-scale outdoor scenes, our method shows significant improvement in the high-quality rendering results. Additionally, Fig. 4.16 shows a qualitative comparison of our method with the SRN [93], NV [71], NeRF [75] and the newly published NSVF [69] methods. The experimental results show that the proposed method achieves much better performance than previous methods (SRN, NV, and NeRF) and comparable performance with NSVF.

Occlusion Handling. The reference color image is calculated based on the projected pixels. If the scene contains occluded regions, incorrect projection pixels

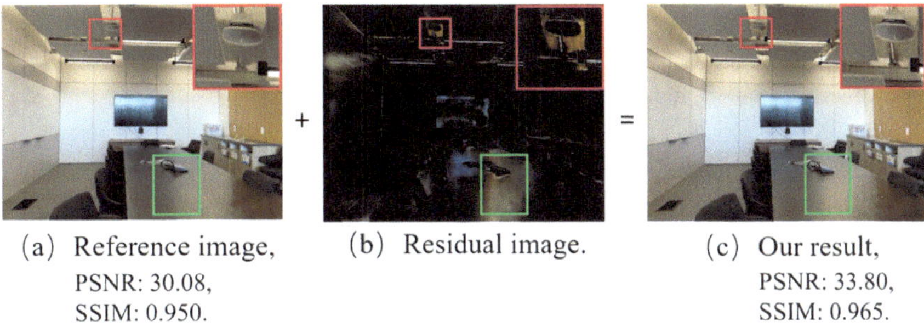

(a) Reference image, PSNR: 30.08, SSIM: 0.950. (b) Residual image. (c) Our result, PSNR: 33.80, SSIM: 0.965.

Fig. 4.17 Comparison of rendering images with and without feature filtering and joint training. From left to right (**b**, **c**): reference images, residual images, and result images. The reference color image in (**b**) has obvious artifacts caused by occluded projection pixels, whereas the reference color image in (**c**) is more accurate. As a result, when feature filtering and joint training are applied, the residual color image has less errors, and the resulting image is improved in the occlusion regions

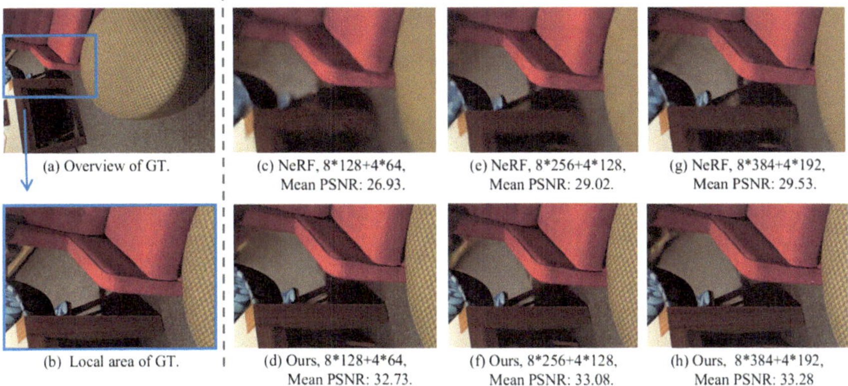

(a) Overview of GT. (c) NeRF, 8*128+4*64, Mean PSNR: 26.93. (e) NeRF, 8*256+4*128, Mean PSNR: 29.02. (g) NeRF, 8*384+4*192, Mean PSNR: 29.53.

(b) Local area of GT. (d) Ours, 8*128+4*64, Mean PSNR: 32.73. (f) Ours, 8*256+4*128, Mean PSNR: 33.08. (h) Ours, 8*384+4*192, Mean PSNR: 33.28

Fig. 4.18 **a** is the result of the original NeRF method. **b** uses residual learning to apply spatial color priors without introducing the additional radiance color loss. This method achieves better performance than the NeRF approach, but the red boxes show that the occlusion area is blurry (top box) or has ghosting artifacts. **c**, **d** introduce the radiance color loss in addition to the residual learning scheme. (**c**) uses a feature filter without joint training, which means that the radiance and residual colors have unique density predictions. (**d**) uses feature filtering and joint training. The comparison shows that the performance improvement is mainly due to the residual learning scheme. Feature filters must be combined with joint training to remove artifacts caused by occlusion

may influence the quality of the reference image. A feature filter and joint training strategy are applied to address this limitation. Figure 4.17b shows that without occlusion detection, the reference color image contains obvious artifacts due to the incorrect pixel projections. The reference color image also shows clear distortion due to the incorrect pixel projections. Residual color prediction can partially correct these issues; however, the results are not perfect, as the resulting images still have certain artifacts. With the patch feature filter and joint training scheme, the calculated reference color image is not affected by the occluded regions, as shown

Fig. 4.19 Comparison of the Lambertian surface (red block) and reflective surface (green block) results. This figure shows that the proposed residual-based method outperforms the NeRF method for points on the Lambertian surface, whereas for the reflective surface, the proposed method performs similarly to the NeRF approach

in Fig. 4.17c. Moreover, as demonstrated in Fig. 4.18, the performance improvement is mainly due to the residual learning scheme, while the feature filter addresses the occlusions. Thus, the feature filter needs to be incorporated with the joint training scheme (Fig. 4.19).

4.3 RGBD-Based Geometry Reconstruction

To enable accurate depth prediction and real-time performance, RGBD cameras are widely used in dense 3D reconstruction systems. For example, ElasticFusion [114] uses dense frame-to-model camera tracking and windowed surfel-based fusion. DPI-SLAM [47, 63] uses a loosely coupled IMU constraint with visual estimation and point clouds for mapping. RGBD Inertial SLAM [63] achieves real-time reconstruction and tightly coupled visual-inertial optimization based on GPUs. [140] used an extended Kalman filter to fuse visual and inertial information and achieved stable tracking under fast camera motions.

Due to the popularity of RGBD dense 3D reconstruction, some researchers investigated large-scale reconstruction using inexpensive RGBD sensors to allow more agile reconstruction results. Steinbrucker et al. [81] reconstructed scenes with 9 rooms using an RGBD SLAM system with a multiresolution tree structure. CHISEL [77] used a dynamic spatially hashed truncated signed distance field (TSDF) to enable CPU-only reconstruction of areas larger than 300 square meters. However, for large building-scale environments, the scanning process was time-consuming and required high concentration. A few works have proposed multiagent collaborative scanning methods. Reference [107] achieved multirobot reconstruction using centralized bundle adjustment and loop closure detection (LCD). References [42, 76, 103] used cloud-based methods to allow distributed scanning. Reference [40] proposed an autonomous strategy for optimized robot collaborative reconstruction. Furthermore, [114] used a server-client architecture and an interagent LCD approach to achieve collaborative scanning.

The following section introduces state-of-the-art RGBD-based geometry reconstruction techniques, including FlashFusion [40], a real-time high-resolution system with CPU computing; iDFusion [138], a dense volumetric reconstruction method with global consistency; and BuildingFusion [136], a building-scale structural reconstruction technique.

4.3.1 FlashFusion: Real-Time Reconstruction

Real-time 3D reconstruction, a fascinating topic in computer vision, robotics, path planning, machine perception, and other fields, has increased the number of practical human-robot interaction applications. First, [78] considered high-resolution dense 3D reconstruction methods using generic depth sensors. Then, [135] used truncated signed distance fields (TSDFs) and multiple depth observations from various viewing angles. Despite KinectFusion's limited volume and the need for graphics processing units to achieve real-time performance, the high-quality reconstruction results led to hundreds of follow-up studies, addressing the scalability [82, 130], efficiency [55, 61, 96], and global consistency [15], among other issues.

The state-of-the-art BundleFusion [15] approach recently demonstrated superior performance in obtaining high-resolution and globally consistent real-time 3D reconstruction results. This was accomplished by using frame-based localization methods for increased robustness and reintegrating frames to correct surfaces online when previous poses were updated based on loop closure constraints and bundle adjustment optimization methods.

However, this method needs two powerful GPUs for computing. The cutting-edge CHISEL [61], on the other hand, only uses CPU computations and portable devices for real-time 3D reconstruction. Nevertheless, CHISEL cannot achieve global consistency due to localized drift and ineffective TSDF fusion, and its real-time implementation can only support a voxel resolution of $20mm$. The current state-of-the-art is still an efficient reconstruction method that operates on portable devices but lacks robustness and quality or a globally consistent high-quality system that needs expensive GPUs to achieve real-time performance, preventing its application in more general cases such as mobile robots or wearable technology.

Han et al. [40] proposed **FlashFusion**, a **F**ast yet **LA**rge-**S**cale **H**igh-resolution (subcentimeter level) dense 3D reconstruction system without GPU computing. FlashFusion has two modules: a very effective reconstruction module and a globally consistent yet fast localization module. For the former module, to obtain globally consistent pose estimates, we adopt and develop our FastGO scheme. For the latter module, we use CPU computing to quickly perform TSDF fusion and mesh extraction.

With the above technical contributions, Han et al. [40] demonstrated **the first CPU-based globally consistent real-time 3D reconstruction system for portable devices.**

System Architecture

The front-end tracking thread, the back-end optimization thread, and the meshing thread are all included in the FlashFusion system design. The tracking thread uses unprocessed RGBD raw data as input, and the drift is removed by MILD [38], a high-precision yet very effective loop closure detector that does not require any training. The following section describes how the localization- and reconstruction-related back-end threads are implemented.

Localization—Due to the high efficiency requirements, the frames are divided into keyframes and local frames based on the disparity criteria; new keyframes are added when the average disparity of the comparable features between the current frame and its associated keyframe exceeds a pre-defined threshold. The only thing that changes during the global optimization process is the pose in the keyframes; the relative transformation between a local frame and the previous keyframe is fixed, and the pose in the local frame is updated based on the pose in the optimized keyframe.

Reconstruction—The world is represented as a two-level tree, resembling the methods applied in [61, 82], where a cube with a certain number of voxels is organized as a chunk and each chunk is spatially hashed into a dynamic hash map.

The reconstruction process can be separated into TSDF fusion and mesh extraction depending on how the keyframe and local frame are organized. To ensure the quality of the reconstructed 3D model, the TSDF is precisely updated based on the depth observations for each incoming frame, and the depth observations from the local frames and keyframes are fused in the TSDF. However, the meshes are updated only when a new keyframe is added. This technique is applied due to the substantial overlap between the keyframe and its corresponding local frames, and there is no need to update the meshes each time a local frame is added.

Real-Time Globally Consistent Dense 3D Reconstruction using CPU Computing

Robust Globally Consistent Localization

There is a distinction between local and global frame registration, which are frequently employed in pose estimation. The local registration process denotes the frame's relationship with its previous keyframe and the current frame. The global registration process recognizes the current local frame as a new keyframe and compares it to all prior keyframes if the keyframe update condition is satisfied. Here, the relative transformation is estimated by minimizing the following cost function on a manifold using the Lie group, and frame registration is accomplished by locating the corresponding ORB features between the two frames, denoted as F_i and F_j (approximately 1000 ORB features are extracted for each frame) [104],

$$E_{i,j}(T_{i,j}|T_{i,j} \in SE3) = \sum_{k=0}^{|C_{i,j}|-1} ||\boldsymbol{p}_i^k - T_{i,j}\boldsymbol{p}_j^k||^2, \tag{4.26}$$

where $C_{i,j} = \{(\boldsymbol{p}_i^k, \boldsymbol{p}_j^k)|k = 0, 1, \cdots, |C_{i,j}| - 1\}$, $(\boldsymbol{p}_i^k, \boldsymbol{p}_j^k)$ indicates the k-th feature correspondence between frame F_i and F_j, and $\boldsymbol{p}_i^k \in F_i$, $\boldsymbol{p}_j^k \in F_j$. $T_{i,j}$ is the relative transformation from F_j to F_i in Euclidean space.

Given the pairwise frame correspondence, we minimize the reprojection error of all corresponding keyframe pairs based on the global registration results to obtain globally consistent pose estimation as follows:

$$E(\boldsymbol{\xi}) = \sum_{i=0}^{N-1} \sum_{j \in \Phi(i)} E_{i,j} = \sum_{i=0}^{N-1} \sum_{j \in \Phi(i)} \sum_{k=0}^{|C_{i,j}|-1} ||\boldsymbol{p}_i^k - T_{i,j}\boldsymbol{p}_j^k||^2, \tag{4.27}$$

where $\Phi(i)$ represents the top 5 most similar images selected via the rapid yet highly accurate loop closure detection technique MILD [38], as a traverse search scheme [15] is impractical for CPU computing.

Suppose there are M feature correspondences in $E(\boldsymbol{\xi})$; then, we have

$$E(\boldsymbol{\xi}) = \boldsymbol{r}(\boldsymbol{\xi})^T \boldsymbol{r}(\boldsymbol{\xi}), \tag{4.28}$$

where $r(\boldsymbol{\xi}) = \left[\boldsymbol{r_0}^T, \boldsymbol{r_1}^T, \cdots, \boldsymbol{r_{M-1}}^T\right]$, and $\boldsymbol{r_l} = \boldsymbol{p}_i^k - T_{i,j}\boldsymbol{p}_j^k$ represents the reprojection vector of the l-th feature correspondence among the M features. Nonlinear Gauss-Newton optimizations are used to iteratively solve this problem on the Lie manifold in $SE3$ space:

$$\boldsymbol{\delta} = -(J(\boldsymbol{\xi})^T J(\boldsymbol{\xi}))^{-1} J(\boldsymbol{\xi})^T \boldsymbol{r}(\boldsymbol{\xi}), \tag{4.29}$$

where $J(\boldsymbol{\xi})$, with size $3M \times 6(N-1)$, is the Jacobian of $\boldsymbol{r}(\boldsymbol{\xi})$. Instead of computing $J(\boldsymbol{\xi})$ directly, $J(\boldsymbol{\xi})^T J(\boldsymbol{\xi})$ and $J(\boldsymbol{\xi})^T \boldsymbol{r}(\boldsymbol{\xi})$ are computed directly for efficiency since they are more compact:

$$\begin{aligned} J(\boldsymbol{\xi})^T J(\boldsymbol{\xi}) &= \sum_{i=0}^{N-1} \sum_{j\in\Phi(i)} \sum_{k=0}^{|C_{i,j}|-1} (J_{i,j}^k)^T J_{i,j}^k, \\ J(\boldsymbol{\xi})^T \boldsymbol{r}(\boldsymbol{\xi}) &= \sum_{i=0}^{N-1} \sum_{j\in\Phi(i)} \sum_{k=0}^{|C_{i,j}|-1} (J_{i,j}^k)^T \boldsymbol{r_l}, \end{aligned} \tag{4.30}$$

where $J_{i,j}^l$ is the Jacobian of $\boldsymbol{r_l}$. As noted in FastGO [41], $\sum_{k=0}^{|C_{i,j}|-1} (J_{i,j}^k)^T J_{i,j}^k$ and $\sum_{k=0}^{|C_{i,j}|-1} (J_{i,j}^k)^T \boldsymbol{r_l}$ in Eq. 4.27 can be efficiently computed with complexity $O(1)$ instead of $O(|C_{i,j}|)$ by integrating the second-order statistics of the feature correspondences from $C_{i,j}$. Thus, Eq. 4.27 can be solved in real time using CPU computing.

Although the FastGO approach is very efficient, the results suffer in the face of outliers. Therefore, we replace the l_2 norm with the more reliable Huber norm in the cost function. This, however, necessitates updating the weight of each feature pair based on the most recent pose estimates, which means that the formulation cannot be well represented using the second-order statistics, as in FastGO. After each Gauss-Newton update, an exhaustive computation would be required to explore all the feature pairings in Eq. 4.27. An online correction approach is used to solve the Huber norm-based reduction in real time by selecting only the top 10 frame pairings whose relative poses vary the most noticeably among all the considered frame pairs. With this correction, in practice, only a small portion of the relative poses of the frame pairs need to be updated following the loop closure detection and global bundle adjustment operations. The reprojection error of the features depends on the relative poses of the frame pairs.

Efficient TSDF Fusion. Recall that the surface can be represented implicitly using the TSDF, where each 3D point in space is mapped based on the distance (denoted as the signed distance function (SDF)) to the nearest surface, and the SDF values of the points on the surface are close to 0. Continuous space can be represented as voxels, with each voxel storing the SDF of its center point. To efficiently index valid voxels near surfaces, hashing—[82] and octree—[130] based data structures employing the sparsity of valid voxels are proposed. In voxel hashing-based reconstruction systems, $8 \times 8 \times 8$ voxels are organized as chunks. Each chunk is mapped to an address based on its spatial hash function [102]. Depth observations are integrated chunkwise using a projective mapping approach. Each voxel center is projected onto a 2.5D depth map, and the difference between the projected distance and the depth map reading is recognized as the surface distance d and fused into the TSDF. Since depth readings have different covariance values due to the applied sensor model, the observations are fused with different weights w, which are determined directly by the depth readings as follows:

$$sdf_n = \frac{sdf_o * w_o + sdf_i * w_i}{w_i + w_o}, \quad w_n = w_i + w_o, \tag{4.31}$$

where sdf_o and w_o represent the original SDF and weight value, sdf_i and w_i denote the future SDF and weight observations, and sdf_n and w_n represent the newly updated SDF and weight.

Previously, all the chunks that fall into the frustum of the camera view were selected as candidate chunks and processed equally during the fusion process [61]. Then, the invalid chunks (the blue grids in Fig. 4.20), whose voxels have no valid SDF, are removed, and the valid chunks (the yellow grids in Fig. 4.20) that are within the truncated band of a surface are retained. However, the valid chunks account for only a small portion of the candidate chunks, and most computational resources are wasted on the invalid chunks.

We suggest a sparse voxel sampling method to quickly identify acceptable chunks by considering the chunk's eight corners and determining whether the minimum absolute SDF value exceeds a predetermined threshold. In general, surfaces are continuous, meaning that there is a strong likelihood that surfaces will cross at least one of the chunk's six faces and that at least one of the eight corners can be projected onto the surface if the chunk includes the surface, as illustrated in Fig. 4.21. For example,

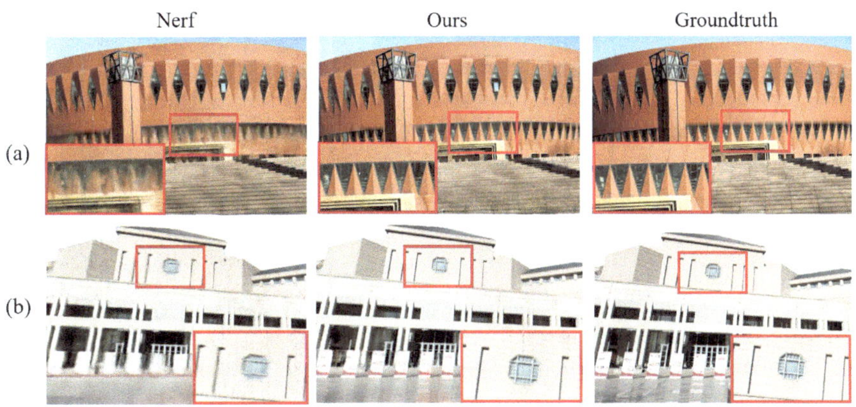

Fig. 4.20 Illustration of chunk selection, showing that FlashFusion effectively handles chunks (yellow grids), while CHISEL processes all the candidate chunks (both the blue and yellow grids)

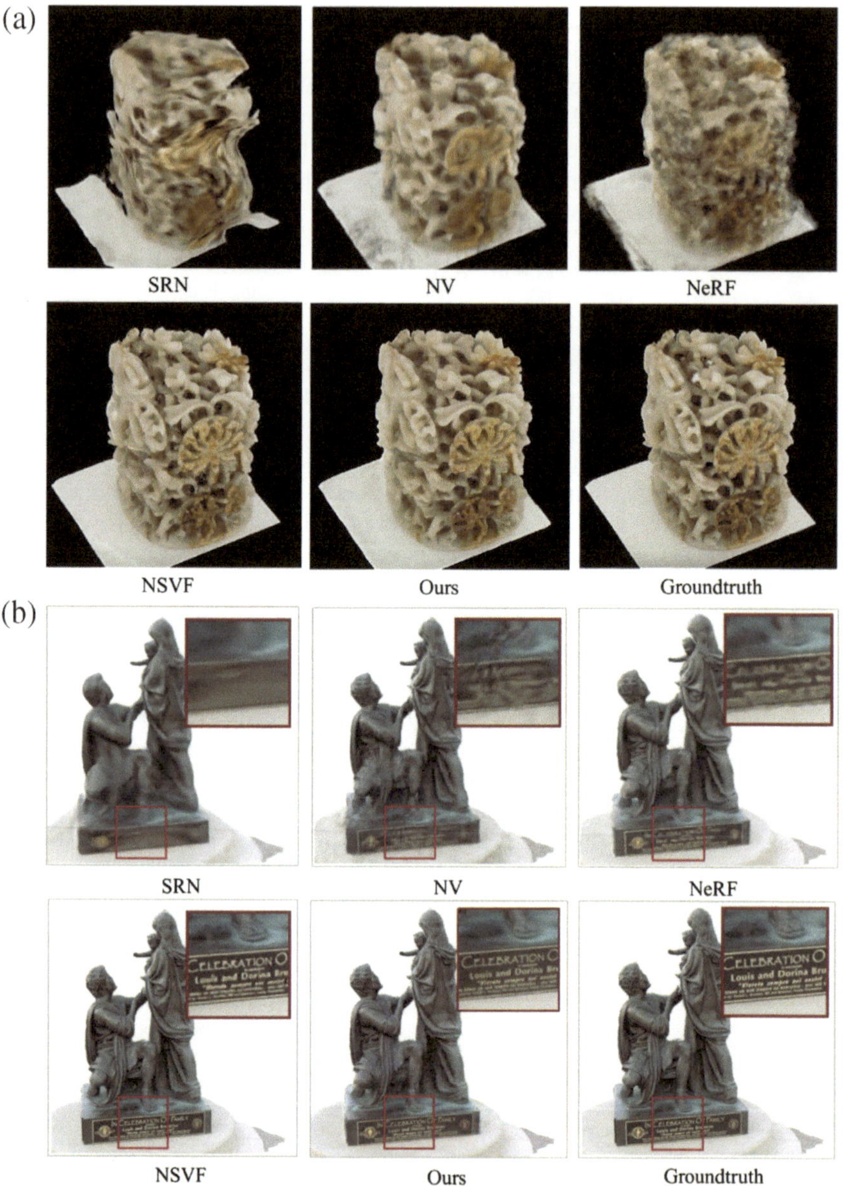

Fig. 4.21 Illustration of sparse voxel sampling for chunk filtering, which calculates the SDF values at the eight corners in each chunk

suppose that the voxel resolution is r and that there are N_v^3 voxels in each chunk; if one voxel falls in the truncated surface, the SDF value d_v of that voxel should satisfy $|d_v| < T_{truncation}$. Accordingly, the SDF value of a corner voxel d_c should satisfy $|d_c| \leq |d_v| + \sqrt{2}N_v \times r < T_{truncation} + \sqrt{2}N_v \times r$. In contrast, if the SDF values of all corners are less than $T_{truncation} + \sqrt{2}N_v \times r$, the chunk is unlikely to contain the surface. The eight corners can be computed in parallel using SIMD operations; however, when the voxel size is less than 1 cm, the computation is laborious. To identify the legitimate chunks, the chunks are first investigated at a resolution of 20 mm using sparse voxel sampling. If the identified SDF values suggest that the large chunk contains a surface, the chunks are traversed at a resolution of 5 mm.

As a result, the valid chunks can be accurately estimated using the aforementioned sparse voxel sampling method. However, a thorough investigation of the stored chunks is required to identify the previous valid chunks that are now invalid due to object movement, resulting in a linear increase in the computational time with the size of the reconstructed 3D model. We use a second hash table and coarse hash chunks to address this issue. Each $8 \times 8 \times 8$ chunk is arranged into a cube, which is then spatially hashed based on its central location, and only the chunk IDs that are currently present are stored. In other words, we traverse the existing chunks stored in the selected cubes to determine if they are visible with the current camera view after selecting the cubes that are within the frustum of the camera view.

Figure 4.22 presents the results of experiments performed based on the fr3office dataset to evaluate the effectiveness of our sparse voxel sampling method in comparison with traditional candidate chunk selection methods. The horizontal axis indicates the frames in the sequence, the left vertical axis indicates the precision of the estimated valid chunks, and the right vertical axis indicates the computational complexity. The results show that the proposed method considerably accelerates the chunk selection process, retaining more than 98% of the valid chunks with a 2% increase in computational complexity on average.

The local frames are then combined in accordance with the keyframe selection based on the valid chunks. In other words, each keyframe stores the IDs of the chosen chunks, which can then be utilized again if the keyframe needs to be reintegrated. The experiments also demonstrate that waiving valid voxel selection for local frames significantly reduces the complexity of TSDF fusion, enabling reintegration using CPU computing. The experiments also demonstrate that each frame's TSDF integration takes approximately 2 ms, and the valid voxel selection takes 6 ms.

Similar to the method in [61, 82], color observations are fused into the TSDF model following the average framework shown in Eq. 4.31. The difference with FlashFusion is that we implicitly save the color values using the multiplication result (expressed by an unsigned short for each channel; the maximum weight is set to 255). With this method, the color values for each created vertex can be simply computed in the rendering stage. Thus, we do not need to explicitly calculate an updated color value for each voxel. In contrast to previous methods, which required two addition, two multiplication, and one division operation to obtain the updated color value and weight, the fusion of new observations requires only a single integer addition operation to update the multiplication result. It should be noted that adding integers

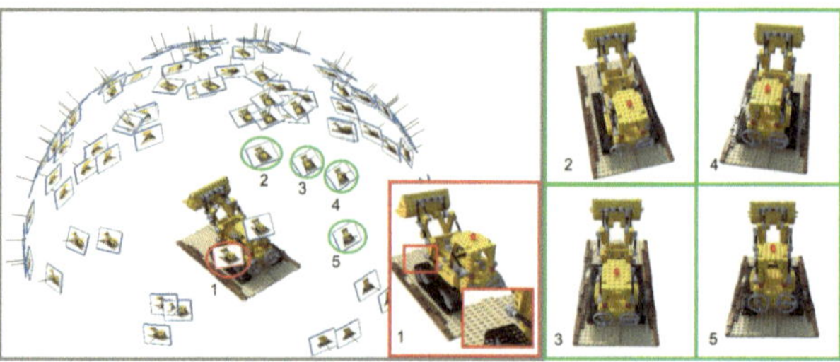

(a) Input scene. The image in the red box shows the inference viewpoint. The images in the green box show some of the adjacent views. The zoomed-in area is the occlusion region, which is visible in the inference viewpoint and not visible in the presented adjacent views

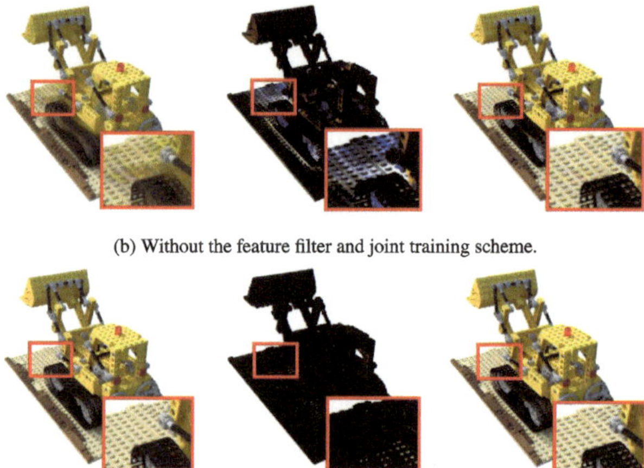

(b) Without the feature filter and joint training scheme.

(c) With the feature filter and joint training scheme.

Fig. 4.22 Performance analysis of our sparse voxel sampling method and comparison with conventional candidate chunk selection strategy based on the fr3office dataset

costs only one clock cycle, while the division operation in Intel SSE requires 38 clock cycles (including latency).

Mesh Extraction. In general, given the TSDFs, we can use the incremental marching cubes algorithm [25] to estimate the surfaces represented by triangles (meshes); however, this method is impractical to implement via CPU computing. For example, the mesh extraction in the state-of-the-art CHISEL method [61] takes approximately 100 ms and 1500 ms at voxel resolutions of $20mm$ and $5mm$, respectively. We thus investigate the major computational costs related to mesh extraction, i.e., polygon generation and normal extraction, achieving highly accelerated mesh extraction in 60 ms at a voxel resolution of $5mm$.

Recall that for each valid voxel v_0, its seven neighboring voxels $v_i, i = 1, 2, \cdots, 7$ are extracted based on a cube c, as shown in Fig. 4.23. Polygons inside this cube can be calculated based on the SDF values of the 8 corners with the classic marching cube algorithm, i.e., vertices exist only on the edges of c when the signs of the two

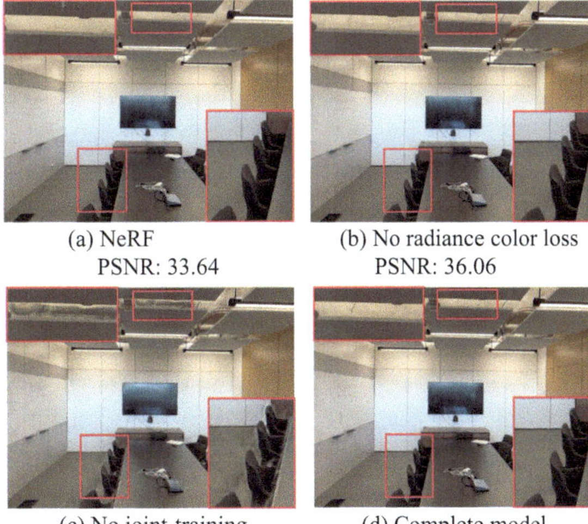

Fig. 4.23 Illustration of generating polygons using 8 neighboring voxels [25]

endpoints' SDF values are different. The exact location v of each vertex is determined by linear interpolation of the two endpoints v_a and v_b based on their SDF values s_a and s_b as follows:

$$\mathbf{v} = \frac{s_a \times \mathbf{v_b} - s_b \times \mathbf{v_a}}{s_a - s_b}. \tag{4.32}$$

Therefore, it is imprudent to predict if a polygon exists in a given voxel's neighborhood by considering all of voxels. Although the voxels with small SDF values are closer to the surface than the voxels with large SDF values, since we use projective SDFs rather than Euclidean SDFs, it is impractical to set a constant threshold to filter out voxels with high SDF values. For instance, voxels that are adjacent to a surface may have high SDF values when the surface is seen parallel to its normal. We use a dynamic threshold strategy to allocate various chunks with different thresholds to address this issue. The highest absolute SDF value of voxels near surfaces is used as the threshold when constructing the mesh. Voxels with SDF values greater than the threshold are disregarded when new observations are added and the mesh has to be modified. Assuming that only 10% of the meshes are new and that the remaining 90% of the meshes must be updated in each mesh extraction stage, the technique successfully accelerates the polygon generation process by a factor of 2.

Moreover, by analyzing the specific structure of the cube c generated in the mesh extraction process, we find that linear interpolation is essentially unnecessary since relative to its origin corner v_0, c is a fixed cube whose side length equals the voxel resolution. The generated vertices are expected to be on the edge of c, allowing one degree of freedom for the position of each vertex, which is linear with respect to the interpolation coefficient, $\frac{s_a}{s_a - s_b}$. The remaining two degrees of freedom depend on the edge and can be determined through a small look-up table.

The normal n of each vertex is computed via the derivative of the TSDF as [78]:

$$\mathbf{n} = \begin{bmatrix} \delta_x \\ \delta_y \\ \delta_z \end{bmatrix} = \begin{bmatrix} s_{i+1,j,k} - s_{i-1,j,k} \\ s_{i,j+1,k} - s_{i,j-1,k} \\ s_{i,j,k+1} - s_{i,j,k-1} \end{bmatrix}, \tag{4.33}$$

where i, j, k indicates the index of voxel v_n that is closest to vertex v. Note that v_n can be determined directly in the polygon generation step, where v is close to v_a if $fabs(s_a) < fabs(s_b)$ or v_b if $fabs(s_a) > fabs(s_b)$. Three of the neighboring voxels required for the normal estimation of v_n (Eq. 4.33) are in the cube c, while the other three voxels can be found in nearby chunks; thus, the normal can be estimated directly in the mesh extraction step with limited computational cost.

Voxel access, which is heavily used in existing methods, is the final bottleneck that must be addressed in the rendering stage. For example, 6 adjacent voxels are accessed during the normal extraction for each vertex, whereas 7 nearby voxels are accessible during polygon formation for each voxel candidate. For interchunk voxel access, we must first determine whether the corresponding chunk exists before locating it in the large hash table that contains all of the chunks. As chunks are added to or withdrawn from the FlashFusion program, a look-up table is stored for each chunk. By storing the addresses of neighboring chunks in the look-up table, the chunk hash map does not need to be accessed for each voxel in the polygon formation stage or for each vertex in the normal estimation step.

Reintegration. We can incorporate live depth observations to realize real-time 3D reconstruction with CPU computations by using the accelerated TSDF fusion and mesh extraction approach. We recombine up to 10 prior keyframes for every keyframe. One keyframe may correspond to multiple local frames, whereas we select at most 10 local frames to guarantee that reintegration can be performed within a certain number of clock cycles and that this process does not limit the overall system pipeline. We choose 10 local frames to ensure both efficiency and diversity because too many keyframe linkages between local frames may suggest that the camera is stationary or moving slowly, and successive local frames are less informative.

Experiments

To verify the performance of the FlashFusion approach, extensive experiments are performed based on both the synthetic ICL-NUIM [44] dataset (with noise) and the real-world TUM RGBD dataset [97] using an Intel Core i7 7700 @3.6 GHz CPU. The ICL-NUIM dataset provides ground truth data for both localization and reconstruction, while the TUM RGBD dataset provides ground truth data for only localization. State-of-the-art methods, including BundleFusion [15], ElasticFusion [114], InfiniTAM [56], RGB-D SLAM [22], FastFusion [96] and CHISEL [61], are used for comparisons.

Table 4.6 Accuracy comparison of localization results based on the TUM RGBD dataset (cm)

	fr1/desk	fr2/xyz	fr3/office	fr3/nst
RGBD SLAM	2.3	0.8	3.2	1.7
ElasticFusion	2.0	1.1	**1.7**	1.6
BundleFusion (online)	1.7	1.4	2.9	1.6
BundleFusion (offline)	**1.6**	**1.1**	2.2	**1.2**
FlashFusion	1.9	1.3	2.5	1.8

Table 4.7 Accuracy comparison of the localization results based on the ICL-NUIM dataset (cm)

	kt0	kt1	kt2	kt3
RGBD SLAM	2.6	0.8	1.8	43.3
ElasticFusion	0.9	0.9	1.4	10.6
BundleFusion (online)	0.8	0.5	1.1	1.2
BundleFusion (offline)	**0.6**	**0.4**	**0.4**	**1.1**
FlashFusion	0.7	0.8	1.1	1.4

Moreover, live scanning via an Asus Xtion sensor using FlashFusion implemented via CPU computing based on the portable tablet Surface Pro[1] is demonstrated. Here, FlashFusion operates at subcentimeter resolution (5 mm) and achieves global consistency in real time without the use of GPU computing.

Accuracy. The localization accuracy is evaluated using the RMSE of the absolute trajectory error (ATE) Sturm et al. [97], as shown in Table 4.6 for the TUM RGBD dataset and Table 4.7 for the ICL-NUIM dataset, where a smaller ATE value implies higher accuracy. Regarding the surface reconstruction accuracy, we compute the difference between the reconstructed meshes and the ground truth 3D model using the SurfReg tool [44], as depicted in Table 4.8 for the ICL-NUIM dataset.

BundleFusion, which uses two powerful GPUs, shows the best overall performance, obtaining the maximum localization and surface reconstruction accuracy. This method compares each keyframe to all prior keyframes utilizing dense registration and stronger feature descriptors. FlashFusion uses CPU processing for both localization and reconstruction, achieving accuracy that is comparable to that of BundleFusion; however, FlashFusion is somewhat less accurate.

The representative qualitative results of FlashFusion are presented in Fig. 4.24 for the fr3office dataset [97] at 5 mm resolution. Figure 4.24a shows the reconstructed model under phong-shaded rendering, Fig. 4.24b shows the normal map, and Fig. 4.24c shows the color map. To acquire this dataset, the camera was moved in a circle around the desk. Reintegration after loop closure is a FlashFusion technique that successfully addresses localization drift-related misalignments and generates a

[1] https://www.microsoft.com/en-us/surface.

Table 4.8 Accuracy comparison of the surface reconstruction results based on the ICL-NUIM Dataset (cm)

	kt0	kt1	kt2	kt3
RGBD SLAM	4.4	3.2	3.1	16.7
FastFusion	5.6	7.5	7.0	6.6
ElasticFusion	0.7	0.7	0.8	2.8
InfiniTAM	1.3	1.1	**0.1**	2.8
BundleFusion	**0.5**	**0.6**	0.7	**0.8**
FlashFusion	0.8	0.8	1.0	1.3

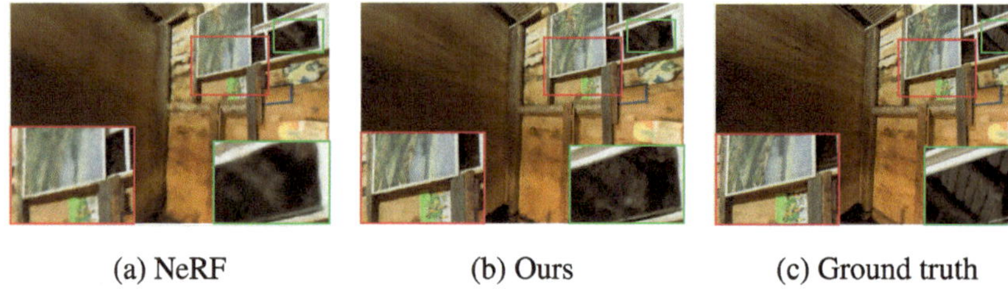

(a) NeRF (b) Ours (c) Ground truth

Fig. 4.24 Qualitative illustration of the FlashFusion results based on the fr3office dataset [97]. **a** Phong-shaded rendering, **b** normal map, and **c** color map

3D model that is globally consistent in real time using CPU processing. The supplemental video shows the live scanning results.

Efficiency. The detailed computing components used in FlashFusion are presented in Fig. 4.25a. The tracking thread, serving as the front-end thread for pose estimation, runs continuously at 30 Hz. An optimization thread is used for global posture optimization based on all the prior keyframes when a new keyframe F_i is inserted. Valid chunk selection is used to choose the chunk candidates to be updated depending on the current keyframe once the pose of the keyframe has been established. When the next keyframe is added, F_i and its related local frames are fused into the TSDF, and the meshing thread updates the meshes based on the updated chunks. All the above computations are accomplished via CPU computing. The reconstructed model, including the vertices, colors, and normals, is then transmitted to the GPU for visualization. To further demonstrate the efficiency of FlashFusion for large-scale datasets, quantitative measurements are conducted based on the representative Dyson_lab [114] dataset, which includes scans of an entire lab over 6400 frames, as presented in Fig. 4.25b. The corresponding qualitative illustration of the reconstructed model is shown in Fig. 4.26. Note that the other 5 components are only called at the keyframe rate while running the back-end threads, excluding the processing time for each frame. The real-time performance of FlashFusion via CPU computing is guaranteed by the clearly efficient completion of all computing components.

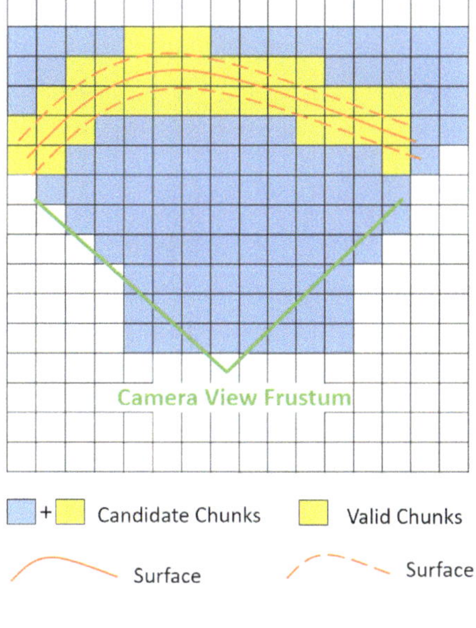

Fig. 4.25 The thread implementation of FlashFusion and its computational complexity analysis

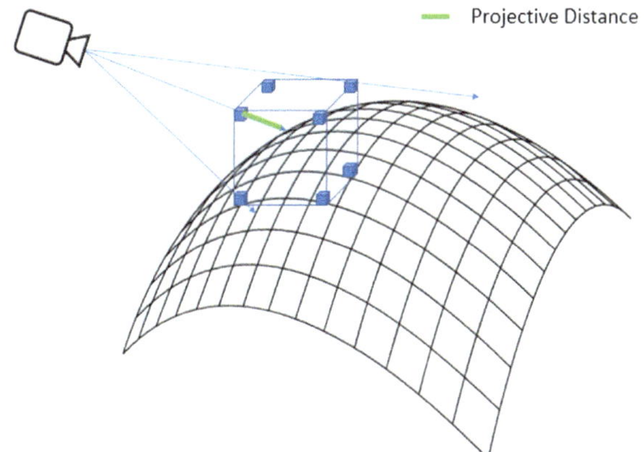

Fig. 4.26 Reconstructed 3D model based on the Dyson_lab [114] dataset

As an efficiency comparison between GPU and CPU computing cannot be performed, we compare our method with state-of-the-art CPU-based systems FastFusion [96] and CHISEL [61] based on the fr3office dataset. FastFusion employs an octree data structure, with the voxel size varying from 20 mm to 5 mm. FastFusion requires 20 ms for TSDF fusion for each frame on average and 300 ms to 1000 ms for mesh extraction, which is approximately 6 times slower than FlashFusion. The detailed efficiency comparison for the TSDF fusion and mesh extraction processes at different voxel resolutions (5 mm, 10 mm and 20 mm) is presented in Table 4.9. While CHISEL can be implemented in real time at a resolution of 20 mm, this becomes computationally expensive as the resolution increases. In contrast, FlashFusion achieves a subcentimeter resolution in real time, or 5 mm resolution at 300 Hz

Table 4.9 Efficiency comparison between CHISEL and FlashFusion (ms)

	CHISEL		FlashFusion	
	TSDF fusion	Mesh extraction	TSDF fusion	Mesh extraction
5 mm	483	1518	3.6	38.4
10 mm	86	312	1.1	19.7
20 mm	15	70	0.7	6.5

for TSDF fusion and 25 Hz for mesh reconstruction, ensuring improved performance in applications such as VR/AR and mobile robots that interact closely with the environment. Moreover, FlashFusion realizes global consistency, while CHISEL cannot due to its localization drift and slow TSDF fusion, as shown in the supplemental video.

Conclusions and Limitations

In this study, we introduce FlashFusion, a real-time 3D reconstruction system that is CPU-based and globally consistent. Based on the suggested method, TSDFs can be updated at $300Hz$ and $5mm$ voxel resolution or $900Hz$ and 1 cm voxel resolution using CPU computing. This is advantageous for real-world applications such as high-resolution dense 3D reconstruction with wearable technology and onboard path planning for flying vehicles or robot arms.

Although the proposed keyframe-based fusion technique significantly reduces the complexity of TSDF fusion, it may reduce the quality of the results by inducing minor artifacts at the borders of keyframes, as shown in Fig. 4.27. Since we only select valid voxels for each keyframe and update the TSDFs of these voxels based on the incoming local frames, surfaces that are detected in local frames but beyond the scope of the keyframe are neglected in the TSDF fusion process. Surfaces in the first and second keyframes are fused with $N_1 + N_2$ observations upon the arrival of the following keyframe, whereas surfaces in later keyframes are fused with only N_2 observations. Artifacts due to uneven fusion arise only at the keyframe edges, and these artifacts quickly disappear when more observations are added.

In our upcoming work, we will attempt to address these issues while considering the potential benefits of combining 3D reconstruction methods with semantic understanding to improve how robots interact with their environment.

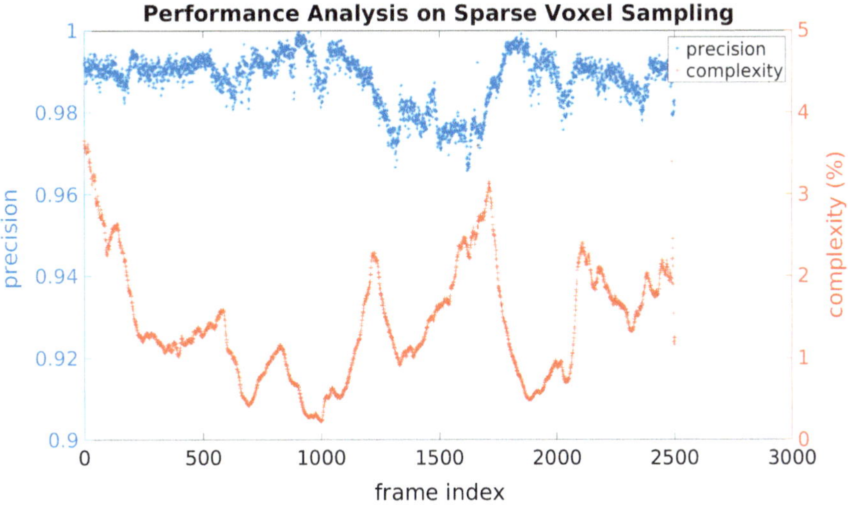

Fig. 4.27 Artifacts introduced due to the keyframe-based integration scheme

4.3.2 IDFusion: Robust Reconstruction

As a key element in spatial AI computing, globally consistent dense 3D reconstruction based on RGBD and inertial measurements has received considerable attention [17]. Due to the increasing use of RGBD sensors, numerous studies have aimed to reduce the registration errors of RGBD scans. However, environments without textures and with motion blur may cause the reconstruction algorithm to fail or produce unpleasant artifacts if the algorithm relies only on visual input. By combining visual and inertial observations for camera state estimation, visual-inertial navigation systems can obtain high precision. For scalability, most visual-inertial fusion systems rely on Kalman filters or window-based nonlinear optimization methods. Because of the marginalization process used in these systems, they obtain great local accuracy but cannot guarantee global consistency.

iDFusion [138] was proposed to investigate the joint benefit, i.e., global consistency and high local accuracy, of visual observations and inertial measurements, with the goal of developing a practical dense 3D reconstruction system that is reliable, globally consistent, and can be implemented in real time. The posture estimation issue is defined as the joint optimization of all camera and IMU data. This issue can be formulated as a nonlinear Gauss-Newton optimization problem, which guarantees a globally consistent estimate of the camera positions and can be immediately solved via max a posteriori estimation. Because this nonlinear optimization approach necessitates traversing every visual and IMU observation, it has significant computational complexity. As demonstrated in FastGO, the complexity of the visual portion is reduced to improve efficiency by decreasing the number of visual observations according to the number of keyframes based on the depth information of each point.

In addition, the IMU preintegration technique is used to group the high-frequency inertial measurements [28]. As a result, in contrast to previous sliding window-based optimization or filtering-based approaches, all previous observations can be optimized in real time, leading to a state-of-the-art performance with public datasets.

Moreover, loop closure [38] can be applied to eliminate the built-up localization drift. When two locations appear to be the same, loop closure based on visual observations can easily produce false-positive loops, resulting in inaccurate camera pose estimates and inconsistently constructed 3D models. False positives can lead to inconsistent posture estimates, making it difficult to differentiate false and true positives because of the accumulated drift in the frame-to-frame registration results. Thus, a unique loop-validity detector based on the IMU bias states was developed. This detector differed from earlier approaches, which used robust cost functions to reduce the impact of erroneous loop closures as the computational complexity increased. To determine whether the loop closure observations are consistent, the detector uses the estimated bias of the IMU state. A truly positive loop closure maintains prior bias estimates when a new loop closure restriction is applied, whereas a false-positive loop closure dramatically changes these estimates to minimize the global energy function. The truncated signed distance field (TSDF) is used at sub-centimeter resolution to fuse the depth observations, with VoxelHashing [81] used for scalability and FlashFusion [40] used for efficiency, given the globally consistent camera pose estimates. The experimental results show that iDFusion can achieve real time, reliable, and globally consistent dense 3D reconstruction results with portable devices for a variety of scenes, including scenes with no texture or motion-blurred backgrounds.

Localization System with Globally Consistent IMU Information

The proposed reconstruction system is shown in Fig. 4.28. For camera localization, the inertial and visual data are combined. To achieve globally consistent localization, all historical observations, including the extrinsic parameters of the camera-IMU transformation, are incorporated into the global optimization process. Erroneous loop detection operations are eliminated by the use of a loop closure validation module. We combine the depth images and localization findings using TSDF fusion to obtain the 3D model. The multithreading architecture speeds up the system and is module independent.

Globally Consistent Pose Estimation

The correctness of the localization model is greatly impacted by the global optimization results. We use the visual information, plane detection information, and IMU observation constraints to further increase the localization robustness. As shown in Fig. 4.29, in addition to the feature matching information, IMU integration is another

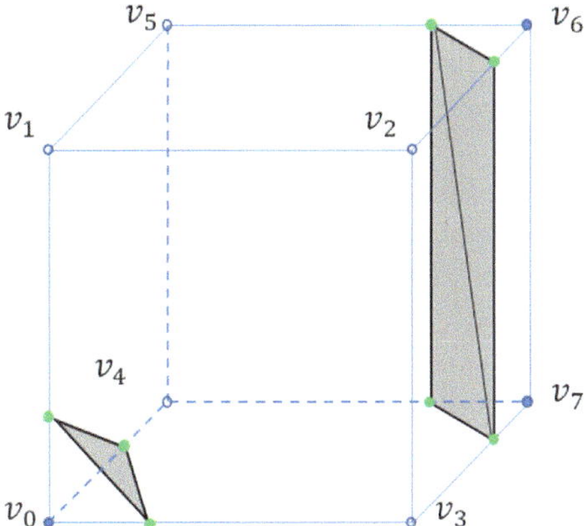

Fig. 4.28 Illustration of the framework of our RGBD reconstruction system. The multithreading architecture achieves modularization and real-time capacity

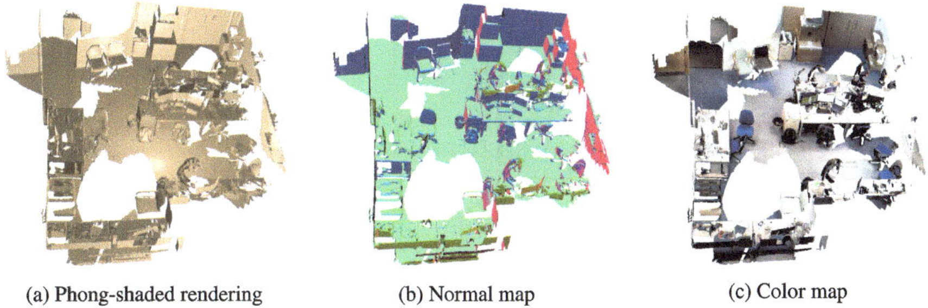

(a) Phong-shaded rendering (b) Normal map (c) Color map

Fig. 4.29 Illustration of our optimization framework. The visual and inertial measurement constraints are shown by the dotted lines

limitation for the adjacent frames. Visual limitations are used to optimize the IMU data. Then, we can retrieve the camera-IMU alignment matrix and the robust camera position.

Next, we discuss the notation definitions. F_i indicates the i_{th} frame, which is a combination of color image I_i and depth image D_i. $\tilde{\boldsymbol{\omega}}$ and $\tilde{\boldsymbol{\alpha}}$ are the angular velocity and acceleration, respectively. The IMU constraint $C^I_{i,i+1}$ is integrated with the sequential information between keyframes. The state variables of the selected keyframes are optimized in the global optimization process. The general frames are located based on the last keyframe. The most recent keyframe is assessed by the loop closure detection module to obtain the matching keyframes. We calculate the feature correspondence $C^F_{i,j}$ and plane correspondence $C^P_{i,j}$ to enhance the geometry constraint. The keyframe correspondence set is denoted by Ω.

The state variables of frame F_i are the rotation R_{ci}, translation P_{ci}, velocity V_i, angular velocity bias b^g_i, and acceleration bias b^a_i. The rotation and translation of the IMU frame are denoted by R_{Ii} and P_{Ii}, respectively. The rotation and translation between the camera and IMU are denoted by R_0 and P_0, respectively. The default gravity is g, while R_g denotes the gravity direction after optimization.

As the situation develops, minor inaccuracies will accumulate. The transformation between the camera, IMU, and initial gravity inevitably deviates, leading to accumulated localization errors. To eliminate these accumulated errors, we optimize the camera-IMU transformation and gravity direction via a global optimization process. The global variables R_0, P_0, and R_g and frame status variables R_{ci}, P_{ci}, V_{Ii}, b_i^g, and b_i^a are considered in one global optimization function. $\mathbf{S}$ denotes all optimization variables, where $\mathbf{S} = \{\mathbf{R}, \mathbf{P}, \mathbf{V}, \mathbf{b^g}, \mathbf{b^a}, R_0, P_0, R_g\}$ and $\{\mathbf{R}, \mathbf{P}, \mathbf{V}, \mathbf{b^g}, \mathbf{b^a}\} = \{R_{ci}, P_{ci}, V_{Ii}, b_i^g, b_i^a, i = 0, \cdots, N-1\}$. $\mathbf{S}$ is optimized by the following function:

$$\mathbf{S}^* = \arg\min_{\mathbf{S}} \sum_{i=0}^{N-1} C_{i,i+1}^I(\mathbf{S}) + \sum_{(i,j)\in\Omega} [\lambda_1 C_{i,j}^P(\mathbf{S}) + \lambda_2 C_{i,j}^F(\mathbf{S})]. \tag{4.34}$$

Feature Constraint. $C_{i,j} = \left\{C_{i,j}^k = \left(p_i^k, p_j^k\right)\right\}$ is the point correspondence based on the frame correspondence (F_i, F_j), which is extracted by the feature matching scheme. P_i^k is a point in the i-th frame. T_{ci} is the transformation matrix of the i-th frame. The corresponding Lie algebra on SE3 is ξ_i. The rotation matrix on SE3 is R_{ci}, while P_{ci} is the translation vector. The visual features are extracted by the ORB extraction algorithm. Then, we formulate the residual functions for point correspondence as follows:

$$r_{i,j}^k(\boldsymbol{\xi}) = T(\xi_i)\, P_i^k - T\left(\xi_j\right) P_j^k. \tag{4.35}$$

By combining all the Jacobian matrices of the corresponding points, we can obtain the feature correspondence constraint as follows:

$$C_{i,j}^F = \sum_{k=0}^{\|C_{i,j}\|-1} \left\|T_i P_i^k - T_j P_j^k\right\|^2. \tag{4.36}$$

The homogeneous format of the point position p_i^k is P_i^k. T_0 denotes the pose of the initial frame, which is not considered in the optimization process. There are N keyframes. Pose initialization before global optimization is achieved by minimizing Eq. (4.36).

Plane Constraint. The plane extraction algorithm with agglomerative hierarchical clustering [24] is used to rapidly extract the planes in the depth images. The plane formulation is

$$X^T P_c + D = 0, \tag{4.37}$$

where X is the plane normal vector, D is the vertical distance from the plane to the origin, and P_c includes all plane points in the depth images. We use the point coordinates to fit the plane equation and obtain the parameters X and D. Then, we transform the plane parameters to the world frame based on the pose of the current depth frame R, P as follows:

$$(RX)^T Pw + [D - (RX)^T T] = 0. \tag{4.38}$$

where P_w is the transformed position of plane points in the world frame.

X_i and D_i are the fitting plane parameters in the i-th camera frame. The pose of this frame is R_i, P_i. The corresponding normal vector and distance in the world frame are $m_i R_i X_i$ and $m_i[D_i - (R_i X_i)^T T_i]$, respectively, while m_i is determined based on the following conditions:

$$m_i = \begin{cases} 1, \; D_i - (R_i X_i)^T T_i > 0, \\ -1, \; D_i - (R_i X_i)^T T_i < 0. \end{cases} \tag{4.39}$$

This parameter is used to adjust the distance between the pose and the positive value. The k-th detected plane $C^k_{P_{i,j}}$ is detected in frames i and j simultaneously. The plane constraint $C_{P_{i,j}}$ is formulated based on the Euclidean distance of the plane parameters. We formulate the residual equation of the plane correspondence as follows:

$$\begin{aligned} r^k_{N_{i,j}} &= m^k_i R_{ci} X^k_i - m^k_j R_{cj} X^k_j, \\ r^k_{D_{i,j}} &= m^k_i [D^k_i - (R_{ci} X^k_i)^T P_{ci}] - m^k_j [D^k_j - (R_{cj} X^k_j)^T P_{cj}]. \end{aligned} \tag{4.40}$$

The plane constraint is the combination of the residuals of the plane correspondences.

$$C^P_{i,j} = \sum_{k=0}^{\left\|C_{P_{i,j}}\right\|-1} (\left\|r^k_{N_{i,j}}\right\|^2 + \left\|r^k_{D_{i,j}}\right\|^2). \tag{4.41}$$

IMU Constraint. Based on the known pose R_{ci}, P_{ci} of keyframe F_i, we can calculate the pose R_{cj}, P_{cj} of keyframe F_j with the extracted point correspondences. V_{Ii} is initialized by the speed calculation. The transformation matrix between the camera and IMU is R_O, P_O. The IMU pose can be transformed based on the camera pose using the following equation:

$$\begin{aligned} R_{Ii} &= R_{ci} R_O, \\ P_{Ii} &= P_{ci} + R_{ci} P_O. \end{aligned} \tag{4.42}$$

The pose variation between keyframes F_i and F_j can be determined based on the calculated IMU pose as follows:

$$\begin{aligned}\Delta R_{ij} &= R_{Ii}^T R_{Ij},\\ \Delta V_{ij} &\doteq R_{Ii}^\top \left(V_{Ij} - V_{Ii} - g\Delta t_{ij}\right),\\ \Delta P_{ij} &\doteq R_i^\top \left(P_{Ij} - P_{Ii} - V_{Ii}\Delta t_{ij} - \frac{1}{2}\mathbf{g}\Delta t_{ij}^2\right).\end{aligned} \tag{4.43}$$

In addition, the IMU observations are integrated to obtain the motion information $\Delta\tilde{R}_{ij}$, $\Delta\tilde{V}_{ij}$ and $\Delta\tilde{P}_{ij}$ based on the motion integral model. The integration is calculated only once since the changed camera poses have no effect on the IMU integration results. The motion residual is the alignment error between the transformed IMU poses and the integrated IMU motion:

$$\begin{aligned}r_{\Delta R_{ij}} &= Log(\Delta R_{ij}\Delta\tilde{R}_{ij}),\\ r_{\Delta V_{ij}} &= \Delta V_{ij} - \Delta\tilde{V}_{ij},\\ r_{\Delta P_{ij}} &= \Delta P_{ij} - \Delta\tilde{P}_{ij}.\end{aligned} \tag{4.44}$$

The motion constraint based on the IMU observations is the combination of the terms between adjacent keyframes:

$$C_{i,i+1}^I = \left\|r_{\Delta R_{ij}}\right\|^2 + \left\|r_{\Delta V_{ij}}\right\|^2 + \left\|r_{\Delta P_{ij}}\right\|^2. \tag{4.45}$$

Efficient Solution. The global optimization process is formulated as the minimization of Eq. (4.34). To solve the optimization problem, we apply the following Gauss-Newton iteration method:

$$\boldsymbol{\delta} = -\left(J(\boldsymbol{\xi})^T J(\boldsymbol{\xi})\right)^{-1} J(\boldsymbol{\xi})^T \boldsymbol{r}(\boldsymbol{\xi}). \tag{4.46}$$

The large number of variables in the optimization problem leads to high computational complexity. We simplify the calculation of constructing the matrix in the Gauss-Newton iterations to achieve real-time performance. In the following section, we show the simplification of constructing $J(\boldsymbol{\xi})^T J(\boldsymbol{\xi})$, and $J(\boldsymbol{\xi})^T \boldsymbol{r}(\boldsymbol{\xi})$ is processed in a similar manner.

The matrix used in the Gauss-Newton iteration can be decomposed, as illustrated in Fig. 4.30. We separate the Jacobian matrix into small blocks that can be calculated separately to combine $J(\boldsymbol{\xi})^T J(\boldsymbol{\xi})$. The matrix manipulation can be separated into block manipulation. The computational complexities of calculating C^F, C^P and C^I with Eqs. (4.36), (4.41) and (4.45) are $O(k * N_{corr})$, $O(m * N_{corr})$ and $O(N)$, respectively. N and N_{corr} represent the number of keyframes and the number of frame correspondences, respectively. k and m denote the average number of feature correspondences and plane correspondences per image correspondence, respectively. The number of feature correspondences is much larger than the number of other correspondences. Thus, the efficiency bottleneck is the constraint of C^F. As a result, the calculation cost of $J^T J$ in Fig. 4.30 is mainly from the calculation of $J_A^T J_A$. Therefore, the key to the accumulation is the simplification of the $J_A^T J_A$ calculation. We use

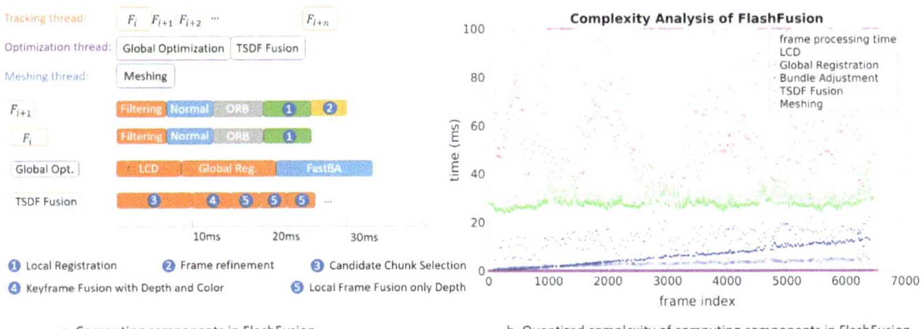

Fig. 4.30 Jacobian in the Gauss-Newton function. The number of corresponding points is K. The number of keyframes is N, and the number of corresponding planes is M. The Jacobian of the feature point constraint is J_A. The Jacobian of the IMU constraint is J_B, J_C, J_D. The Jacobian of the plane constraint is J_E

the improved FastGO [42] algorithm to reduce the number of repeated integration calculations. $J_A^T J_A$ is combined with the statistics of the feature correspondences, which simplifies the complexity of C^F from $O(k * N_{corr})$ to $O(N_{corr})$. As a result, the time consumption is significantly reduced.

FastGO [42] is based on the minimization of feature matching residuals (Eq. 4.35). The derivative matrix of $r_{i,j}^k(\boldsymbol{\xi})$ is

$$J_i^k(\xi_i) = [I_{3*3} \quad -[T(\xi_i)P_i^k]^\wedge]. \tag{4.47}$$

The optimization update of the variables is consistent with the derivation on SE3:

$$exp(\xi^\wedge) = exp(\delta\xi^\wedge) * exp(\xi^\wedge). \tag{4.48}$$

This is equivalent to

$$\begin{aligned} P &= P + exp(\delta\Phi^\wedge) * \delta P, \\ R &= exp(\delta\Phi^\wedge) * R. \end{aligned} \tag{4.49}$$

R and P are the rotation matrix and transformation vector in $T(\xi)$. $\delta\Phi$ and δP are the rotation vector and the transformation part of $\delta\xi$, respectively.

We use the SO3 derivation since the SE3 derivation is more complex. R and P are computed separately and updated as follows:

$$\begin{aligned} P &= P + \delta P, \\ R &= exp(\delta\Phi^\wedge) * R. \end{aligned} \tag{4.50}$$

Then, R and P are differentiated to obtain their Jacobian matrix:

$$J_i^k(\xi_i) = [I_{3*3} \quad -[R(\Phi_i)P_i^k]^\wedge]. \tag{4.51}$$

In contrast to Eqs. (4.47) and (4.51) has no translation term, which simplifies the calculation.

The submatrix of the feature correspondence between frames i and j is

$$J_m = [0 \cdots J_i^k(\xi_i) \cdots 0 \cdots -J_j^k(\xi_j) \cdots 0]. \tag{4.52}$$

The submatrix $J_{C_{i,j}}^T J_{C_{i,j}}$ is the combination of all feature correspondence submatrices:

$$J_{C_{i,j}}^T J_{C_{i,j}} = \sum\nolimits_{m \in C_{i,j}} J_m^T J_m. \tag{4.53}$$

The whole Jacobian matrix $J_A^T J_A$ is computed based on matrices such as $J_{C_{i,j}}^T J_{C_{i,j}}$. The calculation is shown as follows:

$$J_A^T J_A = \sum J_{C_{i,j}}^T J_{C_{i,j}}. \tag{4.54}$$

Equation (4.52) shows that J_m is constructed with only two nonzero matrices ($3 * 6$). As a result, $J_m^T J_m$ ($m \in C_{i,j}$) is constructed with only four nonzero matrices ($6 * 6$). $J_{C_{i,j}}^T J_{C_{i,j}}$ is the integration of $J_m^T J_m$ ($m \in C_{i,j}$), which is also constructed with only four nonzero matrices ($6 * 6$). This matrix is calculated by combining the second-order statistics of the feature information. Then, $J_A^T J_A$ is constructed based on $J_{C_{i,j}}^T J_{C_{i,j}}$, which means that this matrix can be directly calculated by integrating the second-order statistics of the feature information. This approach simplifies the complexity of the C^F calculation from $O(k * N_{corr})$ to $O(N_{corr})$

As shown in Fig. 4.30, the main complexity is the calculation of $J^T J$. We greatly reduce the calculation complexity. The $J^T r$ calculation is simplified by a similar integration approach. Then, Eq. (4.46) is solved using the sparse matrix solver for efficiency.

Self-Calibration and Gravity Optimization

For visual-inertial localization, the accuracy of the transformation between the camera and IMU is crucial. Small transformation inaccuracies accumulate during the optimization process. The use of plug-ins is not appropriate for offline calibration tools such as Kalibr [30] because they require calibration instruments and laborious procedures. We proposed an online calibration method to fit the nonfixed IMU device to eliminate these laborious procedures and improve the overall consistency. The calibration was carried out during the global optimization to eliminate the accumulated errors. To achieve tightly coupled calibration, $R_{ci} R_0$ and $P_{ci} + R_{ci} P_0$ are used to replace R_{Ii} and P_{Ii} in the optimization formulation. The camera pose information R_{ci} and P_{ci} and transformation matrix information R_0 and P_0 are optimization variables, which are jointly optimized in the global optimization process.

Fig. 4.31 The camera-IMU rotation vector rapidly converges to the accurate value after visual initialization

The gravity constant is crucial for IMU integration since it must be removed from IMU acceleration observations. This value can be approximated by optimization approaches. [85] represented the gravity vector with an orthogonal basis in the optimization process, whereas [63] explicitly used the gravity vector in the optimization function. Both techniques require a normalization procedure and modify the gravity magnitude during the optimization process.

R_0, P_0 and R_g are introduced in the globally consistent optimization process to achieve self-calibration. The gravity optimization is achieved by adjusting the direction of R_g. The default R_0, P_0 and R_g are arbitrarily set values that different considerably from the real values. The optimization process may not converge due to a poor initial value. The first 20 keyframes are utilized to estimate the calibration data based on the tracked camera postures, which helps to enhance the initialization. We use the global optimization approach to increase the accuracy using precise calibration data. The convergence of this approach is robust. Figure 4.31 shows the rapid convergence of the rotation vector between the camera and IMU after the proposed initialization.

Robust Loop-Validity Detector

The accuracy of the detection findings cannot be guaranteed by the loop closure detection method based on visual cues. The identified images look identical, suggesting that additional areas should be considered. Consider the placement of two identical posters on two separate white walls. To eliminate accumulated errors, loop detection accuracy is crucial. The reconstructed model is severely deformed by false loop detection results. We utilize the following information to reject false loop detections:

(1) The suggested global optimization formulation obtains accurate IMU bias, which varies only slightly as the optimization process progresses. (2) Erroneous loop closures add poor constraints to the overall optimization. The anticipated IMU bias values in the global optimization are noticeably altered to fit these inconsistencies. Consistent constraints are added by appropriate loop closures, which are obtained based on the accumulated errors. The global optimization process maintains the stability of the predicted IMU bias.

We develop a robust loop closure validator based on the predicted IMU bias fluctuation in the optimization process based on the above findings. To evaluate the state of the incoming keyframes, Gauss-Newton iterations are carried out. We calculate the summation of the acceleration bias variation $\delta \mathbf{b}^{\mathbf{a}}$ and angular velocity bias variation $\delta \mathbf{b}^{\mathbf{g}}$. T_{b_a} and T_{b_g} are the thresholds. When a loop closure result is introduced in the global optimization process, if $[\delta \mathbf{b}^{\mathbf{a}}] > T_{b_a}$ and $[\delta \mathbf{b}^{\mathbf{g}}] > T_{b_g}$, the loop closure is considered to be false. We remove the false detected loops and recover the optimization variables.

Online Texture Mapping

TSDF fusion and the marching cube technique can be used to create mesh models based on registered depth images. The extracted mesh color image, on the other hand, is the average color image in the TSDF, which is blurrier than the original color image. Foley *et al.* [27] used texture mapping to incorporate the high-definition image details into the model. However, label selection in texture mapping is a complex, time-consuming optimization problem that limits the application of online texture mapping methods.

In this section, we streamline the texture mapping calculation and combine this calculation with the real-time dense reconstruction technique. The marching cube algorithm and TSDF fusion are used to construct the model. The texture model is then created using our streamlined online texture mapping algorithm, which has considerably better visual performance than previous models. TSDF-based light adjustment and chunk grouping in view selection are the two primary components of the simplified texture mapping approach. All of the meshes in a chunk are treated as one optimization object by the chunk grouping method. The same observation image is used for texture mapping across all of the chunked meshes. As a result, this method has significantly fewer optimization variables than previous approaches. Texture mapping may obtain blocky results because the various images were collected with different exposure settings. To alter the lighting in the texture results, we refer to the TSDF color. We can create a high-quality textured model in real time using the suggested chunk grouping and TSDF-based light adjustment methods.

Chunk Grouping in View Selection

The mesh model $M = \{m_i\}_{i=0}^{N}$ has N triangles. The mapping calculation is to select a suitable image from all observation images $\mathcal{F} = \{f_i\}_{i=0}^{M}$ for every triangle. The mesh m_i selects the images f_i, which means that the projection patch on f_i is used for the mapping of m_i.

With the reconstructed model, we can construct a graph $G_t(\mathcal{V}_t, \mathcal{E}_t)$ based on the chunk structure, including the chunks $\mathcal{V}_t = \{v_i\}_{i=1}^{N_t}$ and the edges $\mathcal{E}_t \subseteq \{(v_i, v_j)\}_{i \neq j; v_i, v_j \in \mathcal{V}_t}$. The edges re geometrically adjacent chunks. $l(v_i)$ indicates the selected image of mesh v_i. As discussed in [64], the view selection can be modeled as the best label selection, which can be optimized by combining the data term $E_{data}(l^t(v_i))$ and smoothing term $E_{smooth}(l(v_i), l(v_j))$. The data term constrains the sharpness and angle of the mapped image:

$$E_{data}(l(v_i)) = -\int_{\Phi(v_i)} ||\nabla I(p)||_2 \cdot \cos\theta dp. \tag{4.55}$$

$\nabla I(p)$ is the pixel gradient, which indicates the sharpness of the image patch. $\Phi(v)$ refers to the mapping pixels of v_i on $l(v_i)$. θ is the angle between the mesh normal and ray direction. The smoothing term ensures that the images selected by adjacent meshes are as similar as possible:

$$E_{smooth}(l(v_i), l(v_j)) = -[l^t(v_i) = l^t(v_j)] \cdot [(v_i, v_j) \in \mathcal{E}_t]. \tag{4.56}$$

The smoothing term uses the Iverson bracket, which is efficient and effective for encouraging mapping continuity. The optimization loss is the combination of the data term and smoothing term:

$$\min_{l}[\Sigma_i E_{data}(l(v_i)) + \lambda \Sigma_{i,j} E_{smooth}(l(v_i), l(v_j))]. \tag{4.57}$$

The target is to obtain a suitable mapping $l : \mathcal{V} \to \mathcal{F}$, which can be optimized by the MRF solver [105]. However, the prior texture mapping task has a very long computational time, and each mesh is considered an independent variable to improve the meshing efficiency. The mesh models in online reconstruction systems regularly change due to altered camera postures and new keyframes. Real-time performance requires a more effective online mapping system. The mesh in a chunk is approximately continuous for indoor dense reconstruction tasks, which typically use the same image for texture mapping. This implies that considering each mesh as an independent variable results in many unnecessary computations. By treating the entire collection of meshes as a single optimization variable, we can decrease the number of optimization variables. As a result, the problem has significantly fewer optimization variables, thereby enabling real-time mapping calculations. Additionally, the mapping performance is very similar to that of the original offline mapping scheme.

At time t, an undirected graph $G_t(\mathcal{V}_t, \mathcal{E}_t)$ is constructed based on the chunk set $\mathcal{V}_t = \{v_i\}_{i=1}^{N_t}$ and the edge set $\mathcal{E}_t \subseteq \{(v_i, v_j)\}_{i \neq j; v_i, v_j \in \mathcal{V}_t}$. Here, an edge exists between a chunk and its 6 neighbors when their submeshes form a continuous geometric surface. Following the labeling procedure in [64], the overall target is to assign labels, denoted as $l^t : \mathcal{V}_t \rightarrow \mathcal{F}_t$, by optimizing the following two-term objective function:

$$\min_{l^t}[\Sigma_i E^t_{data}(l^t(v_i)) + \lambda \Sigma_{i,j} E^t_{smooth}(l^t(v_i), l^t(v_j))], \tag{4.58}$$

where the first data term can be represented as

$$E_{data}(l(v_i)) = -\int_{\Phi(v_i)} ||\nabla I(p)||_2 \cdot \cos\theta dp. \tag{4.59}$$

This is similar to the "gradient magnitude", i.e., the integration of intensity gradients in an image [34]. In Eq. 4.59, $\Phi(v)$ refers to the mapping between triangles in v and the corresponding image area in $l(v)$. $I(p)$ represents the image intensity at pixel p, and θ denotes the angle between the view direction and the surface normal. In the gradient magnitude task, image components with larger gradients (finer details) are more suitable as high-resolution candidates. A *cosine* factor is included in the integral since the view angle increases the original texture area by $\cos^{-1}\theta$ and dilutes texture details.

The second smoothing term can be represented as

$$E_{smooth}(l(v_i), l(v_j)) = -[l^t(v_i) = l^t(v_j)] \cdot [(v_i, v_j) \in \mathcal{E}_t], \tag{4.60}$$

using the Potts model, where $[\cdot]$ denotes the Iverson bracket. The simple submodular pairwise cost term improves efficiency [23] while ensuring large homogeneous texture patches. Moreover, this term is not influenced by illumination differences.

The optimization procedure is executed via the MRF solver introduced in [105]. The conventional texture stitching frameworks tend to directly handle millions of triangles for the sake of quality. Additionally, because the frame poses are chosen to achieve globally consistent dense 3D reconstruction results, the geometry frequently changes. Real-time texture mapping is more challenging because the texture must be updated due to the changing geometry. For indoor conditions, the surfaces are typically continuous, suggesting that nearby triangles frequently have the same label. In other words, it would be superfluous for each triangle m_i to have its own label. Recall that the triangles are grouped by chunks $\mathcal{V} = \{v_i\}$ based on spatial locations. We exploit the chunk structure and enforce that triangles belonging to the same chunk share the same corresponding frame. Following this insight, the chunk structure allows us to easily manipulate the triangles and reduce the problem scale. Triangles in the same chunk are grouped as an atomic node (submesh) in the optimization process. This approach substantially reduces the number of variable, and the effect on the visual appearance is limited compared with the results of other offline frameworks.

TSDF-Based Light Adjustment

Due to the varying lighting and exposure settings of different images, textured models typically have an undesirable blocky look. Even if the camera exposure remains unchanged and the differences in light sensitivity in different pixel regions are calibrated, the blocky effect is still visible since the illumination of the image changes over time. The mesh color can be changed using the optimization-based approach as a minimization problem. However, the real-time mapping calculation is constrained due to the increased computational overhead. To achieve online texture mapping and color adjustment with only CPU computations, we propose a TSDF-based light adjustment method. Suppose a chunk has K vertices $\{ver_i\}_{i=1}^{K}$. Based on the existing TSDF model, we can obtain the relevant TSDF color $\{c_i^t\}_{i=1}^{K}$. The mean TSDF color is C_{mean}^t. Based on the view selection results, we can obtain the relevant mapping pixel color $\{c_i^p\}_{i=1}^{K}$. The mean pixel color is C_{mean}^p. Then, we can obtain the calibration color $C_{cal} = C_{mean}^p - C_{mean}^t$. The color of all texture mapping pixels for the chunk can be adjusted as $c_i^{map} = c_i^{map} - C_{cal}$. The fusion of the many images produces the TSDF color result. Due to the fuzzy effect of the fusion process, the color extracted from the TSDF model is useful as a reference to bridge the gap between various images. The TSDF-based color adjustment strategy is particularly effective since we already have the TSDF model, which enables real-time processing.

Experiments

Localization with Globally Consistent IMU Information. The localization accuracy is evaluated based on the ICL-NUIM [43] and TUM RGBD [97] datasets. Similar to RGB-D-Inertial SLAM system [63], we used synthetic IMU data. The experimental hardware included an ASUS Xtion depth camera and SC-AHRS-100D2 IMU. We collected challenging real-world data, including fast motion and textureless scenes. We used ElasticFusion [114] and FlashFusion [40] and compared their reconstruction performance with that of our approach. The proposed system was run on an Intel Core i7-9700K CPU without the need for parallel GPU calculations.

Localization Accuracy. The RMSE of the trajectory distance [97] is used as an evaluation metric. Table. 4.10 shows the comparison results based on the public ICL-NUIM and TUM RGBD datasets. A larger RMSE value indicates a worse localization result. The methods used for comparison are CPA-SLAM [73], ElasticFusion [114], NIO [109], BundleFusion [15], Noniterative SLAM [110], RGB-D SLAM [22], RGB-D-Inertial SLAM [63], and FlashFusion [40]. The experiments show that our method is comparable to parallel acceleration methods using GPUs and is more accurate than lightweight methods using only CPUs.

Complexity Evaluation. The online performance of our method is effectively ensured by the delicate multithread structure. Global optimization is conducted in a loop following the initialization stage. The amount of time to optimize different datasets is displayed in Fig. 4.33. The added IMU and plane limitations do not

Table 4.10 Quantitative evaluations based on public datasets in terms of the absolute trajectory error (cm)

	kt0	kt1	kt2	kt3	fr1/desk	fr2/xyz	fr3/office
ElasticFusion (GPU) [114]	0.9	0.9	1.4	10.6	2.0	1.1	**1.7**
BundleFusion online (GPU) [15]	0.8	**0.5**	1.1	**1.2**	**1.7**	1.4	2.9
CPA-SLAM (GPU) [73]	1.8	1.4	2.5	1.6	–	–	–
RGB-D-Inertial SLAM (GPU) [63]	0.9	1.2	**0.9**	1.9	–	–	–
RGBD SLAM (CPU) [22]	2.6	0.8	1.8	43.3	2.3	**0.8**	3.2
NIO (CPU) [109]	–	–	–	–	2.5	1.2	3.3
Noniterative SLAM (CPU) [110]	–	–	–	–	2.5	1.1	–
FlashFusion (CPU) [40]	**0.7**	0.8	1.1	1.4	1.9	1.3	2.5
Ours (CPU)	**0.7**	0.7	**0.9**	**1.2**	2.0	1.0	**1.7**

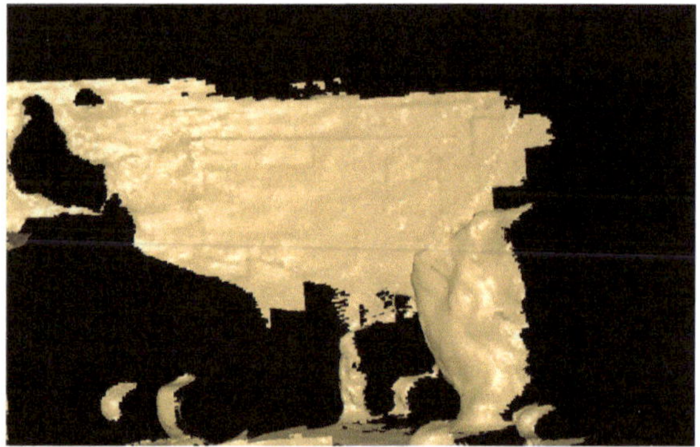

Fig. 4.32 Reconstruction with and without IMU optimization. **a** shows the fusion of the IMU information, and the accurate localization results ensure a smooth reconstructed model. However, the incorrect localization in **b** leads to obvious artifacts in the reconstructed model

increase the optimization time. The global optimization process takes an average of 32.08 milliseconds for 2488 frames. Moreover, the tracking thread is not blocked by the optimization thread.

Reconstruction in Repetitive Environments. Figure 4.34 illustrates the variation in the IMU bias estimation. When the position loop closure approach is used, the variation is mild. However, the introduction of the false loop closure scheme significantly alters the ariation results. We validate the loop identification results using the change in the IMU bias based on this phenomenon.

Figures 4.35a, b illustrate the loop detection validation results. The test scene contains comparable sites that are simple to identify as the same place. Without a loop limitation, ElasticFusion clearly accumulates errors. FlashFusion introduces a

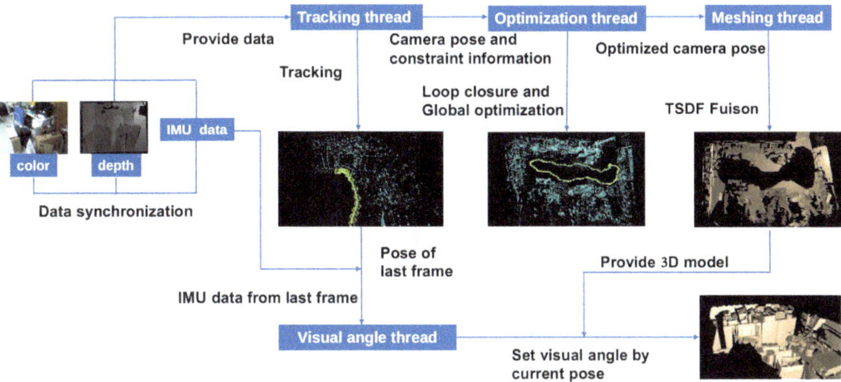

Fig. 4.33 Efficiency evaluation based on the TUM dataset fr3/officet [97]. The global optimization time increases linearly with the number of frames

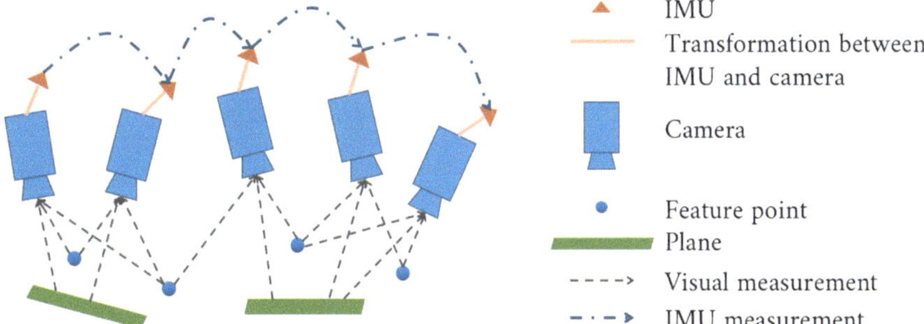

Fig. 4.34 The norm of bias variation. False loop causes severe bias variation while true loop will not lead to sudden change

false loop, which causes the model to be severely warped. Both methods cannot determine the validity of the discovered loop based on only the visual observations. However, we can use the variation in the high-quality predicted IMU bias to validate the loop identification results. To guarantee global consistency, the correct filtered loop is added during the global optimization process.

Reconstruction in Scenes with Challenging Movements. Fast motions and textureless areas are two situations in which camera tracking systems relying on visual information are prone to failure. When visual tracking is unsuccessful, IMU integration can be used to localize the camera. The direction of gravity, the camera and IMU transformation settings, and the velocity and bias of the IMU all have significant impacts on IMU integration. Figure 4.32a shows the localization results based on only IMU integration, demonstrating the accuracy of our globally consistent state optimization results. Figure 4.32b shows the localization results based on an IMU without state optimization, which has obvious artifacts due to the inaccurate state variables.

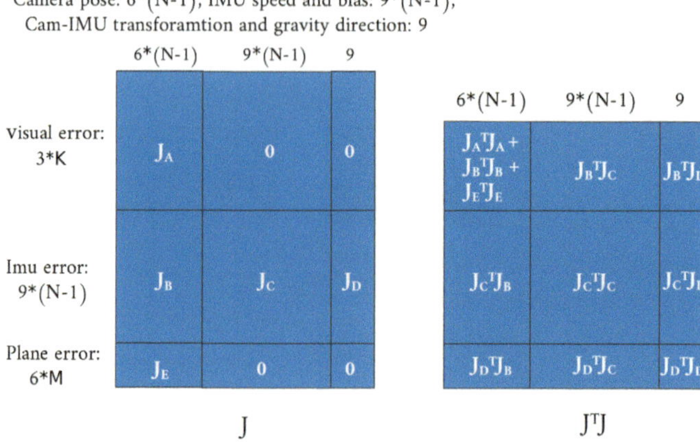

Fig. 4.35 Quantitative evaluation. The concerned methods are the state-of-the-art ElasticFusion and FlashFusion. **a, b** present the scenario with two repetitive appearances (highlighted by circle). iDFusion discards false loop, achieving the global consistency. FlashFusion is misled by the false loop, producing a disordered result. ElasticFusion discards the true loop, causing the accumulated error. **c** shows the scenario under fast motion. iDFusion tracks robustly, assuring complete and consistent reconstruction. FlashFusion suffers failed tracking, causing incomplete reconstruction. ElasticFusion applies wrong visual tracking, leading to overlapped reconstruction. **d, e** show the zoomed local geometry and color details under fast motion. FlashFusion has a more uneven surface and a blurry color appearance. Both the surface and color of ElasticFusion appear more sparse losing many details. On the contrary, iDFusion achieves higher quality reconstruction by fusing visual and inertial observations for higher localization accuracy

Visual tracking fails for blurred images showing fast motion. Figure 4.35c shows the performance for a scene with fast motion. The results demonstrate that FlashFusion's failed tracking limits the rebuilding procedure. ElasticFusion's chaotic reconstruction outcomes are caused by incorrect tracking results. Using the additional IMU information, our technique provides strong localization results when the visual tracking system fails. The authenticity and integrity of the rebuilt model are guaranteed by the precise camera poses. Figure 4.35d, e show that our proposed method achieves higher quality details in scenes with fast motion than ElasticFusion and FlashFusion.

Online Texture Mapping. The quantitative evaluation was performed using the structural similarity index measure (SSIM). We compared the quality of the rendered images with the rendered images generated by other online reconstruction schemes, including the surfel-based method SurfelMeshing [91] and the volume-based method FlashFusion [40]. We used public datasets [15, 44] for the comparison. As shown in Table 4.11, our proposed method obtains better texture mapping performance than the other approaches. In the scene with complex illumination conditions (BundleFusion office2), our method is slightly inferior due to the blocky effect.

Then, we compared the texture mapping efficiency of the different methods. We used chunk grouping in the view selection calculation. The number of optimization

Table 4.11 SSIM evaluation

	Ours	FlashFusion	SurfelMeshing [91]
ICL-NUIM kt1	**0.978**	**0.978**	0.971
ICL-NUIM kt0	**0.943**	0.933	0.871
Copyroom	**0.886**	0.846	0.859
BundleFusion office2	0.967	**0.979**	0.954

Table 4.12 Efficiency comparison based on public datasets

	Graph node number of ours	Time of ours (ms)	Graph node number of [108]	Time of [108] (s)
ICL-NUIM kt1	10482	112	1211399	18.6
ICL-NUIM kt0	11677	163	1325307	36.8
BundleFusion office2	43844	633	2259375	63.0
Copyroom	16817	276	1271893	35.8

variables was greatly reduced, accelerating the speed of the MRF solver by more than 100 times. Table 4.12 shows a comparison of the texture mapping efficiency of our approach and an offline scheme [108].

Figure 4.36 shows a qualitative evaluation of our method with a volume-based method [40] and an offline texture mapping method [108]. The comparison methods use the same mesh model and images as input. As shown in Fig. 4.36, our proposed method has much better rendering performance than the volume-based method [40] and has comparable performance with the offline method [108].

4.3.3 *UnstructuredFusion: Real-Time 4D Reconstruction*

Virtual reality (VR) and augmented reality (AR) technologies have advanced over the past ten years, offering creative approaches to present information in ways that were unimaginable only a few years ago. These technologies have a wide range of uses, including entertainment, business, gaming, education, military, and the arts. In particular, live 4D (3D spatial plus 1D time) content generation and reconstruction have become cutting-edge but bottleneck techniques in VR/AR applications. These techniques are constrained by imperfect 3D sensing systems using existing RGBD sensors and imperfect reconstruction algorithms, especially when handling challenging dynamic scenes such as nonrigid human motions. Computer vision and computer graphics researchers have recently investigated how to recreate 4D models of human actions for improved VR/AR experiences.

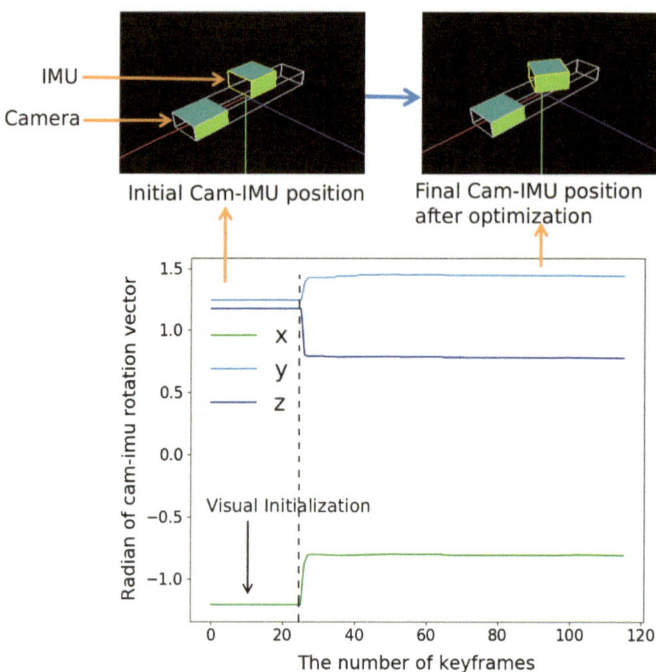

Fig. 4.36 Qualitative comparisons of FlashFusion [40] (top), our online texture mapping method (middle) and an offline texture mapping method [108] (bottom). The local contents highlighted in the red bounding boxes are magnified for better visualization

Recent technological advances using RGBD sensors for 4D reconstruction have achieved considerable progress in terms of both effectiveness and efficiency by leveraging high-end GPUs. Reconstruction techniques using single-depth camera setups [36, 49, 79, 94, 127] usually adopt temporal fusion pipelines to address incomplete observation challenges; however, the reconstruction results still suffer from inherent self-occlusion issues due to lack of camera resources. Although the cutting-edge DoubleFusion [128] method uses human shape priors and achieves robust dynamic reconstruction results for innocuous human body motions, it still only produces geometric reconstruction results and cannot generate interesting textures because of input limitations. One approach is to use collaborative multicamera systems such as Fusion4D and Motion2Fusion to reconstruct high-quality 4D geometries and textures at the same time. However, due to the need for noncommercial depth cameras, which include tens of RGB/infrared cameras integrated with structured lighting systems, these systems are expensive and challenging to deploy. More crucially, the requirement that all cameras and illumination systems be synchronized and precalibrated severely restricts the range of daily use applications.

We propose *UnstructuredFusion*, which allows real time, high-quality, complete reconstruction of 4D textured models of human body motions with only three commercial RGBD cameras. The three depth cameras cover the whole human body in a compensated yet flexible way, i.e., they are allocated in an unstructured manner without any precalibration or synchronization. Compared with [19, 20], our system is significantly simpler to set up and less expensive; however, this simplicity leads to several issues for the algorithms used to reconstruct the high-quality 4D geometries and textures. (1) The RGBD camera array is sparse and unsynchronized, making traditional precalibration very time-consuming and challenging. (2) The camera array collects internal asynchronous RGB and depth videos; this unsynchronized data capture involves not only the RGBD cameras but also each individual camera. The multicamera system needs to register the unstructured depths and movies based on human performance online to address these issues.

According to the above analysis, the six streams, including the three depth streams and three RGB video streams, are all completely unstructured, meaning that they are all geographically and temporally unsynchronized. Our main strategy for resolving this problem is to identify the best anchor for aligning the depth and color streams. We use the human skeleton and surface information for depth alignment, and we propose a coarse-to-fine alignment technique by using skeleton data and a nonrigid optimization algorithm to adjust the multiview depths. We suggest fusing a canonical texture atlas as an anchor to guide the updating of the temporal dynamic texture for texture reconstruction. The following is a summary of the technical contributions of our UnstructuredFusion system.

- We propose an unstructured, multiple RGBD camera system using only three commercial RGBD cameras for real-time human performance capture.
- We propose a skeleton warping-based nonrigid tracking scheme for unstructured multiview depth alignment. This scheme can be used in both the online calibration step and the tracking step.
- We propose a dynamic atlas texturing scheme for warping and updating the texture based on the fused geometry, achieving high-quality appearance reconstruction in real time.

Because of the uniqueness of our scheme, UnstructuredFusion acts as a good middle ground between an excessively demanding hardware setup and a high-quality reconstruction system, encouraging 4D reconstruction in immersive telepresence scenarios and potentially achieving more immersive and interactive experiences.

Overview of UnstructuredFusion

The proposed UnstructuredFusion system aims to achieve real-time human performance capture in unstructured and sparse multiview environments. To achieve this, we developed a new pipeline that is not only resistant to unstructured misalignment across views but also fully utilizes the depth and color information of several views for accurate reconstruction while still operating at real-time speeds.

Figure 4.37 illustrates the high-level components of our system pipeline, which obtains considerably more vivid results than previous real-time performance capture

Fig. 4.37 The system pipeline of UnstructuredFusion. We initialize our system in the first frame by performing online multicamera calibration. Then, for each frame, we sequentially perform the next three steps: skeleton warping-based nonrigid tracking, geometric fusion, and atlas texturing based on temporal blending. Finally, live textured meshes with geometric details are obtained

systems under the sparse multiview setting. Our system uses RGBD images from sparse and unstructured views as input and output textured meshes. Specifically, three Kinect v2 sensors are utilized to generate three uncalibrated and unsynchronized RGBD streams with 512×424 resolution at 30 fps. A temporal fusion strategy to accumulate the 3D reconstruction data is adopted, and the truncated signed distance function (TSDF) [14] volume is utilized as the underlying data structure. Similar to [128, 137], the embedded deformation (ED) model [99] and a linear human body (SMPL) model [72] are combined for nonrigid motion representation. Furthermore, the main components of our pipeline are briefly reviewed as follows.

Initialization. We propose a novel online multiview calibration scheme that simultaneously optimizes the camera poses, initial pose, and shape parameters of the SMPL model to initialize the first frame. This aligns the uncalibrated RGBD streams and ensures that the SMPL human prior model is automatically embedded. Even when taking into account the lack of suitable overlap regions between close views in the sparse multiview setting, our online calibration approach can reliably align the various RGBD streams. It is important to note that no additional equipment, such as a checkerboard or an IMU, nor any manual rigging or prescanning processes, are required during the online calibration. The user only needs to initialize the system with an approximate A-pose.

Nonrigid Tracking. We use our pipeline to determine the nonrigid alignment parameters based on the canonical frame with respect to the camera views of the present RGBD inputs. By using the distinctive and solid human shape to divide the nonrigid tracking problem into two subproblems, we offer a novel skeleton warping-based nonrigid tracking scheme. This method uses an iterative flip-flop strategy to optimize the fit skeletons and the hybrid motion fields. Our skeleton warping approach effectively and reliably models nonrigid motion and addresses the misalignment problem in unstructured sparse multiview settings.

Geometry Fusion. We combine the depth observations in a global canonical TSDF volume after estimating the nonrigid movements and aligning the unstructured RGBD input streams. This volume is kept constant to obtain temporally coherent reconstruction results. When updating the canonical volume, similar to previous works [19, 20], we discard the data of the voxels that are warped in invalid areas in the current inputs and explicitly detect overlapping voxels to prevent erroneously fused geometry. The body shape and pose are also optimized in the fused SDF using the efficient volumetric shape-pose optimization algorithm developed in previous work [128] to obtain better canonical body fitting and skeleton embedding results. Finally, marching cubes are used to extract triangle meshes.

Atlas Texturing. In the sparse multiview setting, we propose a unique atlas texturing method to obtain more vivid performance capture results. We build a highly effective projective atlas map with virtual camera views related to the canonical volume to ensure real-time performance. The reconstructed mesh based on the canonical volume is textured using the projective atlas. Our approach is the first single-view-input compatible real-time atlas texturing solution for dynamic reconstruction in sparse multiview environments.

Technique Details of UnstructuredFusion

In contrast to traditional structured multicamera systems, which require rigid allocation and meticulous offline calibration of cameras, our UnstructuredFusion system enables unstructured and even manual allocation of cameras. In other words, we provide a unique online calibration approach to eliminate the cumbersome overhead of camera array systems and propose a nonrigid tracking algorithm based on skeleton warping. A temporal blending-based online atlas texturing approach is suggested to obtain high-quality appearance results that vividly convey human motions. In the following subsections, we first describe how our method represents motion. Then, we discuss the online calibration, nonrigid tracking, and online atlas texturing methods.

Human Motion Representation. Our approach aims to capture human motion; hence, we apply the effective and reliable [128] double-layer surface representation, which fuses the SMPL model with the embedded deformation (ED) model, to represent the motion. We provide a brief description of these two motion parameterization techniques and define the mathematical notations used in our method in this subsection.

The embedded deformation model is represented by a nonrigid motion field $G = \{\mathbf{dq}_j, \mathbf{x}_j\}$ consisting of the dual quaternions $\{\mathbf{dq}_j\}$ and the corresponding sparse ED nodes $\{\mathbf{x}_j\}$. Let $SE3(\mathbf{dq}_j)$ denote the rigid transformation associated with the j-th dual quaternion. Neighboring ED nodes are connected to form an ED graph. For any 3D vertex $\mathbf{v}_c$ in the canonical volume, the ED warping operation is formulated as follows:

$$\tilde{\mathbf{v}}_c = ED(\mathbf{v}_c; G) = SE3(\sum_{i \in \mathcal{N}(v_c)} w(\mathbf{x}_i, \mathbf{v}_c)\mathbf{dq}_i)\mathbf{v}_c, \tag{4.61}$$

where $\mathcal{N}(v_c)$ represents a set of node neighbors of $\mathbf{v}_c$, and $w(\mathbf{x}_i, \mathbf{v}_c) = \exp(-\|\mathbf{v}_c - \mathbf{x}_i\|_2^2/(2r_k^2))$ is the influence weight of the i-th node $\mathbf{x}_i$ to $\mathbf{v}_c$. The influence radius r_k is set as 0.075 m for all the ED nodes. Similarly, $\tilde{\mathbf{n}}_{v_c} = ED(\mathbf{n}_{v_c}; G)$ denotes the warped normal of $\mathbf{v}_c$ using the ED motion field G.

The SMPL model [72] has $N = 6890$ vertices and a skeleton with $K = 24$ joints. Before posing, the body model $\bar{\mathbf{T}}$ deforms into a morphed model $T(\boldsymbol{\beta}, \boldsymbol{\theta})$ with shape parameters $\boldsymbol{\beta}$ and pose parameters $\boldsymbol{\theta}$ to accommodate different identities and pose-dependent deformations. Mathematically, the body shape $T(\boldsymbol{\beta}, \boldsymbol{\theta})$ is morphed according to

$$T(\boldsymbol{\beta}, \boldsymbol{\theta}) = \bar{\mathbf{T}} + B_s(\boldsymbol{\beta}) + B_p(\boldsymbol{\theta}), \tag{4.62}$$

where $B_s(\boldsymbol{\beta})$ and $B_p(\boldsymbol{\theta})$ represent the shape and pose, respectively. Let $T(\bar{\mathbf{v}}; \boldsymbol{\beta}, \boldsymbol{\theta})$ denote the morphed 3D position of any vertex $\bar{\mathbf{v}} \in \bar{\mathbf{T}}$. The pose function of the SMPL model can be formulated as $W(T(\boldsymbol{\beta}, \boldsymbol{\theta}), J(\boldsymbol{\beta}), \boldsymbol{\theta}, \mathcal{W})$, which is a general blend skinning function in terms of the morphed body $T(\boldsymbol{\beta}, \boldsymbol{\theta})$, pose parameters $\boldsymbol{\theta}$, joint locations $J(\boldsymbol{\beta})$ and skinning weights $\mathcal{W}$. Then, for any 3D vertex $\mathbf{v}_c$, the linear blend skinning (LBS) operation with the SMPL skeleton motions is formulated as follows:

$$\begin{aligned} \hat{\mathbf{v}}_c =& \mathbf{G}(\mathbf{v}_c, \boldsymbol{\theta})\mathbf{v}_c, \mathbf{G}(\mathbf{v}_c, \boldsymbol{\theta}) = \sum_{i \in \mathcal{B}} w_{i,v_c} \mathbf{G}_i, \\ \mathbf{G}_i =& \prod_{k \in \mathcal{K}_i} \exp(\theta_k \hat{\xi}_k), \end{aligned} \tag{4.63}$$

where $\mathbf{G}(\mathbf{v}_c, \boldsymbol{\theta})$ is the posed rigid transformation of $\mathbf{v}_c$, $\mathcal{B}$ is the bone index set, $\mathbf{G}_i$ is the cascaded rigid transformation of the i-th bone, $\mathcal{K}_i$ are the parent indices of the i-th bone in the backward kinematic chain, $\exp(\theta_k \hat{\xi}_k)$ is the exponential map of the twist associated with the k-th bone, and w_{i,v_c} is the skinning weight associated with the i-th bone and point $\mathbf{v}_c$. For the w_{i,v_c} setting, if $\mathbf{v}_c$ is on the SMPL model, w_{i,v_c} is pre-defined in $\mathcal{W}$. If $\mathbf{v}_c$ is on the fused surface, w_{i,v_c} is determined based on the weighted average of the skinning weights of its KNN nodes. If $\mathbf{v}_c$ is on the depth input, we find its KNN nodes by transforming the ED nodes to the camera view.

Online Multicamera Calibration

Recall that due to the unstructured multicamera setting of our system, the RGBD cameras are uncalibrated and unsynchronized. Our online calibration scheme is illustrated in Fig. 4.38. For the first frame, the user needs to start with a rough A-pose to initialize the system. Then, we jointly optimize the initial camera poses $\boldsymbol{T} = \{\boldsymbol{T}_i\}, i = 1, 2, 3$, initial skeleton pose $\boldsymbol{\theta}_0$ and shape parameters $\boldsymbol{\beta}_0$ of the SMPL model as follows:

$$\begin{aligned} E_{\text{init}}(\boldsymbol{T}, \boldsymbol{\beta}_0, \boldsymbol{\theta}_0) =& \lambda_{\text{vdata}} E_{\text{vdata}} + \lambda_{\text{sdata}} E_{\text{sdata}} + \\ & \lambda_{\text{pdata}} E_{\text{pdata}} + \lambda_{\text{prior}} E_{\text{prior}}. \end{aligned} \tag{4.64}$$

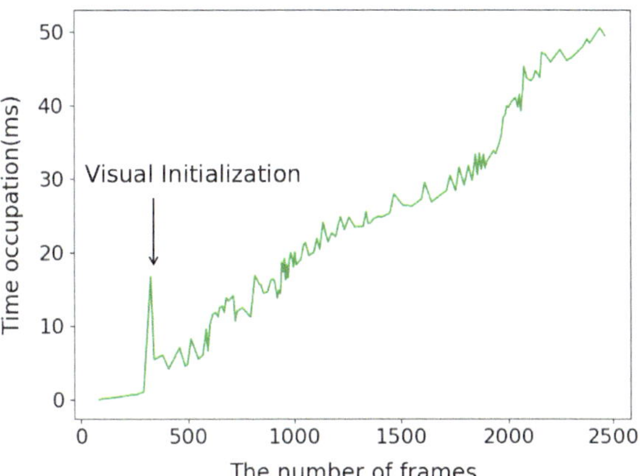

Fig. 4.38 Illustration of the online calibration scheme. **a**, **b**, **c** are the depth inputs with the initial camera poses, the camera poses after solving Eq. 4.67, and the camera poses after solving Eq. 4.64. **d** The final output of the online calibration scheme. From left to right: the aligned color frames, the aligned depth frames and the embedded SMPL model

Here, the volumetric data term E_{vdata} measures the misalignment error between the SMPL model and the reconstructed mesh in the reference volume [128]:

$$E_{\text{vdata}}(\boldsymbol{\beta}_0, \boldsymbol{\theta}_0) = \sum_{\bar{\mathbf{v}} \in \bar{\mathbf{T}}} \psi(\mathbf{D}(W(T(\bar{\mathbf{v}}; \boldsymbol{\beta}_0, \boldsymbol{\theta}_0); J(\boldsymbol{\beta}_0), \boldsymbol{\theta}_0)), \tag{4.65}$$

where $\mathbf{D}(\cdot)$ takes a point in the canonical volume as input and outputs the bilinear interpolated TSDF, and $\psi(\cdot)$ represents the robust Geman-McClure penalty function.

The dense projective data term E_{pdata} forces the warped vertices on the SMPL model to move to the depth of the input depth data based on a point-to-plane distance metric, which is formulated as

$$E_{\text{pdata}}(\boldsymbol{T}) = \sum_{i=1}^{3} \sum_{(\bar{\mathbf{v}}, \mathbf{u}_i) \in C_i} \psi(\mathbf{n}_{\bar{\mathbf{v}}}^T(\boldsymbol{T}_i W(T(\bar{\mathbf{v}}; \boldsymbol{\beta}_0, \boldsymbol{\theta}_0)) - \mathbf{u}_i)), \tag{4.66}$$

where $(\bar{\mathbf{v}}, \mathbf{u}_i)$ is a pair in the i-th camera view found via a projective look-up method; $\mathbf{u}_i$ is a sampled point in the depth map; and $\bar{\mathbf{v}}$ is a vertex in the SMPL model. In addition to the dense alignment scheme, we detect the global human skeleton using the Kinect SDK. Let $\mathbf{J}_{p,i}$ denote the p-th 3D joint position of the detected skeleton in the i-th camera view. This term includes additional global constraints for fitting the SMPL model to the present depth maps, which can be formulated with the following sparse feature term:

$$E_{\text{sdata}}(\boldsymbol{T}) = \sum_{1 \le i < j \le 3} \sum_{p=1}^{N_p} \tau(p; i, j) \|\boldsymbol{T}_i^{-1}\mathbf{J}_{p,i} - \boldsymbol{T}_j^{-1}\mathbf{J}_{p,j}\|_2^2, \tag{4.67}$$

where N_p is the number of 3D skeleton joints and $\tau(p; i, j)$ is the indicator function, which equals 1 only if the p-th joint is observable in both the i-th and j-th camera views.

Similar to [10, 128], we utilize a pose prior to penalize unnatural poses, which is defined as

$$E_{\text{prior}}(\boldsymbol{\theta}_0) = -log(\sum_j w_j N(\boldsymbol{\theta}_0; \mu_j, \delta_j)). \tag{4.68}$$

This term is formulated as a Gaussian mixture model (GMM), where w_j, μ_j and δ_j denote the mixture weight, mean and variance of the j-th Gaussian model, respectively.

Skeleton Warping-Based Nonrigid Tracking

It is important to note that the RGBD input streams in our unstructured configuration suffer from errors due to the lack of synchronization, in addition to extrinsic camera localization errors and depth map distortion errors, because of the asynchronous nature of the consumer-oriented hardware. These problems lead to the misalignment of the raw RGBD inputs, which complicates the nonrigid tracking problem and produces strange reconstruction outcomes. To address these misalignment challenges while creating a nonrigid tracking approach, we propose aligning all asynchronous raw RGBD streams to a global reference using skeleton warping, with the unique human shape prior serving as a reliable reference. With this approach, the skeletal motions and nonrigid ED deformations are both investigated by jointly optimizing the current SMPL skeleton pose $\boldsymbol{\theta}$ and the ED motion field G in a frame-by-frame manner, given that the ED node graph is bound tightly to the SMPL model.

Mathematically, we introduce the "*fit skeleton*" $\hat{\boldsymbol{\theta}}_i$, denoting the SMPL skeleton pose fit to current live RGBD image input $\mathcal{D}_i$ in the i-th camera view, which corresponds to subframe-level asynchronous capture time in the unstructured setting. Let $\boldsymbol{\theta}$ denote the optimized skeleton for the current frame without the influence of unstructured errors. Then, we can warp each pixel $\mathbf{u}_i \in \mathcal{D}_i$ from the fit skeleton pose $\hat{\boldsymbol{\theta}}_i$ to the global pose $\boldsymbol{\theta}$ using the "skeleton warping" operation, which is formulated as

$$\mathbf{u}_i' = \mathbf{G}(\mathbf{u}_i, \boldsymbol{\theta})\mathbf{G}(\mathbf{u}_i, \hat{\boldsymbol{\theta}}_i)^{-1}\mathbf{u}_i, \tag{4.69}$$

where $\mathbf{G}(\cdot)$ is the linear blend skinning (LBS) rigid transformation of the SMPL skeleton. Note that the skinning weight of $\mathbf{u}_i$ is determined based on the weighted average of the skinning weights of its KNN nodes. To align all the raw RGBD streams using the skeleton warping scheme, we combine a dense data term and the pose prior term [10] to optimize the fit skeleton $\hat{\boldsymbol{\theta}}_i$, which is formulated as

$$E_{\text{skwarp}}(\hat{\boldsymbol{\theta}}_i) = \lambda_{\text{fit}} E_{\text{fit}}(\hat{\boldsymbol{\theta}}_i) + \lambda_{\text{prior}} E_{\text{prior}}(\hat{\boldsymbol{\theta}}_i). \tag{4.70}$$

Here, the definition of the pose prior term E_{prior} resembles Eq. 4.68 in terms of $\hat{\boldsymbol{\theta}}_i$ instead of $\boldsymbol{\theta}_0$. The dense data term measures the skeleton fit between the reconstructed double-layer surface and the depth map:

$$E_{\text{fit}}(\hat{\boldsymbol{\theta}}_i) = \sum_{(v_c,u_i)\in\mathcal{P}_i} \tau_1(\mathbf{v}_c)\psi(\hat{\mathbf{n}}_{v_c}^{\mathbf{T}}(\mathbf{G}(\mathbf{v}_c,\hat{\boldsymbol{\theta}}_i)\mathbf{v}_c - \mathbf{u}_i)) + \tau_2(\mathbf{v}_c)\psi(\tilde{\mathbf{n}}_{v_c}^{\mathbf{T}}(\mathbf{G}(\mathbf{v}_c,\hat{\boldsymbol{\theta}}_i)\mathbf{G}(\mathbf{v}_c,\boldsymbol{\theta})^{-1}\tilde{\mathbf{v}}_c - \mathbf{u}_i)), \tag{4.71}$$

where $\mathcal{P}_i$ is the correspondence set of the i-th camera view, u_i is a sampled point in the depth map, and its closest point v_c can be on either the body shape or the fused surface. $\tau_1(\cdot)$ and $\tau_2(\cdot)$ are correspondence indicator functions, where $\tau_1(\mathbf{v}_c)$ equals 1 only if $\mathbf{v}_c$ is on the body shape, and $\tau_2(\mathbf{v}_c)$ equals 1 only if $\mathbf{v}_c$ is on the fused surface. We follow the same correspondence search scheme on the double-layer surface as in [128]; please refer to [128] for more details.

After obtaining all the fit skeletons $\hat{\boldsymbol{\theta}}_i$ of the current frame, we jointly optimize the global skeleton pose $\boldsymbol{\theta}$ and current ED nonrigid motion field G as follows:

$$E_{\text{mot}}(G,\boldsymbol{\theta}) = \lambda_{\text{data}}E_{\text{data}} + \lambda_{\text{bind}}E_{\text{bind}} + \lambda_{\text{reg}}E_{\text{reg}} + \lambda_{\text{prior}}E_{\text{prior}} + \lambda_{\text{skele}}E_{\text{skele}}. \tag{4.72}$$

Similarly, the pose prior term E_{prior} resembles Eq. 4.68. Following [128], the binding term E_{bind} constrains both motions to be consistent, while the geometry regularity term E_{reg} produces locally as-rigid-as-possible (ARAP) motions to prevent overfitting based on the depth inputs. These two terms are described in more detail in [36, 128].

The dense projective data term E_{data} is formulated as the sum of the point-to-plane distances in our multiview setting:

$$E_{\text{data}}(G,\boldsymbol{\theta}) = \sum_{i=1}^{3} \sum_{(v_c,u_i)\in\mathcal{P}_i} (\tilde{\mathbf{n}}_{v_c}^{\mathbf{T}}(\tilde{\mathbf{v}}_c - \mathbf{u}_i'))^2, \tag{4.73}$$

where $\mathbf{u}_i$ is a sampled point in the depth map, and $\mathbf{v}_c$ denotes its closest point on the fused surface. $\mathcal{P}_i$ is the set of correspondences in the i-th camera view found via a projective local search [79, 117]. $\mathbf{u}_i'$ denotes the aligned depth pixel after skeleton warping using Eq. 4.69. To align the fit skeletons $\hat{\boldsymbol{\theta}}_i$ and current global skeleton $\boldsymbol{\theta}$, we introduce the following skeleton term:

$$E_{\text{skele}}(\boldsymbol{\theta}) = \sum_{i=1}^{3} \sum_{u_i\in\mathcal{P}_i} \| W_{\mathbf{u}_i}(\boldsymbol{\theta} - \hat{\boldsymbol{\theta}}_i) \|_2^2, \tag{4.74}$$

where $W_{\mathbf{u}_i}$ is the LBS weight vector of the depth point $\mathbf{u}_i$. Note that we first warp the ED nodes to the i-th camera view to identify the KNN nodes of $\mathbf{u}_i$; then, $W_{\mathbf{u}_i}$ is determined based on the weighted average of the skinning weights of the KNN nodes.

We solve the optimization problem in Eq. 4.72 under the ICP framework. The Levenberg-Marquardt (LM) method is used to resolve the nonlinear least squares issue. The bone and node transformations in each iteration use twist representations, and the transformations are approximated using first-order Taylor expansions around the most recent values. The resulting linear system is solved using an extremely effective GPU-based preconditioned conjugate gradient (PCG) solver [36].

Temporal Blending-Based Atlas Texturing

High-quality texturing plays a critical role in reconstructing vivid appearances for human motion capture. Prior works [19, 36] have adopted per-vertex colors in the final output; however, simply associating the color sampling with the geometry resolution leads to unfavorable trade-offs and blurred results. Recent works [12, 20, 21] have proposed atlas texturing to alleviate the geometry resolution constraints, leading to sharper results. However, these methods perform atlas texturing independently in a frame-by-frame manner, implying that the texture atlas changes for each live frame. Thus, sufficient camera views must be used to ensure that the color information covers all vertices for all possible motions of the model in the live frame (i.e., 8 and 106 camera views have been used in previous works [12, 20]). In view of this issue, we propose a novel temporal blending-based atlas texturing scheme that generates a sharp and realistic textured model in the sparse multiview setup and is even applicable for single-view inputs. More specifically, considering the real-time computational costs, we choose to construct a highly efficient projective atlas map. To utilize the intraframe information and establish a global atlas, the projective poses are all bound to the global canonical model. In particular, to capture more details in the head region, we build two kinds of virtual projective poses in the first frame, $\boldsymbol{T}_{H,i}$ and $\boldsymbol{T}_{B,i}$, for texturing the head and body regions, respectively, where $i \in 1, 2, 3$ denotes the index of the virtual projective views. Three views are sufficient to cover the static canonical model in the A-pose setting.

As shown in Fig. 4.39, for every input color frame, we first project all the visible canonical vertices with $\boldsymbol{T}_{H,i}$ and $\boldsymbol{T}_{B,i}$. Then, we write the color at the projected position based on the corresponding texture coordinates to build the partial projective textures $\boldsymbol{a}_i$ of the i-th virtual view for the head and body regions. This partial texture

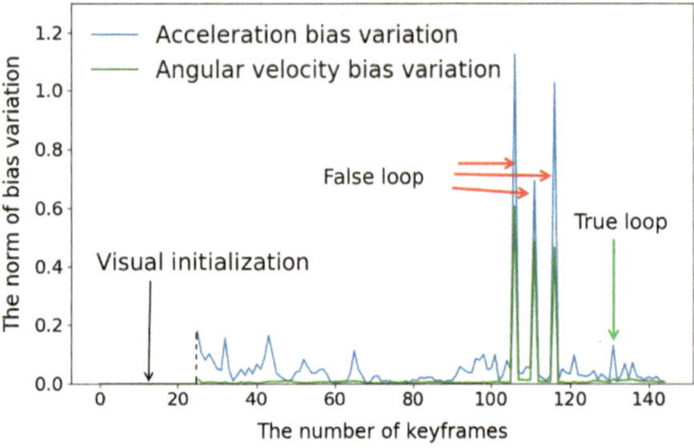

Fig. 4.39 Illustration of our projective atlas scheme. **a** Example RGBD image inputs. **b** The corresponding partial atlas $\{\boldsymbol{a}_i\}$. **c** The global blended atlas $\{\boldsymbol{A}_i\}$. **d** The textured mesh output in the live frame

generation can be easily achieved using the OpenGL rasterization pipeline. We then temporally blend all the partial texture images $\{\boldsymbol{a}_i\}$ into a complete and global atlas $\{\boldsymbol{A}_i\}$ in a frame-by-frame manner, as illustrated in Fig. 4.39b, c. To texture the non-covered and occluded areas in the canonical model, we also fuse all the color frames into a global color volume **C**, similar to the TSDF volume. This hybrid texturing scheme can easily be achieved with the OpenGL pipeline. The faces in the canonical model with valid projected UV coordinates are textured using $\{\boldsymbol{A}_i\}$, while the per-vertex color values for faces without valid UV coordinates are extracted directly from the color volume **C**.

Dynamic Projected Texture Blending. To obtain a sharp and complete atlas from the partial projective atlas, we perform texture blending in a frame-by-frame and per-pixel manner. Similar to TSDF fusion in the volume space, we perform temporal atlas blending as follows. For each pixel p, $\boldsymbol{a}_i(p)$ and $\boldsymbol{A}_i(p)$ denote the color values in the i-th virtual camera view in the partial and blended atlases, respectively; $\mathbf{W}_i(p)$ denotes the accumulated blending weight; and $w_i(p) = cos(\theta)$ denotes the view-dependent weight of the current frame, where θ is the angle between the projected normal in the camera view and the view direction of the camera. Finally, the projected atlas is dynamically blended with the weight truncation scheme as follows:

$$
\begin{aligned}
\boldsymbol{A}_i(p) &\leftarrow \frac{\boldsymbol{A}_i(p)\mathbf{W}_i(p) + \boldsymbol{a}_i(p)w_i(p)}{\mathbf{W}_i(p) + w_i(p)}, \\
\mathbf{W}_i(p) &\leftarrow min(\mathbf{W}_i(p) + w_i(p), w_{max}).
\end{aligned}
\tag{4.75}
$$

The maximum blending weight w_{max} enables the moving average texture blending scheme to support dynamic texturing. This weight is set to 4 for the head and 8 for the body region to obtain more dynamic texture details in the face region.

We next examined the blurry effect in the atlas texturing process. The motion blur caused by rapid motions can be removed by discarding bad color frames by selecting views using the blurriness measure developed by Crete *et al.* [13]. For the atlas blur caused by the nonrigid misalignment between the live mesh and the color images, a 2D as-similar-as-possible (ASAP) grid-based warping scheme, denoted as grid-based warping, is adopted between the partial atlas $\boldsymbol{a}_i(p)$ and the temporally blended atlas $\boldsymbol{A}_i(p)$ during atlas blending using Eq. 4.75. Let $\{p_A, p_a\}$ denote all the feature pairs between $\boldsymbol{A}_i$ and $\boldsymbol{a}_i$ based on *ORB* descriptors and the *GMS* matching method [7]. Regular grid cells are sampled in $\boldsymbol{A}_i$, and each cell is split into two triangles. Similar to [48, 68, 70], the 2D warping from $\boldsymbol{A}_i$ to $\boldsymbol{a}_i$ is modeled based on the positions of the deformed grid vertices, denoted as $\hat{V}$. Mathematically, we optimize the 2D deformation using the following energy function:

$$
E(\hat{V}) = E_{\mathrm{d}}(\hat{V}) + \alpha E_{\mathrm{s}}(\hat{V}). \tag{4.76}
$$

The data term $E_{\mathrm{d}}(\hat{V})$ that sums the distances of all the feature pairs in the atlas domain after warping $\hat{V}$ is formulated as

$$E_{\mathrm{d}}(\hat{V}) = \sum_{p_A} \| w_{p_A} \hat{V}_{p_A} - p_a \|_2^2, \tag{4.77}$$

where $\hat{V}_{p_A}$ are the warped grid vertices enclosing p_A, and w_{p_A} is the corresponding bilinear interpolation weight.

The ASAP term $E_{\mathrm{s}}(\hat{V})$ is formulated as

$$E_{\mathrm{s}}(\hat{V}) = \sum_{\hat{v}} \tau(\hat{v}) \| \hat{v} - \hat{v}_1 - s R_{90}(\hat{v}_0 - \hat{v}_1) \|_2^2, \; R_{90} = \begin{bmatrix} 0 & 1 \\ 1 & 0 \end{bmatrix}, \tag{4.78}$$

where v, v_0, and v_1 are the neighboring triangle vertices in the clockwise direction, and $s = \|v - v_1\| / \|v_0 - v_1\|$ is a known scalar in the initial grid. $\tau(\hat{v})$ equals 1 only if $\hat{v}$ is in the valid area of the blended atlas $\boldsymbol{A}_i$. For more details on ASAP warping, please refer to [48, 68, 70]. Before blending $\boldsymbol{A}_i$ and $\boldsymbol{a}_i$ using Eq. 4.75, we warp $\boldsymbol{A}_i$ using bilinear interpolation with the grid deformation $\hat{V}$, which can be efficiently performed via the OpenGL rasterization pipeline. The proposed texturing scheme produces sharper and more realistic textured results than the per-vertex scheme, as shown in Fig. 4.40.

Experimental Results

In this section, we first describe how our UnstructuredFusion system was implemented. We next assess our primary technological contributions and qualitatively and quantitatively compare our method with existing state-of-the-art approaches. The final subsection describes the limitations of the UnstructuredFusion system.

Implementation Details. UnstructuredFusion was implemented on a single NVIDIA GeForce GTX TITAN X GPU and a 3.2 GHz 4-core Xeon E3-1230 CPU with 16 GB of memory. The input live RGBD streams were captured by three Kinect One sensors at 30 fps with 512×424 resolution. The entire pipeline runs at 33 ms per frame, with the skeleton warping-based motion tracking requiring approximately 16 ms with 5 ICP iterations, TSDF fusion requiring approximately 6 ms, atlas texturing requiring approximately 8 ms, and the remaining computations requiring 3 ms. For online calibration, the parameters λ_{vdata}, λ_{sdata}, λ_{pdata} and λ_{prior} were set as 1.0, 2.0, 1.0, and 0.01, respectively. For motion tracking, we set $\lambda_{\mathrm{fit}} = 1.0$, $\lambda_{\mathrm{data}} = 1.0$, $\lambda_{\mathrm{bind}} = 1.0$, $\lambda_{\mathrm{reg}} = 5.0$ and $\lambda_{\mathrm{skele}} = 10.0$. Note that these parameters were set empirically to balance the cost of each term. For the ED model, we used the 4 nearest node neighbors for ED warping and the 8 nearest node neighbors to construct the ED graph following previous work [19, 36]. The TSDF voxel size was set to 4 mm in each dimension to preserve sufficient geometric details of the target.

Evaluation. Several representative sequences reconstructed by UnstructuredFusion are illustrated in Fig. 4.41. The results show that challenging motions and high-quality textures were both reconstructed by our method. In addition, we evaluate the contributions of our technique, including the skeleton warping-based nonrigid

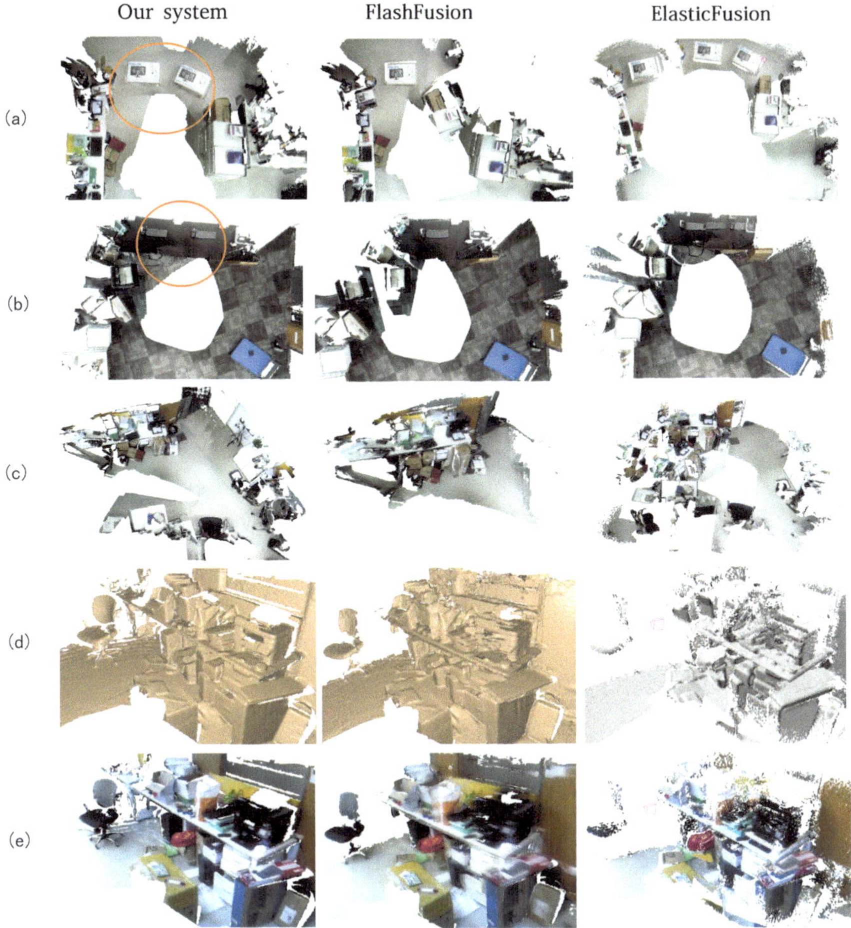

Fig. 4.40 Our atlas texturing scheme (right) compared to the per-vertex color scheme (left). The per-vertex color scheme leads to block artifacts, while our method obtains sharper textures with more dynamic facial expressions

tracking algorithm, temporal blending-based atlas texturing approach, and online multicamera calibration method, in the following sections.

We use three sequences as examples to qualitatively demonstrate the usefulness of the proposed skeleton warping strategy during the nonrigid tracking process, as shown in Fig. 4.42. Due to the mismatch among the unstructured RGBD streams at various camera viewpoints, the fused model suffers from severe cumulative errors without skeleton warping, particularly for the regions highlighted with red circles. In contrast, our method aligns the unstructured sequences using the proposed skeleton warping scheme, producing reconstructions of 4D geometries and textures that are aesthetically pleasing.

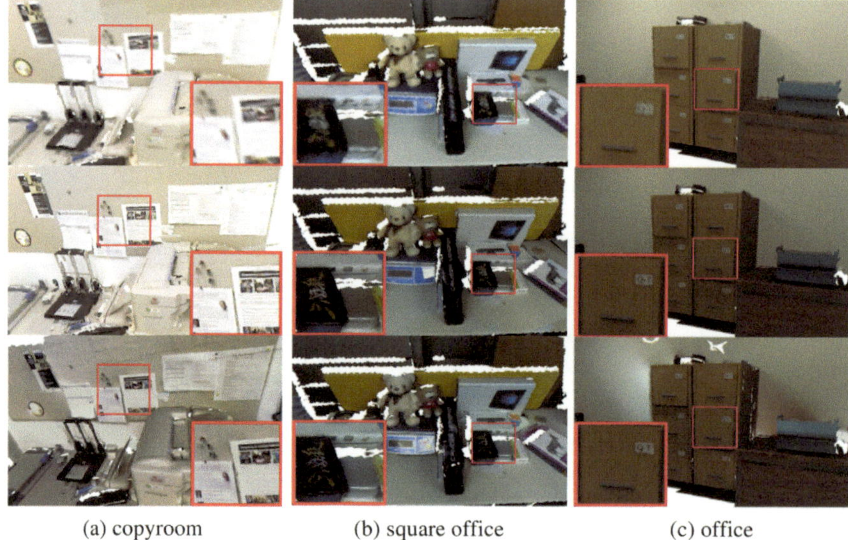

Fig. 4.41 Several examples that demonstrate the quality and fidelity of the 4D geometry and texture results reconstructed with the proposed UnstructuredFusion system

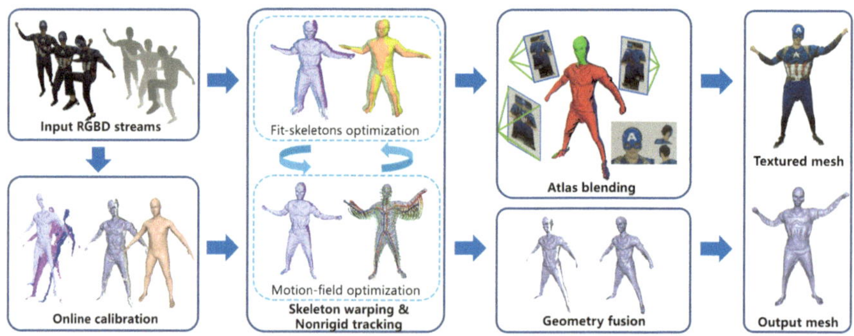

Fig. 4.42 Evaluation of the proposed skeleton warping scheme. **a** Geometric and textured results without skeleton warping. **b** The corresponding results with skeleton warping

Furthermore, we evaluated the effectiveness of our method using three sequences captured by manually moved cameras, as illustrated in Fig. 4.43. To record these three sequences, the cameras were moved slightly in roughly fixed places, forward in roughly straight lines, and around the target to keep the target performer in the capture views. With this manual adjustment approach, the multicamera movement makes it more challenging to consistently register unstructured camera movements. Because of accumulated misalignment errors, the reconstruction method without the skeleton warping algorithm fails and produces results with strange geometries, as indicated in the areas with red circles in Fig. 4.43. The misalignment errors caused

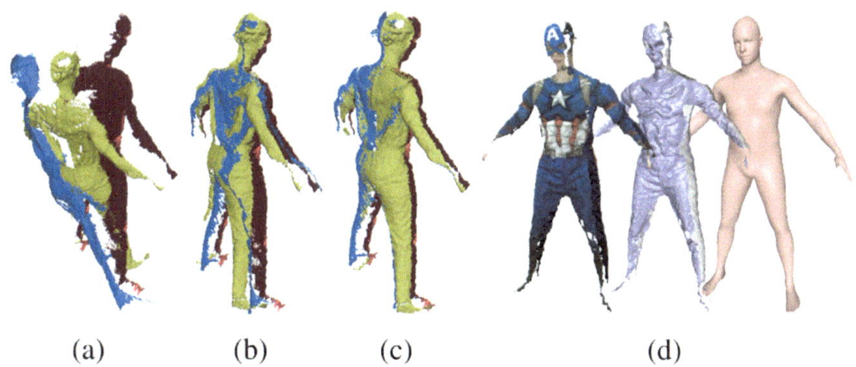

Fig. 4.43 Evaluation of our method using moving cameras. **a**, **b**, **c** Three example sequences, in which the three cameras move in roughly fixed positions, move forward in a roughly straight line and circle around the captured target, respectively. From left to right: the captured scene; the reconstruction results without skeleton warping and our results

by the highly unstructured inputs from the moving cameras with various types of motions, however, were addressed by our method with skeleton warping, resulting in high-quality reconstructed meshes.

Recall that in Fig. 4.40, the representative sequences qualitatively illustrate that the proposed atlas texturing method outperforms the traditional per-vertex scheme in terms of producing sharper and more realistic textured results. To further evaluate the effectiveness of the grid-based warping procedure in our atlas texturing method, we take one representative sequence as an example to demonstrate the results of our atlas texturing scheme without the proposed grid-based warping procedure, as shown in Fig. 4.44b, c. The result produced by the per-vertex texturing scheme [36] is also shown in Fig. 4.44d. The color-coded residual produced by contrasting the textured output with the input color image is represented in the blue map. As anticipated, the proposed atlas texturing method performs better than the per-vertex scheme, producing considerably less residue and sharper textured outputs. Moreover, the specific grid-based warping procedure significantly enhances the sharpness of the final textured result.

Furthermore, the corresponding quantitative error curves of our atlas texturing scheme and the per-vertex scheme are depicted in Fig. 4.44e. The residuals are determined as the per-pixel Euclidean distances of the RGB values between the textured results and the color image inputs, where each color channel is normalized to [0,1]. The results demonstrate the effectiveness of our atlas blending technique and the grid-based warping optimization method, with the proposed atlas texturing scheme achieving an average normalized error of approximately 0.33, while the method without grid-based warping and the per-vertex scheme obtain normalized errors of 0.42 and 0.54, respectively. Note that the blended atlas technique performs worse than the proposed atlas texturing technique; thus, the per-vertex approach is the best method to apply at the beginning of the sequence.

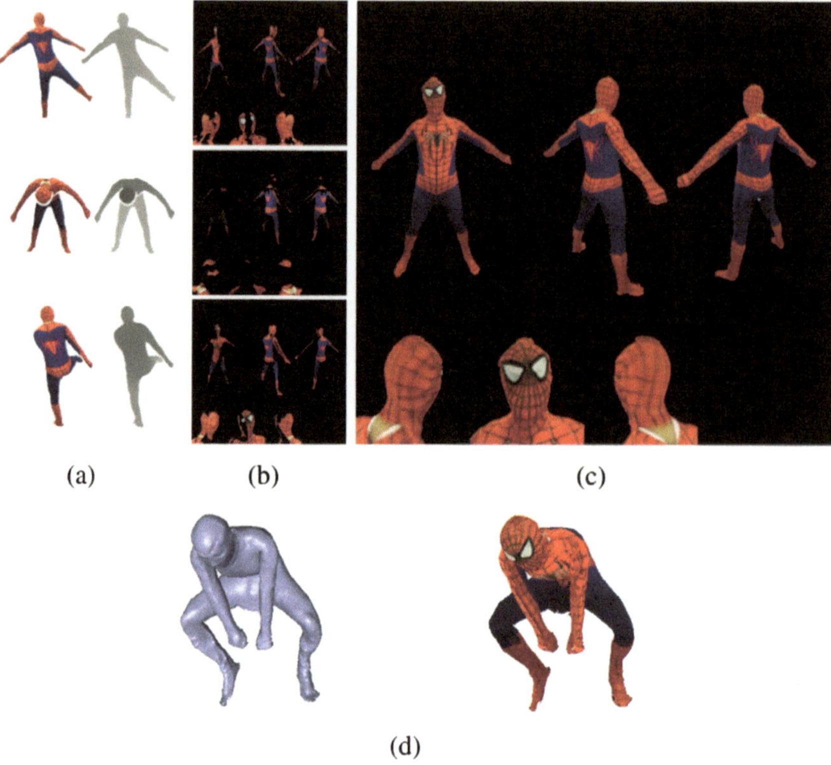

Fig. 4.44 Evaluation of the atlas blending scheme. **a** Input depth and color image. **b** The reconstructed result of our atlas texturing method with grid-based warping, where the blue map indicates the color-coded residual compared with the input color image. **c** The reconstructed result of our atlas texturing scheme without grid-based warping. **d** The reconstructed result using the per-vertex scheme. **e** The corresponding quantitative error curves

Moreover, to further illustrate the effectiveness of the proposed atlas texturing scheme, we artificially demonstrate the atlas texturing result using a sequence from DoubleFusion [128] with only single-view RGBD inputs in Fig. 4.45. The results show that our atlas texturing method generates sharp and realistic textured results for both single-view inputs and commodity sparse multiview inputs.

To evaluate the proposed online multicamera calibration scheme, the state-of-the-art global registration methods 4PCS [5] and Go-ICP [122] were adopted for comparison. Go-ICP [122] combines a local ICP method with a branch-and-bound search to find the global minimum, and 4PCS [5] performs global registration by constructing a congruent set of 4 points between the range images. Figure 4.46a presents the original depth inputs from the unstructured three camera views, and Fig. 4.46b–d shows the registration results of 4PCS [5], Go-ICP [122] and our online multicamera calibration scheme, respectively, for different render views for qualitative visualization. As highlighted by the circles and boxes, the 4PCS and Go-ICP methods fail to

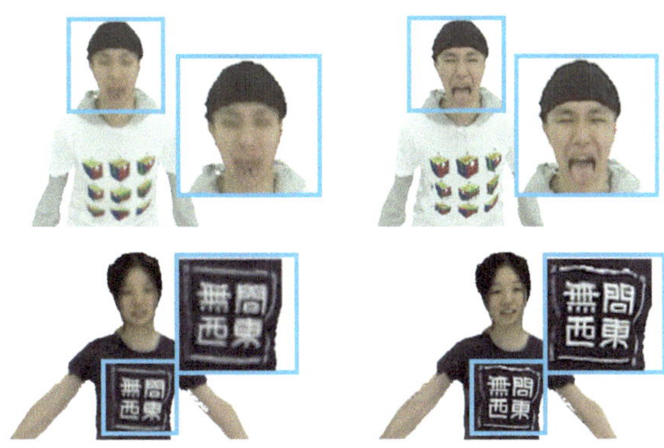

Fig. 4.45 Evaluation of the atlas blending scheme in terms of the number of input views. **a** The textured and relighting results with atlas blending using a single-view sequence from DoubleFusion [128]. **b** The results using the three-view sequence captured by UnstructuredFusion

align the three unstructured views and lead to overlap for the partial meshes from different camera views, especially for the head and limb regions, due to the small overlap between different camera views. In contrast, our proposed method obtains considerably better registration results by utilizing the solid human shape prior, as shown in Fig. 4.46d.

In contrast to existing high-end multiview capture systems [12, 20], our system utilizes only 3 commercial RGBD cameras to cover the entire target. Note that the reconstruction algorithm in our system can be extended to more cameras, but the current lightweight sparse multiview setup is more convenient for daily usage. In Fig. 4.47, we assess the performance of our system with different numbers of cameras. Our technique allows for high-fidelity reconstructions with fewer cameras. Unfortunately, in scenarios with fewer cameras, our system fails to track challenging motions in regions such as the elbows and knees because of the lack of adequate acceptable constraints provided by the input depth streams. With more camera resources, target motions can be tracked more precisely with fewer cumulative misalignment errors.

Comparison. In the previous subsections, we evaluated the contributions of each technique. In this subsection, we demonstrate the overall performance of the proposed UnstructuredFusion system by qualitatively and quantitatively comparing our system with other state-of-the-art methods.

While the recently developed DoubleFusion [128] also utilizes the SMPL model to regularize the embedded deformation, it is a single-view method. For a fair comparison of the real-time dynamic reconstruction performance in unstructured and sparse multiview settings, we extended DoubleFusion [128] to the sparse multiview setting by directly formulating the data for DoubleFusion [128] as multiview depth inputs; we denoted this model as Multi-DoubleFusion. In this basic extension, we adopt the same online calibration scheme that is used in our method to obtain the initial camera poses and the embedded SMPL model. Note that similar to other state-of-the-art multiview methods [19, 20], in Multi-DoubleFusion, we estimate the global rigid motion using the rigid-ICP algorithm for each frame and the nonrigid parameters based on the fixed global rigid motion parameters.

Fig. 4.46 Evaluation of the online calibration scheme. **a** The original depth inputs. **b**, **c**, **d** The registration results of 4PCS [5], Go-ICP [122] and our online multicamera calibration scheme, respectively. The red circles highlight the misaligned regions, while the corresponding boxes visualize the misalignment in different render views. Note that different colors of the boxes represent different body regions

Figure 4.48 shows a qualitative comparison of our UnstructuredFusion method with the comparison methods. Due to the limited capture view resources and inadequate geometry, the geometry results of DoubleFusion [128] suffer from rapid self-occluded motions and difficult loop closure operations. Furthermore, although Multi-DoubleFusion tends to be more resilient to occlusions, the Multi-DoubleFusion results still include accumulated misalignment errors between different views, which causes unnatural reconstruction results, particularly in the head and limb regions, as highlighted in the red circled regions. In contrast, our approach can handle unstructured inputs and produce outcomes that are loop-closed and have fine geometric

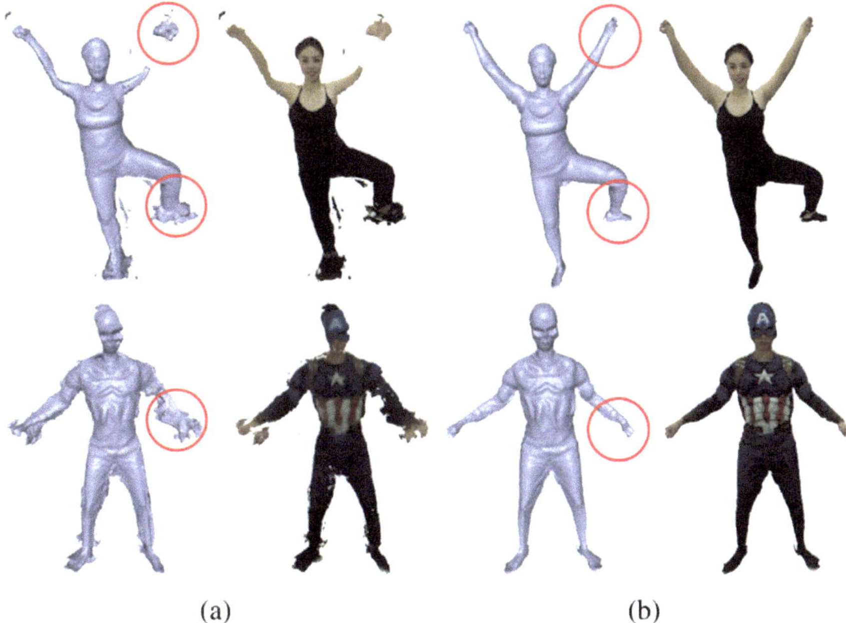

Fig. 4.47 Evaluation of the number of cameras. **a–c** show the reconstructed geometry and texture results using a single camera, 2 cameras, and 3 cameras, respectively

details. Moreover, our technique conducts comprehensive and dynamic atlas texturing, ensuring more realistic reconstruction results.

For a quantitative comparison, the reconstructed geometry is rendered as a 2D depth map in the camera view, and the mean absolute error (MAE) is calculated by using the depth input as the reference only in the visible surface regions. The MAE metric includes the reconstruction errors for both the rigid and nonrigid ICP processes of each approach. This enables an accurate quantitative comparison, even without ground truth reconstruction data. As shown in Fig. 4.49, our method achieves high-quality reconstruction results with fewer accumulated artifacts. The MAE of our method is approximately 17.47 mm, compared with MAEs of 37.52 mm for DoubleFusion [128] and 36.03 mm for Multi-DoubleFusion. Moreover, the MAEs of all the captured sequences in our experiments are listed in Table 4.13, where the errors were computed in the visible surface regions only. Our method has considerably less error, with an average MAE of 22.34 mm, compared with average MAEs of 44.48 mm for DoubleFusion [128] and 39.04 mm for Multi-DoubleFusion. These quantitative comparisons reveal the effectiveness of our method for nonrigid tracking during dynamic reconstruction in unstructured and sparse multiview settings.

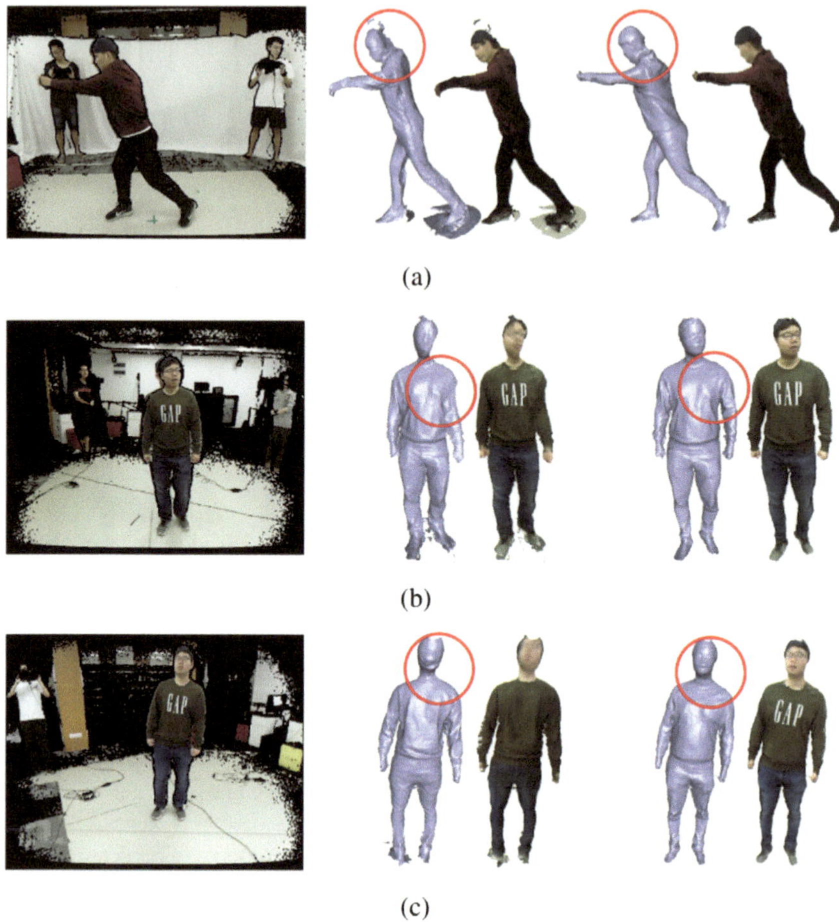

Fig. 4.48 Qualitative comparison. **a**–**c** show the reconstructed geometry/texture results of DoubleFusion [128], Multi-DoubleFusion [128] and our UnstructuredFusion, respectively

We specifically compare our results with the marker-based motion capture data obtained using the OptiTrack system to statistically assess our method. It should be noted that the infrared LED blinks to synchronize the capture sequences between our system and the OptiTrack system. Similar to [128, 137], the two systems are calibrated using manually chosen and preselected matched pairs. Following calibration, the detected marker positions in the OptiTrack coordinates system are converted to the camera coordinates of the first frame. We then use the reconstructed motion field to monitor the motions of these markers in the subsequent frames, and we compare the positions in each frame with the ground truth data determined by OptiTrack. Figure 4.50 presents the numerical curves of the per-frame maximum error of DoubleFusion [128], Multi-DoubleFusion, and our method based on one

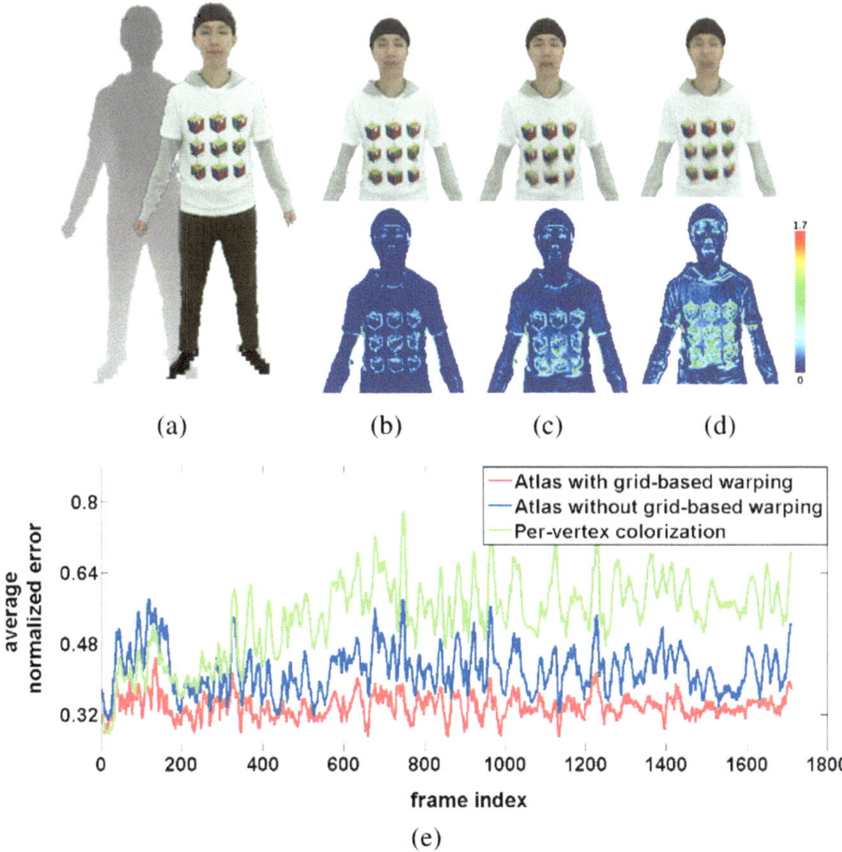

Fig. 4.49 Quantitative comparison. **a–c** show the reconstructed geometry/texture results of DoubleFusion [128], Multi-DoubleFusion [128] and our UnstructuredFusion, respectively. The color-coded maps in the bottom row indicate the projective error maps

sequence, in which the performer circles periodically as a challenging loop closure motion. DoubleFusion cannot track the nonrigid motions in the self-occluded regions in the side view; thus, the numerical error of DoubleFusion gradually increases, notably in the performer's side view. Until approximately frame 250, when the target motion is still slow and the accumulated error is still acceptable for reconstruction, Multi-DoubleFusion yields results that are equivalent to those of our method and is more resistant to self-occlusion issues than DoubleFusion. However, the Multi-DoubleFusion results still contain accumulated errors caused by the misalignment of the unstructured inputs. In contrast, our method can handle the misalignment of the unstructured inputs without the self-occlusion problem and achieves the smallest numerical error with the ground truth obtained by the OptiTrack system. The maximal and average errors of DoubleFusion [128], Multi-DoubleFusion, and our method

Table 4.13 Average numerical errors of all the captured sequences for the DoubleFusion [128], Multi-DoubleFusion [128] and UnstructuredFusion methods

	DoubleFusion (mm)	Multi-DoubleFusion (mm)	UnstructuredFusion (mm)
Sports	51.01	45.15	21.13
Yoga	54.68	48.07	27.71
Captain America	39.22	30.47	25.49
Dancing	43.18	40.54	16.38
Kicking	37.52	36.03	17.47
Walking	34.13	37.10	20.96
Spiderman	50.36	43.27	22.07
Waving	45.83	41.23	23.74
Crossing	44.38	47.52	26.09

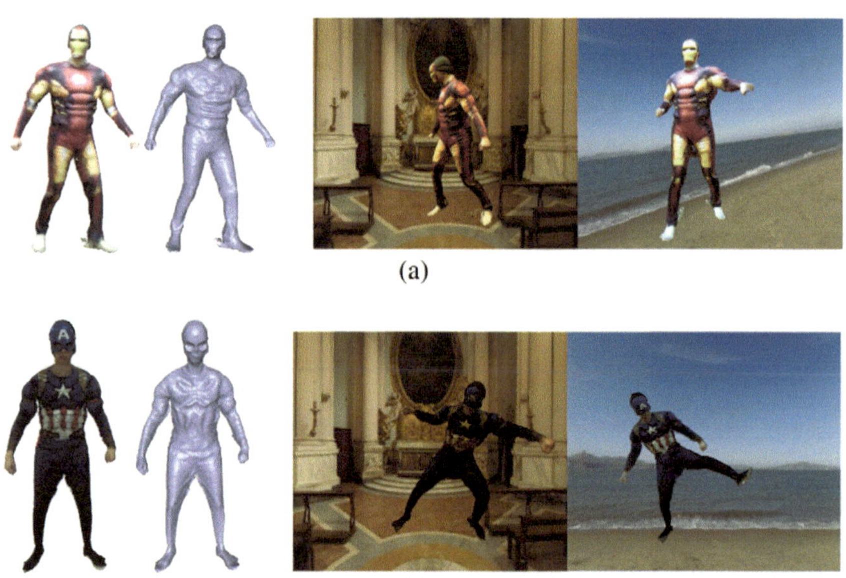

Fig. 4.50 Numerical error curves of our method, DoubleFusion [128] and Multi-DoubleFusion. Note that the ground truth is obtained via the Vicon system

for the sequences captured via the OptiTrack system are listed in Table 4.14 for a better comparison. Our numerical findings show that our method obtains more accurate tracking results than previous schemes in difficult unstructured and sparse multiview environments. We strongly advise watching the accompanying video for a more thorough analysis of our methodology and future outcomes.

Table 4.14 Average and maximal numerical errors for the entire sequence compared with the ground truth observations by the OptiTrack system for DoubleFusion [128], Multi-DoubleFusion and our method

	DoubleFusion (m)	Multi-DoubleFusion (m)	Our method (m)
Max error	0.2001	0.1195	0.0231
Average error	0.0976	0.0368	0.0107

Limitations and Discussion. As the first work to explore the problem of real-time dynamic reconstruction for both geometries and textures in unstructured sparse multiview settings, the proposed UnstructuredFusion system still has several limitations.

The highly fine details of the target, particularly in the facial region, cannot be reconstructed by our method based on the reconstructed geometry due to the low resolution of the depth input. To produce more synthetic geometric details in certain model-specific locations, data-driven approaches can be used. Moreover, our system cannot adapt to changes in surface topology; however, we hope to address this issue in future work by utilizing the key-volume update technique [19]. Moreover, human-object interactions must be modeled for many real-world applications, and our method is limited to human reconstruction. We intend to introduce the static object reconstruction approach [39] into our present framework. Moreover, the reconstructed meshes exhibit jittery behavior in areas where feet and fittings provide invalid depth data. Further postprocessing techniques, such as temporal filtering, could address the jitter in the 4D meshes. For atlas texturing, our technique relies on the fused color volume for occluded regions when projecting onto the camera views, which creates discrete colors around the boundaries of the occluded regions. As a postprocessing step, we intend to introduce the atlas processing method [84] in the gradient domain to obtain more seamless blending results. Using generative models and data-driven techniques to develop a more comprehensive and precise atlas is another promising approach. Moreover, even if the projective atlas is effective, it is not sufficiently compact for our texturing approach. For more effective applications, a texture atlas compression approach is required to streamline all of the textured meshes. Due to the limitations of the available commercial RGBD sensors, our system cannot be configured to operate in outdoor environments. To improve the quality of the acquired raw data, we intend to integrate the binocular solution with the available learning technology [120]. Our system's reliance on the A-pose initialization of the performer is another issue that should be addressed. To recognize the human shape and stance during initialization, we will use a data-driven technique. A human shape detector can also be applied in our nonrigid tracking pipeline to prevent cumulative tracking errors.

Conclusion

We proposed using unstructured commercial RGBD cameras to develop UnstructuredFusion, a real-time human performance capture system without markers. Our system alleviates the stringent requirements of previous systems (such as highly structured multicamera setups and laborious calibration and synchronization procedures) for generating high-quality 4D geometry and texture results representing human activities. We resolved the challenging issues of online multicamera calibration, nonrigid tracking, and atlas texturing based on numerous asynchronous movies with our flexible hardware architecture utilizing only three unstructured RGBD cameras. The proposed approach is supported by the strong global restrictions of the human body and motion, which are represented by the skeleton and warped skeleton, respectively. We carried out numerous tests to assess UnstructuredFusion's efficiency in high-quality 4D geometry and texture reconstruction without time-consuming calibration, even when flexibly and manually arranging the three cameras. Our UnstructuredFusion system alleviates the onerous hardware and software constraints of typical structured multicamera systems while addressing the occlusion problems that arise with single camera arrangements.

4.3.4 Multimodal Material Perception with a Structured Light Camera

In scientific fields such as computer vision, computer graphics, and robotics, it is advantageous to infer an object's underlying material qualities based on acquired images. For instance, despite looking comparable, a porcelain mug is significantly more delicate than a plastic mug, necessitating cautious manipulation by a robot. Unfortunately, image-based material recognition is ill posed since material, geometry, and lighting characteristics are entangled in the image perception process. The structured light camera offers helpful detangling elements needed for material recognition due to its depth acquisition capability and active illumination settings.

The structured light camera offers an effective probing function to detect subsurface scattering effects. This function depicts how light penetrating the material surface is scattered and how the light is reflected by the surface at a different location. This is the main reason to employ the structured light camera. Subsurface scattering, which occurs frequently with everyday translucent surfaces such as skin, plastic, marble, and wax, serves as a defining characteristic for material classification. However, due to spatial integration, it is challenging to determine the distinctive point spread function (PSF) when the surface is illuminated by a diffusive light source. For system diagnostics, a probing function is therefore needed. In related works such as Su et al. [98], Tanaka et al. [100], incident light was temporally modulated to implicitly reconstruct the temporal PSF. In [95], the incident light was spatially modulated for binary material classification, showing the feasibility of this approach. Similar to [95], a speckle dot pattern projected by a structured light camera was used in this work as a spatial probing function. We intend to manage geometry and distance variance, unlike their contrast-based method, which is restricted to a constant distance and flat surface. In addition, we can observe reflection features in the infrared (IR) spectrum due to the diffusive illumination settings, a byproduct of the diffractive optical element (DOE) in the structured light projector.

In addition to the active IR sensor, common off-the-shelf structured light cameras also include RGB sensors for acquiring visual information. RGB images illuminated by ambient light are more sensitive to the texture of the material, which severely impacts the dot pattern in IR images. The material discrimination ability can be improved by fusing both modalities because the scattering and reflection features learned based on the IR image are orthogonal to the texture features learned based on the RGB image. Synthesized and acquired datasets were both used for assessment to validate the suggested methodology. This work makes two significant contributions.

- We show how the surface properties, including surface reflection and subsurface scattering, are related to the different components of the actively illuminated IR image.
- A multimodal fusion method using all the images acquired by an off-the-shelf structured light camera is proposed based on the orthogonality of the corresponding learned features.

Theoretical analyses and experimental validations are both presented. The experimental results show that the multimodal fusion method is superior to state-of-the-art appearance-based methods.

Structured Light Camera and Material Properties

In this section, we relate the observed IR images captured by the structured light camera to different material optical properties. Specifically, subsurface scattering and direct reflection properties are revealed based on the mixed illumination pattern projected by the structured light camera. As shown in Fig. 4.51, by observing the subsurface scattering effect, the plastic bottle cap can be distinguished from the upper-right part of the notebook in the raw IR image, although both are red in the RGB image. By observing the reflection effect, the chair cushion can be distinguished from the chair back in the "smoothed" IR image, termed IR_{dif} in Subsection **IR Image Processing**, although both are black in the RGB image. However, when the textures are considered material features, RGB images are better for discrimination since these images are illuminated by low-frequency ambient light. The physical IR imaging process (involving material optical properties, geometry, and lighting) is analyzed in this section.

Material Optical Properties. Following Jensen et al. [51], the general bidirectional surface scattering distribution function (BSSRDF) models the relationship between the outgoing radiance and incoming radiant flux from different angles and locations,

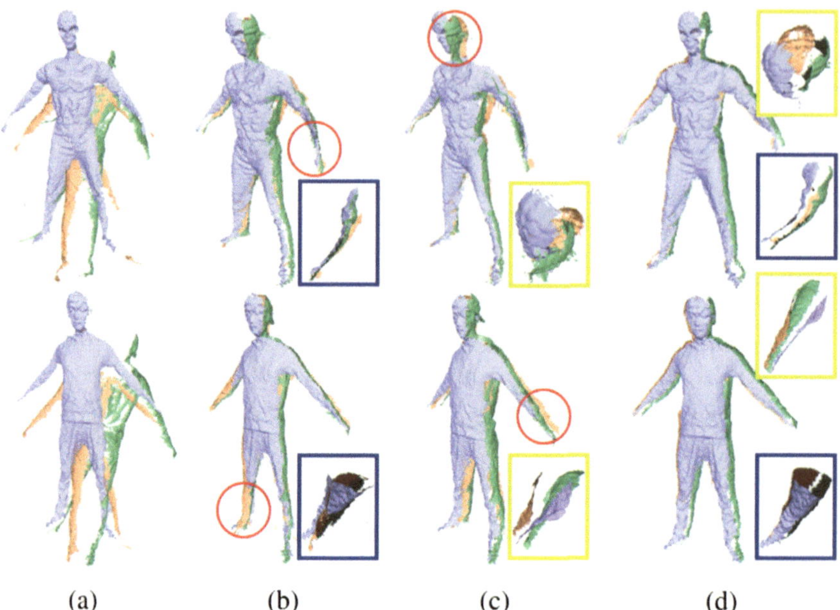

Fig. 4.51 **a** The physical imaging process. Infrared (IR) light is actively projected by the IR projector and received by the IR sensor after being reflected and scattered (red). Visible ambient light is received by the RGB sensor (yellow). **b** Different material properties are related to different images. The top shows the raw images directly obtained by the structured light camera. Note that the reflection property is revealed in the IR_{dif} image, which is obtained based on the IR image by using the method described in Subsection **IR Image Processing**

accounting for direct reflection and subsurface scattering. The outgoing radiance L_o at point $\mathbf{x}$ in direction $\mathbf{w_o}$ can be split into two terms: the direct reflection term L_r and the subsurface scattering term L_s [29].

$$L_o(\mathbf{x}, \mathbf{w_o}) = L_r(\mathbf{x}, \mathbf{w_o}) + L_s(\mathbf{x}, \mathbf{w_o}). \tag{4.79}$$

Jensen et al. [51] decomposed subsurface scattering into single and multiple (diffuse) terms. Since single-scattering terms for optically dense materials decrease much faster than multiple scattering terms as the distance between the incident point $\mathbf{x_i}$ and outgoing point $\mathbf{x_o}$ increases, these terms contribute little to the overall outgoing radiance. The widely adopted dipole method models the outgoing radiance of multiple scattering targets as a spatial convolution of the PSF and incoming irradiance at the object surface. where the material-related PSF $R_d(||\mathbf{x} - \mathbf{x_i}||; \sigma_a, \sigma_s, p)$ is a shift-invariant function of the material's absorption coefficient σ_a, scattering coefficient σ_s and phase function p. The final subsurface scattering term can be integrated as

$$L_s(\mathbf{x}, \mathbf{w_o}) = L_1(\mathbf{x}, \mathbf{w_o}) + \int_A R_d(||\mathbf{x} - \mathbf{x_i}||)E(\mathbf{x_i})dA(\mathbf{x_i}) \\ \approx R_d(\mathbf{x}) * E(\mathbf{x}), \tag{4.80}$$

where $E(\mathbf{x})$ is the incident irradiance at surface point $\mathbf{x}$, and L_1 is the single-scattering term. Following Schmitt et al. [89], we ignored the Fresnel term in the above equation because such Fresnel effects cannot be observed under our settings because the incident and outgoing light directions are approximately the same, i.e., $\mathbf{w_i} \approx \mathbf{w_o}$.

The direct reflection component can be well described by the bidirectional reflectance distribution function (BRDF), which describes direct reflection without penetration, i.e., $\mathbf{x_i} = \mathbf{x}$. The well-known physical-based Cook-Torrance model parameterizes the BRDF with the diffuse albedo k_d, specular albedo k_s and surface roughness k_r as follows:

$$L_r(\mathbf{x}, \mathbf{w_o}) = k_d E(\mathbf{x}) + k_s \int_{2\pi} S_p(\mathbf{n}, \mathbf{w_i}, \mathbf{w_o}; k_r) L_i(\mathbf{x_i}, \mathbf{w_i}) d\mathbf{w_i}, \tag{4.81}$$

where S_p, parameterized by the material roughness k_r, is the specular term describing the specular reflectance at the surface normal $\mathbf{n}$, and L_i is the incident radiance. The structured light camera cannot determine the true depth of the pixels corresponding to intense specular reflection angles due to overexposure. This severely impacts the local geometry information. To address the incomplete depth information, we concentrate on materials with reasonably high roughness in this work.

Finally, suppose that the projection of a surface point $\mathbf{x}$ on the image is $\mathbf{x^p}$. With sufficiently high resolution, the intensity of this pixel, denoted as $I(\mathbf{x^p})$, is approximately proportional to the outgoing radiance.

$$I(\mathbf{x^p}) \propto L_o(\mathbf{x}, \mathbf{w_o}). \tag{4.82}$$

Structured Light Camera. Structured light cameras are widely utilized in 3D object recognition tasks. The depth of a scene can be determined with off-the-shelf structured light cameras by projecting known patterns and examining the distorted patterns that are reflected back. As shown in Fig. 4.51, these cameras are equipped with a projector emitting patterned infrared light and a sensor with a filter in the corresponding spectrum (usually the infrared spectrum, as shown by the red-tinted region in Fig. 4.51). Usually, an additional calibrated RGB sensor (yellow-tinted region in Fig. 4.51) is equipped for general visual perception under ambient light conditions.

The most commercially successful projected pattern is the collimated dot pattern created by Primesense, although the pattern may differ among various cameras. This design is commonly used by the iPhone, Xtion (Asus), and Kinect-V1 (Microsoft) (Apple). This spatial multiplexing speckle dot pattern can be produced by diffracting the laser light using a diffractive optical component (DOE). This pattern has a useful spatial PSF probing capability. The diffuse illumination and 0-th order pattern are also produced by a DOE as a result of the pseudorandom dot pattern.

For subsurface scattering, dot pattern illumination provides multiple probing functions for the diffusive PSF kernel R_d. Specifically, the radiance of each collimated laser beam can be modeled as a δ function with fixed intensity L_{dot}. Therefore, the irradiance at the illuminated material surface point $\mathbf{x}$ is only related to the incident direction:

$$E_{dot}(\mathbf{x}) = L_{dot} M(\mathbf{x})\langle \mathbf{n}, \mathbf{w_i} \rangle, \tag{4.83}$$

where $M(\mathbf{x})$ is a sparse mask indicating whether the surface point $\mathbf{x}$ is illuminated by the dot pattern illumination, $\mathbf{w_i}$ is the direction from the surface point to the projector center, $\mathbf{n}$ is the normal of the surface at point $\mathbf{x}$, and $\langle \cdot, \cdot \rangle$ denotes the inner product operator. Since the designed energy of the outgoing light exceeds the dynamic range of the image sensor, the sensor cannot distinguish different reflection intensities.

In contrast, diffusive isotropic illumination is suitable for direct reflection feature learning. The isotropic illumination can be modeled as a point light source with fixed radiant intensity I_{dif}. Then, the radiance at surface point $\mathbf{x}$ is attenuated following an inverse square law based on its distance to the light source, denoted as r:

$$E_{dif}(\mathbf{x}) = \frac{I_{dif}}{r^2}\langle \mathbf{n}, \mathbf{w_i} \rangle. \tag{4.84}$$

A simplified model considering only the diffuse subsurface scattering and diffuse direct reflection near a local region on the surface can be formulated as

$$\begin{aligned} L_o(\mathbf{x}, \mathbf{w_o}) \approx& R_d(\mathbf{x}) * E(\mathbf{x}) + k_d E(\mathbf{x}) \\ =& (R_d(\mathbf{x}) + k_d \delta(\mathbf{x})) * (E_{dot}(\mathbf{x}) + E_{dif}(\mathbf{x})) \\ \approx& (R_d(\mathbf{x}) + k_d \delta(\mathbf{x})) * E_{dot}(\mathbf{x}) \\ &+ (R_d * 1(\mathbf{x}) + k_d)\frac{I_{dif}}{r^2}\langle \mathbf{n}, \mathbf{w_i} \rangle \\ =& R'_d(\mathbf{x}) * M(\mathbf{x}) L_{dot}\langle \mathbf{n}, \mathbf{w_i} \rangle + k'_d \frac{I_{dif}}{r^2}\langle \mathbf{n}, \mathbf{w_i} \rangle, \end{aligned} \tag{4.85}$$

where $1(\mathbf{x})$ is a function with a constant value of one used to approximate local diffusive illumination, $k'_d = R_d * 1(\mathbf{x}) + k_d$ denotes the augmented diffuse albedo, and $R'_d(x) = R_d(\mathbf{x}) + k_d\delta(\mathbf{x})$ denotes the augmented PSF. The first part of the formula involves a convolution of the augmented local material PSF $R'_d(\mathbf{x})$ with the dot pattern illumination mask $M(\mathbf{x})$, while the second part is simply the diffuse radiance amplified by the augmented diffuse albedo k'_d.

To better understand these equations, we can decompose images based on the illumination conditions. Consider an infrared image IR_{dif} acquired under diffuse illumination conditions, i.e., $E(\mathbf{x}) = E_{dif}(\mathbf{x})$. It is difficult to distinguish the material's subsurface scattering properties because in the second term, the spatial PSF information is lost due to the convolution with the similar local illumination term; however, it is easy to distinguish the augmented diffuse albedo k'_d (which is dominated by the diffuse albedo k_d value). In contrast, for an infrared image IR_{dot} acquired under dot pattern illumination conditions, i.e., $E(\mathbf{x}) = E_{dot}(\mathbf{x})$, the diffuse albedo contributes to the pixel's intensity only for the directly illuminated points, whose values are usually overexposed in a limited dynamic range. While it is simple to capture the spatial PSF by probing with an illumination mask, it is challenging to differentiate albedo features under dot pattern illumination conditions alone. Hence, combining the two schemes improves the system's ability to recognize different materials.

The PSF kernel cannot be accurately retrieved by deconvolution alone since the spatial convolution occurs on the material surface rather than in the image plane. The point spreading effect seen in the image is deformed by the surface's shape and distance, as well as the anisotropic single-scattering term. In particular, the projected PSF kernel decreases proportionately with increasing distance (i.e., approximately at $1/d$, where d is the depth from the surface point to the sensor image plane), and the PSF is tilted by different shapes.

Method

Our method uses reflection, scattering, and texture features for material classification. To separate the geometry-induced fluctuations in the observed image, a geometry term is created based on the lighting model and calculated based on the depth image.

Image Preprocessing. As analyzed in Subsection ***Structured Light Camera***, under fixed lighting conditions, the material and geometry features of the objects are entangled in the observed IR image. To classify materials regardless of their geometry, several factors are needed for disentanglement, including the inverse of the depth $1/d$, the inverse square between each point and the center of the projector $1/r^2$, and the inner product between the surface normal and incident light direction $\langle \mathbf{n}, \mathbf{w_i} \rangle$. The incident light direction needs to be calculated for the last two factors. Since commercial structured light cameras have relatively small baselines (several centimeters), according to the scale of the scene (several meters), it is reasonable to approximate the incident light direction $\mathbf{w_i}$ as the outgoing light direction $\mathbf{w_o}$. The latter is easy to calculate given the camera's FOV and pixel location. We term the collection of such information the geometry modality and use this information as input for disentangling the geometry factors for material classification.

Specifically, we employ the conventional definition of the camera space, with the origin located at the camera's optical center, the Z axis pointing toward the image center, and the X, Y axes aligned with the camera's edges. The projection of the material surface point $\mathbf{x}(x, y, z)$ on the IR image is $\mathbf{x^p}(u, v)$, where (x, y, z) and (u, v) are the coordinates in the camera space and image plane, respectively. With a stereo vision calculation, each pixel in the IR image is assigned a depth value $d = z - f$ representing the distance between its corresponding surface point $\mathbf{x}$ and the image plane, where f is the camera's focal length. Based on the ratio between the focal length and the pixel width and height of the IR sensor, termed f_x, f_y, the unnormalized surface normal direction at surface point $\mathbf{x}$ can be calculated based on its projected location $\mathbf{x^p}$ as follows:

$$\mathbf{n}'(u, v) = \left[\frac{dz}{dx}, \frac{dz}{dy}, -1\right]^T = \left[\frac{f_x}{z}\frac{dz}{du}, \frac{f_y}{z}\frac{dz}{du}, -1\right]^T. \tag{4.86}$$

Since d and z differ only by a constant value, their differential is the same. Therefore, $\frac{dz}{dv}$ and $\frac{dz}{du}$ can be calculated by applying a differential operator to the whole depth image. In our implementation, we use vertical and horizontal Sobel operators to calculate the differentials. The unnormalized outgoing direction from surface point $\mathbf{x}$ to the camera sensor can be calculated based on its projected location $\mathbf{x^p}$ as follows:

$$\mathbf{w_o}'(u, v) = \left[\frac{x}{z}, \frac{y}{z}, 1\right]^T = \left[\frac{u}{f_x}, \frac{v}{f_y}, 1\right]^T. \tag{4.87}$$

Since we approximate $\mathbf{w_i}$ with $\mathbf{w_o}$, we calculate the desired $\langle \mathbf{n}, \mathbf{w_i} \rangle$ for every pixel $\mathbf{x^p}$ in the image plane by first normalizing the above unnormalized term with the L2 norm and then taking the inner product.

$$\langle \mathbf{n}, \mathbf{w_i} \rangle \approx \langle \mathbf{n}, \mathbf{w_o} \rangle = \langle \frac{\mathbf{n}'}{||\mathbf{n}'||_2}, \frac{\mathbf{w_o}'}{||\mathbf{w_o}'||_2} \rangle. \tag{4.88}$$

Similarly, we approximate the IR camera's origin as the light source point to calculate the distance from surface point $\mathbf{x}$ to the diffusive light source point as $r(u, v) \approx ||z\mathbf{w_o}'(u, v)||_2$. By taking the elementwise inverse square for all pixels in the image, we obtain the radiance attenuation factor $1/r^2$. Finally, the PSF kernel shrinking factor $1/d$ is calculated by elementwise inversion of the depth image.

Note that every factor is calculated based on a single-channel image with the same size as the IR image, and we stack these factors in a channelwise manner to generate the three-channel geometry modality image. Specifically, if the IR image has dimensions (H, W), the stacked geometry modality image is a $(3, H, W)$ tensor, where the $\langle \mathbf{n}, \mathbf{w_i} \rangle$, $1/r^2$ and $1/d$ images are the channels.

IR Image Processing. As analyzed in Subsection **Structured Light Camera**, the observed infrared image IR is a superposition of the diffuse illuminated image IR_{dif} (termed diffusion modality) and dot pattern illuminated image IR_{dot} (termed dot modality). Since the dot pattern dominates the overall illumination, its corresponding image also dominates the observed image, i.e.,

$$IR = IR_{dot} + IR_{dif} \approx IR_{dot}. \tag{4.89}$$

To better distinguish the material's albedo properties, the weak diffusion modality IR_{dif} needs to be explicitly recovered. We consider the observed infrared image as having strong salt-and-pepper noise distortion as a straightforward solution to this issue. Unfortunately, because the noise is not randomly distributed but rather spatially connected due to the PSF, the noise cannot be completely eliminated by median filtering alone. To determine the edges in the grayscale image, we employ guided image filtering [46] to the median filtered IR image (Fig. 4.52, middle). This

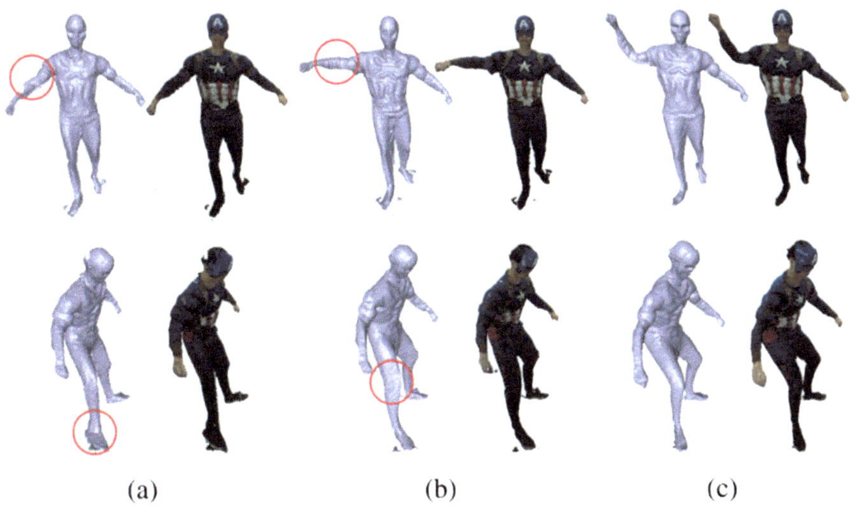

Fig. 4.52 Recovering IR_{dif} by filtering (median filtered) the IR image based on the (aligned) grayscale image. Note that the intensity in the median filtered IR image and IR_{dif} image were both enhanced 4x for better visualization

approach allows us to remove the dot pattern while preserving the sharp edges. We directly use the observed IR image as the dot modality since it is the dominant image component and validate this choice experimentally.

RGB Image Processing. To learn texture features, we utilize a grayscale RGB image (also referred to as the gray modality). To prevent color variations caused by ambient lighting and dyeing, which are invariant to materials, we employ the gray modality image rather than the raw RGB image.

Because the images were acquired by different cameras, the RGB and IR images are not perfectly matched. Therefore, we must align the images from RGB image space to IR image space to address this mismatch. Since the depth is determined in IR image space, we perform the alignment in this direction. In more detail, using the depth value and RGB-IR camera calibration settings, we first determine the coordinates of each pixel in the IR image. Then, we assign its value by interpolating the surrounding RGB values. The average of the aligned RGB image yields the final gray modality image.

Material Classification Model. The overall material classification process is depicted in Fig. 4.53. We first recover the diffusion modality image from the IR image and directly use the IR image as the dot modality image, as described in Subsection **IR Image Processing**. As analyzed in Subsection ***Structured Light Camera***, direct reflection and subsurface scattering are deeply entangled with the geometry features, producing the corresponding diffusion and dot modality images. We disentangle the geometry variance by taking it as an input for the corresponding feature extractors. Specifically, we stack the diffusion and geometry modality images (denoted as

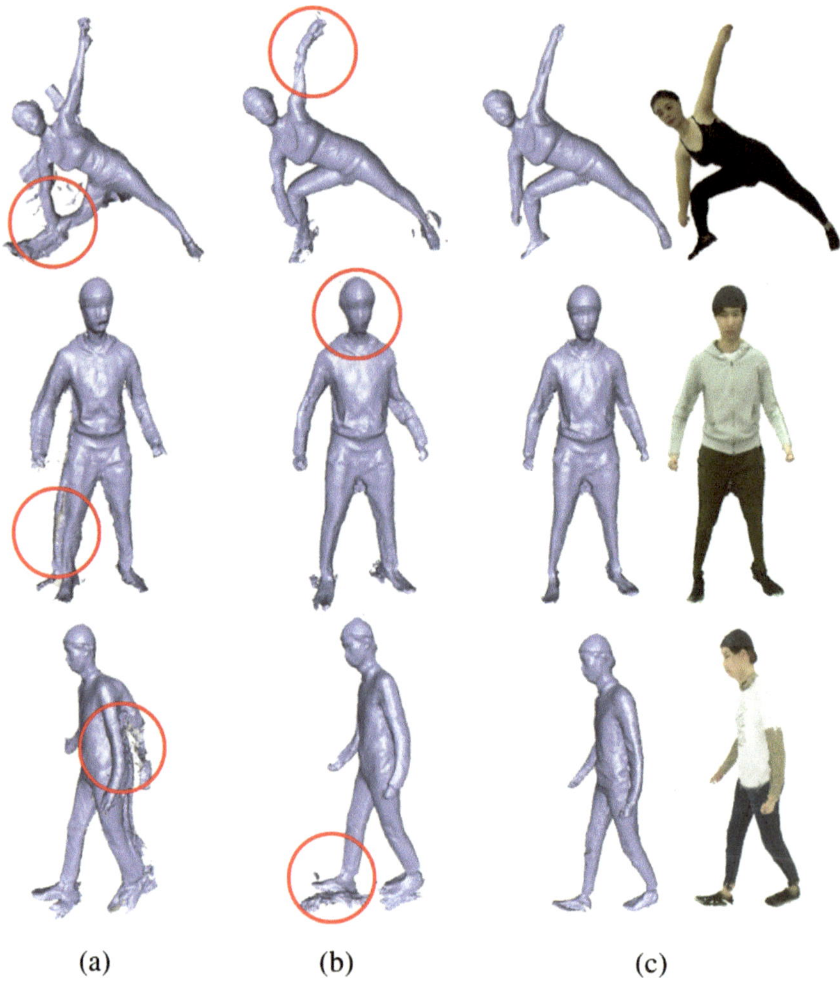

Fig. 4.53 An overview of the material classification method

diffusion-geometry modality in Fig. 4.53) in a channelwise manner. Then, we extract the material reflection features using ResNet-based convolutional neural networks (composed of four residual blocks followed by 2×2 max-pooling layers). Similarly, the material scattering features are extracted from the channelwise stacked dot and geometry modality images (denoted as dot-geometry modality in Fig. 4.53). For texture feature learning, the aligned gray modality image is directly fed into the feature learning networks described above.

The resulting reflection, scattering, and texture features are then combined to carry out the final material classification using a fully connected network with two hidden layers with 512 and 128 neurons. The material dimension is included in the

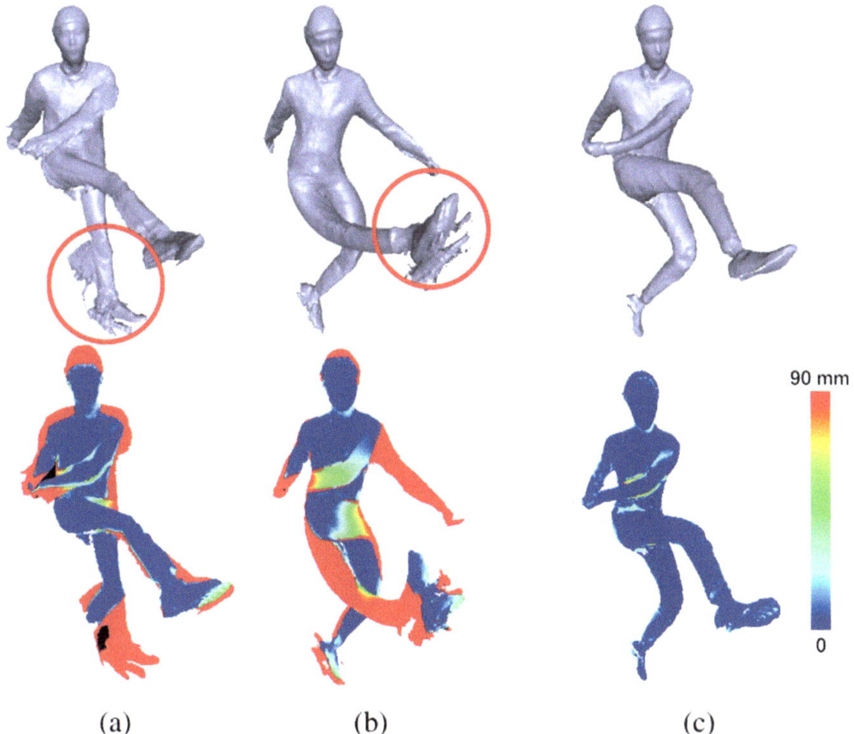

Fig. 4.54 Different fusion methods. The CF and FF methods require explicit recovery of the diffusion modality image

final output layer. We utilize a local patch-based classification method for material recognition to consider materials with minimal specularity. Patches with size $P \times P$ at the same location in the aligned modality images are cropped and fed into the networks.

In addition to the feature fusion (FF) method described above, we also implemented image fusion (IF) and channel fusion (CF) methods in the experiment, and the performance of the different methods was compared. As shown in Fig. 4.54, the image fusion method adds the diffusion and dot modality images in an elementwise manner to generate the synthesized "ir" image. Then, the method concatenates the ir modality (1 channel) and geometry modality (3 channels) images in a channelwise manner. A CNN is then used for categorization. The channel fusion method employs the early fusion paradigm, which concatenates all modality images in a channelwise manner before sending the data to a CNN classifier. Before sending the features to a CNN classifier, the feature fusion approach employs the late fusion paradigm, which concatenates the features collected from the discrete and dot-geometry modality images in a channelwise manner. While we employ an artificial "raw IR" image in the image fusion process, we do not explicitly recover the diffusion modality image. An explicitly recovered diffusion image is necessary for the other two fusion techniques.

Data Collection

We use both synthesized and captured datasets to validate the performance of the proposed material classification method.

Synthesized Dataset. First, a synthetic dataset is produced using Blender, a physical-based rendering program. This dataset is used to verify that the reflection and scattering properties are learned under the known active illumination settings. We discuss the lighting conditions, camera settings, and material attribute modeling results in detail.

The illumination is modeled by assembling an isotropic point light source and a dot pattern projector that projects the reverse engineered Primesense pattern. Since the Primesense pattern is patent protected, we adopt the binary pattern proposed by Reichinger [87]. As shown in Fig. 4.55, the pattern is composed of 3×3 blocks, with a bright dot in the center of each block. Each block is composed of 211×165 holographic orders (the counterpart of pixels in diffractive optics), with a total of 633×495 holographic orders. This pattern is believed to be created by a double-layer DOE that first creates a single block and then repeats the block 9 times, as discussed in [92]. The final image was processed to approximate each dot as a Gaussian shape, and the core dots were extended to simulate artifacts to synthesize the final pattern. Furthermore, we alter the UV mapping of the projection to match the pin-cushion distortion of the actual Xtion camera. The IR camera was positioned exactly 1 m away from a white wall to obtain the reference images. We adjusted the relative intensities

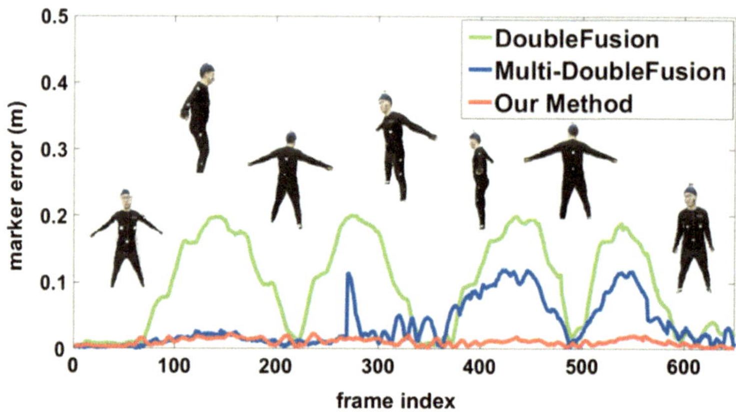

Fig. 4.55 Structured light cameras that project the Primesense pattern (left): Xtion (top), Kinect V1 (middle, showing the key components inside) and Occipital (bottom)

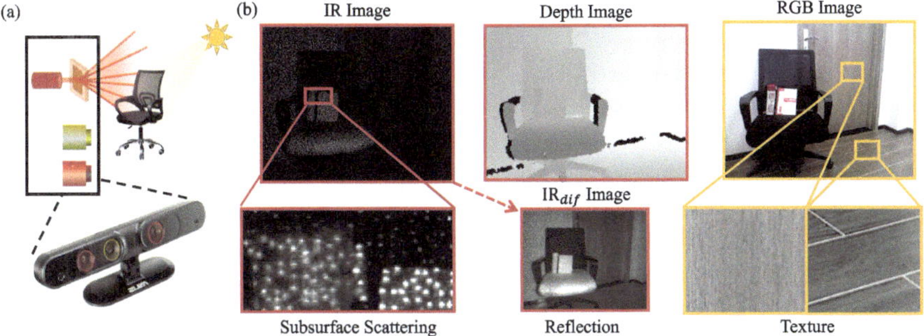

Fig. 4.56 Settings for the synthesized dataset. **a** The modeled IR camera (red frustum with triangle pointed the upwards direction), IR projector (orange dotted ball with direction), and random bumpy planes in Blender. **b** The projected light is modeled to simulate the real pattern shown in Fig. 4.55

of the dot pattern and isotropic light based on the assumption that the white wall is a perfect Lambertian surface without subsurface scattering effects. We produced two images using isotropic and dot pattern illuminations by using the customizable lighting settings.

As shown in Fig. 4.56, a camera with the same FOV (vertical 45°, horizontal 58°) and resolution (1280 × 1024) as Xtion's IR camera was coupled rigidly with the projector at a fixed baseline of 6 cm on the left. In the model, we applied the same translation and rotation to the camera and projector model to keep their relative position fixed. To align with the actual dynamic range of the Xtion sensor, we clipped the received intensity in the rendered image to prevent overexposure.

With a physical-based rendering engine, Blender has a material shader named "principled BSDF", which has base layers for users to control the diffuse, specular, subsurface, and transmission properties. This allows modification of object material properties such as the diffuse albedo, roughness, and optical depth for the subsurface. In the experiments, materials with 5 different albedo values were uniformly sampled in the range 0.2-1, and materials with 5 different optical depth values (subsurface scattering parameters) were exponentially sampled in the range 0.001-0.02 (Fig. 4.57a).

To disentangle geometry differences, the scene includes both flat and random bumpy planes (Fig. 4.57b). For each material in the training set, images were rendered at 4 different distances within the range of 1–4 m, which is a typical working range for indoor structured light cameras such as Xtion. Additionally, 9 different angles were sampled uniformly in the range of ±30°. In the testing set, two distances and angles were randomly selected within the described range.

Captured Dataset. To test the performance of the proposed method for real-world materials, we captured images with the Xtion sensor. The IR camera of Xtion is equipped with a bandpass filter that is only sensitive to emitted 830 nm laser light while filtering out the visible spectrum. The raw image resolution for the RGB and IR cameras is 1280 × 1024, and the depth resolution is 640 × 480. The PSF spans

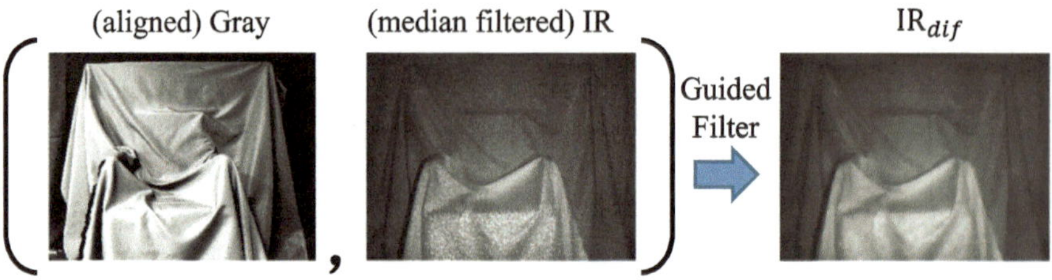

Fig. 4.57 Different generated material properties. **a** Materials with different properties are illuminated by various lighting conditions. The different rows show various subsurface scattering properties, and the different columns show various reflection albedo properties. **b** IR_{dot} and IR_{dif} images rendered under their corresponding illumination conditions; the surfaces are either flat or randomly bumpy. The diffusion images are enhanced for better visualization

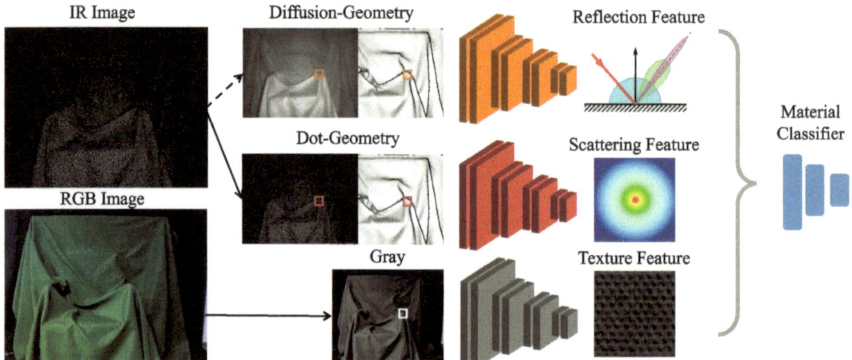

Fig. 4.58 Settings for the captured dataset. **a** Images in the training set were acquired at 4 different depth ranges and three different angles. Additionally, images in the testing set were acquired at two random angles in the depth range. **b** The Xtion sensor model and image capture settings

only limited space (typically approximately 4×4 patches for the highest resolution), and the highest resolution images were captured in our experiment (Fig. 4.58).

As shown in Fig. 4.59, we acquired images of 40 different objects made of 30 different types of materials. Several samples of various materials, including cloth, carpet, and sponge, were considered. These samples had various hues and textures. The materials that were sampled varied widely in terms of texture, transparency, reflectance, and subsurface scattering characteristics. All materials were imaged at 4 different depth ranges between 1–3 m, each with three different viewing directions (roughly left, middle, and right) and flat and deformed shapes (if deformable), resulting in 24 images for each material in the training set. This approach was used to disentangle geometry differences. The testing set was acquired using the same approach, with the exception that each material was sampled twice at random.

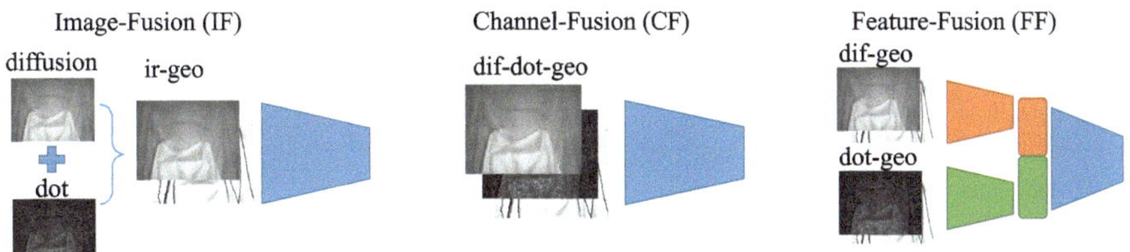

Fig. 4.59 Materials in the captured dataset. Forty different objects made of 30 distinct materials were included

Experiment

Experiments with the Synthesized Dataset. To validate the connections between the different modality images and material properties in a controlled manner, we tested the classification performance of the proposed method based on the synthesized dataset. We abbreviate the diffusion and geometry modalities as "dif" and "geo", respectively, when the modalities are channelwise stacked as input (see Subsection ***Material Classification Model***). We abbreviate the model taking "x" modalities as inputs as the "x model", e.g., the dif-geo model represents the model taking diffusion and geometry modality images as inputs. As shown in Table 4.15, although the overall accuracy of the dif-geo model is 22.8%, the model achieves an accuracy of 97.7% for the albedo property, indicating that these modalities have good discrimination ability for the material's albedo property but are confused by the subsurface property. In contrast, the dif-geo model shows the opposite behavior, showing the ability to distinguish different subsurface properties.

The confusion matrices (Fig. 4.60) show the grouping effects of different modalities. Note that the vertical and horizontal axes are the ground truth and predicted classes, respectively. A total of 800 patches from the testing set were sampled for each material (each with a unique albedo and subsurface parameter), with darker patches representing higher confusion values. The materials can be separated into five groups along each axis, each with a different albedo value and continually varying subsurface characteristics within the group. The confusion matrix of the dif-geo model has a diagonal distribution. Except for the diagonal line, which indicates the accurate prediction results, the most likely false prediction has an offset of ±5 elements relative to the diagonal line, corresponding to materials with the same subsurface property

Table 4.15 Performance based on the synthesis dataset w/o geometry features

Metric	dif-geo	dot-geo	diffusion	dot
Overall ac. (%)	22.8	**58.9**	4.3	34.0
Subsurface ac. (%)	23.3	**96.9**	21.5	86.2
Albedo ac. (%)	**97.7**	60.8	19.9	39.5

Fig. 4.60 Confusion matrix of different modalities based on the synthesized dataset. **a** The confusion matrix of the dot-geo model has a diagonal distribution with a 5-element shift. **b** The confusion matrix of the dif-geo model has a clustering effect along the diagonal line

and albedo values that are either slightly smaller or larger than the actual albedo value. In contrast, the confusion matrix of the dot-geo model is distributed along block diagonals. Every five classes are clearly grouped together, demonstrating that while materials with varying albedo values are distinct, the classifier tends to favor particular subsurface features within the group.

This finding was validated by visualizing the learned features using principal component analysis (PCA). Here, the features refer to the flattened output vectors of the feature learning networks described in Subsection ***Material Classification Model***, which are fed to the downstream classifier. As shown in Fig. 4.61, when projecting the dif-geo feature to the largest principal axes, similar albedo properties are clustered together. Within each group, the subsurface properties are distributed almost uniformly. An exactly opposite behavior can be observed in the dot-geo feature space. This validates the connection between the learned features and material properties.

To validate the necessity of introducing the geometry term in the dot and diffusion modalities, an ablation study was performed. The overall accuracies of the models using only the dot and dif modalities are 34.0% and 4.3%, respectively, and both models show significant performance reductions compared to the geometry stacking model. This performance decrease validates the necessity of accounting for geometry variance in the model. When using only the diffusion modality, the overall accuracy is approximately the same as using a random guess (4.3% vs. 4.0%). Based on only the reflection of the diffusely illuminated surface, it is difficult to attribute the variance in the intensity to different albedo properties because surfaces at different distances or tilted at different angles can also have the same variance in the intensity around the pixel. In contrast, the model using the dot modality distinguishes materials mainly

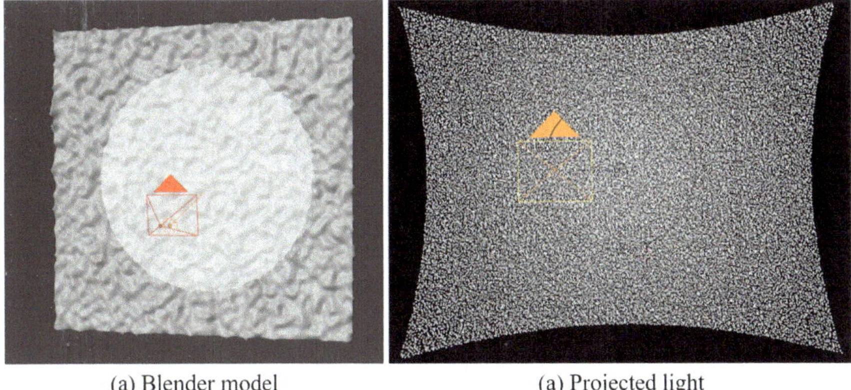

(a) Blender model (a) Projected light

Fig. 4.61 PCA visualizations of learned features of different material properties. The features learned from dif-geo modalities, denoted as the "dif-geo features", exhibit a clear clustering structure for the material albedo value but not for the subsurface albedo value in the major PCA directions. The "dot-geo features" show the opposite behavior

Table 4.16 Performance of different fusion methods based on the synthesized dataset

Metric	IF	CF	FF
Overall ac. (%)	65.6	**96.0**	**96.0**
Subsurface ac. (%)	98.3	**99.5**	99.0
Albedo ac. (%)	66.7	96.5	**97.0**

based on the spatial PSF feature. This feature varies mainly due to material variations and not depth and orientation variation and is thus more robust when used alone.

In the following experiments, we use the geometry-stacked versions of the diffusion and dot modalities by default.

To validate the necessity of explicitly recovering the diffusion modality, we evaluated different fusion methods, as described in Subsection ***Material Classification Model***. As shown in Table 4.16, since the image fusion method is dominated by the dot pattern, it has a similar performance to the dot-geo model. In contrast, the channel fusion and feature fusion methods both benefit from the explicit recovered diffusion modality, leading to better albedo accuracy when paired with the dif-geo model (97.7%).

Experiments Based on the Captured Dataset. We validated the physical connections between the different modalities and corresponding properties based on the synthesized dataset. We next examined the proposed method based on the captured dataset for real-world applications. Since we sampled 40 objects with 30 different materials in the captured dataset, we trained our model based on only the object label and tested the accuracy based on the corresponding object samples and material labels. This training method tends to "overclassify" the materials but is useful for verifying whether the learned models classified materials based on material-invariant

Table 4.17 Performance based on the captured dataset w/o the geometry modality. The accuracies are calculated separately for the 40 samples and 30 materials, and the inner-material similarity gain measures the difference between these two accuracies

Metric	Without geo			With geo			Fusion (with geo)			
	Dot	Dif	Gray	Dot	Dif	Gray	Dot-Dif	Dot-Gray	Dif-Gray	All
Sample ac. (%)	51.9	28.0	**64.9**	62.4	35.8	**67.8**	64.2	90.7	81.7	**90.8**
Material ac. (%)	59.3	33.4	**66.9**	**70.0**	41.4	69.9	71.70	92.2	83.8	**92.5**
Inner-mat. gain (%)	**7.4**	5.3	2.0	**7.6**	5.7	2.2	**7.5**	1.5	2.1	1.7

features, e.g., color and geometry bias. We denoted the difference between the material accuracy and sample accuracy as the inner-material similarity gain. For similar sample accuracy, the larger the gain is, the better the samples with the same materials are clustered together. This indicates that the learned model tends to perform classification based on the joint features of the same material instead of the inner-material difference.

As shown in Table 4.17, when the model was trained using single modalities without geometry stacking, the model using the gray modality has the highest sample accuracy, which we attribute to the fact that the texture difference is significant in the captured dataset. However, models using the dot modality have the highest inner-material similarity. This indicates that the model using the dot modality is more robust in terms of the inner-material variance and thus captures good material-related properties. With geometry stacking, models using the dot-geo and dif-geo modalities exhibit significant inner-material gain compared to models using the gray-geo modalities. These results were consistent with the synthesized experiment results and theoretical analysis, where the dot and diffusion modalities derived from the IR image were illuminated by known patterns, and the material's optical properties could be accurately inferred by disentangling the geometry factors involved in the imaging process. On the other hand, the grayscale image is illuminated by uncontrolled ambient light, making it difficult to disentangle the material's optical properties, even with known geometry factors. We attribute the insignificant performance gain achieved by stacking the geometry and gray modalities to the fact this stacking only helps the model explain the depth-related texture variation.

To further validate that IR modalities can handle difficult textureless cases while gray modality cannot, we select 10 textureless white materials from the captured materials, including materials such as white plastic, paper, sponge, wall, cloth, and foam. The model using gray modality achieves 63.5% material accuracy, compared with 92.4% for the model using dot-dif-geo. It seems that the model using gray modal struggles to find useful texture differences for these materials, while the optical subsurface and reflection differences among sampled materials are more significant.

Table 4.18 Comparison with state-of-the-art methods

Methods	Sample ac. (%)	Material ac. (%)
DEP w/o pretrain	75.0/67.3	78.4/71.1
DeepTEN w/o pretrain	72.9/67.1	75.1/70.5
Gray (ours)	64.9	66.9
Fusion-all (ours)	90.8	92.5

Since the features learned from the three modalities lie in different dimensions, we fuse them in a channel fusion fashion to evaluate the classification performance. As shown in Table 4.17, on average, the pairwise fusion exhibits performance gain compared with the single modal. The method fusing all modalities achieves the highest 92.5% on material accuracy.

Comparison with State-of-the-Art Methods. We compare the proposed method with state-of-the-art material classification methods such as DeepTEN [132] and DEP [118, 119]. Since we are the first to use the structured light camera and none of these methods incorporate the dot and diffusion modalities described above, we compare the performance using grayscale images only. Both of these state-of-the-art models use pretrained ResNet18/50 networks as backbones, and we validate their performance with and without using the pretrained weights for a fair comparison with our method. The comparisons are summarized in Table 4.18. With a predesigned texture feature recognition network and a pretrained backbone, DEP and DeepTEN outperform our method, showing better texture recognition ability. However, the performance using only texture information (78.4%) is not comparable with that of our fusion method, which incorporates texture, reflection, and scattering information (92.5%), indicating the significance of these additional modalities. The ablation study results indicate that the superior performance of DEP and DeepTEN relies largely on pretraining for better texture feature extraction rather than dedicated network design. It is worth noting that the purpose of this work is not to fully exploit the texture feature but to utilize additional optical material features, as shown by the fusion-all model. Additionally, the performance improvement achieved by using pretrained weights suggests that a larger real-world dataset that incorporates both RGB and IR images is beneficial for multimodal feature learning.

Segmentation Experiment. In addition to the patch-based CNN classification, a sliding window can be applied to the full image to achieve material segmentation. In this experiment, we trained our classifier in a fully convolutional manner and used a single fully convolutional network (FCN) for testing. Since patch-based training methods lack local consistency in the outputs, a conditional random field (CRF) was attached at the back-end to obtain the final output. The results are shown in Fig. 4.62.

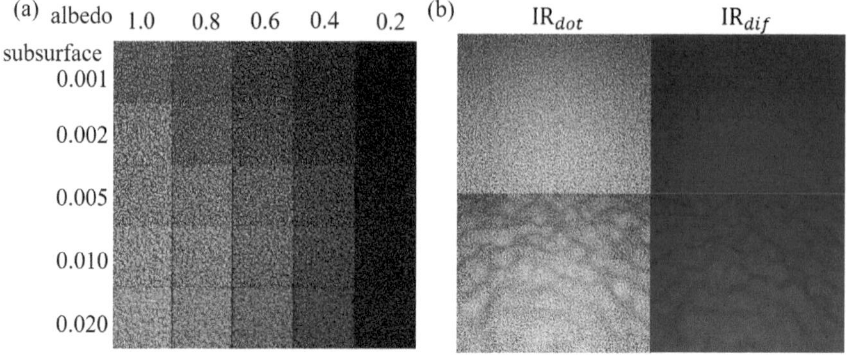

Fig. 4.62 Segmentation results of the proposed method. The raw segmentation result (middle) lacks local consistency, which is mitigated by the CRF back-end (right)

References

1. RayTrix. Avaliable: http://www.raytrix.de/, 2010. 2018. [Online].
2. Lytro. Avaliable: https://www.lytro.com/, 2011. 2018. [Online].
3. *Casual 3D Photography*, volume 36, 2017. ACM.
4. Henrik Aanæs, Rasmus Ramsbøl Jensen, George Vogiatzis, Engin Tola, and Anders Bjorholm Dahl. Large-scale data for multiple-view stereopsis. *International Journal of Computer Vision*, 120(2):153–168, 2016.
5. D. Aiger, N. J. Mitra, and D. Cohen-Or. 4-points congruent sets for robust surface registration. *ACM SIGGRAPH 2008 papers*, 27(3), 2008.
6. Robert Anderson, David Gallup, Jonathan T Barron, Janne Kontkanen, Noah Snavely, Carlos Hernández, Sameer Agarwal, and Steven M Seitz. Jump: virtual reality video. *ACM Transactions on Graphics (TOG)*, 35(6):1–13, 2016.
7. JiaWang Bian, Wen-Yan Lin, Yasuyuki Matsushita, Sai-Kit Yeung, Tan Dat Nguyen, and Ming-Ming Cheng. Gms: Grid-based motion statistics for fast, ultra-robust feature correspondence. In *IEEE Conference on Computer Vision and Pattern Recognition*, 2017.
8. Tom E. Bishop and Paolo Favaro. The light field camera: Extended depth of field, aliasing, and superresolution. *IEEE Transactions on Pattern Analysis and Machine Intelligence*, 34:972–986, 2012.
9. Tom E Bishop, Sara Zanetti, and Paolo Favaro. Light field superresolution. In *2009 IEEE International Conference on Computational Photography (ICCP)*, pages 1–9. IEEE, 2009.
10. Federica Bogo, Angjoo Kanazawa, Christoph Lassner, Peter Gehler, Javier Romero, and Michael J. Black. Keep it smpl: Automatic estimation of 3d human pose and shape from a single image. In Bastian Leibe, Jiri Matas, Nicu Sebe, and Max Welling, editors, *Computer Vision – ECCV 2016*, pages 561–578, Cham, 2016. Springer International Publishing.
11. Neill DF Campbell, George Vogiatzis, Carlos Hernández, and Roberto Cipolla. Using multiple hypotheses to improve depth-maps for multi-view stereo. In *European Conference on Computer Vision*, pages 766–779. Springer, 2008.
12. Alvaro Collet, Ming Chuang, Pat Sweeney, Don Gillett, Dennis Evseev, David Calabrese, Hugues Hoppe, Adam Kirk, and Steve Sullivan. High-quality streamable free-viewpoint video. *ACM Transactions on Graphics (TOG)*, 34(4):69, 2015.
13. F. Crete, T. Dolmiere, P. Ladret, and M. Nicolas. The blur effect: perception and estimation with a new no-reference perceptual blur metric. In *Human Vision and Electronic Imaging XII*, volume 6492 of *SPIE*, page 64920I, February 2007. https://doi.org/10.1117/12.702790.

14. Brian Curless and Marc Levoy. A volumetric method for building complex models from range images. In *Proceedings of the 23rd Annual Conference on Computer Graphics and Interactive Techniques*, SIGGRAPH '96, pages 303–312, New York, NY, USA, 1996. ACM. ISBN 0-89791-746-4. https://doi.org/10.1145/237170.237269. URL https://doi.org/10.1145/237170.237269.
15. Angela Dai, Matthias Nießner, Michael Zollhöfer, Shahram Izadi, and Christian Theobalt. Bundlefusion: Real-time globally consistent 3d reconstruction using on-the-fly surface reintegration. *ACM Transactions on Graphics (ToG)*, 36(4):1, 2017.
16. Yuchao Dai, Zhidong Zhu, Zhibo Rao, and Bo Li. Mvs2: Deep unsupervised multi-view stereo with multi-view symmetry. In *2019 International Conference on 3D Vision (3DV)*, pages 1–8. IEEE, 2019.
17. Andrew J Davison. Futuremapping: The computational structure of spatial ai systems. *arXiv preprint*arXiv:1803.11288, 2018.
18. Chao Dong, Chen Change Loy, Kaiming He, and Xiaoou Tang. Learning a deep convolutional network for image super-resolution. In *ECCV*, 2014.
19. Mingsong Dou, Sameh Khamis, Yury Degtyarev, Philip Davidson, Sean Fanello, Adarsh Kowdle, Sergio Orts Escolano, Christoph Rhemann, David Kim, Jonathan Taylor, Pushmeet Kohli, Vladimir Tankovich, and Shahram Izadi. Fusion4D: Real-time Performance Capture of Challenging Scenes. In *ACM SIGGRAPH Conference on Computer Graphics and Interactive Techniques*, 2016.
20. Mingsong Dou, Philip Davidson, Sean Ryan Fanello, Sameh Khamis, Adarsh Kowdle, Christoph Rhemann, Vladimir Tankovich, and Shahram Izadi. Motion2fusion: Real-time volumetric performance capture. *ACM Trans. Graph.*, 36(6):246:1–246:16, November 2017. ISSN 0730-0301.
21. Ruofei Du, Ming Chuang, Wayne Chang, Hugues Hoppe, and Amitabh Varshney. Montage4d: Interactive seamless fusion of multiview video textures. In *Proceedings of the ACM SIGGRAPH Symposium on Interactive 3D Graphics and Games*, I3D '18, pages 5:1–5:11, New York, NY, USA, 2018. ACM. ISBN 978-1-4503-5705-0.
22. Felix Endres, Jürgen Hess, Nikolas Engelhard, Jürgen Sturm, Daniel Cremers, and Wolfram Burgard. An evaluation of the rgb-d slam system. In *Robotics and Automation (ICRA), 2012 IEEE International Conference on*, pages 1691–1696. IEEE, 2012.
23. Pedro F. Felzenszwalb and Daniel P. Huttenlocher. Efficient belief propagation for early vision. *International Journal of Computer Vision*, 70(1):41–54, 2006.
24. C. Feng, Y. Taguchi, and V. R. Kamat. Fast plane extraction in organized point clouds using agglomerative hierarchical clustering. In *2014 IEEE International Conference on Robotics and Automation (ICRA)*, pages 6218–6225, May 2014. https://doi.org/10.1109/ICRA.2014.6907776.
25. Randima Fernando. *GPU Gems: Programming Techniques, Tips and Tricks for Real-Time Graphics*. Pearson Higher Education, 2004.
26. John Flynn, Michael Broxton, Paul Debevec, Matthew DuVall, Graham Fyffe, Ryan Overbeck, Noah Snavely, and Richard Tucker. Deepview: View synthesis with learned gradient descent. In *IEEE Conference on Computer Vision and Pattern Recognition*, pages 2367–2376, 2019.
27. James D Foley, Foley Dan Van, Andries Van Dam, Steven K Feiner, John F Hughes, J HUGHES, and EDWARD ANGEL. *Computer graphics: principles and practice*, volume 12110. Addison-Wesley Professional, 1996.
28. Christian Forster, Luca Carlone, Frank Dellaert, and Davide Scaramuzza. Imu preintegration on manifold for efficient visual-inertial maximum-a-posteriori estimation. In *arXiv preprint*. Georgia Institute of Technology, 2015.
29. Jeppe Revall Frisvad, Toshiya Hachisuka, and Thomas Kim Kjeldsen. Directional dipole model for subsurface scattering. *ACM Transactions on Graphics*, 34(1):1–12, 2014.
30. P. Furgale, J. Rehder, and R. Siegwart. Unified temporal and spatial calibration for multi-sensor systems. In *2013 IEEE/RSJ International Conference on Intelligent Robots and Systems*, pages 1280–1286, Nov 2013. https://doi.org/10.1109/IROS.2013.6696514.

31. Y. Furukawa and C Hernández. *Multi-View Stereo: A Tutorial*, volume 9. Now Publishers Inc., 2015a.
32. Yasutaka Furukawa and Carlos Hernández. Multi-view stereo: A tutorial. *Found. Trends Comput. Graph. Vis.*, 9:1–148, 2015b.
33. Yasutaka Furukawa and Jean Ponce. Accurate, dense, and robust multiview stereopsis. *IEEE Transactions on Pattern Analysis and Machine Intelligence*, 32(8):1362–1376, 2010.
34. Ran Gal, Yonathan Wexler, Eyal Ofek, Hugues Hoppe, and Daniel Cohen-Or. Seamless montage for texturing models. *Eurographics 2010*, 29(2):479–486, 2010.
35. Silvano Galliani, Katrin Lasinger, and Konrad Schindler. Massively parallel multiview stereopsis by surface normal diffusion. In *IEEE International Conference on Computer Vision*, pages 873–881, 2015.
36. Kaiwen Guo, Feng Xu, Tao Yu, Xiaoyang Liu, Qionghai Dai, and Yebin Liu. Real-time geometry, albedo and motion reconstruction using a single rgbd camera. *ACM Transactions on Graphics (TOG)*, 2017.
37. Martin Habbecke and Leif Kobbelt. A surface-growing approach to multi-view stereo reconstruction. In *IEEE Conference on Computer Vision and Pattern Recognition*, 2007.
38. Lei Han and Lu Fang. Mild: Multi-index hashing for appearance based loop closure detection. In *2017 IEEE International Conference on Multimedia and Expo (ICME)*, pages 139–144. IEEE, 2017.
39. Lei Han and Lu Fang. Flashfusion: Real-time globally consistent dense 3d reconstruction using cpu computing. In *Proceedings of Robotics: Science and Systems*, Pittsburgh, Pennsylvania, June 2018a. https://doi.org/10.15607/RSS.2018.XIV.006.
40. Lei Han and Lu Fang. Flashfusion: Real-time globally consistent dense 3d reconstruction using cpu computing. In *Robotics: Science and Systems*, volume 1, page 7, 2018b.
41. Lei Han, Lan Xu, Dmytro Bobkov, Eckehard Steinbach, and Lu Fang. Real-time global registration for globally consistent rgbd slam. *http://www.luvision.net/FastGO*, 2018.
42. Lei Han, Lan Xu, Dmytro Bobkov, Eckehard Steinbach, and Lu Fang. Real-time global registration for globally consistent rgb-d slam. *IEEE Transactions on Robotics*, 35(2):498–508, 2019. https://doi.org/10.1109/TRO.2018.2882730.
43. A. Handa, T. Whelan, J. McDonald, and A. J. Davison. A benchmark for rgb-d visual odometry, 3d reconstruction and slam. In *2014 IEEE International Conference on Robotics and Automation (ICRA)*, pages 1524–1531, May 2014. https://doi.org/10.1109/ICRA.2014.6907054.
44. Ankur Handa, Thomas Whelan, John McDonald, and Andrew J Davison. A benchmark for rgb-d visual odometry, 3d reconstruction and slam. In *Robotics and automation (ICRA), 2014 IEEE international conference on*, pages 1524–1531. IEEE, 2014.
45. Richard Hartley and Andrew Zisserman. *Multiple view geometry in computer vision*. Cambridge university press, 2003.
46. Kaiming He, Jian Sun, and Xiaoou Tang. Guided image filtering. In *European Conference on Computer Vision*, pages 1–14, Crete, 2010. Springer.
47. Ming Hsiao, Eric Westman, and Michael Kaess. Dense planar-inertial slam with structural constraints. In *2018 IEEE International Conference on Robotics and Automation (ICRA)*, pages 6521–6528. IEEE, 2018.
48. Takeo Igarashi, Tomer Moscovich, and John F. Hughes. As-rigid-as-possible shape manipulation. In *ACM SIGGRAPH 2005 Papers*, SIGGRAPH '05, pages 1134–1141, New York, NY, USA, 2005. ACM.
49. Matthias Innmann, Michael Zollhöfer, Matthias Nießner, Christian Theobalt, and Marc Stamminger. Volumedeform: Real-time volumetric non-rigid reconstruction. In *Computer Vision–ECCV 2016: 14th European Conference, Amsterdam, The Netherlands, October 11-14, 2016, Proceedings, Part VIII 14*, pages 362–379. Springer, 2016.
50. Sergey Ioffe and Christian Szegedy. Batch normalization: Accelerating deep network training by reducing internal covariate shift. In *International conference on machine learning*, pages 448–456. PMLR, 2015.
51. Henrik Wann Jensen, Stephen R Marschner, Marc Levoy, and Pat Hanrahan. A practical model for subsurface light transport. In *Proceedings of the 28th Annual Conference on Computer Graphics and Interactive Techniques*, pages 511–518, Los Angeles, 2001.

52. Hae-Gon Jeon, Jaesik Park, Gyeongmin Choe, Jinsun Park, Yunsu Bok, Yu-Wing Tai, and In-So Kweon. Accurate depth map estimation from a lenslet light field camera. *2015 IEEE Conference on Computer Vision and Pattern Recognition (CVPR)*, pages 1547–1555, 2015.
53. Mengqi Ji, Juergen Gall, Haitian Zheng, Yebin Liu, and Lu Fang. Surfacenet: An end-to-end 3d neural network for multiview stereopsis. In *Proceedings of the IEEE International Conference on Computer Vision*, pages 2307–2315, 2017.
54. Mengqi Ji, Jinzhi Zhang, Qionghai Dai, and Lu Fang. Surfacenet+: An end-to-end 3d neural network for very sparse multi-view stereopsis. *IEEE Transactions on Pattern Analysis and Machine Intelligence*, 43(11):4078–4093, 2020.
55. Olaf Kähler, Victor Adrian Prisacariu, Carl Yuheng Ren, Xin Sun, Philip Torr, and David Murray. Very high frame rate volumetric integration of depth images on mobile devices. *IEEE transactions on visualization and computer graphics*, 21(11):1241–1250, 2015.
56. Olaf Kähler, Victor A Prisacariu, and David W Murray. Real-time large-scale dense 3d reconstruction with loop closure. In *European Conference on Computer Vision*, pages 500–516. Springer, 2016.
57. James T Kajiya and Brian P Von Herzen. Ray tracing volume densities. *ACM SIGGRAPH computer graphics*, 18(3):165–174, 1984.
58. Nima Khademi Kalantari, Tingxian Wang, and Ravi Ramamoorthi. Learning-based view synthesis for light field cameras. *ACM Transactions on Graphics (TOG)*, 35:1–10, 2016.
59. Tejas Khot, Shubham Agrawal, Shubham Tulsiani, Christoph Mertz, Simon Lucey, and Martial Hebert. Learning unsupervised multi-view stereopsis via robust photometric consistency. *arXiv preprint*arXiv:1905.02706, 2019.
60. Jiwon Kim, Jung Kwon Lee, and Kyoung Mu Lee. Accurate image super-resolution using very deep convolutional networks. In *Proceedings of the IEEE conference on computer vision and pattern recognition*, pages 1646–1654, 2016.
61. Matthew Klingensmith, Ivan Dryanovski, Siddhartha Srinivasa, and Jizhong Xiao. Chisel: Real time large scale 3d reconstruction onboard a mobile device using spatially hashed signed distance fields. In *Robotics: science and systems*, volume 4, page 1. Citeseer, 07 2015. https://doi.org/10.15607/RSS.2015.XI.040.
62. Arno Knapitsch, Jaesik Park, Qian-Yi Zhou, and Vladlen Koltun. Tanks and temples: Benchmarking large-scale scene reconstruction. *ACM Transactions on Graphics*, 36(4), 2017.
63. Tristan Laidlow, Michael Bloesch, Wenbin Li, and Stefan Leutenegger. Dense rgb-d-inertial slam with map deformations. In *2017 IEEE/RSJ International Conference on Intelligent Robots and Systems (IROS)*, pages 6741–6748. IEEE, 2017.
64. V. Lempitsky and D. Ivanov. Seamless mosaicing of image-based texture maps. In *2007 IEEE Conference on Computer Vision and Pattern Recognition*, pages 1–6, 2007.
65. Marc Levoy and Pat Hanrahan. Light field rendering. *Proceedings of the 23rd annual conference on Computer graphics and interactive techniques*, 1996.
66. Xing Lin, Jiamin Wu, Guoan Zheng, and Qionghai Dai. Camera array based light field microscopy. *Biomedical optics express*, 6 9:3179–89, 2015.
67. Zhouchen Lin and Harry Shum. A geometric analysis of light field rendering. *International Journal of Computer Vision*, 58:121–138, 2004.
68. Feng Liu, Michael Gleicher, Hailin Jin, and Aseem Agarwala. Content-preserving warps for 3d video stabilization. In *ACM SIGGRAPH 2009 Papers*, SIGGRAPH '09, pages 44:1–44:9, New York, NY, USA, 2009. ACM. ISBN 978-1-60558-726-4.
69. Lingjie Liu, Jiatao Gu, Kyaw Zaw Lin, Tat-Seng Chua, and Christian Theobalt. Neural sparse voxel fields. *Advances in Neural Information Processing Systems*, 33:15651–15663, 2020.
70. Shuaicheng Liu, Lu Yuan, Ping Tan, and Jian Sun. Bundled camera paths for video stabilization. *ACM Trans. Graph.*, 32(4):78:1–78:10, July 2013. ISSN 0730-0301.
71. Stephen Lombardi, Tomas Simon, Jason Saragih, Gabriel Schwartz, Andreas Lehrmann, and Yaser Sheikh. Neural volumes: Learning dynamic renderable volumes from images. *arXiv preprint*arXiv:1906.07751, 2019.
72. Matthew Loper, Naureen Mahmood, Javier Romero, Gerard Pons-Moll, and Michael J. Black. Smpl: A skinned multi-person linear model. *ACM Trans. Graph.*, 34(6):248:1–248:16, October 2015. ISSN 0730-0301.

73. Lingni Ma, Christian Kerl, Jörg Stückler, and Daniel Cremers. Cpa-slam: Consistent plane-model alignment for direct rgb-d slam. In *Robotics and Automation (ICRA), 2016 IEEE International Conference on*, pages 1285–1291. IEEE, 2016.
74. Ben Mildenhall, Pratul P Srinivasan, Rodrigo Ortiz-Cayon, Nima Khademi Kalantari, Ravi Ramamoorthi, Ren Ng, and Abhishek Kar. Local light field fusion: Practical view synthesis with prescriptive sampling guidelines. *ACM Transactions on Graphics (TOG)*, 38(4):1–14, 2019.
75. Ben Mildenhall, Pratul P. Srinivasan, Matthew Tancik, Jonathan T. Barron, Ravi Ramamoorthi, and Ren Ng. Nerf: Representing scenes as neural radiance fields for view synthesis. In *ECCV*, pages 405–421. Springer, 2020.
76. Anastasios I Mourikis, Stergios I Roumeliotis, et al. A multi-state constraint kalman filter for vision-aided inertial navigation. In *ICRA*, volume 2, page 6, 2007.
77. Raul Mur-Artal and Juan D Tardós. Orb-slam2: An open-source slam system for monocular, stereo, and rgb-d cameras. *IEEE transactions on robotics*, 33(5):1255–1262, 2017.
78. Richard A Newcombe, Shahram Izadi, Otmar Hilliges, David Molyneaux, David Kim, Andrew J Davison, Pushmeet Kohi, Jamie Shotton, Steve Hodges, and Andrew Fitzgibbon. Kinectfusion: Real-time dense surface mapping and tracking. In *2011 10th IEEE International Symposium on Mixed and Augmented Reality*, pages 127–136. IEEE, 2011.
79. Richard A. Newcombe, Dieter Fox, and Steven M. Seitz. DynamicFusion: Reconstruction and Tracking of Non-Rigid Scenes in Real-Time. In *IEEE/CVF Conference on Computer Vision and Pattern Recognition (CVPR)*, June 2015.
80. Michael Niemeyer, Lars Mescheder, Michael Oechsle, and Andreas Geiger. Differentiable volumetric rendering: Learning implicit 3d representations without 3d supervision. In *Proceedings of the IEEE/CVF Conference on Computer Vision and Pattern Recognition*, pages 3504–3515, 2020.
81. M. Nießner, M. Zollhöfer, S. Izadi, and M. Stamminger. Real-time 3d reconstruction at scale using voxel hashing. *ACM Transactions on Graphics (TOG)*, 32(6):169, 2013a.
82. Matthias Nießner, Michael Zollhöfer, Shahram Izadi, and Marc Stamminger. Real-time 3d reconstruction at scale using voxel hashing. *ACM Transactions on Graphics (TOG)*, 32(6):169, 2013b.
83. Eric Penner and Li Zhang. Soft 3d reconstruction for view synthesis. *ACM Transactions on Graphics (TOG)*, 36(6):1–11, 2017.
84. Fabián Prada, Misha Kazhdan, Ming Chuang, and Hugues Hoppe. Gradient-domain processing within a texture atlas. *ACM Trans. Graph.*, 37(4):154:1–154:14, July 2018. ISSN 0730-0301. https://doi.org/10.1145/3197517.3201317. URL https://doi.org/10.1145/3197517.3201317.
85. T. Qin, P. Li, and S. Shen. Vins-mono: A robust and versatile monocular visual-inertial state estimator. *IEEE Transactions on Robotics*, 34(4):1004–1020, Aug 2018. ISSN 1552-3098. https://doi.org/10.1109/TRO.2018.2853729.
86. Nasim Rahaman, Aristide Baratin, Devansh Arpit, Felix Draxler, Min Lin, Fred Hamprecht, Yoshua Bengio, and Aaron Courville. On the spectral bias of neural networks. In *International Conference on Machine Learning*, pages 5301–5310. PMLR, 2019.
87. Andreas Reichinger. Kinect pattern uncovered, 2022. URL https://azttm.wordpress.com/2011/04/03/kinect-pattern-uncovered/.
88. Ruslan Salakhutdinov and Geoff Hinton. Learning a nonlinear embedding by preserving class neighbourhood structure. In *Artificial Intelligence and Statistics*, pages 412–419. PMLR, 2007.
89. Carolin Schmitt, Simon Donné, Gernot Riegler, Vladlen Koltun, and Andreas Geiger. On joint estimation of pose, geometry and svbrdf from a handheld scanner. In *Proceedings of the IEEE/CVF Conference on Computer Vision and Pattern Recognition*, pages 3493–3503, 2020.
90. Johannes Lutz Schönberger, Enliang Zheng, Marc Pollefeys, and Jan-Michael Frahm. Pixelwise view selection for unstructured multi-view stereo. In *European Conference on Computer Vision (ECCV)*, pages 501–518. Springer, 2016.

91. Thomas Schöps, Torsten Sattler, and Marc Pollefeys. Surfelmeshing: Online surfel-based mesh reconstruction. *IEEE transactions on pattern analysis and machine intelligence*, 42(10):2494–2507, 2019.
92. Alexander Shpunt. Optical designs for zero order reduction, 2009.
93. Vincent Sitzmann, Michael Zollhöfer, and Gordon Wetzstein. Scene representation networks: Continuous 3d-structure-aware neural scene representations. In *Advances in Neural Information Processing Systems*, pages 1121–1132, 2019.
94. M. Slavcheva, M. Baust, D. Cremers, and S. Ilic. KillingFusion: Non-rigid 3D Reconstruction without Correspondences. In *IEEE Conference on Computer Vision and Pattern Recognition (CVPR)*, 2017.
95. Jürgen Steimle, Andreas Jordt, and Pattie Maes. Flexpad: highly flexible bending interactions for projected handheld displays. In *Proceedings of the SIGCHI Conference on Human Factors in Computing Systems*, pages 237–246, Paris, 2013.
96. Frank Steinbrücker, Jürgen Sturm, and Daniel Cremers. Volumetric 3d mapping in real-time on a cpu. In *Robotics and Automation (ICRA), 2014 IEEE International Conference on*, pages 2021–2028. IEEE, 2014.
97. Jürgen Sturm, Nikolas Engelhard, Felix Endres, Wolfram Burgard, and Daniel Cremers. A benchmark for the evaluation of rgb-d slam systems. In *Intelligent Robots and Systems (IROS), 2012 IEEE/RSJ International Conference on*, pages 573–580. IEEE, 2012.
98. Shuochen Su, Felix Heide, Robin Swanson, Jonathan Klein, Clara Callenberg, Matthias Hullin, and Wolfgang Heidrich. Material classification using raw time-of-flight measurements. In *Proceedings of the IEEE Conference on Computer Vision and Pattern Recognition*, pages 3503–3511, Las Vegas, 2016.
99. Robert W Sumner, Johannes Schmid, and Mark Pauly. Embedded deformation for shape manipulation. *ACM Transactions on Graphics (TOG)*, 26(3):80, 2007.
100. Kenichiro Tanaka, Yasuhiro Mukaigawa, Takuya Funatomi, Hiroyuki Kubo, Yasuyuki Matsushita, and Yasushi Yagi. Material classification using frequency-and depth-dependent time-of-flight distortion. In *Proceedings of the IEEE Conference on Computer Vision and Pattern Recognition*, pages 79–88, Hawaii, 2017.
101. Michael W. Tao, Sunil Hadap, Jitendra Malik, and Ravi Ramamoorthi. Depth from combining defocus and correspondence using light-field cameras. *2013 IEEE International Conference on Computer Vision*, pages 673–680, 2013.
102. Matthias Teschner, Bruno Heidelberger, Matthias Müller, Danat Pomerantes, and Markus H Gross. Optimized spatial hashing for collision detection of deformable objects. In *Vmv*, volume 3, pages 47–54, 2003.
103. Sebastian Thrun. Probabilistic robotics. *Communications of the ACM*, 45(3):52–57, 2002.
104. Sebastian Thrun, Wolfram Burgard, and Dieter Fox. *Probabilistic robotics*. MIT press, 2005.
105. Daniel Thuerck, Michael Waechter, Sven Widmer, Max von Buelow, Patrick Seemann, Marc E. Pfetsch, and Michael Goesele. A fast, massively parallel solver for large, irregular pairwise Markov random fields. In *Proceedings of High Performance Graphics 2016*, pages 173–183, 2016.
106. Engin Tola, Christoph Strecha, and Pascal Fua. Efficient large-scale multi-view stereo for ultra high-resolution image sets. *Machine Vision and Applications*, pages 1–18, 2012.
107. Jonathan Ventura, Clemens Arth, Gerhard Reitmayr, and Dieter Schmalstieg. Global localization from monocular slam on a mobile phone. *IEEE transactions on visualization and computer graphics*, 20(4):531–539, 2014.
108. Michael Waechter, Nils Moehrle, and Michael Goesele. Let there be color! large-scale texturing of 3d reconstructions. In *ECCV 2014*, pages 836–850, 2014.
109. Chen Wang, Minh-Chung Hoang, Lihua Xie, and Junsong Yuan. Non-iterative rgb-d-inertial odometry. In *arXiv preprint*arXiv:1710.05502, 2017a.
110. Chen Wang, Junsong Yuan, and Lihua Xie. Non-iterative slam. *2017 18th International Conference on Advanced Robotics (ICAR)*, pages 83–90, 2017b.
111. Fangjinhua Wang, Silvano Galliani, Christoph Vogel, Pablo Speciale, and Marc Pollefeys. Patchmatchnet: Learned multi-view patchmatch stereo. In *Proceedings of the IEEE/CVF Conference on Computer Vision and Pattern Recognition*, pages 14194–14203, 2021a.

112. Tingxian Wang, Alexei A. Efros, and Ravi Ramamoorthi. Occlusion-aware depth estimation using light-field cameras. *2015 IEEE International Conference on Computer Vision (ICCV)*, pages 3487–3495, 2015.
113. Zhaoqiang Wang, Lanxin Zhu, Hao Zhang, Guo Li, Chengqiang Yi, Yi Li, Yicong Yang, Yichen Ding, Mei Zhen, Shangbang Gao, Tzung K. Hsiai, and Peng Fei. Real-time volumetric reconstruction of biological dynamics with light-field microscopy and deep learning. *Nature methods*, 18:551–556, 2021b.
114. Thomas Whelan, Renato F Salas-Moreno, Ben Glocker, Andrew J Davison, and Stefan Leutenegger. Elasticfusion: Real-time dense slam and light source estimation. *The International Journal of Robotics Research*, 35(14):1697–1716, 2016.
115. Gaochang Wu, Yebin Liu, Lu Fang, and Tianyou Chai. Revisiting light field rendering with deep anti-aliasing neural network. *IEEE Transactions on Pattern Analysis and Machine Intelligence*, 44:5430–5444, 2022.
116. Zhaolin Xiao, Qing Wang, Guoqing Zhou, and Jingyi Yu. Aliasing detection and reduction in plenoptic imaging. *2014 IEEE Conference on Computer Vision and Pattern Recognition*, pages 3326–3333, 2014.
117. L. Xu, Y. Liu, W. Cheng, K. Guo, G. Zhou, Q. Dai, and L. Fang. Flycap: Markerless motion capture using multiple autonomous flying cameras. *IEEE Transactions on Visualization and Computer Graphics*, PP(99):1–1, 2017. ISSN 1077-2626. https://doi.org/10.1109/TVCG.2017.2728660.
118. Jia Xue, Hang Zhang, and Kristin Dana. Deep texture manifold for ground terrain recognition. In *Proceedings of the IEEE Conference on Computer Vision and Pattern Recognition*, pages 558–567, 2018.
119. Jia Xue, Hang Zhang, Ko Nishino, and Kristin Dana. Differential viewpoints for ground terrain material recognition. *IEEE Transactions on Pattern Analysis and Machine Intelligence*, 2020.
120. Shi Yan, Chenglei Wu, Lizhen Wang, Feng Xu, Liang An, Kaiwen Guo, and Yebin Liu. Ddrnet: Depth map denoising and refinement for consumer depth cameras using cascaded cnns. In Vittorio Ferrari, Martial Hebert, Cristian Sminchisescu, and Yair Weiss, editors, *Computer Vision – ECCV 2018*, pages 155–171, Cham, 2018. Springer International Publishing.
121. Fengting Yang, Qian Sun, Hailin Jin, and Zihan Zhou. Superpixel segmentation with fully convolutional networks. *2020 IEEE/CVF Conference on Computer Vision and Pattern Recognition (CVPR)*, pages 13961–13970, 2020.
122. J. Yang, H. Li, D. Campbell, and Y. Jia. Go-icp: A globally optimal solution to 3d icp point-set registration. *IEEE Transactions on Pattern Analysis and Machine Intelligence*, 38(11):2241–2254, 2016.
123. Yao Yao, Zixin Luo, Shiwei Li, Tian Fang, and Long Quan. Mvsnet: Depth inference for unstructured multi-view stereo. In *Proceedings of the European Conference on Computer Vision (ECCV)*, pages 767–783, 2018.
124. Yao Yao, Zixin Luo, Shiwei Li, Jingyang Zhang, Yufan Ren, Lei Zhou, Tian Fang, and Long Quan. Blendedmvs: A large-scale dataset for generalized multi-view stereo networks. In *Proceedings of the IEEE/CVF Conference on Computer Vision and Pattern Recognition*, pages 1790–1799, 2020.
125. Youngjin Yoon, Hae-Gon Jeon, Donggeun Yoo, Joon-Young Lee, and In-So Kweon. Learning a deep convolutional network for light-field image super-resolution. *2015 IEEE International Conference on Computer Vision Workshop (ICCVW)*, pages 57–65, 2015.
126. Fisher Yu and Vladlen Koltun. Multi-scale context aggregation by dilated convolutions. *arXiv preprint*arXiv:1511.07122, 2015.
127. Tao Yu, Kaiwen Guo, Feng Xu, Yuan Dong, Zhaoqi Su, Jianhui Zhao, Jianguo Li, Qionghai Dai, and Yebin Liu. Bodyfusion: Real-time capture of human motion and surface geometry using a single depth camera. In *The IEEE International Conference on Computer Vision (ICCV)*. ACM, October 2017.
128. Tao Yu, Zerong Zheng, Kaiwen Guo, Jianhui Zhao, Qionghai Dai, Hao Li, Gerard Pons-Moll, and Yebin Liu. Doublefusion: Real-time capture of human performances with inner body shapes from a single depth sensor. In *The IEEE International Conference on Computer Vision and Pattern Recognition(CVPR)*. IEEE, June 2018.

129. Kaan Yucer, Changil Kim, Alexander Sorkine-Hornung, and Olga Sorkine-Hornung. Depth from gradients in dense light fields for object reconstruction. In *2016 Fourth International Conference on 3D Vision (3DV)*, pages 249–257. IEEE, 2016.
130. Ming Zeng, Fukai Zhao, Jiaxiang Zheng, and Xinguo Liu. A memory-efficient kinectfusion using octree. In *Computational Visual Media*, pages 234–241. Springer, 2012.
131. Cha Zhang and Tsuhan Chen. Spectral analysis for sampling image-based rendering data. *IEEE Trans. Circuits Syst. Video Technol.*, 13:1038–1050, 2003.
132. Hang Zhang, Jia Xue, and Kristin Dana. Deep ten: Texture encoding network. In *Proceedings of the IEEE conference on computer vision and pattern recognition*, pages 708–717, 2017.
133. Jinzhi Zhang, Mengqi Ji, Guangyu Wang, Xue Zhiwei, Shengjin Wang, and Lu Fang. Surrf: Unsupervised multi-view stereopsis by learning surface radiance field. *IEEE Transactions on Pattern Analysis and Machine Intelligence*, 2021.
134. Richard Zhang, Phillip Isola, Alexei A Efros, Eli Shechtman, and Oliver Wang. The unreasonable effectiveness of deep features as a perceptual metric. In *IEEE/CVF Conference on Computer Vision and Pattern Recognition (CVPR)*, pages 586–595, 2018.
135. Zhengyou Zhang. Microsoft kinect sensor and its effect. *IEEE multimedia*, 19(2):4–10, 2012.
136. Tian Zheng, Guoqing Zhang, Lei Han, Lan Xu, and Lu Fang. Building fusion: semantic-aware structural building-scale 3d reconstruction. *IEEE Transactions on Pattern Analysis and Machine Intelligence*, 2020.
137. Zerong Zheng, Tao Yu, Hao Li, Kaiwen Guo, Qionghai Dai, Lu Fang, and Yebin Liu. Hybridfusion: Real-time performance capture using a single depth sensor and sparse imus. In *European Conference on Computer Vision (ECCV)*, Sept 2018.
138. Dawei Zhong, Lei Han, and Lu Fang. idfusion: Globally consistent dense 3d reconstruction from rgb-d and inertial measurements. In *Proceedings of the 27th ACM International Conference on Multimedia*, pages 962–970, 2019.
139. Tinghui Zhou, Richard Tucker, John Flynn, Graham Fyffe, and Noah Snavely. Stereo magnification: Learning view synthesis using multiplane images. *arXiv preprint*arXiv:1805.09817, 2018.
140. Zhu Zunjie and Feng Xu. Real-time indoor scene reconstruction with rgbd and inertia input. *arXiv preprint*arXiv:1812.03015, 2018.

Chapter 5
Toward Large-Scale Plenoptic Reconstruction

Reconstructing real-world scenes with unparalleled levels of realism and detail has been a long-standing goal in the fields of computer vision and graphics. Achieving this goal necessitates coordinated efforts in both sensing techniques and plenoptic reconstruction algorithms. In the previous chapters, the breakthroughs in gigapixel-level sensing have been presented, which lays the solid foundations for large-scale plenoptic reconstruction with ultra-high resolution. In this chapter, we introduce state-of-the-art plenoptic reconstruction methods which fully exploits the sensation resolution for hyper realistic plenoptic 3D reconstruction.

To achieve gigapixel-level plenoptic reconstruction, several pivotal issues in existing reconstruction methods must be addressed: (1) Large-scale datasets (e.g., the ultra-large-scale, gigapixel-level 3D reconstruction dataset GigaMVS introduced in Sect. 5.1), which are essential for advancing and analyzing state-of-the-art reconstruction methods, are needed; (2) sparse view reconstruction methods (Sect. 5.2), which are more practical and more cost-efficient but increase the occlusion difficulty, are necessary; (3) high-resolution neural reconstruction techniques (Sect. 5.3), which exploit the underlying manifold sparsity to unleash the expressivity for structural and textual details; and (4) knowledge-based reconstruction methods (Sect. 5.4), which combine geometric structure priors and semantic knowledge to expedite large-scale geometry recovery, should be developed. The following chapter investigates these aspects to expand the scale of reconstruction methods, including detailed descriptions of the algorithms and applications.

L. Fang, *Plenoptic Imaging and Processing*, Advances in Computer Vision and Pattern Recognition, https://doi.org/10.1007/978-981-97-6915-5_5

5.1 GigaMVS: A Benchmark for Gigapixel-Level 3D Reconstruction

Multiview stereopsis (MVS) techniques are widely used to recover 3D geometries and texture details based on multiple images. However, existing benchmark algorithms do not provide high-resolution observations of small details for scenes with large-scale geometries.

To address this issue, we introduce a new benchmark called HighResMVS, which supports ultrahigh-resolution imagery for reconstructing large-scale 3D models. HighResMVS was designed to evaluate both the geometry and texture of the reconstructed models. We call this benchmark the ***Structure & Texture*** model, which has the following characteristics: **(1)** Multiscale: The dataset includes both *structure*-scale scenes and *texture*-scale details, captured by up to 10 gigapixel images; **(2)** Large scale: The captured scenes cover an area of up to 50000 m^2, which is ten times larger than the area covered by current state-of-the-art benchmarks [70]; **(3)** High-resolution: the large number of gigapixel-level images (with both wide field-of-view and high spatial resolution) provides extremely high-resolution details for multiscale scenarios, with approximately $10\times$ higher resolution than the existing high-resolution 3D reconstruction benchmark [112].

For evaluation, the benchmark provides precise ground truth 3D models based on a laser scanner, with noisy point clouds and unavoidable moving objects either carefully addressed or eliminated during postprocessing. During the evaluation, the geometric performance and visual quality both are assessed, which can serve as a complimentary evaluation protocol for the 3D reconstruction benchmark. Considering that the sparse viewpoints [60] and image resolution affect the 3D reconstruction performance, various spatio-angular resolutions are set to investigate the performance of existing MVS algorithms (Fig. 5.1).

Popular Reconstruction Benchmarks

The evaluation of computer vision algorithms has long been an important topic in the field. The Middlebury benchmark [113] has been widely used as the standard dataset to evaluate stereo reconstruction algorithms. However, due to the limited size and lighting conditions of the dataset, the Middlebury benchmark is not sufficient to assess the performance of stereo algorithms in real-world scenarios. Recently, Aanæs et al. [1] introduced the DTU dataset, which contains images of dozens of indoor objects with various materials collected under different lighting conditions, with the images captured by a robotic arm. Although the DTU dataset includes realistic, high-resolution (1600×1200) images, the controlled laboratory environment and the fixed camera trajectory limit the complexity of the dataset. Moreover, deep learning-based models trained based on the DTU dataset may not be robust enough to handle real-world scenarios.

Fig. 5.1 Illustration of the high-resolution (gigapixel-level) property of our GigaMVS benchmark algorithm. The captured gigapixel-level images have both wide FoVs and high resolution, supporting multiscale observations, such as the *palace*-scale scenes and *relief*-scale local details

The recent benchmarks for evaluating 3D reconstruction algorithms have focused on outdoor scenes representing real-world scenarios. The EPFL benchmark [120] includes images of outdoor scenes captured at a high resolution of 6.2 megapixels. However, the benchmark lacks sufficient data for evaluation, as the images include only three building facades. Merrell et al. proposed the UNC dataset [85], which includes realistic images captured under conditions without controlled lighting or set camera poses. However, this dataset includes only one scene, limiting its applicability.

The Tanks and Temples benchmark [70] by Knapitsch et al. includes many indoor and outdoor scenes and offers many video images and 3D models. However, camera poses are not included in this benchmark because this benchmark is aimed at evaluating full reconstruction pipelines, including structure-from-motion and multiview stereo methods. Although the benchmark is supposedly aimed at large-scale scene reconstruction, only one scene has an area of approximately 30,000 m^2, and only three scenes are larger than 300 m^2.

We introduce a novel benchmark for MVS known as the MegaMVS benchmark, which can be used to evaluate both the 3D geometry and texture details of reconstructed scenes. In contrast to the BlendedMVS dataset [144], which uses an online platform to generate 3D models, our benchmark provides ground truth models generated by precise scanners.

In some previous benchmarks, the reconstruction results were evaluated without ground truth 3D models. For example, Chil et al. [22] evaluated the quality of the reconstruction results using human observers, and Waechter et al. [128] compared rendered images with corresponding real images. Although these methods can be used to evaluate both the geometry and texture, the geometry evaluation may not be precise due to the limited views available for the evaluation. In our benchmark, we evaluate the texture details in the reconstruction results using both ground truth 3D models and image-based methods.

High-Resolution Images. We adopt a high-resolution imaging technique to collect our dataset, as this approach provides detailed information for the 3D reconstruction of large-scale outdoor scenes with distant details and wide fields of view. In recent years, with the development of advanced recording systems such as DSLR and array cameras, superhigh-resolution images have become easier to capture. For instance, the Atlas dataset is a human-centric image dataset that contains high-resolution images with resolutions up to 81 megapixels that were collected using a custom-designed camera array. However, the increased resolution poses a significant challenge for existing computer vision algorithms, particularly learning-based methods. Our dataset was inspired by the high-resolution PANDA dataset [132], which contains gigapixel-level images with more than 25000 × 14000 resolution captured by a special camera array, presenting a significant challenge for existing algorithms.

To evaluate the performance of 3D reconstruction algorithms with high-resolution images, we propose a benchmark dataset that contains images captured based on a gigapixel imaging pipeline. This pipeline increases the difficulty of training algorithms based on our benchmark, as the scalability and robustness of the algorithms may be influenced by the high-resolution images with rich details. In contrast, other benchmark datasets such as ETH3D [112] and Yang et al. [141] contain high-resolution images with resolutions up to 24 and 5.07 megapixels, respectively, for stereo matching. Nonetheless, our dataset is unique, containing high-resolution images with wide fields of view, making it an ideal benchmark to evaluate the performance of 3D reconstruction algorithms with superhigh-resolution images.

Ground Truth Geometry

We utilized a high-precision Faro Focus3D laser scanner to obtain the ground truth 3D geometry. With a range of 120 m, 320 × 240 pixel resolution, and an accuracy of 2 mm, the scanner can capture highly detailed point clouds. To ensure full coverage and minimize occlusions, we scanned each scene from multiple viewpoints. For example, for a typical scene such as MoCap Studio, we scanned the scene from 12 different positions.

The point clouds obtained by the scanner were initially noisy and contained a considerable number of outliers. To clean the data, we employed a series of filters, including a statistical noise filter, a voxel grid filter, and an outlier removal filter, which removed most of the noise; however, some artifacts and noise remained.

To remove the remaining noise and artifacts, we developed a manual labeling pipeline that included several steps. First, we clustered the point clouds to separate the data into groups corresponding to individual objects. Then, we carefully examined each cluster and manually removed any outlier points. Finally, we labeled each cluster based on the object that the cluster represented.

To evaluate the quality of our ground truth data, we computed various statistics based on the point clouds. For instance, in the MoCap Studio scene, the average distance between nearest neighbor points was 2.35 mm, while the median distance was 1.93 mm. Additionally, 95% of the points had a nearest neighbor with a distance of less than 5.5 mm, which demonstrated the high density of the ground truth data.

Gigapixel Texture

Conventional high-resolution images are limited by the trade-off between the FoV and resolution, resulting in a lack of large-scale global structures and high-resolution details. In contrast, the gigapixel images of large-scale structures in our dataset contain local high-resolution details and globally wide FoVs. Thus, the gigapixel images include both local details and global structures, demonstrating their effectiveness for various computer vision applications.

To demonstrate the importance of such global structures in the gigapixel images, we conducted experiments using small FoV high-resolution images as inputs. As shown in Fig. 5.2, the state-of-the-art MVS algorithm COLMAP [111] failed to provide globally consistent results due to the lack of global structure in the inputs. COLMAP cannot handle repeated local patterns, resulting in mismatches among local patches. In contrast, when stitched gigapixel images were used as inputs, COLMAP eliminated many inaccurate local matching points, resulting in globally consistent results.

Previous researchers have proposed Gigapan [11, 71] as a method to generate gigapixel images. However, our benchmark dataset faces challenges such as stitching artifacts caused by inaccurate pose estimation and parallax caused by camera displacements, especially for complex scenes captured from close viewpoints. Therefore, we adopted a hybrid method to generate high-quality gigapixel images.

To ensure the accuracy and robustness of our gigapixel images, we used both GigaPan and Canon EOS R6 cameras. GigaPan is suitable for capturing images of distant content, while Canon EOS R6 shows better performance for nearby content due to its high resolution and subpixel scanning technology. We chose a 70 mm lens for normal distance viewpoints and a 400 mm lens for extremely far viewpoints.

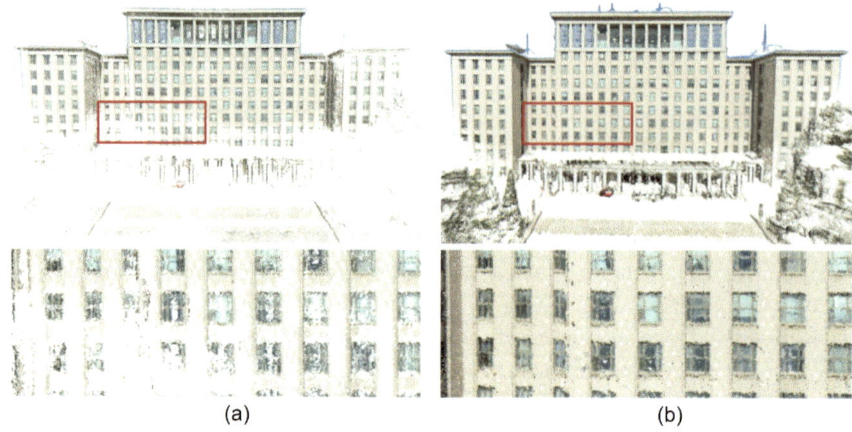

Fig. 5.2 Comparison of COLMAP reconstruction results using **a** the captured standard images and **b** the stitched gigapixel images as inputs

Specifically, for the GigaPan camera (capturing content at distances >5 m), we generated gigapixel images by projecting the images onto the same plane using homography [11]. To improve the stitching accuracy, we employed the state-of-the-art structure-from-motion (SfM) method [110] by COLMAP [111] to estimate the camera matrices and calculate the homography using all images. Traditional image stitching methods, such as the off-the-shelf GigaPan method, address each viewpoint independently without considering the geometrical constraints of all viewpoints. As a result, these methods suffer from inaccurate camera matrix estimation, leading to image warping errors and stitching artifacts for complex scenes. In contrast, the proposed stitching approach estimates the camera matrices based on all images collected from different views. As a result, the proposed method obtains more accurate camera matrices and therefore produces more accurately stitched gigapixel images.

Figure 5.3 presents the pipeline for generating gigapixel images using two different cameras, Sony A9 and Nikon Z7II. The Sony A9 camera is used to capture images of distant scenes, while the Nikon Z7II camera is used to capture images of nearby scenes. To ensure color consistency among the images collected by the two cameras, we use an automatic exposure bracketing method and select the optimal image based on the exposure histogram.

For distant scenes captured by the Sony A9 camera, we use a motor-controlled camera head to capture multiple images and stitch the images together using a state-of-the-art structure-from-motion (SfM) algorithm. The SfM algorithm estimates the camera matrices and calculates the homography based on all images. This approach ensures accurate camera matrix estimation and image stitching results, resulting in high-quality gigapixel images. We use a 70 mm lens for normal distance viewpoints and a 400 mm lens for extremely far viewpoints.

For nearby scenes captured by the Nikon Z7II camera, we utilize a pixel-shift mode to capture multiple images with subpixel shifts and combine the different

(a) Off-the-Shelf GigaPan solution. (b) Our solution.

Fig. 5.3 Comparison of the stitched images produced by the GigaPan method and our approach

images to generate a high-resolution image. This approach is robust when capturing images at close viewpoints, as the camera is fixed while collecting multiple images. However, this mode is sensitive to dynamic objects. Therefore, we only used this camera for scenes in which dynamic objects were avoidable.

To evaluate the stitching error, we project the gigapixel images onto the laser-scanned 3D model to obtain the ground truth images. Then, we measure the average displacement of the matched feature points between the stitched images and their corresponding ground truth images. The average displacement is 0.2356 pixels, indicating the high quality of the stitched gigapixel images.

Scalability

The scalability of MVS algorithms is crucial for large-scale outdoor scenes in which different contents have considerably different scales. Our dataset includes a museum scene that contains both a large-scale building and small-scale sculptures, and the scale variation among these components may be as large as 10^4, as shown in Fig. 5.4 However, existing MVS algorithms based on voxel or point cloud representations have difficulty handling significant scale differences, even when the algorithms are based on well-designed coarse-to-fine strategies [16] or octree structures [106, 123]. This difficulty can potentially be addressed by using continuous implicit representations [20], which present the scene with arbitrary high resolution. By combining adaptive sampling strategies such as level of detail (LOD) [122, 159], the representation and reconstruction results can flexibly and efficiently change with different scales. To evaluate the scalability of MVS algorithms, we use a scale-adaptive accuracy measure that assesses the reconstruction error across different scales.

Fig. 5.4 Ground truth models for representative scenes in our benchmark dataset. For better visualization, obstructing trees are rendered transparent

Gigapixel

The high-resolution images in our dataset include fine-grained object details, while the wide field-of-view (FoV) of the gigapixel images allows the global structure of the scene to be captured. As shown in Fig. 5.2, the gigapixel images allow us to reconstruct the scene with higher accuracy and more completeness than when only high-resolution images are used because the gigapixel images contain more context, enabling a better understanding of the overall scene. Recently, there have been efforts to develop novel gigapixel-level scene reconstruction methods that use deep learning-based approaches to generate high-quality gigapixel-level 3D models.

Sparsity

Capturing images of large-scale outdoor scenes can be a tedious task. Using a sparse camera setup is a practical and cost-effective alternative; however, the complex occlusions and varied scales in the scene can increase the difficulty of using these images as input for reconstruction algorithms. Figure 5.5 shows an evaluation of the performance of different methods for scenes with various sparsities. Our results indicate that existing algorithms have difficulty reconstructing ultralarge-scale scenes with sparse image inputs. However, recent studies on very sparse MVS algorithms [60] have shown promising results. These studies used geometry-aware view-selection strategies to address the challenges posed by sparse setups. We propose the GigaMVS dataset as a benchmark to evaluate and improve the performance of these algorithms. Since sparse MVS is particularly challenging for ultralarge-scale outdoor scenes, we believe that our dataset is important and may lead to more valuable insights.

Occlusion

Real-world large-scale scenes are known to contain complex occlusions that pose significant challenges to the robustness of reconstruction algorithms. For instance, the representative scenes in our benchmark, such as Haiyan halls and great fountains, contain severe occlusions. Figure 5.6 shows that state-of-the-art learning-based methods such as R-MVSNet [143], which fuses multiview features that contain irrelevant occluded views, produce poorly estimated depth maps. To address this issue, more advanced occlusion-aware MVS methods [82] are urgently needed.

We highlight the importance of identifying occlusions with MVS algorithms and emphasize the need for more advanced algorithms to address severe occlusion issues. The representative scenes in our benchmark can serve as valuable resources for evaluating the performance of such algorithms.

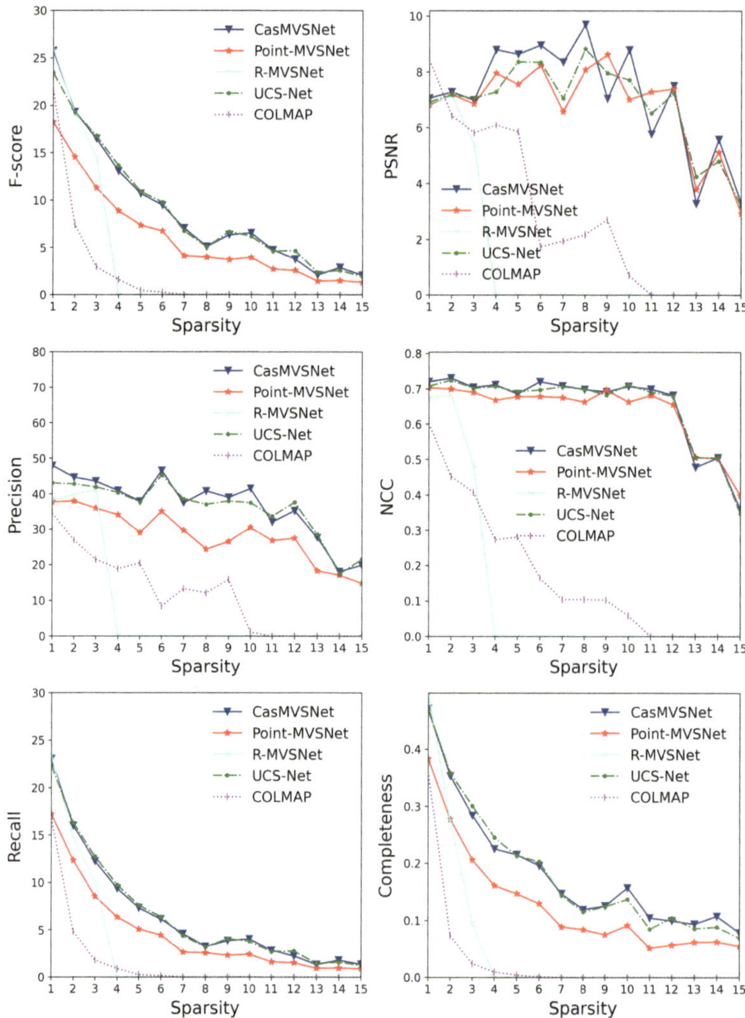

Fig. 5.5 The performance of state-of-the-art methods under different view sparsities. The results are averaged over 13 scenes

Efficiency

Computational efficiency is a major challenge in handling ultralarge-scale scenes and gigapixel images. Traditional methods such as COLMAP require substantial computing time and resources since they iteratively reconstruct 3D scenes in a pixel-by-pixel manner. In contrast, learning-based methods use parallel computations for inference since convolutional networks can simultaneously address many pixels in larger receptive fields. However, these methods have slow training speeds and large

Fig. 5.6 Illustration of failed depth map prediction obtained by a learning-based method (R-MVSNet) compared with that obtained by the traditional method (COLMAP)

memory requirements. To address the high computational complexity of existing algorithms, structure-adaptive reconstruction methods are potential solutions. These approaches allocate fewer computing resources to regions with simple structures, such as walls and floors, and more computing resources to regions with complex structures and fine details, thereby achieving better modeling results [77, 78, 158].

Another promising solution is semantic-aware reconstruction. Region-based stereo matching methods use segmentation and classification [29, 158] algorithms to significantly reduce the number of possible stereo pairs by focusing on semantic-aware regions with strong geometric primitives. These approaches offer better scalability and improved time efficiency since they reduce the computational loads for less significant regions.

We propose that structure-adaptive and semantic-aware reconstruction methods can be combined to obtain better results and improved computational efficiency for ultralarge-scale scenes and gigapixel images. Additional research in this area will be valuable in improving the efficiency and accuracy of 3D reconstruction algorithms.

Failure Cases

While learning-based MVS methods have achieved great success based on some datasets, such as DTU and T&T, these methods often struggle to generate accurate results in complex regions. For example, as shown in Fig. 5.6, R-MVSNet fails to estimate depth maps with reasonable 3D structures for the Ruins of the Old Summer Palace (Great Fountain) dataset. Although R-MVSNet can successfully predict the depth map for the Central Main Building, it fails to generate the thin structure of the flagpole. The lack of diversity in the training data severely affects the robustness and stability of learning-based methods.

Therefore, existing benchmark datasets and algorithms can be improved further. Future studies should focus on improving the robustness and stability of learning-based methods and increasing the diversity of training data to enhance the ability of existing methods to handle complex regions.

5.2 Sparse-View Reconstruction for Large-Scale Scenes

5.2.1 SurfaceNet+: Very Sparse Multiview Stereopsis

Multiview stereopsis under dense observations has gained considerable interest in the fields of light-field imaging and rendering due to its potential to recover accurate 3D geometric models based on 2D image sets. Its benefits, including its resistance to occlusions [134, 149] and its ability to reduce image noise [9, 24], have been extensively researched. Some studies have focused on enhancing the depth map generation ability using semantic embeddings [19, 41, 79] or object-level shape priors [19, 41, 58, 64, 79, 86]. Other approaches [16, 46, 98, 111, 142, 143] have attempted to estimate the depth maps of each input view and fuse the different maps in 3D models. SurfaceNet [59], a representative voxel-based reconstruction methods, improves the 3D geometry by directly learning the volume-wise geometric context based on unprojected 3D color volumes in an end-to-end manner.

However, densely sampled images are usually required to obtain good reconstruction accuracy, which is impracticable for large-scale scenes. The sparser the observations, the more 3D information of the imaged scene is lost during the sensing process, increasing the difficulty of MVS tasks. On the other hand, sparser observations with wider image baselines are more practical and cost-effective; however, these data also increase the difficulty of MVS tasks because dense correspondence matching is more difficult with wider baseline angles.

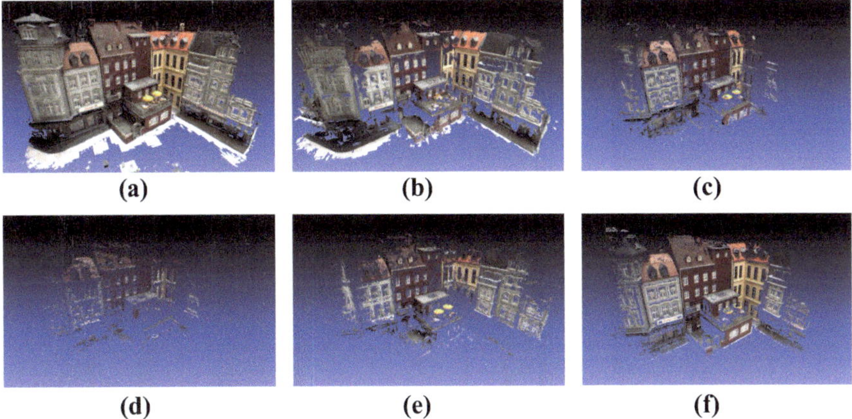

Fig. 5.7 Illustration of a very sparse MVS setting using only one seventh of the camera views, i.e., $\{v_i\}_{i=1,8,15,22,...}$, to recover the model 23 in the DTU dataset [1]. Compared with the state-of-the-art methods, the proposed SurfaceNet+ provides much complete reconstruction, especially around the boarder region captured by very sparse views. **a** SurfaceNet+. **b** SurfaceNet [59]. **c** Gipuma [37], **d** R-MVSNet [143], **e** Point-MVSNet [16], **f** COLMAP [111]

Ji et al. [60] proposed the concept of sparse MVS and developed an approach for addressing general sparse MVS tasks with a large range of baseline angles reaching up to 70°. As illustrated in Figs. 5.7 and 5.8a, under the sparse configuration, the consistency among the images is only marginally satisfied based on depth predictions obtained from two views with a wide baseline angle. Although 3D regularization methods can be directly applied to address the shortcomings of depth map fusion methods, volume-based methods such as SurfaceNet [59] are influenced by noisy surfaces and large holes near regions with repeating patterns and complex geometries.

Ji et al. proposed SurfaceNet+ [60] to address the constraints of previous MVS algorithms. SurfaceNet+ is an end-to-end learning framework that can adapt to various degrees of sparsity. The key contributions include a trainable occlusion-aware view selection technique that considers geometric priors via a coarse-to-fine scheme. This volume-wise view selection strategy considerably improves the performance of learning-based volumetric MVS methods.

In the following sections, the architecture of SurfaceNet+ [60] is presented. Then, the network and optimization strategy are analyzed in detail. Finally, the benchmark based on the sparse-MVS approach is presented, and the performance of the SurfaceNet model is evaluated in detail.

Architecture of the Sparse View Reconstruction Approach

As the sparsity of the observations increases, the number of available consistent image views decreases; hence, the view selection strategy becomes more critical. SurfaceNet+ includes a novel trainable occlusion-aware view selection strategy that

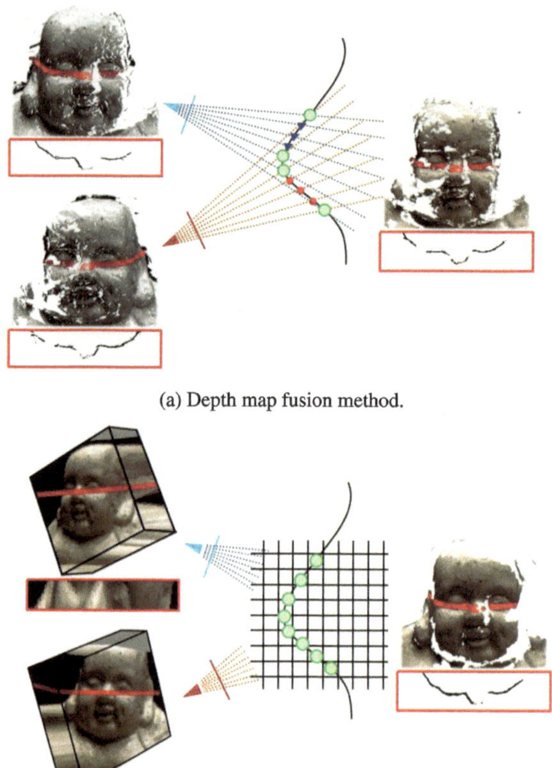

(a) Depth map fusion method.

(b) Volumetric method

Fig. 5.8 Illustration of two types of multi-view reconstruction methods. The front view of the 3D model and the top view of the selected region (red) are shown in pair. The circles (green) indicate the prediction. **a** Because the 2D image unevenly samples the 3D surface, as the baseline angle increases, it is rare for view pair (red and blue) to have intersected rays during depth fusion. The 2D regularization gets less helpful. **b** Volumetric method optimizes the 3D geometry by directly regularizing in 3D space

considers the geometric prior using a coarse-to-fine approach. In brief, the **multiscale inference process** generates the geometric prior required by the **occlusion-aware view selection** strategy. As illustrated in Fig. 5.9, a very coarse 3D surface is projected from a bounding box by examining all input views. Then, the coarse-level geometry is iteratively refined by gradually rejecting the occluded views based on the coarse geometric prior. The structure of the backbone network, which is fully convolutional, is described in depth in the next section.

Multiscale Inference Process For volume-based reconstruction algorithms, the 3D surface usually includes noisy predictions with recurring patterns [59]. In addition, it

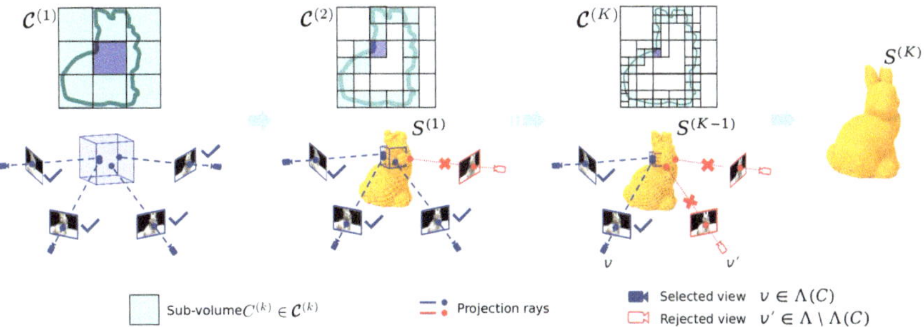

Fig. 5.9 SurfaceNet+ recovers the whole scene $S^{(K)}$ by progressive refinement of the geometric predictions $S^{(k)}$. So that for each sub-volume $C^{(k)} \in \boldsymbol{C}^{(k)}$ (drawn as blue cube) the occlusion-aware view selection is performed on the geometric prior. The occluded projection rays are drawn in red and the blue views are the selected ones for reconstruction. In each scale, the volume-wise algorithm only loops through the region in cyan to boost the precision and efficiency

is computationally intensive to investigate the large amount of empty 3D space. While it may seem obvious to examine the 3D geometry prior during the reconstruction process, standard MVS tasks do not include shape priors for the scenario. Therefore, an effective coarse-to-fine approach is utilized to gradually polish the geometric surface based on the idea that the scene's 3D surface occupies a minority of the space.

During the first stage, the entire scene bounding box is split into coarse subvolumes $\boldsymbol{C}^{(1)}$ with size $l^{(1)} = s \cdot r^{(1)}$, where $r^{(1)}$ is the resolution of the voxel at the coarsest level and the voxelization operation forms a tensor of size $s \times s \times s$. The tensor size depends on the GPU memory; for example, s values of 32, 64, and 128 are chosen here. The estimated surface at the coarsest level is the output of this stage, which is denoted as $S^{(1)}$, where $x \in S$ indicates an occupied voxel in the surface prediction.

Then, the scene is divided into several subvolumes with different resolutions, i.e., $\{\boldsymbol{C}^{(2)}, \ldots, \boldsymbol{C}^{(k)}, \ldots, \boldsymbol{C}^{(K)}\}$, with an iterative approach, with the resolutions defined in a geometric sequence with the common ratio δ, i.e., $r^{(k)} = \delta \cdot r^{(k+1)}$. Usually, $\delta = 4$ is used to balance effectiveness and efficiency. This procedure is executed iteratively until the condition $r^{(K)} \leq r$ is satisfied, where r is the predefined resolution and $r^{(K)}$ is the finest resolution when the iteration stops. The subvolume division strategy is highly dependent on the predicted point cloud at the coarse level $S^{(k-1)}$, where $k = 2, 3, \ldots$, with each subvolume $C^{(k)} \in \boldsymbol{C}^{(k)}$ containing at least one point:

$$\begin{aligned} \boldsymbol{C}^{(k)} = \arg\min_{\boldsymbol{C}} \{ |\boldsymbol{C}| \, | \forall \boldsymbol{C} : \\ (S^{(k-1)} \subseteq \bigcup \boldsymbol{C}) \wedge (\forall C \in \boldsymbol{C} : S^{(k-1)} \cap C \neq \varnothing)\}, \end{aligned} \tag{5.1}$$

where $|\boldsymbol{C}|$ denotes the number of subvolumes, and $\bigcup \boldsymbol{C}$ is represents the union of all the subvolumes, i.e., $\bigcup_{C \in \boldsymbol{C}} C$. To eliminate the border effect in the convolution process, we typically relax the aforementioned restriction and permit a small overlap between adjacent subvolumes. In the following paragraph **Network Architecture**, the point cloud output $S^{(k)}$ of SurfaceNet+ is discussed.

Trainable Occlusion-Aware View Selection Approach As demonstrated in Fig. 5.7, although SurfaceNet [59] prevents artifacts caused by uneven sampling from the 3D surface to the 2D depth, large holes are formed in regions with complicated geometries. Thus, for the sparse MVS problem, view selection methods are extremely important. Using the same annotation as SurfaceNet [59], we introduce a trainable occlusion-aware view selection scheme, which can rank and select the top-N_v most valuable view pairs $\boldsymbol{V}_C$ for each subvolume C based on all the possible view pairs as follows:

$$\boldsymbol{V} = \{(v_i, v_j)|(v_i, v_j \in \Lambda) \wedge (v_i \neq v_j)\}, \tag{5.2}$$

These view pairs are selected based on the learned relative weights $w_C^{(v_i,v_j)}$, which are inferred based on the geometric and photometric priors. Note that the multi-scale scheme provides the crucial geometric prior $S^{(k-1)}$. Consequently, according to Eq. 5.7, the surface for each subvolume C contains $|\boldsymbol{V}_C| = N_v$ predictions.

The geometric prior can be easily determined based on the multiscale predictions. For any camera view v w.r.t. each subvolume $C \in \boldsymbol{C}$, a convex hull $H(C, v) \subset \mathbb{R}^3$ is uniquely defined by a set of points:

$$H(C, v) = Conv(\Gamma(C) \cup \{o_v\}), \tag{5.3}$$

where o_v is the camera center of v, and the set $\Gamma(C) = \{c_1, c_2, \dots, c_8\}$ contains the 8 corners of C.

The more points in the coarse surface prediction $S^{(k-1)}$ that appear in the region between camera view v and subvolume $C^{(k)}$, the more likely it is that view v is occluded. These barrier points are defined in the set

$$B(C^{(k)}, v) = S^{(k-1)} \cap H(C^{(k)}, v) \backslash C^{(k)}. \tag{5.4}$$

As indicated in [59], the end-to-end trainable relative weights not only enhance the algorithm efficiency by filtering out the majority of the less important view pairs for each subvolume but also improve the effectiveness of the surface prediction via the weighted fusion strategy. The number of viable views for each subvolume in a sparse MVS problem may be too small for heuristic occlusion detection. Instead, a view pair selection strategy is proposed that can be trained to consider occlusions by utilizing prior knowledge of both the geometry and light conditions:

$$w_C^{(v_i,v_j)} = p_{C^{(k)}}^{(v_i,v_j)} \cdot r\left(\theta_C^{(v_i,v_j)}, e(C, I_i), e(C, I_j)\right), \tag{5.5}$$

where the photometric priors are the same as those in SurfaceNet [59], i.e., the baseline angle $\theta_C^{(v_i,v_j)} = \angle o_{v_i} o_C o_{v_j}$ and the embeddings $e(\cdot)$ of the cropped patches near the 2D image of o_C in both I_i and I_j. Moreover, the geometric prior is encoded as the probability of a view not being occluded, i.e., :

$$p_{C^{(k)}}^{(v_i,v_j)} = \exp\left(-\alpha \cdot r_k^2 \cdot (|B(C^{(k)}, v_i)| + |B(C^{(k)}, v_j)|)\right), \tag{5.6}$$

where α is a hyperparameter that controls the sensitivity of the occlusion probability term and the coefficient r_k^2 is a normalization term w.r.t. different scales. In Section **Sparse MVS Benchmark**, the effect of α and its effectiveness in improving the reconstruction performance is discussed.

Finally, the same fusion scheme as applied in SurfaceNet [59] is utilized; the scheme ranks and selects a small subset of view pairs $\boldsymbol{V}_C$. Subsequently, p_x represents the confidence that a voxel x is on the surface, which is inferred based on the weighted average of the predictions $p_x^{(v_i,v_j)}$:

$$p_x = \frac{\sum_{(v_i,v_j)\in \boldsymbol{V}_C} w_C^{(v_i,v_j)} p_x^{(v_i,v_j)}}{\sum_{(v_i,v_j)\in \boldsymbol{V}_C} w_C^{(v_i,v_j)}}, \tag{5.7}$$

where $\boldsymbol{V}_C$ denotes the set of selected view pairs with a size of $|\boldsymbol{V}_C| = N_v$, and the relative weight $w_C^{(v_i,v_j)}$ for each view pair is an end-to-end trainable parameter that can be inferred based on Eq. 5.5. A smaller N_v value may lead to more efficient and less effective outcomes, as discussed in the later ablation study.

Network and Optimization

At every reconstruction stage, a 3D convolutional neural network is used to predict whether each voxel in the subvolumes lies on the surface. Specifically, given C^k and the corresponding image view pairs (I_i, I_j), a Gaussian kernel is used to blur each image to propagate the local texture information around the large receptive field to guarantee multiview consistency among the different stages. The 3D subvolume $(I_i^{C^{(k)}}, I_j^{C^{(k)}})$ constructed by projecting a view pair is shown in Fig. 5.8b. This representation implicitly encodes both the camera parameters and the scale information, thereby allowing adaption in the fully convolutional neural network.

The detailed structure of the designed network is demonstrated in Fig. 5.10. The model consists of a UNet-like network and a refinement network. Two subvolumes $(I_i^{C^{(k)}}, I_j^{C^{(k)}})$ that contain two RGB color values are taken as input to generate a six-channel tensor of size $6 \times s \times s \times s$, and the probability that each voxel $p_{x\in C^{(k)}}$ is on the surface is precited, forming a tensor of size $1 \times s \times s \times s$. To extract hierarchical geometry information at different scale levels, a pyramid structure is used to process the receptive field features at various scales. After concatenating different scale features, the features are passed through two 3×3 convolution layers, followed by a one-channel convolution layer with the sigmoid function to better aggregate the multiscale information. Similar to [142], a prediction refinement network is employed at

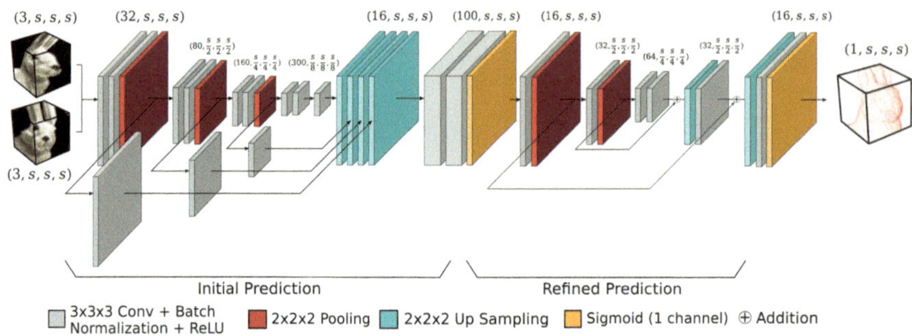

Fig. 5.10 The network design of SurfaceNet+. The network input is two subvolumes with a size of (3, s, s, s) that are not projected from different views. The final prediction is a one-channel tensor containing the probability that each voxel is found on the surface

the end of the previous network. Based on the initial output $\tilde{S}_C^{(k)}$, which is a tensor of size $1 \times s \times s \times s$, the skip connections in each layer are used to learn the residual prediction and produce the final output $S_C^{(k)}$.

Loss The training loss is composed of two components that penalize the initial prediction p_x and the refined prediction p'_x. The first stage involves comparing the discriminative prediction for each voxel p_x with the ground true value $\hat{s}_x$. Since most of the voxels are not on the surface ($\hat{s}_{x \in C^{(k)}} = 0$), a class-balanced cross-entropy function is used, i.e., , for each $C^{(k)}$, we have

$$L_{init} = - \sum_{x \in C^{(k)}} \left\{ \beta^{(k)} \hat{s}_x \log p_x + (1 - \beta^{(k)})(1 - \hat{s}_x) \log(1 - p_x) \right\}, \tag{5.8}$$

where the hyperparameter $\beta^{(k)}$ is the ground truth occupancy ratio at scale k.

In the second stage, the refined prediction p'_x is regressed to the ground truth based on the mean square error (MSE). This is performed to penalize small residues as follows:

$$L_{refine} = \sum_{x \in C} \left\| \hat{s}_x - p'_x \right\|_2, \tag{5.9}$$

where $p_x \in S_C^{(k)}$. Consequently, the training loss is defined as

$$L_{total} = L_{refine} + L_{init}. \tag{5.10}$$

Sparse MVS Benchmark

In this section, the imperative sparse-MVS approach is applied to different datasets, including the DTU dataset [1], the Tanks and Temples (T&T) dataset [69], and the ETH3D low-res dataset [112]. The results are extensively compared with the results of existing MVS methods at various observation sparsity levels.

In addition, it is practical to sample small batches of images at sparse viewpoints, i.e., grouping batches of views with certain batch sizes with the previously defined sparse viewpoints for a certain sparsity level. When $sparsity = 3$ and $batchsize = 1$, the chosen camera indices are 1/4/7/10/$\cdots$. When $sparsity = 3$ and $batchsize = 2$, the chosen camera indices are 1,2/4,5/7,8/10,11/$\cdots$.

Figure 5.11 shows the relationship between the sparsity n and the average baseline angle $\bar{\theta}$ averaged over all the ground truth points for the 22 models in the DTU dataset, 8 models in the Tanks and Temples dataset, and 5 models in the ETH3D low-res dataset. Note that the baseline angle data are calculated using only the nearest view pairs for simplicity.

$$\theta = \{\angle o_{v_i} x o_{v_j} | x \in \hat{S}, v_i \in \Lambda, v_j = \arg\min_{v \in \Lambda} \overline{o_{v_i} o_v}\} \tag{5.11}$$

The average baseline angle $\bar{\theta}$, which is defined based on the intersecting projection rays, steadily increases as the sparsity increases $n = 1, \ldots, 11$. For example, the average baseline angle reaches more than 70° for both the DTU and T&T datasets. The proposed sparse-MVS configuration is reasonable since it not only accounts for various sparsity levels but also includes irregular sampling sites due to the positive correlation between n and $\bar{\theta}$.

DTU Dataset [1] The performance of the different methods based on the DTU dataset [1] with different sparse MVS settings is quantitatively compared. The DTU dataset is a large-scale MVS benchmark that includes 80 different scenes acquired at 49 camera locations under seven different lighting conditions. This dataset includes

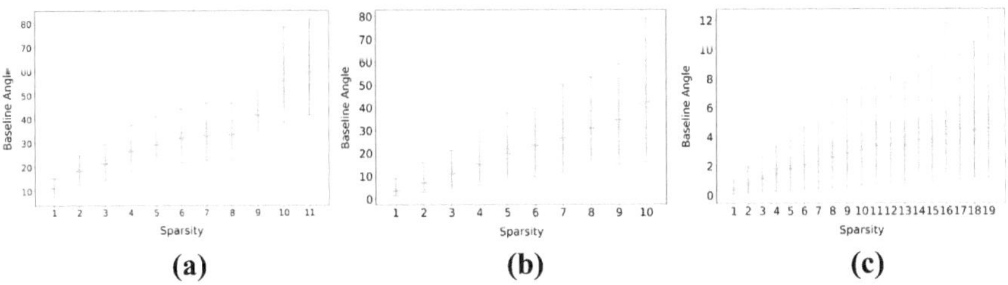

Fig. 5.11 The relationship between sparsity and the average baseline angle over all the models in **a** the DTU dataset [1], **b** the Tanks and Temples dataset [69] and **c** the ETH3D low-res dataset [112]

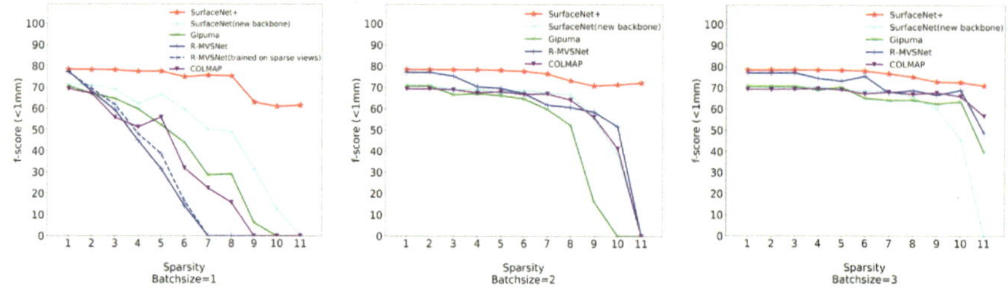

Fig. 5.12 Comparison with the existing methods in the DTU Dataset [1] with different sparsely sampling strategy. When $Sparsity = 3$ and $Batchsize = 2$, the chosen camera indexes are 1,2 / 4,5 / 7,8 / 10,11 / .... SurfaceNet+ constantly outperforms the state-of-the-art methods at all the settings, especially at the very sparse scenario

various items and materials. Following [59], 22 models in the DTU dataset are chosen as the evaluation set.[1]

The performance for various sparsity levels is shown in Fig. 5.12 in terms of the *F1 score* (1 *mm*), which combines the *recall* and *precision*. The results demonstrate that in all sparse environments, the proposed strategy consistently outperforms other algorithms. Surprisingly, SurfaceNet+ consistently performs well with only negligible degradation, especially in the situation with $\bar{\theta} < 40°$. SurfaceNet+ exhibits a slight performance degradation in the extremely sparse scenario when $\bar{\theta} > 50°$. However, the depth fusion approaches only generate a small number of points as outputs. The same depth fusion method is used for depth map-based techniques, including R-MVSNet [143] and Gipuma [37]. To ensure a fair comparison, we adjusted the hyperparameters of the depth fusion algorithm to produce better F1 scores under 1 *mm* at each sparsity level.

Table 5.1 shows the quantitative results, with 3 matrices used for the evaluation. The distance metric [1] and the percentage metric [69] are two measures representing the precision and recall. The F1 score is used to calculate the *overall* percentage score, while the mean precision and mean recall are used to calculate the *overall* distance score. At all levels of sparsity, SurfaceNet+ clearly outperforms existing techniques in terms of the recall and precision. SurfaceNet+ shows nearly constant recall and high precision, in contrast to the other approaches, for which the performance rapidly decreases as the sparsity increases.

Figure 5.13 illustrates a qualitative comparison of SurfaceNet+ with two other approaches, R-MVSNet [143] and Gipuma [37], demonstrating that SurfaceNet+ accurately reconstructs the scenes while showing relatively high recall. When

[1] Following the same dataset split as in SurfaceNet [59]. Training: 2, 6, 7, 8, 14, 16, 18, 19, 20, 22, 30, 31, 36, 39, 41, 42, 44, 45, 46, 47, 50, 51, 52, 53, 55, 57, 58, 60, 61, 63, 64, 65, 68, 69, 70, 71, 72, 74, 76, 83, 84, 85, 87, 88, 89, 90, 91, 92, 93, 94, 95, 96, 97, 98, 99, 100, 101, 102, 103, 104, 105, 107, 108, 109, 111, 112, 113, 115, 116, 119, 120, 121, 122, 123, 124, 125, 126, 127, 128. Validation: 3, 5, 17, 21, 28, 35, 37, 38, 40, 43, 56, 59, 66, 67, 82, 86, 106, 117. Evaluation: 1, 4, 9, 10, 11, 12, 13, 15, 23, 24, 29, 32, 33, 34, 48, 49, 62, 75, 77, 110, 114, 118.

Table 5.1 Quantitative results of reconstruction quality on the DTU dataset in terms of the distance metric (lower is better) and the percentage metric [69] (higher is better) with 1 mm and 2 mm as thresholds. SurfaceNet+ constantly outperforms the state-of-the-arts in all the sparse-MVS settings with $n = 1, 3, 5, 7, 9, 11$

Sparsity	Method	Mean distance (mm)			Percentage (<1mm)			Percentage (<2mm)		
		Precision	Recall	**Overall**	Precision	Recall	**f-score**	Precision	Recall	**f-score**
1	SurfaceNet+	0.385	**0.448**	**0.416**	88.01	**73.01**	**78.44**	92.33	**78.1**	**83.55**
	SurfaceNet [59]	0.450	1.021	0.735	84.49	64.58	71.65	89.10	68.72	76.21
	Gipuma [37]	**0.283**	0.873	0.578	**94.65**	59.93	70.64	**96.42**	63.81	74.16
	R-MVSNet [143]	0.383	0.452	0.417	87.63	72.48	77.09	91.74	76.39	82.01
	COLMAP [111]	0.411	0.657	0.534	82.24	52.48	61.34	88.26	62.20	72.93
3	SurfaceNet+	0.446	**0.482**	**0.464**	86.06	**74.41**	**78.15**	90.87	**78.25**	**82.91**
	SurfaceNet	0.461	0.997	0.729	83.02	61.09	68.87	88.31	66.39	74.41
	Gipuma	**0.267**	1.252	0.759	**95.51**	50.88	64.63	**97.49**	50.33	63.68
	R-MVSNet	0.465	1.012	0.738	89.55	48.03	59.28	96.96	57.92	69.04
	COLMAP	0.467	1.090	0.778	78.45	49.26	59.62	91.44	55.98	65.77
5	SurfaceNet+	0.446	**0.491**	**0.469**	88.58	**71.63**	**77.48**	92.86	**76.04**	**82.28**
	SurfaceNet	0.445	0.948	0.701	81.07	58.62	66.55	85.40	62.76	70.97
	Gipuma	0.460	1.633	1.046	**92.38**	38.53	52.36	**95.10**	48.15	61.78
	R-MVSNet	**0.329**	2.209	1.269	89.26	20.51	31.60	93.99	32.74	46.37
	COLMAP	0.443	1.284	0.863	88.79	42.51	55.94	92.91	54.89	65.77
7	SurfaceNet+	**0.435**	**0.524**	**0.479**	**91.36**	**72.23**	**75.59**	**95.21**	**76.54**	**81.86**
	SurfaceNet	0.688	1.130	0.909	66.86	36.91	50.24	69.21	46.91	61.70
	Gipuma	0.569	1.770	1.169	85.35	17.91	28.66	90.78	28.00	41.31
	R-MVSNet	Empty	Empty	Empty	Empty	Empty	Empty	Empty	Empty	Empty
	COLMAP	0.545	1.756	1.150	59.28	15.14	22.46	80.92	31.56	41.89
9	SurfaceNet+	**0.441**	**0.895**	**0.668**	**85.99**	**53.16**	**63.01**	**89.86**	**57.63**	**67.86**
	SurfaceNet	1.112	2.176	1.644	35.84	29.53	31.47	38.36	34.01	35.49
	Gipuma	Empty	Empty	Empty	Empty	Empty	Empty	Empty	Empty	Empty
	R-MVSNet	Empty	Empty	Empty	Empty	Empty	Empty	Empty	Empty	Empty
	COLMAP	Empty	Empty	Empty	Empty	Empty	Empty	Empty	Empty	Empty
11	SurfaceNet+	**0.445**	**0.880**	**0.663**	**85.81**	**51.52**	**61.54**	**90.05**	**55.41**	**65.99**
	SurfaceNet	Empty	Empty	Empty	Empty	Empty	Empty	Empty	Empty	Empty
	Gipuma	Empty	Empty	Empty	Empty	Empty	Empty	Empty	Empty	Empty
	R-MVSNet	Empty	Empty	Empty	Empty	Empty	Empty	Empty	Empty	Empty
	COLMAP	Empty	Empty	Empty	Empty	Empty	Empty	Empty	Empty	Empty

$sparsity = 7$, SurfaceNet+ produces a point cloud with a high recall value, particularly in regions with complicated geometries and no texture, demonstrating that an accurate 3D model can be fused with correctly selected nonoccluded views.

To perform sparse sampling in a slightly different manner, three $batchsize$ values $\{1, 2, 3\}$ are assessed, as shown in Fig. 5.12. SurfaceNet+ consistently outperforms the other methods, although the performance of the depth-fusion approaches (Gipuma [37], R-MVSNet [143], and COLMAP [111]) improves as the $batchsize$ increases. Furthermore, SurfaceNet+ outperforms the other approaches as the dis-

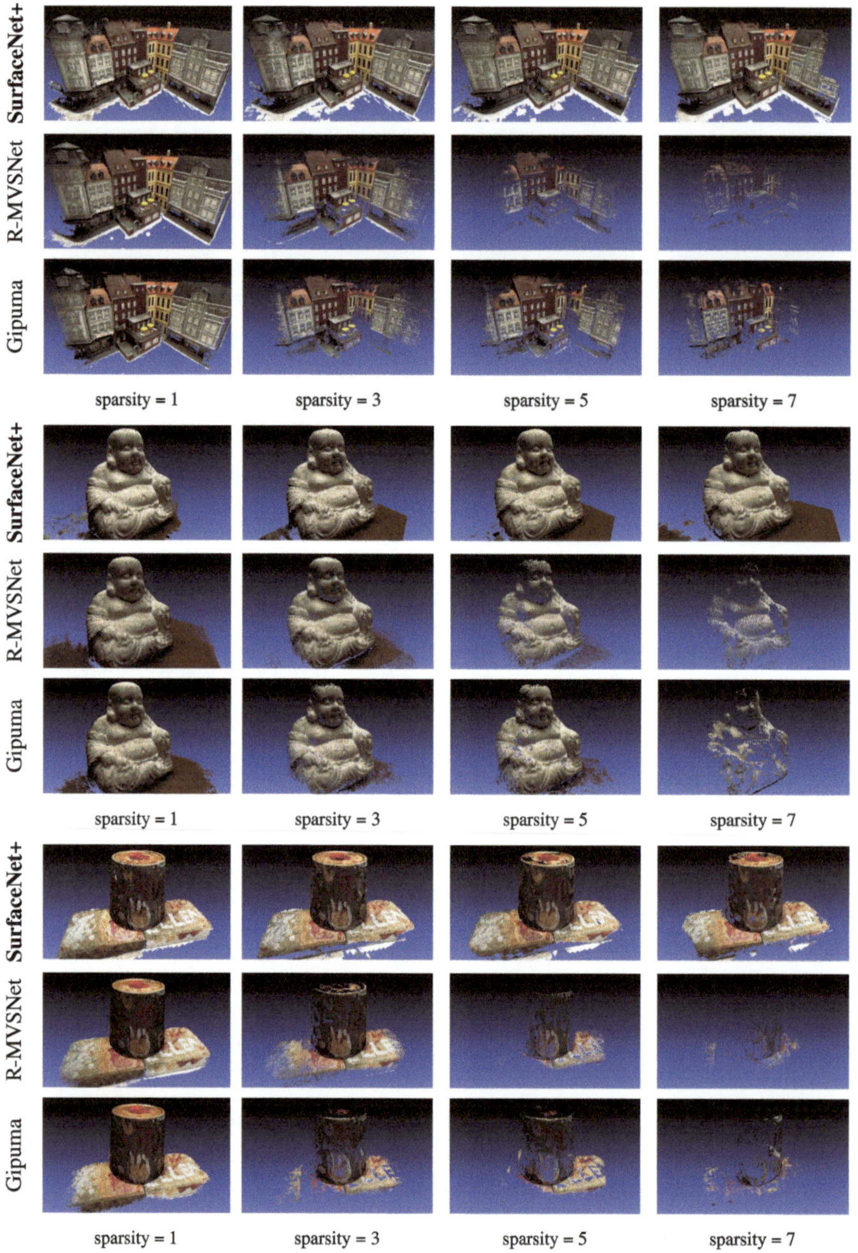

Fig. 5.13 Quanlitative results of three scans 1, 23 and 114 of the DTU dataset compared with R-MVSNet [143] and Gipuma [37]. SurfaceNet+ shows superior performance, particularly with its stable recall quality in sparse cases. Note that the reconstruction of SurfaceNet+ corresponds to the highest completeness and overall quality as seen in Fig. 5.12 and Table 5.1

parity increases. R-MVSNet was retrained for sparse MVS tasks using randomly chosen nonoccluded view pairs at a $batchsize$ of 1. The improvement in the F1 score is minimal, as demonstrated in Fig. 5.12. The wide baseline angles in the very sparse MVS problem lead to drastically skewed matching patches across views that significantly complicate the dense correspondence problem since the depth fusion-based MVS methods (R-MVSNet) rely more on the photometric consistency in the input observations. By immediately assuming a 3D surface based on each unprojected 3D subvolume, learning-based volumetric MVS systems such as SurfaceNet+ bypass the 2D correspondence searching problem. This may explain why learning-based volumetric methods outperform depth fusion-based systems for highly sparse MVS tasks. In the experiments, the depth fusion strategy and hyperparameters of the R-MVSNet and Gipuma methods are tuned to obtain better performance in terms of the F1 score < 1 mm at all sparsity levels. More specifically, as there is a trade-off between accuracy and completeness, we select depth fusion parameter values that lead to high accuracy at sparsity values of 1 and 2 and good completeness at sparsity values ≥ 3, as noted for Gipuma [37]. The remaining portion is unchanged from the R-MVSNet [143] and Gipuma [37] papers. All parameters in COLMAP [111] were implemented with their default settings.

Tanks and Temples Dataset [69] The Tanks and Temples dataset [69] includes many real-world scenes with complicated lighting conditions. The qualitative results of R-MVSNet [143] and COLMAP [111] based on the Tanks and Temples dataset [69] were compared, as shown in Fig. 5.14. The experimental results show that the proposed strategy is effective at various levels of sparsity. The SurfaceNet+ results ($sparsity = 1$) were evaluated based on the online leaderboard. Overall, SurfaceNet+ outperforms R-MVSNet [143], MVSNet [142], COLMAP [111], and Point-MVSNet [16], as shown in Table 5.2. The following table lists all of the top non-anonymous methods on the leaderboard for comparison.

Generalization Based on the ETH3D Dataset [112] In addition, the generalizability of the proposed technique was evaluated based on the ETH3D dataset [112]. The model was trained based on the DTU training dataset, and no network fine-tuning was performed. Figure 5.15 displays the results for the low-resolution scenes. Because the images were captured by the camera with minimal camera movement, the baseline angle in the ETH3D dataset is small among the different camera views. The relationship between the sparsity and the average baseline angle for all the models in the ETH3D low-resolution training set is illustrated in Fig. 5.11c. It should be noted that the ETH3D dataset might not be appropriate for the sparse MVS benchmark because the average baseline angle is much less than $8°$ (Fig. 5.16).

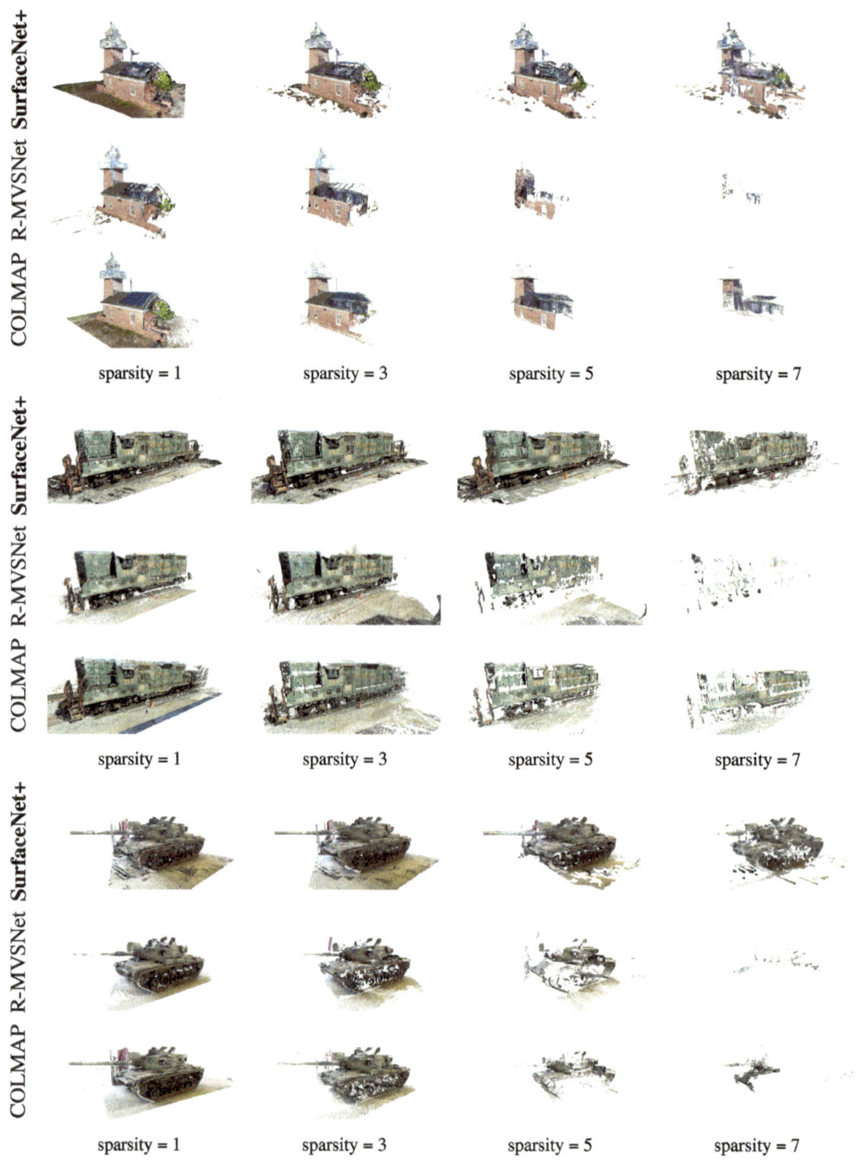

Fig. 5.14 Results of three models in Tanks and Temples 'intermediate' set [69] compared with R-MVSNet [143] and COLMAP [111], which demonstrate the power of SurfaceNet+ of high recall prediction in sparse-MVS

Table 5.2 The top and non-anonymous methods on the Tanks and Temples (T&T) dataset [69] leaderboard. The average rank of SurfaceNet+ is higher than that of R-MVSNet [143], MVSNet [142], COLMAP [111], and Point-MVSNet [16]

Method	Average rank	Mean	Family	Francis	Horse	Lighthouse	M60	Panther	Playground	Train
ACMM [137]	14 00	57.27	69.24	51.45	46.97	63.20	55.07	57.64	60.08	54.48
CasMVSNet [46]	15 75	56.84	76.37	58.45	46.26	55.81	56.11	54.06	58.18	49.51
ACMH [137]	22 25	54.82	69.99	49.45	45.12	59.04	52.64	52.37	58.34	51.61
UCSNet [21]	22 62	54.83	76.09	53.16	43.03	54.00	55.60	51.49	57.38	47.89
PLC [75]	24 38	54.56	70.09	50.30	41.94	58.86	49.19	55.53	56.41	54.13
SurfaceNet+	36 12	49.38	62.38	32.35	29.35	62.86	54.77	54.14	56.13	43.10
Dense R-MVSNet [143]	41 00	50.55	73.01	54.46	43.42	43.88	46.80	46.69	50.87	45.25
VisibilityAwarePointMVSNet [17]	43 88	48.70	61.95	43.73	34.45	50.01	52.67	49.71	52.29	44.75
Point-MVSNet [16]	44 38	48.27	61.79	41.15	34.20	50.79	51.97	50.85	52.38	43.06
R-MVSNet [143]	46 88	48.40	69.96	46.65	32.59	42.95	51.88	48.80	52.00	42.38
MVSNet [142]	57 50	43.48	55.99	28.55	25.07	50.79	53.96	50.86	47.90	34.69
COLMAP [111]	60 50	42.14	50.41	22.25	25.63	56.43	44.83	46.97	48.53	42.04

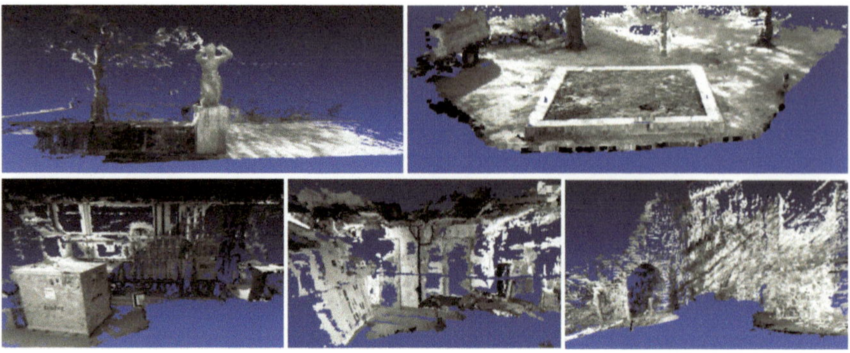

Fig. 5.15 Point cloud reconstructions of the ETH3D low-res dataset [112]

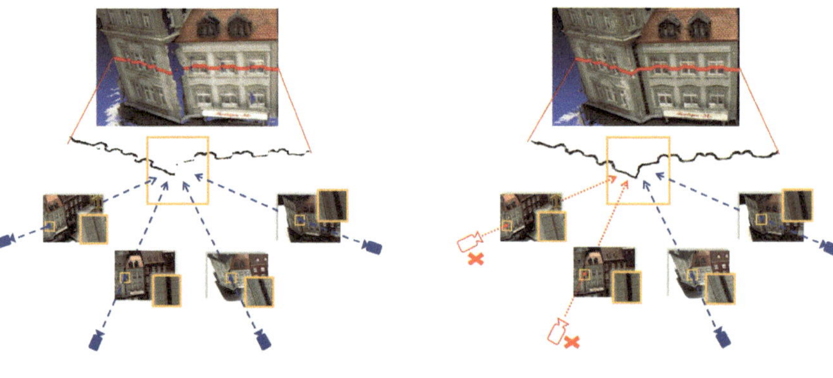

(a) Without occlusion detection

(b) **SurfaceNet+(with occlusion detection)**

Fig. 5.16 Qualitative analysis on occlusion detection module. Top: predicted 3D model with selected region (red). Middle: top view of the selected region. Bottom: illustration of the selected (red)/rejected (blue) views. **a** the algorithm without occlusion detection leads to large hole around complex geometry, bounded by a yellow square. **b** occlusion-aware view selection is performed by considering geometric prior and significantly improves the recall (completeness)

5.2.2 ParseMVS: Structure-Aware Scene Parsing for Sparse-View Reconstruction

There is rich semantic information in macro scenes, which is of great significance for realizing scene reconstruction with higher accuracy and completeness. Semantic-based reconstruction results are also more consistent with human recognition of scenes than MVS methods based on depth fusion [111, 142], points [35] or voxels [59, 60]. Existing works have utilized object classes or model libraries to achieve more compact reconstructions [15, 72]. However, these methods are limited to objects with fixed categories and cannot be easily generalized to large-scale scene reconstruction. In addition to the object-level category information, the lower-level

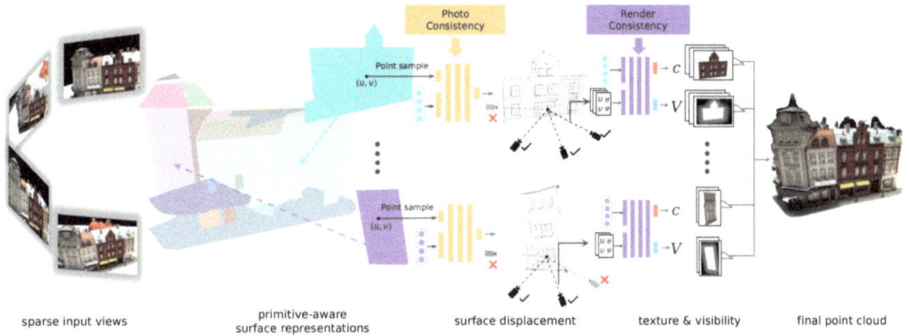

Fig. 5.17 An overview of ParseMVS (**P**rimitive-**A**wa**Re** **S**urface r**E**presentation) and the learning framework for sparse multiview stereopsis. The local point-level geometry, texture, and visibility are learned from two multilayer perceptrons (MLPs) defined in the global primitive-level spaces

geometric semantic/primitive information is meaningful for reconstructing complete scenes, even in scenarios with sparse observations; this topic has been well studied in recent decades [36, 38, 78, 116, 138]. For instance, in global 3D space, lines [116] and planes [36, 78] have been employed as photometric consistency cues. To address more complicated scenes, more sophisticated priors are utilized, including nonplanar regions [38] or polygonal regions [138] (Fig. 5.17).

Unfortunately, due to the lack of pointwise geometric modeling methods, purely primitive-based techniques frequently have low geometric accuracy. On the other hand, recent developments in the field of implicit representations, such as deepSDF [97] and the surface radiance field (SurRF) [154], have enabled the high-quality preservation of fine *local* geometric details.

Cross-view correspondence methods require both strong global primitive-level regularization and local point-level geometric modeling, especially for sparse observations, as noted by Ying et al. [147]. Ying et al. [147] proposed improving sparse MVS performance by learning primitive-aware surface representations, which encode geometry, texture, and visibility details in an integrated manner; this approach is introduced in detail in this section.

Representation

The scene is initially parsed into various geometric primitives, such as planes, spheres, cylinders, and cones, as shown in Fig. 5.18. The method recovers the scene by gradually 'embossing' the primitives based on the local surface information, such as the geometry, texture, and visibility, as opposed to crudely fitting the primitives to the surface. To address the incompleteness and accuracy issues, the locally changed surface attributes in the 2D parametric space are used to preserve global primitive structures while optimizing local details. We first explain how we encode the geometry, color, and visibility information in the scene in a differential manner. Then, we discuss how to address the primitive boundaries before assembling the entire scene.

A scene or object is initially deconstructed into M geometric primitives, each representing a fundamental parametric shape such as a plane, sphere, cylinder, or cone. Each primitive m is assigned a parametric mapping function $P_m : (u, v) \in R^2 \to R^3$ that maps each of the primitive's 2D coordinates (u, v) to a 3D point.

Because primitive m is inherently unbounded, we establish an initial boundary Ω_m for this primitive based on the primitive's supporting point set. Specifically, for each primitive, the output of the efficient RANSAC algorithm is a supporting point set [109] (as discussed in Section **Optimization**). We project the point set onto the 2D parametric space of each primitive and then compute the 2D bounding box of the projected 2D points as the initial boundary Ω_m.

Each primitive m is also assigned an embedding grid γ_m to represent the geometry, texture, and visibility information of the primitive. The embedding grid γ_m consists of k_m embedding vectors (each with vector length l) that are evenly spaced along the initial boundary Ω_m. The whole scene can be represented by M primitives and a set of embedding grids $\Gamma = \{\gamma_1, \gamma_2, \ldots, \gamma_M\}$.

Geometry. To represent the geometry information of the primitives, a multilayer perceptron (MLP) network $r_{\Theta_s} : R^2 \times R^l \to R$ is defined to decode the pointwise surface displacement along the normal vector of the primitives. The following mapping function can be used to represent the geometry of a primitive m:

$$x = P_m(u, v) + r_{\Theta_s}(u, v; \gamma_m^k)\mathbf{n}, \tag{5.12}$$

where γ_m^k is the feature embedding vector calculated by bilinear interpolation based on embedding grid γ_m (Fig. 5.18a). The normal vector can be inferred as

$$\mathbf{n} = \frac{\partial_u P_m \times \partial_v P_m}{\|\partial_u P_m \times \partial_v P_m\|}. \tag{5.13}$$

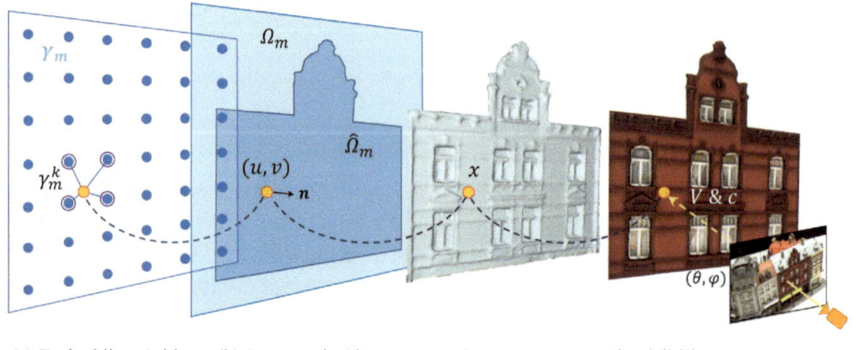

Fig. 5.18 Illustration of our primitive-aware surface representation

Texture. A texture MLP is designed to predict the color of each point based on the primitive. The inputs to the texture MLP are a 2D coordinate (u, v), view direction (θ, ϕ), and the interpolated embedding vector γ_m^k:

$$\begin{aligned} \mathbf{C}_{\Theta_\mathbf{c}} : R^2 \times R^2 \times R^l &\rightarrow R^3 \\ (u, v; \theta, \phi; \gamma_m^k) &\rightarrow (c) \end{aligned} \tag{5.14}$$

Visibility. An additional visibility representation is implemented to address the extreme occlusions that may occur during sparse view sampling. The probability that a certain point on the primitive is visible is represented by the visibility value along each ray. This allows the model to identify the occlusion relations between each primitive and its boundary. The visibility is also modeled by an MLP with the same inputs as the texture MLP:

$$\begin{aligned} \mathbf{V}_{\Theta_\mathbf{v}} : R^2 \times R^2 \times R^l &\rightarrow R^1 \\ (u, v; \theta, \phi; \gamma_m^k) &\rightarrow (V), V \in [0, 1]. \end{aligned} \tag{5.15}$$

Surface Formation The texture and visibility information of each primitive under specific view directions have been modeled, enabling the framework to learn correct surface properties. However, both are defined as radiance fields, which need to be combined to produce a view-independent surface.

The visibility values under all views are combined to determine the exact boundary of each primitive. Essentially, if a point can be seen from at least N_v different input viewpoints, the point is valid. More precisely, we characterize the visible view set $\mathbf{B_m}(u, v)$ of each point as the set of all input view directions for which the visibility is greater than a threshold τ:

$$\mathbf{B_m}(u, v) = \left\{ (\theta_n, \phi_n) \middle| \mathbf{V}_{\Theta_\mathbf{v}}(u, v; \theta_n, \phi_n; \gamma_m^k) \geqslant \tau, n \in N \right\}. \tag{5.16}$$

Then, the boundary of primitive m can be defined as

$$\hat{\Omega}_m = \left\{ (u, v) \middle| (u, v) \in \Omega_m, |\mathbf{B_m}(u, v)| \geqslant N_v \right\}. \tag{5.17}$$

The final color value of the point is determined by taking a weighted average of the colors seen in each of the views in the view set $\mathbf{B_m}(u, v)$:

$$c_m(u, v) = \frac{\sum_{(\theta,\phi)\in\mathbf{B_m}(u,v)} \mathbf{V}_{\Theta_\mathbf{v}}(u, v; \theta, \phi; \gamma_m^k)\mathbf{C}_{\Theta_\mathbf{c}}(u, v; \theta, \phi; \gamma_m^k)}{\sum_{(\theta,\phi)\in\mathbf{B_m}(u,v)} \mathbf{V}_{\Theta_\mathbf{v}}(u, v; \theta, \phi; \gamma_m^k)}. \tag{5.18}$$

The surface $\mathbf{S_m}$ of primitive m can be represented as

$$\mathbf{S_m} = \left\{ (x, c) \middle| x = P_m(u, v) + r_{\Theta_s}(u, v; \gamma_m^k)\mathbf{n}, \right. \\ \left. c = c_m(u, v), (u, v) \in \hat{\Omega}_m \right\}, \tag{5.19}$$

and the union of M surfaces is exactly the final result: $\mathbf{S} = \bigcup_M \mathbf{S}_m$.

Optimization

In this section, a comprehensive overview of the whole pipeline is described. Given a set of sparsely sampled images as input, an incomplete point cloud with a normal distribution is generated by the PMVS model [35]. Then, rough primitives with geometric parameters and sets of supporting points are extracted from the initial point cloud by running the efficient RANSAC algorithm [109] (Fig. 5.19).

During model training, 2D points (u, v) and embedding vectors γ_m^k are sampled based on each primitive m as the input to the geometry network Θ_s. The network Θ_s

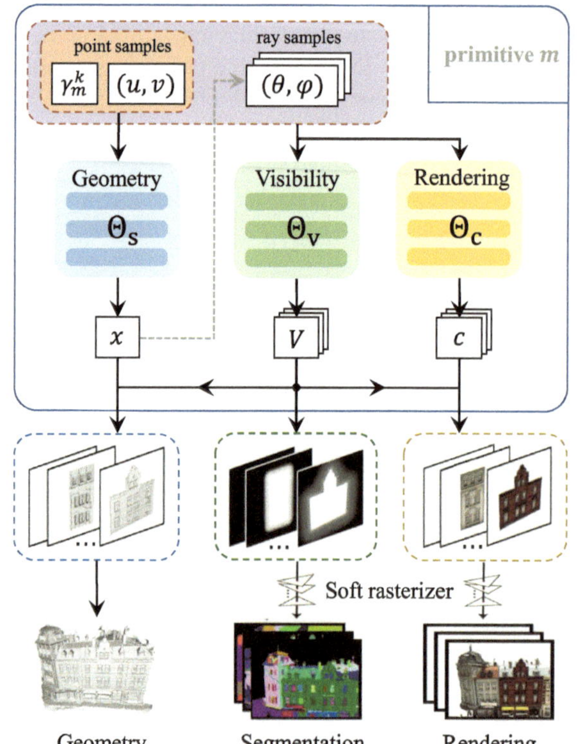

Fig. 5.19 The architecture of ParseMVS. The geometry network predicts the per-point displacement along the normal vector. The other two networks estimate the color and visibility information to produce segmentation and rendering results based on the sparse view inputs

outputs the point displacement x, which can be optimized by verifying multiview photometric consistency. For each input image with a pose, the view direction (θ, ϕ), 2D coordinate (u, v) and embedding vector γ_m^k are passed to networks $\Theta_\mathbf{c}$ and $\Theta_\mathbf{V}$ to generate the per-point RGB color c and visibility V values. These networks are supervised with a render consistency loss via a soft rendering scheme.

Photometric Consistency. The goal of the multiview stereopsis method proposed in this work is to learn the surface deformation r_{Θ_s} defined in Eq. 5.12, which is the optimal offset along the normal direction of the primitive point (u, v) that maximizes the photometric consistency score:

$$\Theta_s = \text{argmax}_{\Theta_s} \mathcal{NCC}(x, \hat{\mathbf{n}}), \tag{5.20}$$

where

$$x = P_m(u, v) + r_{\Theta_s}(u, v; \gamma_m^k)\mathbf{n}, \quad \hat{\mathbf{n}} = \frac{\partial_u x \times \partial_v x}{\|\partial_u x \times \partial_v x\|}. \tag{5.21}$$

Here, $\mathcal{NCC}(x, \hat{\mathbf{n}})$ is the photometric consistency score [35] based on the local tangent plane of point x with normal $\hat{\mathbf{n}}$. Given two images, the photometric consistency score is defined as the normalized cross correlation (NCC) value of the projections onto the two images. Note that only the views in the visible view set $\mathbf{B_m}(u, v)$ are selected (Eq. 5.16). Since it is difficult to optimize the photometric function defined in Eq. 5.20 using gradient descent, the problem can be simplified by dividing it into two subproblems: photometric consistency optimization and offset fitting. Specifically, the nonlinear patch optimization algorithm used in the PMVS model [35] is utilized to find all points $\mathcal{P}$ that maximize the photometric consistency. Then, the objective of the optimization is to minimize the squared distance between the points $\tilde{x}$ in $\mathcal{P}$ and the optimal offset location.

$$\Theta_s = \text{argmin}_{\Theta_s} \left\{ \sum_{(\tilde{u}, \tilde{v}) \in \Omega_m} \left\| \tilde{x} - \left(P_m(\tilde{u}, \tilde{v}) + r_{\Theta_s}(\tilde{u}, \tilde{v}; \gamma_m^k)\mathbf{n} \right) \right\|^2 \right\}, \tag{5.22}$$

where $(\tilde{u}, \tilde{v})$ is the projection of point $\tilde{x}$ onto the primitive.

Render Consistency. To optimize the parameters of networks $\Theta_\mathbf{c}$ and $\Theta_\mathbf{V}$, it is necessary to build the scene appearance based on a collection of primitives using an appropriate sampling approach.

To render an image, the textures of each primitive can be combined based on the visibility along each view direction. Each pixel in a single image may traverse several primitives. Therefore, the visibility defined based on each point of the primitives can be seen as a soft boundary. Specifically, for each pixel p in one of the N images, we select only N_f visible points (denoted as $D_{n,p}$), eliminating the occluded points.

The rendered pixel color value is defined as the weighted average of these point colors:

$$\overline{c}_{n,p} = \frac{\sum_{i \in D_{n,p}} V_i \cdot \mathbf{C}_{\Theta_c}(u_i, v_i; \theta_i, \phi_i; \gamma_i^{k_i})}{\sum_{i \in D_{n,p}} V_i}, \tag{5.23}$$

where V_i is the predicted visibility value. The texture and visibility can be learned by minimizing the differences between the rendered color $\overline{c}_{n,p}$ and the ground truth color $\tilde{c}_{n,p}$ in the image. We also average the instance label of the primitives:

$$\overline{S}_{n,p} = \frac{\sum_{i \in D_{n,p}} V_i \cdot S_{k_i}}{\sum_{i \in D_{n,p}} V_i}, \tag{5.24}$$

where S_{k_i} is the one-hot encoding vector of primitive m. We use compact watershed [92], an unsupervised semantic segmentation algorithm, to generate the segmentation $\tilde{S}_{n,p}$ of the primitives under each view.

The final optimization objective is to minimize the sum of the error:

$$(\Theta_c, \Theta_v) = \text{argmin}_\Theta \left\{ \sum_{n \in N,p} \left(\beta \cdot E(\overline{S}_{n,p}, \tilde{S}_{n,p}) + \|\overline{c}_{n,p} - \tilde{c}_{n,p}\|^2 \right) \right\}, \tag{5.25}$$

where $E(\cdot, \cdot)$ is the cross-entropy function.

Experiments

Dataset and Evaluation. The performance of the proposed method and several baseline methods based on the DTU dataset [1] was evaluated and compared under different sparse MVS settings. As a standard large-scale MVS benchmark, the DTU dataset includes a variety of objects and materials and contains 80 different scenes observed from 49 camera positions under seven different lighting conditions. Our test set consists of 22 scenes in the DTU dataset with complicated surfaces and sophisticated primitives.

The evaluation metrics were used to assess how closely the reconstructed point clouds match the ground truth laser scans. A distance metric [1] and percentage metric [69] were both employed. For the DTU dataset, the distance metric is identical to the standard Chamfer distance [1], whereas the percentage metric indicates the proportion of points that have values less than a given threshold [70]. The mean precision and mean recall were averaged to obtain the *overall* score for the distance metric, while the F1 score was used to calculate the *overall* score for the percentage metric.

Benchmark. We evaluate the results using various sparse MVS settings based on the DTU dataset [1]. In our experiments, the sparsity of the sampled viewpoints is defined the same as in SurfaceNet+ [60]. Only one-tenth of the images (~5 views) in the original DTU dataset are chosen under the severe sparse MVS setting ($sparsity = 10$). In addition to the PMVS method [35], the proposed method is compared with one

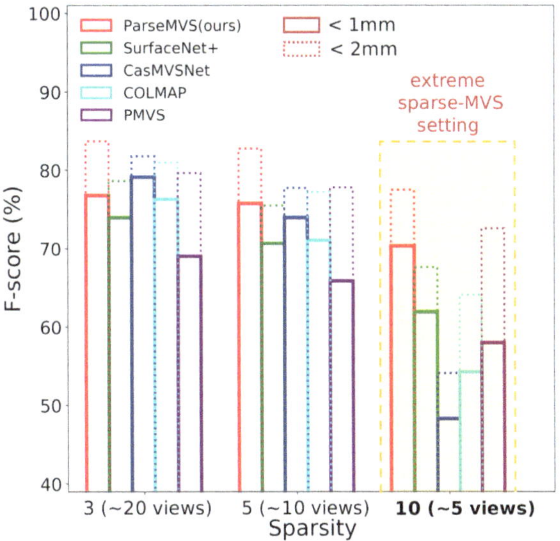

Fig. 5.20 Comparison of the results of the existing approaches based on the DTU dataset [1] with varying input sparsities. Under the severe sparse setting (sparsity = 10), only one-tenth of the images (~5 views) is selected

state-of-the-art sparse MVS algorithm, SurfaceNet+ [60], and two state-of-the-art MVS algorithms, COLMAP [111] and CasMVSNet [46]. The results are compared in terms of the *F1 score* for three levels of sparsity, as reported in Fig. 5.20.

When the distance threshold is set to 2 mm, ParseMVS consistently outperforms the baseline methods for all sparse conditions. The proposed method significantly outperforms the comparison methods for all distance thresholds, especially for the severe situation of $sparsity = 10$ (~5 views).

More detailed quantitative results are shown in Table 5.3, which are presented in terms of three different metrics. Additionally, ParseMVS shows nearly constant recall with good precision, in contrast to the other approaches, for which the recall substantially decreases as the sparsity increases. A qualitative comparison for the $sparsity = 10$ setting is shown in Fig. 5.21. Compared to the baseline methods, ParseMVS precisely reconstructs the scenes while maintaining high recall. Remarkably, ParseMVS can generate much more complete and denser point clouds with higher fidelity textures, indicating the correctness of the occlusion detection and visibility prediction results.

Ablation Studies. This section presents ablation studies that were performed to analyze the advantages of the different designed components. Scan 9 in the DTU dataset, which has a complex surface geometry, multiscale primitives, and spatially varying illumination conditions, was chosen to conduct the experiments.

With the same initial point cloud and the primitive detection results, we train the networks and grid embedding vectors under different settings. Comparisons of the qualitative and quantitative results are presented in Fig. 5.22.

First, the surface displacement function $\mathbf{r}_{\Theta_s}$ is removed from the framework. In this version, only the texture and visibility information of each primitive are

Table 5.3 Quantitative results of reconstruction quality on the DTU dataset in terms of the distance metric [1] (lower is better) and the percentage metric [69] (higher is better) with 1*mm* and 2*mm* as thresholds. Our method constantly outperforms state-of-the-arts in terms of recall and F-score in all the sparse-MVS settings with $n = 3, 5, 10$

Sparsity	Method	Mean distance (mm)			Percentage (<1mm)			Percentage (<2mm)		
		Precision	Recall	**Overall**	Precision	Recall	**F-score**	Precision	Recall	**F-score**
3 (~20 views)	**ParseMVS (ours)**	0.450	**0.589**	**0.520**	84.01	**71.65**	76.76	92.78	**77.08**	**83.64**
	SurfaceNet+ [60]	0.457	0.692	0.575	86.85	71.44	77.93	93.04	75.03	82.59
	CasMVSNet [46]	0.445	1.014	0.730	**94.81**	68.96	**79.12**	**96.61**	71.94	81.76
	COLMAP [111]	**0.438**	1.125	0.782	90.49	66.79	76.25	93.25	72.45	80.94
	PMVS [35]	0.455	1.210	0.832	82.46	60.18	68.99	88.89	73.19	79.63
5 (~10 views)	**ParseMVS (ours)**	0.529	**0.634**	**0.582**	85.73	**68.66**	**75.74**	94.07	**74.65**	**82.73**
	SurfaceNet+	0.543	0.688	0.616	88.67	65.33	74.63	94.31	69.57	79.48
	CasMVSNet	**0.343**	1.423	0.883	**95.63**	61.65	73.94	**97.51**	65.95	77.71
	COLMAP	0.356	1.985	1.171	93.94	57.97	71.04	95.91	65.46	77.18
	PMVS	0.413	1.540	0.977	84.19	54.78	65.85	90.32	69.30	77.78
10 (~5 views)	**ParseMVS (ours)**	0.497	**0.754**	**0.626**	79.48	**64.11**	**70.29**	87.56	**70.64**	**77.48**
	SurfaceNet+	0.511	1.108	0.810	88.72	53.31	65.90	94.50	58.47	71.59
	CasMVSNet	0.329	2.985	1.657	96.99	34.45	48.28	**98.54**	39.79	54.10
	COLMAP	**0.315**	3.342	1.829	**97.43**	38.52	54.25	98.48	48.39	64.06
	PMVS	0.405	1.985	1.195	85.72	44.33	57.98	92.08	60.66	72.53

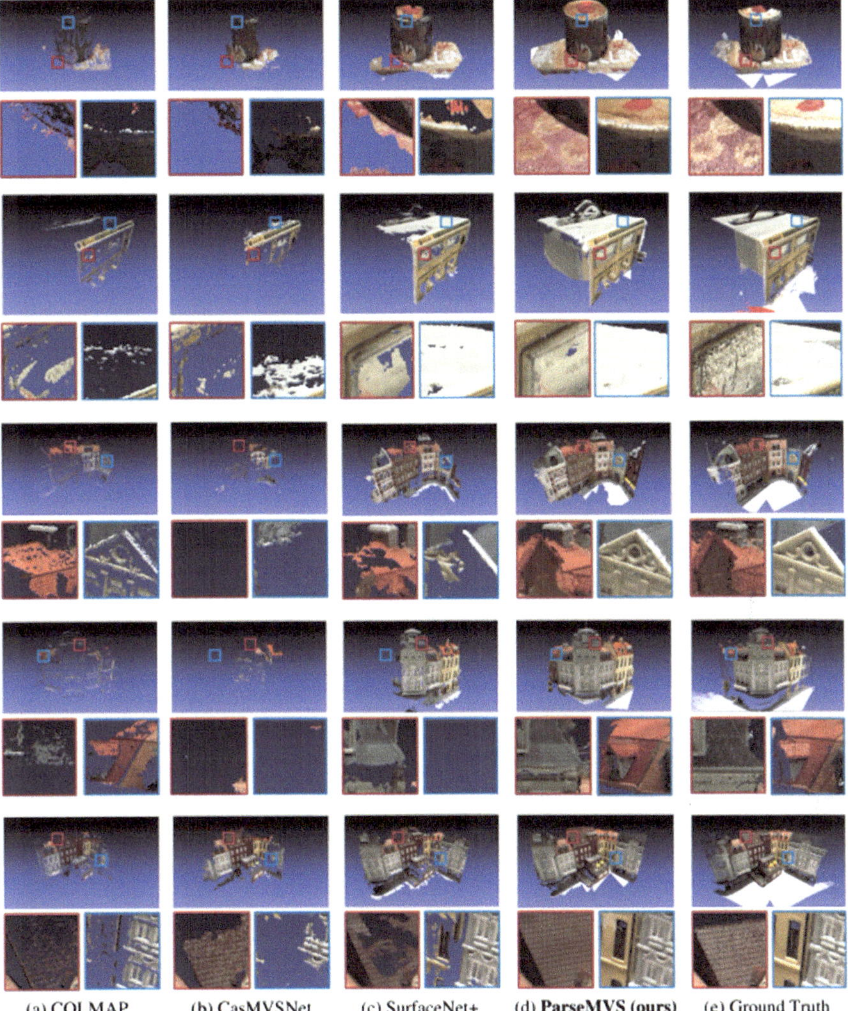

Fig. 5.21 Qualitative comparison for $sparsity = 10$, i.e., for an order of magnitude fewer images. We choose 5 scenes from the DTU dataset [1] for visualization

predicted by the algorithm. The point cloud reconstruction only uses the global geometric structures and ignores local details, leading to a significantly reduced precision and F1 score. As depicted in Fig. 5.22b, this model obtains low geometric accuracy in intricate regions, emphasizing the importance of displacement modeling in improving accuracy.

Fig. 5.22 Ablation study based on the different designed components. The results without the surface displacement (**b**), visibility (**c**), and segmentation modules (**d**) are shown

Then, the visibility information obtained by each viewing camera ray was removed. In Eqs. 5.17 and 5.19, the final boundary $\hat{\Omega}_m$ is replaced with the original coarse boundary Ω_m. Furthermore, Fig. 5.22c shows that this model obtains rough geometries and inconsistent textures. The ambiguous definition of the boundary based on the primitives leads to noise in the reconstruction results. These results illustrate that visibility modeling is important for detecting occlusions, choosing views, and generating accurate boundaries.

Table 5.3 demonstrates that the recall decreases with a near precision value when the segmentation loss is removed (i.e., $\beta = 0$ in Eq. 5.25. The qualitative results are displayed in Fig. 5.22d, showing that discontinuous and incomplete surfaces appear in the regions with repeating textures and specular surfaces. These results indicate that the primitive prior improves regularization in regions with extremely uncertain correspondence during the segmentation process, leading to more complete reconstruction results.

***Robustness to Initialization*.** To evaluate ParseMVS's robustness to initialization, experiments were performed with various initial primitive detection results. The experimental results suggest that ParseMVS is robust to noisy primitive initializations for certain noise levels. Specifically, Gaussian perturbations of various scales were added to the primitive parameters using the original primitive detection results. Each plane's normal vector was normalized to length 1. The noise level was defined as the difference in the angle after this random perturbation was added to the normal of the original primitive.

The F1 score remained constant when the noise level was small (less than 10°C), as demonstrated in Fig. 5.23. This indicates that the displacement field learned based on the primitives has relatively high tolerance to noisy primitive parameters. As expected, the F1 score decreases when the noise reaches an unacceptably high level (greater than 20 °C).

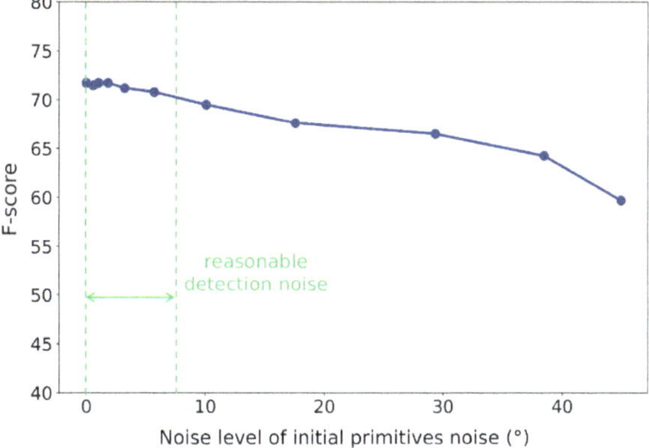

Fig. 5.23 Performance w.r.t. noise level for different primitive parameter initializations

Table 5.4 Comparison with Nerf [87], IDR [145] and SurRF [154]. The table reports the average results on 15 models used in the IDR evaluation set

Method	Training time (h)	No. of parameters	GPU Mem. (MB) (~100,000 rays)
Nerf [87]	25.5	162,959	182,324
IDR [145]	17.3	2,915,258	42,465
SurRF [154]	10.0	369,078	10,324
ParseMVS (ours)	1.0	131,072	2048

Training Efficiency. The ParseMVS representation naturally has an advantageous fast training speed because the surface attributes are modeled and optimized locally in each primitive's 2D space. Quantitative time and memory efficiency comparisons were performed based on the per-scene optimization approaches (Nerf [87], IDR [145], and SurRF [154]), and the results are reported in Table 5.4. The total number of hours spent optimizing all input views (with a size of 600 × 800 pixels) was used to evaluate the training time efficiency. The proposed ParseMVS model is approximately 10 times faster than the other methods during training, as illustrated in Table 5.4. Additionally, ParseMVS exhibits approximately 10x better GPU efficiency than the other methods. Thus, the effective primitive-aware representation proposed in ParseMVS improves network training by a large margin.

5.3 High-Resolution Reconstruction for Large-Scale Scenes

While the development of deep learning has considerably improved computational imaging, the advantages of high-performance imaging [148] in inspiring and promoting more capable reconstruction schemes have not been extensively investigated.

The popularity of high-performance imaging methods poses great challenges but also provides new opportunities for developing downstream reconstruction algorithms. Unfortunately, thus far, there exists a significant gap between the development of sensing and reconstruction algorithms, hindering the full exploitation of the former.

Aiming at bridging the gap between sensing and reconstruction, high-resolution reconstruction techniques with superior capabilities are introduced in this section to better recover textures (Sect. 5.3.2) and geometries (Sect. 5.3.1).

5.3.1 *Surface Radiance Fields for Unsupervised Multiview Stereopsis*

High-resolution geometry reconstructions in the form of explicit point clouds, triangle meshes or, more recently, implicit neural representations are promising for many downstream industries, such as autonomous driving, AR/VR, and meta-universes. Compared with depth sensors with limited sensing resolution or large acquisition costs, recovering dense 3D surfaces based on details in posed 2D images, i.e., multiview stereopsis (MVS), is a more practical solution. Compared with traditional MVS methods [37, 111], recent learning-based works [16, 59, 60, 142, 143] have applied learned geometry and illumination priors to obtain more robust and complete reconstruction results. However, these methods require considerable lased-scanned 3D data for training, which is not always available for outdoor scenes.

To address this issue, some unsupervised MVS methods [28, 68] have used reprojection losses to improve cross-view photometric consistency. However, these methods are susceptible to texture patterns and environmental illumination conditions, leading to incomplete 3D reconstructions near textureless regions. In addition, they are restricted to specific 3D representations (e.g., voxels and point clouds), which may not have smooth and continuous manifold structures for high-fidelity surfaces [45], leading to discretization artifacts.

While continuous mapping defined based on the mesh addresses the discretization issue, it requires fixed geometric positions, which cannot be used for geometry reconstructions. Another solution [93, 117, 145] uses differentiable rendering to progressively update the geometry based on the rendered images. Moreover, implicit shape and texture representations can be learned based on multiple images with a continuous mesh as output [84, 93, 145]. However, dense volumes of 3D space contain redundant radiance field distributions, which are implicitly created using neural networks. This drastically restricts the implicit function's ability to convey detailed geometric content.

To address the above issues, the surface radiance field (SurRF) [154] method was proposed as a novel unsupervised MVS pipeline for learning the surface radiance field, i.e., a radiance field explicitly defined on a continuous 2D surface, which can be optimized without 3D supervision. Specifically, the surface radiance field is a continuous representation of the geometry, texture, and lighting conditions as well

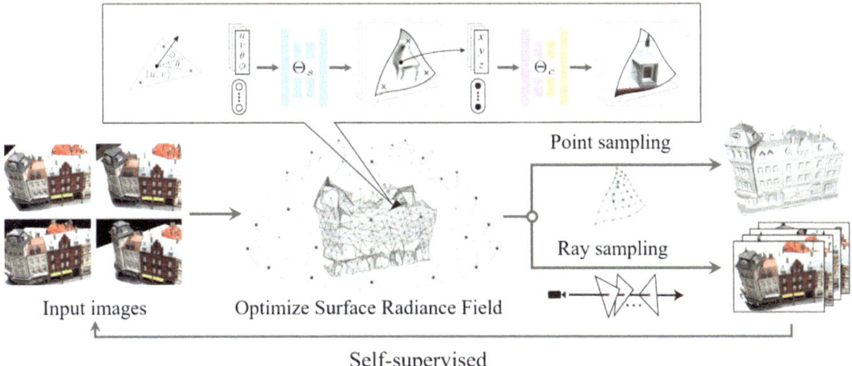

Fig. 5.24 An overview of the surface radiance field (SurRF) and unsupervised optimization procedure. The scene is represented based on a set of surfaces, where each surface includes a combination of 3D shapes, textures, and view directions. The optimization is performed through differentiable rendering by comparing the rendered images with a set of target images

as the explicit surface deformation relative to an initialized coarse-level mesh. As shown in Fig. 5.24, all the information is represented locally on a set of surfaces. In contrast to the height field, which encodes the relative displacement in only one direction, a neural network is used to implicitly represent the geometry as a per-view depth image for each triangle from every viewing direction. Specifically, each of the local surfaces is assigned a geometry and texture embedding, and the geometry MLP encodes all the view-dependent distance information. The distance to the intersection through the target surface of each ray acts as the surface deformation along the view-dependent rays. The geometry embedding is first passed through an MLP network to predict the distance. Then, the color is predicted by another MLP that uses the texture embedding, predicted intersection coordinate, and view direction as input (Fig. 5.25).

Surface Radiance Field

Definition. The scene is represented as a set of surfaces $\mathbf{S}^*$ with continuous geometry and texture information. Each surface $S_i^* \in \mathbf{S}^*$ is a triangle facet containing two learnable embeddings γ_i^s and γ_i^c with code lengths l_s and l_c that describe the surface geometry and texture, respectively. A multilayer perceptron (MLP) network is used to encode the surface indexed by the texture coordinate $\psi^{(x)} = (u, v)$ and the ray direction $\psi^{(d)} = (\theta, \phi)$. Therefore, for a ray with a normalized direction of $\boldsymbol{d}_i^*$ at the 3D point $\boldsymbol{x}_i^*$ of the triangle facet, the surface deformation is determined based a residual prediction as follows:

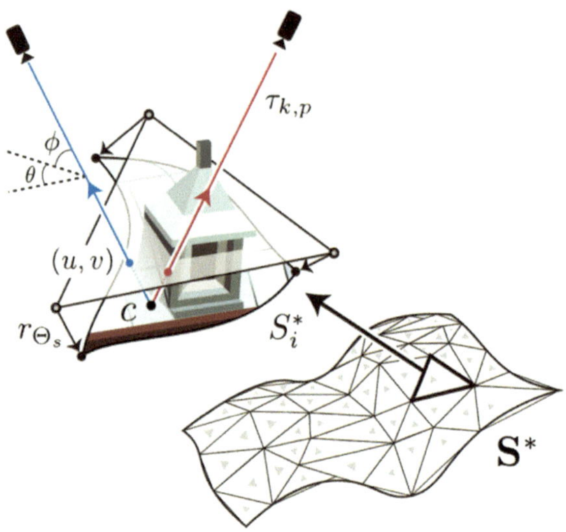

Fig. 5.25 Illustration of the surface radiance field (SurRF) representation

$$s = x_i^* + r_{\Theta_s}(\psi^{(x)}, \psi^{(d)}; \gamma_i^s)d_i^* \tag{5.26}$$

$$r_{\Theta_s} : R^2 \times R^2 \times R^{l_s} \rightarrow R \tag{5.27}$$

encodes the residual length at x_i^* for view direction d_i^* from a point on the triangle facet to the target surface.

The color is predicted as a function of the estimated 3D location of the target surface and the view direction:

$$\begin{aligned} C_{\Theta_c} &: R^3 \times R^2 \times R^{l_c} \rightarrow R^3 \\ &(s - \overline{x}_i^*, \psi^{(d)}; \gamma_i^c) \rightarrow (c) \end{aligned} \tag{5.28}$$

where $\overline{x}_i^*$ is the centroid of the triangle. It is worth noting that unlike the geometry MLP, which maps the point (2D) and the direction (2D) based on the primary triangle facet, the texture MLP maps the 3D point and the texture fields [96]. The different choices of input coordinates to the geometry and texture MLPs enable the reconstruction of a view-consistent 3D geometry. If the geometry MLP produces inconsistent 3D geometry, i.e., different 3D point predictions for different view directions, the texture MLP generates inconsistent color values based on the inconsistent 3D point predictions. This inconsistency constrains the differentiable rendering process, thereby leading to a photometric consistent output.

In general, the full representation can be written as

$$\begin{aligned} L_\Theta &: R^4 \times R^l \rightarrow R^3 \times R^3 \\ &(\psi_i, \gamma_i) \rightarrow (s, c) \end{aligned} \tag{5.29}$$

where $\psi_i = (\psi_i^{(x)}, \psi_i^{(d)})$, $\gamma_i = (\gamma_i^s, \gamma_i^c)$ and $l = l_s + l_c$.

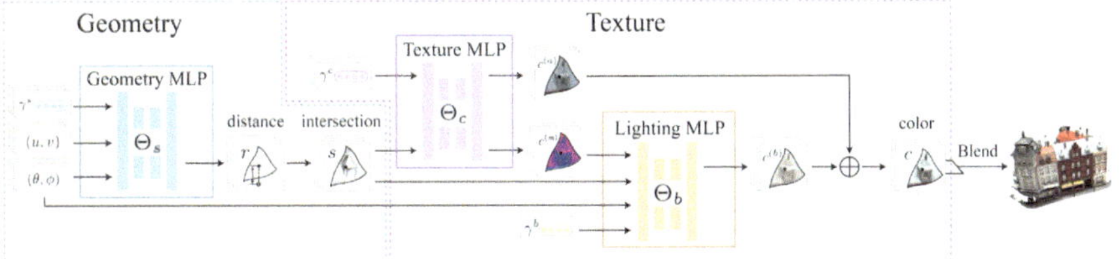

Fig. 5.26 The network architecture of our proposed SurRF method. The surface geometry is first predicted using the 2D texture coordinate, ray direction and an embedding through the geometry multilayer perceptron (MLP) network. Then, the illumination and surface color are estimated through two networks modeling the texture and lighting

Rendering with Lighting Approximation. Although this novel representation allows for the encoding of different colors in various view directions, the direct optimization approach cannot identify the geometry priors, such as the surface normal and multiview consistency, or the view-independent materials, such as the ambient and diffuse lighting conditions.

The detailed network architecture of SurRF is shown in Fig. 5.26, where the geometry and texture of the scene are locally represented and the illumination is globally encoded. To model the illumination conditions, the color prediction C_{Θ_c} in Eq. 5.28 is improved with an enhanced function and a rendering equation B_{Θ_b}:

$$\begin{aligned} C_{\Theta_c} &: R^3 \times R^{l_c} \rightarrow R^3 \times R^6 \\ (s - \overline{x}_i^*; \gamma_i^c) &\rightarrow (c^{(a)}; c^{(m)}), \end{aligned} \tag{5.30}$$

$$\begin{aligned} B_{\Theta_b} &: R^8 \times R^6 \times R^{l_b} \rightarrow R^3 \\ (s, \psi^{(d)}, n; c^{(m)}; \gamma^b) &\rightarrow (c^{(b)}) \end{aligned} \tag{5.31}$$

where $c^{(a)}$ is the view-independent color, and the view-dependent color $c^{(b)}$ is generated based on the diffuse and specular material $c^{(m)}$, the surface normal n computed based on the initial triangle facet, the predicted surface location s, the view direction $\psi^{(d)}$ and a latent lighting embedding γ^b with length l_b. The reason why s is used instead of $s - \overline{x}_i^*$ in Eq. 5.31 is that the learning targets of the texture embedding and the lighting embedding are different. While the texture embedding attempts to recover the local color and material values, the lighting embedding aims to model the rendering equations, which include the surface geometry, material and illumination parameters. As shown in Fig. 5.26 and Eq. 5.31, unlike the texture representation, which assigns a unique embedding γ_i^c to each facet, all the facets have the same lighting embedding γ^b.

The final color for camera k at pixel p is defined as

$$c_{i,k,p} = c_{i,k,p}^{(a)} + c_{i,k,p}^{(b)}. \tag{5.32}$$

Sampling

The goal of the sampling process is to obtain the global scene geometry and color information based on a collection of surface radiance fields. As illustrated in Figs. 5.27 and 5.28, the sampled objects can be surface points, which are used for point cloud generation, or pixel rays from different camera poses, which are used for view synthesis and unsupervised learning.

Point Cloud Generation. As shown in Fig. 5.27, the key component of the point cloud generation method is sampling a height field based on each local triangle facet. Since our geometry is represented as a per-view depth image from every view direction, theoretically, any direction can be selected to generate the corresponding distance to the intersection point. In practice, the surface normal is used as the sampling view direction for convenience. In addition, this approach prevents the network from

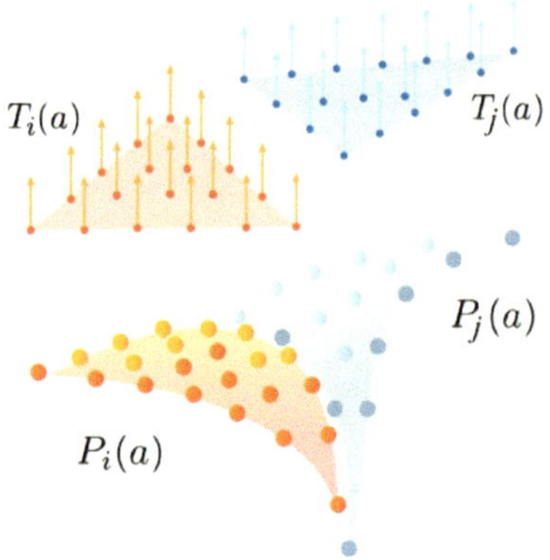

Fig. 5.27 Point sampling for point cloud generation

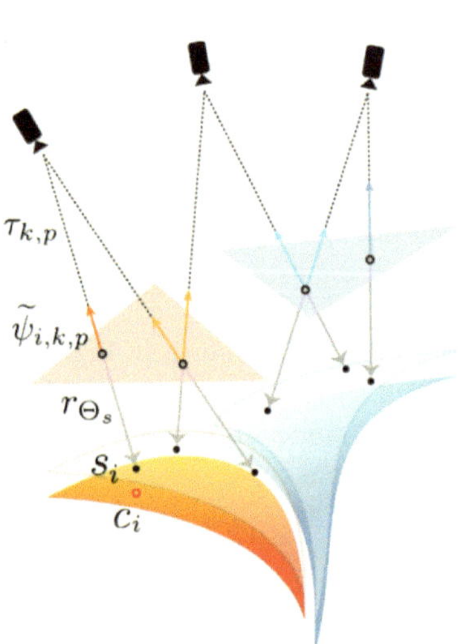

Fig. 5.28 Ray sampling for view synthesis

using singular or overextrapolated coordinate inputs. For each surface, a set of texture coordinates $T(a)$ is first sampled as

$$T(a) = \left\{ \left(u, v, 0, \frac{\pi}{2}\right) | u, v \in \left\{0, \frac{1}{a}, \frac{2}{a}, \ldots, 1\right\}, u + v \leq 1 \right\} \tag{5.33}$$

where $a \in N^*$ is the number of samples on one side. For each surface, the generated point set P_i is

$$P_i(a) = \{(s, c)|(s, c) = L_\Theta(\psi_i, \gamma_i), \psi_i \in T(a)\}, \tag{5.34}$$

and the final point cloud is the union $P(a) = \bigcup_i P_i(a)$.

View Synthesis. The main component of the view synthesis method is a differentiable renderer, such as the soft rasterizer [81], where the deformation and color are predicted, as shown in Fig. 5.28. The predictions are globally aggregated utilizing a z-buffer, as in traditional point-based graphics methods, as shown in Fig. 5.24. Specifically, for images from K views with camera centers t_k, the goal is to generate the color for each pixel.

The ray $\tau_{k,p}$ through each pixel p intersects the triangle facet at $\psi_{i,k,p}$. The triangle facet for which the intersection is inside the triangle is selected. Since a ray may pass through a complex surface several times, only the N_f closest surfaces are selected to prevent unnecessary sampling during the optimization process. The selected facets are denoted as $D_{k,p}$. The sampling number N_f is related only to the training efficiency and not the performance. Then, the rendered pixel color is calculated as the average value of the colors of the selected surfaces:

$$(\overline{s}_{k,p}, \overline{c}_{k,p}) = \frac{\sum_i \mathbf{1}_{i \in D_{k,p}} L_\Theta(\psi_{i,k,p}, \gamma_i)}{\sum_i \mathbf{1}_{i \in D_{k,p}}}. \tag{5.35}$$

Optimization

To learn high-resolution continuous surfaces in complex scenes, two additional strategies are adopted during ray sampling.

Shape Continuity. Unreliable estimations at the surface edge may occur due to surface deformations along each ray in the collection of unique surface radiance fields. Since the latent embedding of each surface can only be optimized based on the sampled rays, directly addressing the discontinuities by reducing the distance to the adjacent surfaces is laborious and necessitates extensive computations to assess the corresponding relationships. Therefore, the variance between the surface location and color $(s^{(\sigma)}_{k,p}, c^{(\sigma)}_{k,p})$ is computed and reduced. Following Eq. 5.35, we have

$$(s_{k,p}^{(\sigma)}, c_{k,p}^{(\sigma)}) = \frac{\sum_i \mathbf{1}_{i \in D_{k,p}}[L_\Theta(\psi_{i,k,p}, \gamma_i) - (\overline{s}_{k,p}, \overline{c}_{k,p})]^2}{\sum_i \mathbf{1}_{i \in D_{k,p}}}. \tag{5.36}$$

Note that this shape consistency optimization approach has the same computational pipeline as the view synthesis method and requires no additional sampling. Since different camera positions and image sizes change the local coordinates on each surface, the continuity is guaranteed by using a sufficient number of training samples.

Progressive Training. The proposed point sampling strategy enables us to progressively refine the triangle facets of each surface to obtain more complex topologies and more detailed self-occlusion information. After generating the previous stage prediction $\mathbf{S}^*$, a set of detailed points $P(a)$ is sampled. In the next stage, detailed triangle facets are obtained based on either the ball pivoting algorithm [7] or by using an equilateral triangle with an appropriate side length whose centroid coincides with the point and whose three sides are perpendicular to the normal of that point.

Before the triangle mesh generation procedure, the point cloud is resampled to obtain uniformly sampled points with resolution v. The resampling rule is defined as follows: The scene space is divided into voxels with size v. If multiple points fall into the same voxel, only the mean of these points is preserved. In the next stage, the point cloud resolution is two times finer than that in the previous stage. The triangle facets in the first five stages are obtained with the ball pivoting algorithm, applying a radius of $2v$. In the final stage, an equilateral triangle with a side length of $1.5v$ is used. The equilateral triangle approach is used for convenience because the ball pivoting method requires a long time to construct a set of triangle facets with no holes as the number of points increases.

The network structure is consistent with that used in the previous stage and uses the same network parameters Θ to accelerate training. In practice, a 6-stage prediction network is used when the number of samples a equals 4. The number of samples a in the finest stage is set to 10 for the final point cloud generation. An illustration of the progressive training scheme is shown in Fig. 5.29.

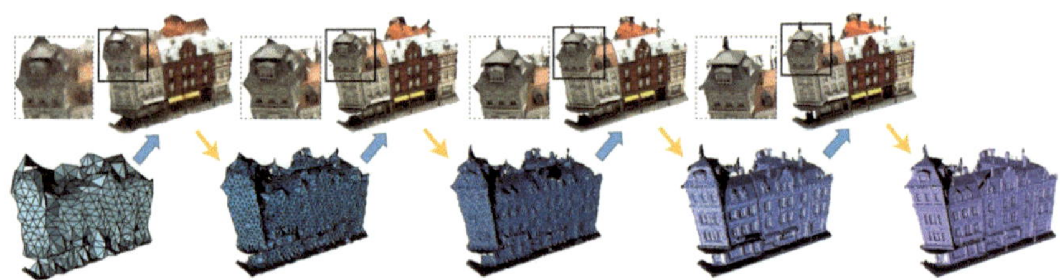

Fig. 5.29 Illustration of the progressive training approach. The SurRF model progressively refines the underlying surface structures via a coarse-to-fine point sampling approach based on a set of posed RGB images

Experimental Results

SurRF significantly outperforms the state-of-the-art unsupervised MVS methods, including DTU [112] and Tanks and Temples [70], and obtains competitive results with the supervised methods, showing high fidelity, high resolution, and many details.

DTU Dataset. The quantitative results based on the DTU dataset are shown in Table 5.5. The distance metric [1] was used to determine the correctness and completeness of the final reconstruction results. The overall distance metric was calculated as the average of the accuracy and completeness. Gipuma [37] achieves the best accuracy; however, our method outperforms all unsupervised methods in terms of the completeness and overall score and produces results that are competitive with state-of-the-art supervised techniques for all distance metrics. A visual comparison including the depth visualization results of SurRF and the other techniques [16, 21, 68] for five scans is shown in Fig. 5.30. SurRF produces finer-grained, smoother point clouds and more continuous surfaces than the other methods.

Tanks and Temples Dataset. The quantitative results based on the Tanks and Temples dataset are shown in Table 5.6. SurRF achieves state-of-the-art performance when compared with state-of-the-art supervised methods and unsupervised multi-view stereo approaches. In particular, SurRF performs noticeably better than all unsupervised learning-based MVS approaches. The reconstructed point clouds are

Table 5.5 Quantitative reconstruction quality results in terms of the distance metric (lower is better) based on the DTU evaluation dataset [1]. SurRF outperforms the state-of-the-art unsupervised methods in terms of the completeness and overall score and obtains competitive results with state-of-the-art supervised methods for all evaluation metrics

Method		Acc. (mm)	Comp. (mm)	Overall (mm)
Supervised	MVSNet [142]	0.396	0.527	0.462
	SurfaceNet+ [59]	0.385	0.448	0.416
	Point-MVSNet [16]	0.342	0.441	0.376
	CasMVSNet [46]	**0.325**	0.385	0.355
	UCS-Net [21]	0.338	0.349	**0.344**
	PVA-MVSNet [146]	0.379	**0.336**	0.357
	D^{2}HC-RMVSNet [139]	0.395	0.378	0.386
Unsupervised	Furu [35]	0.613	0.941	0.777
	Tola [126]	0.342	1.190	0.766
	Gipuma [37]	**0.283**	0.873	0.578
	COLMAP [111]	0.400	0.664	0.532
	Robust consistency [68]	0.881	1.073	0.977
	SurRF [154]	0.388	**0.390**	**0.389**

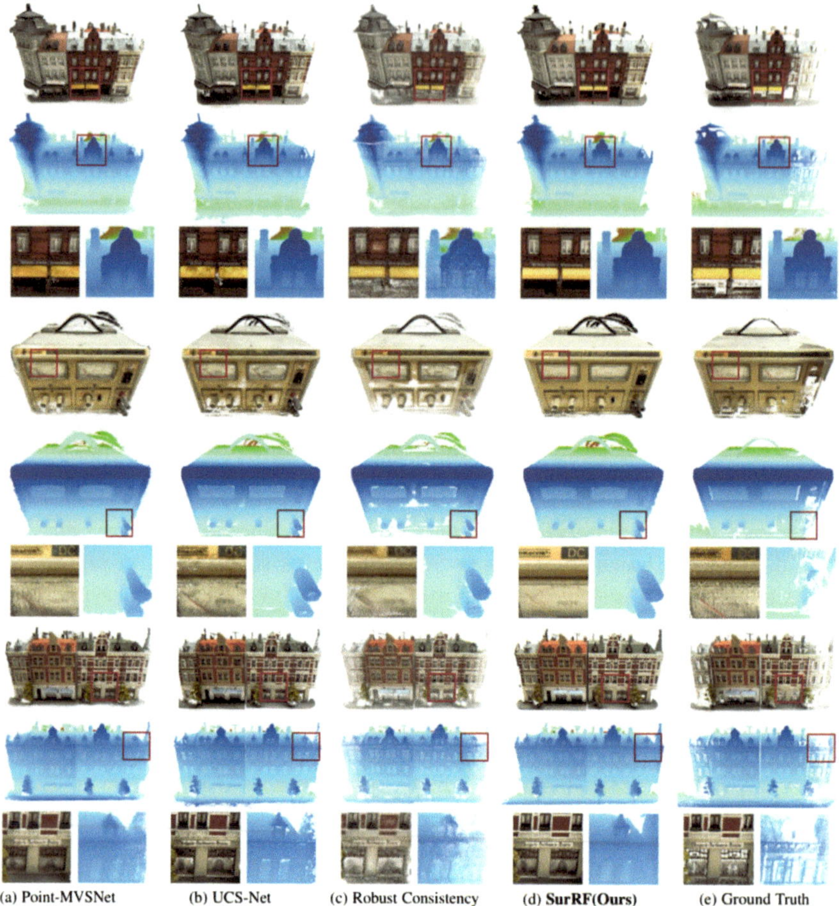

Fig. 5.30 Qualitative results for scans 9, 11, and 15 in the DTU dataset [1] compared with the results of Point-MVSNet [16], UCS-Net [21], Robust Consistency [68] and the ground truth point cloud. Note that SurRF generates denser point clouds with better continuity and colors than the other methods. To ensure a fair comparison, the result with the best completeness is shown

shown in Fig. 5.31 and compared with COLMAP [111], R-MVSNet [143], CasMVSNet [46] and UCS-Net [21]. SurRF produces more thorough, delicate, and accurate surface predictions than modern supervised algorithms, with a competitive F1 score.

The continuous local mapping results do not require laboriously matching the geometry and texture information in the dense volume space, which is required for Nerf [87] and IDR [145], since SurRF can constantly map coarse triangle meshes to finer geometries, leading to compact representations and globally complete shapes for large-scale complex scenes.

In addition, our learned view-dependent function is a clever method for processing the texture and geometry of multiview images that is more generalizable for

Table 5.6 Quantitative results of state-of-the-art learning-based methods based on the Tanks and Temples (T&T) dataset [69]. SurRF achieves state-of-the-art performance among the unsupervised MVS methods and competitive results with the state-of-the-art supervised methods

Method		Rank	Mean
Supervised learning	VisMVSNet [153]	**8.12**	**60.03**
	D^2HC-RMVSNet [139]	9.50	59.20
	CasMVSNet [46]	23.38	56.84
	UCSNet [21]	34.50	54.83
	PVAMVSNet [146]	34.88	54.46
	SurfaceNet+ [60]	51.38	49.38
	Dense R-MVSNet [143]	57.75	50.55
	Point-MVSNet [16]	62.75	48.27
	MVSNet [142]	78.50	43.48
Unsupervised learning	MVS^2 [28]	100.38	37.21
	M^3VSNet [56]	100.38	37.67
	SurRF[154]	**31.88**	**54.36**

outdoor scenes with complex illumination conditions than surface parameterization methods [45, 76], which optimize the geometry based on only a set of parametric surface elements.

Overall, SurRF maintains the advantages of mesh-based (scene manipulation), continuous surface-based (high geometric resolution), and radiance field-based (realistic rendering results) methods.

5.3.2 *Neural Micro Surfaces for Very Large View Synthesis*

A long-standing issue in computer vision and graphics is the creation of unobserved viewpoints based on a group of calibrated images [18, 74, 114, 115]. Recent advances in deep image-based rendering [3, 104, 105, 129] and implicit neural scene representations [6, 87, 127, 145] have enabled the synthesis of photorealistic scenes with complicated contents and view-dependent appearance. However, these methods require inputs with dense coverage of the scene, which are laborious to collect and not always available, especially real-world large-scale scenes with ultrahigh-resolution observations. While the performance of computational imaging methods has substantially improved with the development of recording systems such as array cameras [148], the sensing cost is still considerable due to the difficulty of integrating such systems on mobile devices, leading to an inherent trade-off between the spatial resolution and view resolution. In other words, the development of view synthesis methods that are suitable for sets of sparse but high-resolution input views is a highly challenging but critical task for practical applications such as AR/VR, visual effects, autonomous driving, and robotics.

Fig. 5.31 Point cloud and depth visualization results for Family, M60, and Lighthouse in the intermediate set in the Tanks and Temples dataset [70] compared with the results of COLMAP [111], R-MVSNet [143], CasMVSNet [46] and UCS-Net [21]. Our proposed SurRF method produces more complete, delicate, continuous, and realistic surface predictions than the other methods

The view synthesis quality of state-of-the-art methods degrades drastically when only sparse observations with large baseline angles are used as input. The geometry has a large impact on the view synthesis quality under sparse settings, and existing methods cannot effectively balance precise scene geometries with cross-view consistency. Deep image-based rendering methods [104, 105] warp and aggregate features from nearby views based on fixed or soft geometry proxies. Other methods [3, 107, 124] augment dense geometries with neural descriptors and apply differentiable rendering techniques. When only sparse views are provided, the content in consecutive views may differ significantly, and these methods fail to obtain accurate geometries, resulting in blurry artifacts and poor view-dependent appearance.

Recently, neural volumetric representations [87, 129] and neural implicit surfaces [145] have been used to encode scene geometries based on the volume density or signed distance fields while simultaneously learning radiance fields to produce results with realistic view-dependent appearances. However, due to the inherent ambiguity between the geometry and appearance in these entangled representations, it is easy for the methods to fall into local minima, limiting the representational power of the implicit function, particularly for sparse view settings. Intuitively, an arbitrary incorrect geometry can be well explained by a careful chosen radiance field [155], and existing methods rely on the inductive bias of neural networks to enforce disentanglement. However, as the input observations become sparser, the use of the inductive bias alone cannot obtain good performance, and the underlying geometry severely suffers. In this case, the implicit radiance function becomes less expressive since accurately matching the inputs requires learning a much more complex view-dependent function to compensate for the incorrect geometry, leading to a significantly reduced view synthesis quality.

To address the above limitations, a neural micro surface is introduced in this section. The neural micro surface is a neural implicit field that is continuously defined on explicit surface patches, encapsulating the local geometry and appearance information. In contrast to previous neural scene representations, the neural micro surface is only learned on well-defined surface manifolds instead of in unconstrained and redundant volumes, thus benefiting from the inherent disentanglement property for sparse observations and maintaining superior expressivity to represent complex large-scale scenes. Furthermore, the neural micro surface implicitly reinforces geometric correspondence, making it possible to synthesize realistic fine-grained details. As shown in Fig. 5.32, the neural micro surface is a novel implicit scene representation defined on a coarse triangulation that is composed of surface property mapping, neural shading, and displacement embedding features. For each surface patch, the underlying properties such as the surface normal and BRDF parameters are continuously represented by a coordinate-based MLP network, with the point coordinate and a learnable latent vector used as inputs. To obtain realistic color predictions, the recovered surface properties are interpreted by another shading network that implicitly accounts for the illumination and reflectance conditions. Since precise geometry is crucial to improve the representational power of the implicit shading function, a displacement embedding network is introduced to remap the initial geometry to an embedded latent space, where more detailed geometric correspondence can be implicitly incentivized. To render the neural micro surface with novel views, a differentiable rasterizer is used to efficiently and smoothly sample the implicit field. Notably, the neural micro surface learns a set of implicit functions to continuously represent the surface properties based on mesh parameterizations, which can be easily deployed in any traditional graphics engine. Since the entire framework is fully differentiable, the neural micro surface can be optimized through backpropagation by comparing the synthesized renderings with ground truth observations.

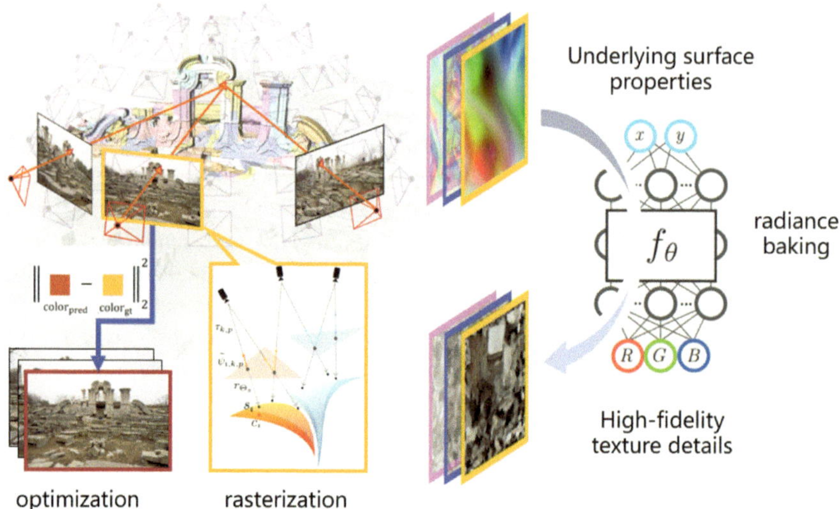

Fig. 5.32 An overview of the neural micro surface representation and the neural rendering framework. The geometric and material properties are locally represented on a set of surface patches using an implicit mapping function modeled by multilayer perceptrons. Novel views can be synthesized by querying another shading function that implicitly accounts for illumination and reflectance. Optimization is performed by matching the synthesized renderings with available observations through a differentiable rendering approach

Neural Micro Surface

Definition. Surface parametrization can be viewed as the process of finding a piecewise continuous function $f : \mathbb{R}^3 \mapsto \mathbb{R}^2$ that corresponds to a planar parameterized domain $\mathcal{U} \in \mathbb{R}^2$ to a surface $\mathcal{S} \in \mathbb{R}^3$. The texture atlas is a more general representation in which the surface is segmented into charts homeomorphic to discs, and each surface is locally parameterized; hence, a local coordinate system for the surface $\mathcal{S}$ can be defined. Texture atlases are important in modern graphics engines. For a specific texture resolution, the texture atlas can explicitly store local appearance information such as the albedo or geometric properties as normals or displacement maps (Fig. 5.33).

We propose to extend the traditional texture atlas and represent scene $\mathcal{M}$ as a compact surface representation with continuous scene properties. Each surface patch $\mathcal{S} \in \mathcal{M}$ is parameterized as an atlas $\mathcal{A}_s$, which is further augmented by a learnable latent grid $\mathcal{Z}_s$ to describe local high-dimensional texture information. We leverage a coordinate-based MLP neural network to represent the continuous underlying the scene properties:

$$\begin{aligned} \Psi_{\Theta_p} &: \mathbb{R}^{Z_s} \mapsto \mathbb{R}^n, \\ z_s &= \psi(\hat{\boldsymbol{x}}; \mathcal{Z}_s), \\ \mathcal{P} &= \Psi_{\Theta_p}(z_s), \end{aligned} \tag{5.37}$$

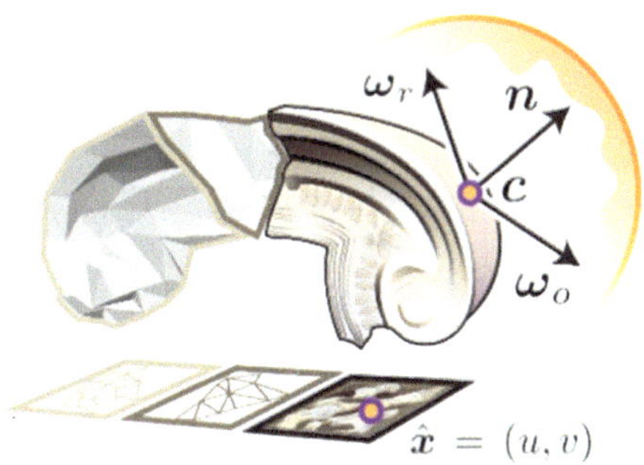

Fig. 5.33 Illustration of the neural micro surface representation

where $\hat{\boldsymbol{x}} = (u, v) \in \mathbb{R}^2$ is the local parametric coordinate in atlas $\mathcal{A}_s$ normalized within the unit square. The latent grid $\mathcal{Z}_s$ is defined by assigning feature vectors to the four corners of the 2D texture atlas, and $z_s \in \mathbb{R}^{Z_s}$ denotes the feature vector at $\hat{\boldsymbol{x}}$ obtained by bilinearly interpolating the corner features. The output $\mathcal{P} \in \mathbb{R}^n$ includes any spatially varying surface properties, such as the normal, albedo, metallic details, roughness, and other BRDF parameters.

The color $\boldsymbol{c} \in \mathbb{R}^3$ is predicted by another shading function Φ_{Θ_c} based on the rendering Eq. [63], where the surface properties $\mathcal{P}$ and ray direction $\boldsymbol{r} \in \mathbb{R}^3$ are used as inputs:

$$\begin{aligned} \Phi_{\Theta_c} &: \mathbb{R}^3 \times \mathbb{R}^n \mapsto \mathbb{R}^3, \\ \boldsymbol{c} &= \Phi_{\Theta_c}(\boldsymbol{r}, \mathcal{P}). \end{aligned} \tag{5.38}$$

Therefore, the view synthesis task can be framed as learning the following mapping $\mathbf{F}_\Theta = \Psi_{\Theta_p} \circ \Phi_{\Theta_c}$:

$$\begin{aligned} \mathbf{F}_\Theta &: \mathbb{R}^{Z_s} \times \mathbb{R}^3 \mapsto \mathbb{R}^3, \\ \boldsymbol{c} &= \mathbf{F}_\Theta(\psi(\hat{\boldsymbol{x}}; \mathcal{Z}_s), \boldsymbol{r}). \end{aligned} \tag{5.39}$$

Unlike neural textures [124], which laboriously store the learned high-dimensional per-texel features based on a global UV parameterization, our representation obtains superior compactness while modeling complex scene properties.

Sampling. The proposed neural micro surface representation continuously represents the local scene properties on the manifold. To render and optimize the neural representation, a sampling mechanism is needed. Similar to [81, 154], we apply a differentiable rendering framework to rasterize the scene surface using the z-buffer and aggregate nearby predictions to produce a smooth scene rendering result.

With this approach, view synthesis is performed by tracing the pixel rays starting at the camera center and querying the neural networks at the intersecting points. For each pixel ray $\boldsymbol{r}$ traced from camera k, the rendered color $\hat{\boldsymbol{c}}_k(\boldsymbol{r})$ is computed as the mean value for a set of rasterized color predictions:

$$\hat{\boldsymbol{c}}_k(\boldsymbol{r}) = \frac{\sum_{i \in \mathcal{M}_k^r} \mathbf{F}_\Theta(\psi(\hat{\boldsymbol{x}}_i; \mathcal{Z}_s), \boldsymbol{r})}{\sum_{i \in \mathcal{M}_k^r} \mathbf{1}}, \tag{5.40}$$

where $\hat{\boldsymbol{x}}_i$ is the texture coordinate of the ray's intersection on the mesh triangle with index i, and $\mathcal{M}_k^r$ denotes the set of triangle indices selected from the z-buffer. Note that only the first $\mathcal{N}_f$ triangles closest to the camera center are selected, and triangles far from the first intersection in terms of the depth value are removed from the selection $\mathcal{M}_k^r$. We empirically find that the number of samples $\mathcal{N}_f$ is important only when the resolution of the rendered image is much higher than that of the initial triangulation.

Detailed Surface Rendering

Displacement Embedding. Although the previously proposed neural micro surface representation outperforms current neural representations by disentangling appearance from geometry and substantially enhancing the representation power of the implicit function, our representation has difficulty rendering realistic micro details with its straightforward optimization approach. The neural micro surface is defined on explicit surface patches obtained from standard multiview stereo and surface reconstruction techniques, which inevitably accumulate errors. As a result, the imprecise surface geometry hinders the representation power of the implicit function, especially for real-world complex scenes with only sparse observations.

Instead of directly optimizing the surface geometry using multiview constraints, another MLP network $\mathbf{G}_{\Theta_r}$ is introduced to compensate for the imperfect geometry by implicitly modeling the micro geometric displacement $\boldsymbol{\delta}_d$ in latent space:

$$
\begin{aligned}
\mathbf{G}_{\Theta_r} &: \mathbb{R}^2 \times \mathbb{R}^3 \times \mathbb{R}^{Z_d} \mapsto \mathbb{R}^{Z_d}, \\
\boldsymbol{\delta}_d &= \mathbf{G}_{\Theta_r}(\hat{\boldsymbol{x}}, \boldsymbol{r}, z_d), \\
\hat{z}_d &= z_d + \boldsymbol{\delta}_d.
\end{aligned} \tag{5.41}
$$

Similar to the texture latent grid $\mathcal{Z}_s$ introduced in Eq. (5.37), each atlas is augmented with a geometric latent grid $\mathcal{Z}_d \in \mathbb{R}^{Z_d}$ to capture the local surface geometry information. Here, $z_d = \psi(\hat{\boldsymbol{x}}; \mathcal{Z}_d)$ denotes the interpolated feature vector at texture coordinate $\hat{\boldsymbol{x}}$, $\psi(\cdot)$ denotes the bilinear interpolation operation, and $\boldsymbol{r} \in \mathbb{R}^3$ represents the camera ray direction. The displacement embedding network $\mathbf{G}_{\Theta_r}$ implicitly remaps the initial geometry along every ray direction $\boldsymbol{r}$ to a shared Z_d-dimensional feature space by with a feature metric-based displacement prediction approach.

Intuitively, the displacement embedding maps disparate coordinates or geometric features on the imperfect initial triangulations based on the ground truth 3D points to the same locations in the latent feature space. By exploiting the inductive bias, the MLP neural network learns a nonlinear, many-to-one mapping that matches the distinct initial features with incorrect geometries but similar colors to nearby latent regions. This approach implicitly incentivizes multiview correspondence, thus allowing the shading function Φ_{Θ_c} to better explain the color deviation among the input views, which improves the expressivity of the implicit function and substantially improves the view synthesis quality.

Unlike traditional displacement mapping methods, which represent the displacement as a 1D scalar along the normal direction or as a global 3D vector, the neural micro surface representation embeds the per-ray geometry into high-dimensional latent space with an implicit function. This approach obtains superior performance since the feature space offers more versatile reparameterizations, potentially allowing for more detailed feature metric-based correspondence.

With this approach, the surface mapping network Ψ_{Θ_p} originally defined in Eq. (5.37) can be reformulated as

$$\begin{aligned} \Psi_{\Theta_p} &: \mathbb{R}^{Z_d} \times \mathbb{R}^{Z_s} \mapsto \mathbb{R}^n, \\ \mathcal{P} &= \Psi_{\Theta_p}(\hat{z}_d, z_s), \end{aligned} \tag{5.42}$$

where $\hat{z}_d \in \mathbb{R}^{Z_d}$ is the remapped latent vector depicting the precise surface geometry.

Reflectance Modeling. Assuming that the surface does not emit light, based on the rendering equation [63], the observed view-dependent color $\mathcal{L}_o(\boldsymbol{x}, \boldsymbol{\omega}_o)$ for a 3D point $\boldsymbol{x}$ with normal vector $\boldsymbol{n}$ can be calculated as an integral of the illumination intensity $\mathcal{L}_i(\boldsymbol{x}, \boldsymbol{\omega}_i) \in \mathbb{R}^3$ over the upper hemisphere $\Omega = \{\boldsymbol{\omega}_i : \boldsymbol{\omega}_i \cdot \boldsymbol{n} > 0\}$:

$$\mathcal{L}_o(\boldsymbol{x}, \boldsymbol{\omega}_o) = \int_{\Omega} \mathcal{L}_i(\boldsymbol{x}, \boldsymbol{\omega}_i)\mathcal{F}_r(\boldsymbol{x}, \boldsymbol{\omega}_o, \boldsymbol{\omega}_i)(\boldsymbol{\omega}_i \cdot \boldsymbol{n})d\boldsymbol{\omega}_i, \tag{5.43}$$

where $\boldsymbol{\omega}_i \in \mathbb{R}^3$ is a unit vector denoting the illumination direction. In addition, $\boldsymbol{\omega}_o \in \mathbb{R}^3$ is the outgoing view direction, which is a unit vector opposite to the ray direction $\boldsymbol{r}$ pointing from a spatial point to the camera center that is defined as $\boldsymbol{\omega}_o = -\boldsymbol{r}$. Furthermore, $\mathcal{F}_r(\boldsymbol{x}, \boldsymbol{\omega}_o, \boldsymbol{\omega}_i)$ is the BRDF parameter composed of diffuse and specular terms, which is defined as

$$\mathcal{F}_r(\boldsymbol{x}, \boldsymbol{\omega}_o, \boldsymbol{\omega}_i) = \boldsymbol{a}_d + \mathcal{F}_s(\boldsymbol{x}, \boldsymbol{\omega}_o, \boldsymbol{\omega}_i). \tag{5.44}$$

With this approach, the formulation in Eq. (5.43) can be decomposed into a view-independent diffuse color $\boldsymbol{d}$ and a view-dependent specular reflectance $\boldsymbol{s}$ as $\mathcal{L}_o(\boldsymbol{x}, \boldsymbol{\omega}_o) = \boldsymbol{d} + \boldsymbol{s}$. Following [10, 65, 67, 127], ignoring self-occlusions and inter-reflections and neglecting exposure variations and tone mapping, the illumination $\mathcal{L}_i(\boldsymbol{x}, \boldsymbol{\omega}_i)$ can be preintegrated, allowing the diffuse and specular components to be approximated as

$$\mathcal{L}_o(\boldsymbol{x}, \boldsymbol{\omega}_o) = \boldsymbol{d} + \boldsymbol{s} \approx \boldsymbol{a}_d\mathcal{L}_d^*(\boldsymbol{x}, \boldsymbol{n}) + \boldsymbol{b}_s\mathcal{L}_s^*(\boldsymbol{x}, \boldsymbol{\omega}_r), \tag{5.45}$$

Where $\boldsymbol{a}_d$ and $\boldsymbol{b}_s$ are the diffuse and specular albedo, respectively. $\mathcal{L}_d^*(\boldsymbol{x}, \boldsymbol{n})$ is the preintegrated illumination component for the diffuse term, and $\mathcal{L}_s^*(\boldsymbol{x}, \boldsymbol{\omega}_r)$ is the preintegrated illumination component for the specular term, which depends on the mirrored view direction $\boldsymbol{\omega}_r$, which is the reflection of the view direction $\boldsymbol{\omega}_o$ about the surface normal $\boldsymbol{n}$:

$$\boldsymbol{\omega}_r = 2(\boldsymbol{\omega}_o \cdot \boldsymbol{n})\boldsymbol{n} - \boldsymbol{\omega}_o. \tag{5.46}$$

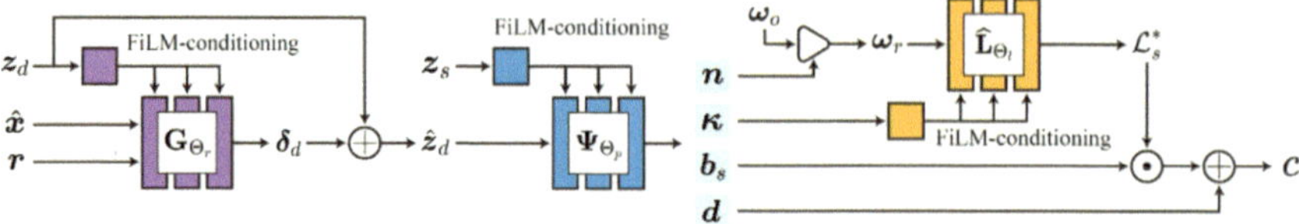

Fig. 5.34 The network architecture of the neural micro surface representation. The feature metric-based multiview correspondence is reinforced by predicting the latent space residual through a displacement embedding-based MLP, whose inputs are the original geometric feature vector, texture coordinate and ray direction. Then, the scene properties are continuously represented by a property mapping-based MLP. An illumination-based MLP is then used to implicitly account for the lighting and reflectance conditions

The detailed network architecture of our rendering framework is shown in Fig. 5.34. The mapping network $\boldsymbol{\Psi}_{\Theta_p}$ uses the local geometric and textural latent vectors as inputs and predicts the spatially varying surface properties, including the diffuse color $\boldsymbol{d} \in \mathbb{R}^3$, normal vector $\boldsymbol{n} \in \mathbb{R}^3$, and specular albedo $\boldsymbol{b}_s \in \mathbb{R}^3$, as well as the high-dimensional spatial feature $\boldsymbol{\kappa} \in \mathbb{R}^F$ encoding the material information:

$$\mathcal{P} = (\boldsymbol{n}, \boldsymbol{d}, \boldsymbol{b}_s, \boldsymbol{\kappa}). \tag{5.47}$$

Note that the diffuse color $\boldsymbol{d}$ is directly predicted by the surface mapping network $\boldsymbol{\Psi}_{\Theta_p}$ since the diffuse albedo and illumination term defined in Eq. (5.45) are determined only by the spatial surface properties.

According to Eq. (5.45), another illumination network $\widehat{\mathbf{L}}_{\Theta_l}$ is used to approximate the preintegrated illumination component of the specularity. This network uses the mirrored view direction $\boldsymbol{\omega}_r$ and spatial feature $\boldsymbol{\kappa}$ as inputs to account for the implicit BRDF reasoning:

$$\begin{aligned} \widehat{\mathbf{L}}_{\Theta_l} &: \mathbb{R}^3 \times \mathbb{R}^F \mapsto \mathbb{R}^3, \\ \mathcal{L}_s^* &= \widehat{\mathbf{L}}_{\Theta_l}(\boldsymbol{\omega}_r, \boldsymbol{\kappa}). \end{aligned} \tag{5.48}$$

As shown in Fig. 5.34, the latent vectors $z_d and z_s$ and the spatial feature $\boldsymbol{\kappa}$ are passed to the neural networks through FiLM conditioning [13, 31, 99]. The latent vector is not a network input and is instead an affine transformation for the intermediate features, thereby enhancing the ability of the network to represent high-frequency details.

Therefore, the overall color originally defined in Eq. (5.38) can be computed based on neural network queries as follows:

$$\boldsymbol{c} = \boldsymbol{d} + \boldsymbol{b}_s \mathcal{L}_s^*. \tag{5.49}$$

Optimization. Similar to other neural rendering frameworks, the L1 loss between the rendered and ground truth pixel color is used as the loss function:

$$l_c = \sum_{k \in \mathcal{V}} \sum_{\boldsymbol{r} \in \mathcal{R}_k} \left\| \hat{\boldsymbol{c}}_k(\boldsymbol{r}) I_k(\boldsymbol{r}) \right\|_1, \tag{5.50}$$

where $\mathcal{V}$ is the set of randomly selected camera views and $\mathcal{R}_k$ is a random set of pixel rays selected from view k.

Furthermore, we propose to guide the learning of the surface properties by constraining the normal prediction of the surface mapping network $\boldsymbol{\Psi}_{\Theta_p}$ using the coarse normal computed based on the input triangle mesh:

$$l_n = \sum_{k \in \mathcal{V}} \sum_{\boldsymbol{r} \in \mathcal{R}_k} \| \boldsymbol{n}_k(\boldsymbol{r}) N_k(\boldsymbol{r}) \|_2^2, \tag{5.51}$$

where $\boldsymbol{n}_k(r)$ denotes the normal prediction of the spatial point at which the pixel ray $\boldsymbol{r}$ traced from camera k intersects with the nearest triangle, and $N_k(\boldsymbol{r})$ is the normal vector derived based on the input geometry.

The color loss and the normal constraint are combined to calculate the final loss:

$$l = l_c + \lambda_n l_n. \tag{5.52}$$

where λ_n denotes a scalar balance weight.

Experiments

We quantitatively (Tables 5.7 and 5.8) and qualitatively (Figs. 5.35 and 5.36) show that our approach outperforms previous methods. First, we demonstrate that our neural micro surface obtains satisfactory view synthesis performance based on the Tanks and Temples [70] dataset. To investigate the scalability of our approach, we validate the performance of the model based on sparse-view high-resolution images in the COLMAP [111], ETH3D [112], and GigaMVS [151] datasets. The neural micro

Table 5.7 Quantitative results of state-of-the-art view synthesis algorithms based on the Tanks and Temples [70] dataset. The neural micro surface outperforms the state-of-the-art algorithms in terms of all evaluation metrics

Sparsity	3			5		
Method	PSNR	SSIM	LPIPS	PSNR	SSIM	LPIPS
Mip-NeRF [6]	19.84	0.758	0.306	17.38	0.730	0.320
Reg-NeRF [94]	19.80	0.755	0.312	17.59	0.737	0.321
NPBG [3]	23.39	0.789	0.141	22.51	0.773	0.189
IBRNet [129]	25.83	0.852	0.154	20.08	0.775	0.278
Neural micro surface	**27.61**	**0.888**	**0.112**	**25.06**	**0.844**	**0.150**

Table 5.8 Quantitative results based on large-scale challenging scenes in the COLMAP [111], ETH3D [112], and GigaMVS [151] datasets. The neural micro surface achieves state-of-the-art performance compared with all prior works for sparse-view high-resolution images

Method	PSNR	SSIM	LPIPS
Mip-NeRF [6]	17.46	0.519	0.539
Reg-NeRF [94]	16.83	0.491	0.603
SVS [105]	17.37	0.585	0.435
NPBG [3]	18.03	0.494	0.333
IBRNet [129]	18.07	0.567	0.402
Neural micro surface	**20.79**	**0.647**	**0.243**

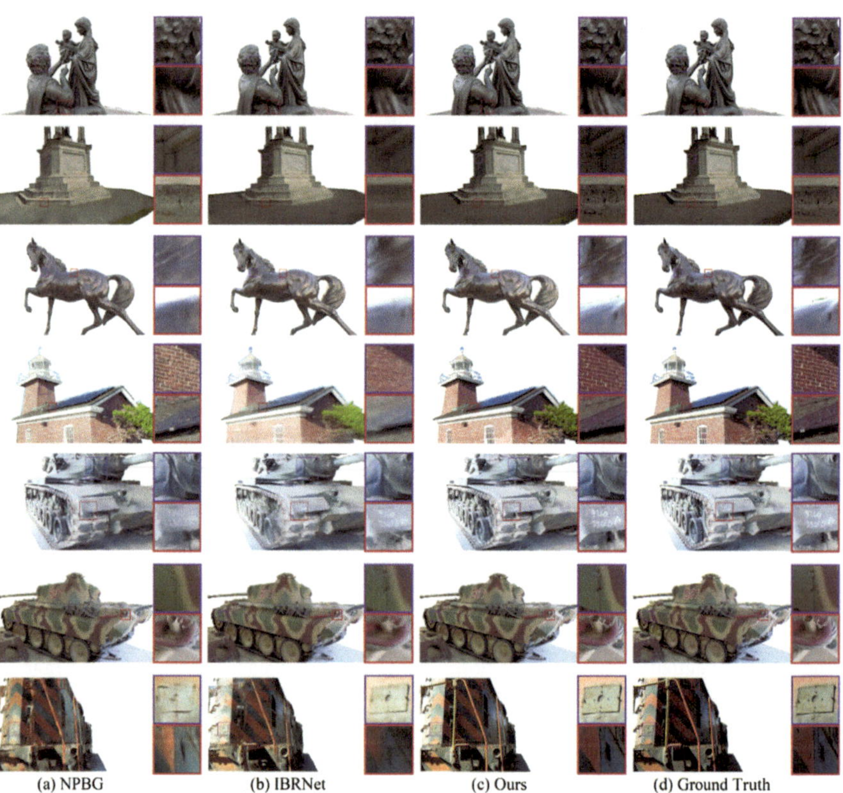

Fig. 5.35 Visual comparisons based on the intermediate set in the Tanks and Temples dataset [70] when the input sparsity is 3. From left to right, the NPBG [3], IBRNet [129], neural micro surface, and ground truth results. The neural micro surface captures fine-grained details with more realistic view-dependent appearance

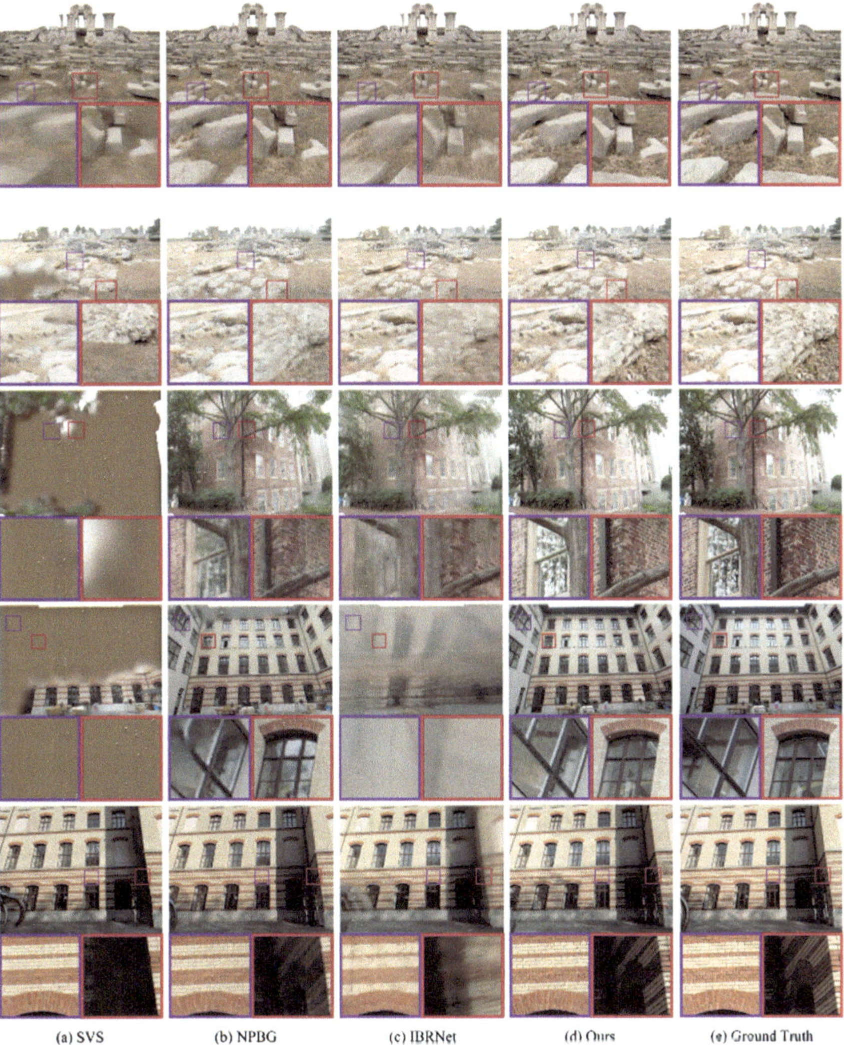

Fig. 5.36 Novel view synthesis of the Great Fountain, Haiyan Hall, South Building, Courtyard, and Facade from the GigaMVS [151], COLMAP [111], and ETH3D [112] datasets, compared with the SVS [105], NPBG [3], and IBRNet [129] results. The neural micro surface synthesizes more detailed and realistic renderings with a complex view-dependent appearance

surfaces show significantly better performance for real-world challenging scenes with only sparse observations than the other methods.

Tanks and Temples Dataset. We compare our method with current state-of-the-art methods, including Mip-NeRF [6], RegNeRF [94], NPBG [3], and IBRNet [129], based on seven complicated real-world scenes in the intermediate set in the Tanks and Temples dataset [70]. A discontinuous collection of target views is used for evaluation, and a set of images is chosen to serve as training views or source images for each scene. Following [60], we select a small proportion of all available training views by consecutively sampling a view from every sparsity $= n$ camera index, i.e., $\{1, n+1, 2n+1, \ldots\}$. Two different sparse-view settings (i.e., sparsity $= 3, 5$) are used, and the corresponding quantitative results are reported in Table 5.7. For scene-specific methods, including Mip-NeRF [6], RegNeRF [94], and our neural micro surface approach, per-scene optimization is performed. For NPBG [3], we finetune the pretrained CNN renderer while optimizing the neural point descriptor from scratch for each scene. For IBRNet [129], we separately finetune the model based on each scene. All methods were assessed based on the widely used PSNR and SSIM metrics (where higher values indicate better performance) and LPIPS metric [156] (where lower values indicate better performance). Note that we run the multiview stereo algorithms using the same sparsity to obtain the input geometries for the NPBG method and our method. We apply the same foreground mask obtained from the dense multiview stereo task before evaluating the metrics, as in [80]. Our neural micro surface approach demonstrates better performance than all baseline methods under the sparse observation setting. Notably, NeRF-based methods such as [6, 94] typically fail in scenarios for which only large baseline observations are available.

The qualitative results for a sparsity of 3 are shown in Fig. 5.35. Note that NPBG [3] often fails to reproduce the view-dependent color with high fidelity, such as the shadows in the Francis (scene 2, Fig. 5.35a) and Lighthouse results (scene 4, Fig. 5.35a) and the specularity in the Horse (scene 3, Fig. 5.35a) and Train results (scene 7, Fig. 5.35a). The results of IBRNet [129] in Fig. 5.35b contain blurry artifacts and lack detailed contents due to the underconstrained geometry and sparse inputs. Our approach more realistically captures fine-grained view-dependent geometric elements, as shown in Fig. 5.35c. The zoomed-in views reveal that our neural micro surface representation more accurately reconstructs high-frequency details, such as the fine-grained texture on the base of the sculpture in the Francis results (scene 2, Fig. 5.35c) and the delicate wrinkles, sharp characters and edges in the M60 result (scene 5, Fig. 5.35c). Moreover, our representation obtains more realistic view-dependent appearance, especially for regions with complex specular reflections, e.g., the specularity in the Horse result (scene 3, Fig. 5.35c) and the roof in the Lighthouse result (scene 4, Fig. 5.35c).

High-Resolution Benchmark. We next show our experimental results for more challenging scenarios, i.e., ultrahigh-resolution view synthesis for large-scale complex scenes based on only sparse observations. The resolution of the sparse-view inputs and target views is typically more than 4K and may even reach the gigapixel level. In this section, the experiments are conducted with a single sparsity setting, i.e., sparsity = 3. We select 5 challenging scenes: South Building in the COLMAP dataset [111], Courtyard and Facade in the ETH3D dataset [112], and Great Fountain, Haiyan Hall, Library, and The Old Gate in the GigaMVS [151] dataset.

Table 5.8 presents the quantitative findings, and qualitative visual comparisons are shown in Fig. 5.36. Our method significantly outperforms previous state-of-the-art techniques for a variety of difficult real-world settings, in terms of both quantitative and qualitative indicators. As shown in Table 5.8, Mip-NeRF [6] and RegNeRF [94] produce severely blurry and distorted synthesis results in all cases, most likely due to their poorly estimated geometry, with significant errors in the radiance function, and the results do not generalize to large-scale 360-degree unbounded scenes with large baseline inputs. SVS [105] shows good performance only when the target contents are well observed in nearby source views; however, SVS generates excessive blur or cracked artifacts when the baseline angles among the selected source views are sufficiently large, as shown in Fig. 5.36a. Similarly, IBRNet [129] produces results with ghosting or blurry artifacts and often fails to reproduce high-frequency details since significant errors are produced by the entangled volumetric geometry. In contrast, NPBG [3] successfully handles large baseline images, and reasonable synthesis results are achieved in all challenging scenarios. However, due to its large dependence on the input point cloud, the resolution of the fine details in the synthesis results is limited, and noisy artifacts are often observed. In addition, NPBG struggles to reconstruct complex view-dependent effects. In contrast to prior works, our method obtains robust, high-resolution view synthesis results with fine-grained details and photorealistic specularity, as shown in Fig. 5.36d. Although IBRNet achieves a slightly better LPIPS score for the Great Fountain case (scene 1, Fig. 5.36), our method consistently produces accurate details with fewer blurry artifacts and more realistic color, e.g., the complicated textures of the grass and stones in the Great Fountain and Haiyan Hall examples (scene 2, Fig. 5.36), the high-resolution structures of the bricks in the South Building (scene 3, Fig. 5.36), Courtyard (scene 4, Fig. 5.36), and Facade cases (scene 5, Fig. 5.36), and the high-fidelity specular reflections that appear on the glass in the South Building and Courtyard cases.

In general, synthesizing fine-grained appearance details and reproducing complex views is quite challenging, particularly for large-baseline sparse observations. The neural micro surface representation successfully addresses these issues by learning implicit fields on local surface manifolds and incentivizing feature metric-based correspondence to enhance micro geometric details. Thus, our approach effectively improves the expressivity of the reflectance model and obtains detailed renderings with realistic color.

The neural micro surface significantly outperforms the current state-of-the-art algorithms for a variety of difficult real-world scenes with sparse observations, including scenes in the GigaMVS [151], ETH3D [112], Tanks and Temples [70], and COLMAP [111] datasets. The neural micro surface representations enable photorealistic rendering of complicated details and are the first attempt toward unprecedented gigapixel-level view synthesis.

5.4 Semantic-Aware Reconstruction for Large-Scale Scenes

Semantics are important information when humans construct their cognition of a scene. Semantic-level 3D scene reconstruction and understanding has received much attention in the computer vision, robotics, and autonomous system communities. By revealing the semantic information underlying an observation, such as the geometric structure and material properties, scene-aware 3D reconstruction methods aim to restore the scene model in a more holistic manner. Based on the scene reconstruction results, the aim of the 3D understanding task is to analyze the 3D model to identify the semantic information of the scene, such as the object category and room structure [27, 50].

From a semantic point of view, the hidden information in the scene is deeply explored, and this section discusses scene reconstruction and scene understanding. We first discuss object-level semantic understanding in indoor scenes, enabling semantic segmentation and instance segmentation. Then, an efficient and incremental instance segmentation architecture is introduced. Finally, we apply semantic segmentation and instance segmentation schemes to large-scale scene reconstruction tasks, achieving the goals of efficient large-scale scene reconstruction and semantic understanding.

5.4.1 Occupancy-Based Semantic Instance Segmentation for Scene Understanding

Computer vision research has covered a wide range of topics on 3D scene semantic and instance segmentation. The most recent deep learning-based methods for semantic segmentation can be divided into two categories based on the type of convolution: point-based [53, 101, 103, 125, 135] and voxel-based [23, 44, 54]. Voxel-based approaches are the main focus in this work. Voxel-based approaches apply 3D convolutions to the voxel grid after receiving voxelized point clouds as input. [25, 52] are two examples of early works that employed dense 3D convolutions. However, the methods proposed in these works could not handle large-scale voxel grids because of the high computational costs of analyzing high-dimensional data. Later, the sparse convolution, which takes advantage of the 3D point cloud's inherent sparsity and obtains state-of-the-art segmentation accuracy, emerged, resolving this crucial computational issue. Hu [54] then suggested training 2D and 3D networks together to achieve the best results. Sparse convolutional networks are also commonly used for segmentation [61, 73]. Lahoud [73] presented a learning-then-clustering strategy, in which mean shift clustering was performed based on the per-point characteristics recovered by a sparse convolutional network. Jiang [61] proposed performing clustering based on both the original and shifted coordinates and used another 3D network to forecast the scores based on the results.

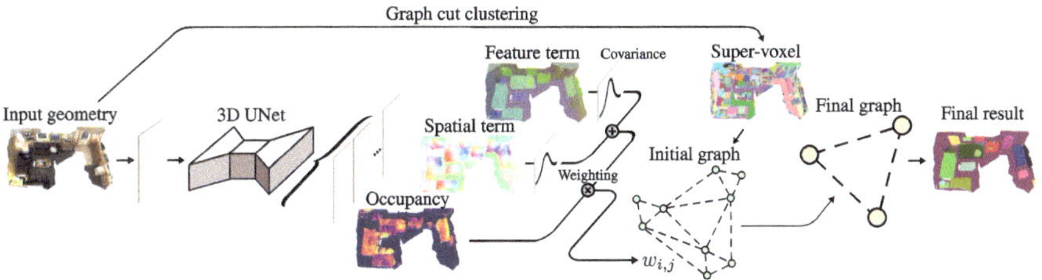

Fig. 5.37 Overview of the occupancy-based semantic instance segmentation pipeline

Although existing 3D segmentation methods have achieved exciting results, they are still sensitive to the incompleteness and noise inherent in 3D input models, leading to unsatisfactory segmentation accuracy. Due to the noise in the data collected by RGBD sensors, some instances may be highly incomplete, leading to irregular and ambiguous geometry. Han et al. [50] argued that existing instance 3D segmentation methods [73, 130] are susceptible to false predictions under these circumstances because they fail to introduce induced biases to constrain the plausibility of the prediction. Typical errors include noisy fractions detected as normal instances.

To address the limitations of existing methods, Han et al. [50] proposed an occupancy term to regularize the instance segmentation process. For each point, the occupancy term predicts the number of points, i.e., the size of its corresponding object. This is possible because 3D models do not have scale and occlusion ambiguities like 2D images, and the sizes of certain classes of objects are predictable.

An overview of the proposed occupancy-based semantic instance segmentation pipeline is shown in Fig. 5.37. The pipeline can be divided into two stages: the network stage and the clustering stage. The network stage uses a voxelized 3D point cloud as input and generates per-point feature embeddings for instance segmentation and semantic classification. A UNet-like 3D sparse convolution network is used in this stage; this network is built using the sparse convolution operations introduced by Graham [44]. In the clustering stage, clustering is performed based on the learned features to generate the final instance proposals.

Network Stage

The purpose of this stage is to learn the per-point features based on the input point cloud using a UNet-like sparse convolution network. Multiple features are required for semantic instance segmentation, including (1) $\mathbf{c}_i$, the semantic label, which provides the per-point classification information; (2) $\mathbf{s}_i$ and $\mathbf{v}_i$, the semantic and spatial embeddings; (3) $\mathbf{b}_i$, the covariance, which is used to fuse $\mathbf{s}_i$ and $\mathbf{v}_i$; and (4) o_i, the occupancy signal, which predicts the size of the target instance, as mentioned above.

Network Architecture. Similar to SCN [44], a UNet-like 3D network with sparse convolutions is adopted. The network architecture is shown in Fig. 5.38. After the input layer transforms the input point cloud into voxels with size 0.02 m, a sparse

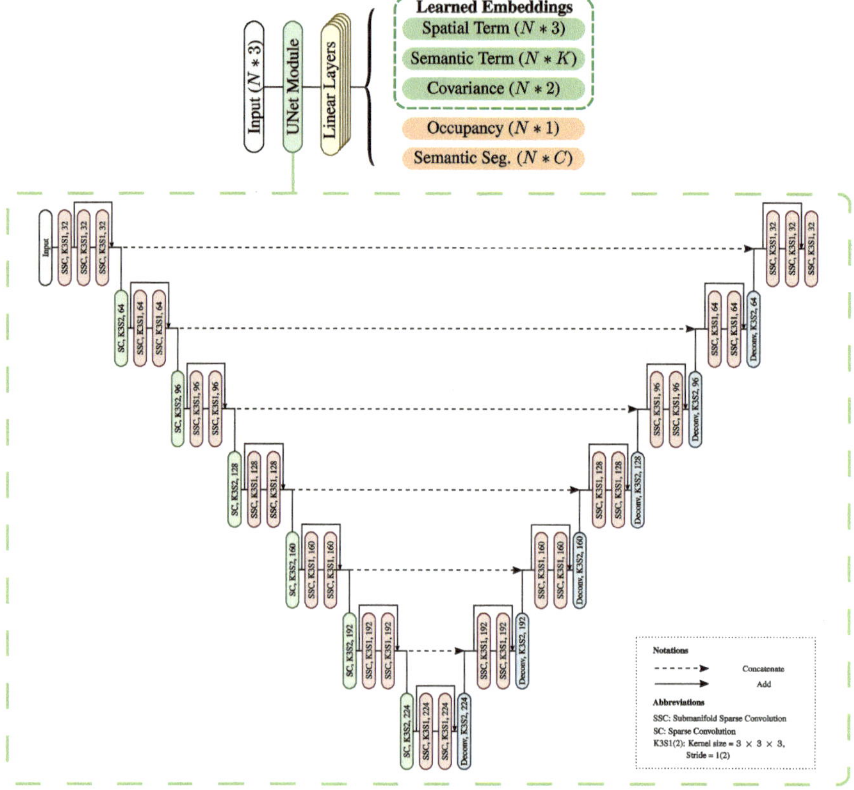

Fig. 5.38 Architecture of the occupancy-based semantic instance segmentation network

convolution UNet module is used to extract multilevel features via the downsampling and upsampling architecture. Different linear layers are used to output the aforementioned terms.

Network Loss. The final loss is composed of three parts:

$$\mathcal{L}_{\text{total}} = \mathcal{L}_{\text{semantic}} + \mathcal{L}_{\text{embedding}} + \mathcal{L}_{\text{occupancy}}, \tag{5.53}$$

where $\mathcal{L}_{\text{semantic}}, \mathcal{L}_{\text{embedding}}, \mathcal{L}_{\text{occupancy}}$ are the semantic classification, embedding learning and occupancy prediction losses, respectively.

The semantic classification loss $\mathcal{L}_{\text{semantic}}$ is used to supervise the semantic classification process. For each point, the network outputs the semantic probabilities of each class. The loss $\mathcal{L}_{\text{semantic}}$ is calculated using the cross-entropy loss, as in SCN [44].

For instance segmentation embedding learning, previous methods such as that by Novotny [95] directly concatenated the semantic and spatial embeddings. However, to independently supervise the learning process of the two embeddings, we propose to explicitly separate them and supervise them with different loss functions.

The motivation is that the spatial and semantic embeddings have different underlying physical meanings and thus inherently have different scales. Therefore, we argue that they should be regularized using different cost functions. The spatial and semantic embeddings are fused using the covariance prediction, which determines which embedding is more reliable for each point. The embedding learning loss $\mathcal{L}_{\text{embedding}}$ is composed of three terms, including the semantic term $\mathcal{L}_{\text{se}}$, the spatial term $\mathcal{L}_{\text{sp}}$, and the covariance term $\mathcal{L}_{\text{cv}}$:

$$\mathcal{L}_{\text{embedding}} = \mathcal{L}_{\text{sp}} + \mathcal{L}_{\text{se}} + \mathcal{L}_{\text{cv}}. \tag{5.54}$$

Spatial Term. Similar to VoteNet [102], we predict the spatial embedding $\mathbf{v}_i \in \mathbb{R}^3$ for each point, which is a vector pointing from the point toward its target object center. This term is supervised with the $L2$ loss as follows:

$$\mathcal{L}_{\text{sp}} = \frac{1}{M} \sum_{m=1}^{M} \frac{1}{N_m} \sum_{k=1}^{N_m} ||\mathbf{v}_k + \mathbf{x}_k - \frac{1}{N_m} \sum_{j=1}^{N_m} \mathbf{x}_j||, \tag{5.55}$$

where M represents the total number of instances in the input 3D model and N_m represents the voxel number of the c-th instance. $\mathbf{x}_k$ denotes the 3D coordinates of the k-th voxel belonging to the m-th instance.

Semantic Term. A semantic embedding $\mathbf{s}_k$ is learned for each point, supervised by the discriminative loss [73], which includes three parts:

$$\mathcal{L}_{\text{se}} = \mathcal{L}_{\text{variance}} + \mathcal{L}_{\text{distance}} + \mathcal{L}_{\text{reg}}, \tag{5.56}$$

where $\mathcal{L}_{\text{variance}}$ is a term that attempts to push the embeddings of an instance toward its center in the embedding space. $\mathcal{L}_{\text{distance}}$ is a distance term that pushes the embeddings of different instances away from each other. Finally, the regularization term $\mathcal{L}_{\text{reg}}$ ensures that the output remains bounded by pushing all instance embeddings toward the origin. The detailed equations for each term are as follows:

$$\mathcal{L}_{\text{variance}} = \frac{1}{M} \sum_{m=1}^{M} \frac{1}{N_m} \sum_{k=1}^{N_M} [||\mathbf{u}_m - \mathbf{s}_k|| - \delta_v]_+^2, \tag{5.57}$$

$$\mathcal{L}_{\text{distance}} = \frac{1}{M(M-1)} \sum_{m_A=1}^{M} \sum_{m_B=m_A+1}^{M} [2\delta_d - ||\mathbf{u}_{m_A} - \mathbf{u}_{m_B}||]_+^2, \tag{5.58}$$

$$\mathcal{L}_{\text{reg}} = \frac{1}{M} \sum_{m=1}^{M} ||\mathbf{u}_m||. \tag{5.59}$$

Here, $\mathbf{u}_m = \frac{1}{N_m}\sum_{k=1}^{N_m} \mathbf{s}_k$ represents the average semantic term of the m-th instance. δ_v and δ_d are two hyperparameters that are set to 0.1 and 1.5, respectively.

Covariance Term. Since the semantic and spatial terms both determine which instance each voxel belongs to, fusing the information of the semantic and spatial terms is important. Instead of using a fixed weight, we propose to predict a covariance term to adaptively determine the clustering range based on the two embeddings. Specifically, we let $\mathbf{b}_k = (\sigma_s^k, \sigma_v^k)$ denote the covariance predicted for the semantic and spatial embeddings of the k-th voxel of the m-th instance. We obtain the covariance for the m-th instance, i.e., (σ_s^m, σ_v^m), by calculating the average of $\mathbf{b}_k$. Then, we calculate the probability of voxel k belonging to instance m, which is denoted as p_k^m, with the following formula:

$$p_k^m = \exp\left(-\left(\frac{||\mathbf{s}_k - \mathbf{u}_m||}{\sigma_s^m}\right)^2 - \left(\frac{||\mathbf{x}_k + \mathbf{v}_k - \mathbf{e}_m||}{\sigma_v^m}\right)^2\right), \tag{5.60}$$

where $\mathbf{e}_m = \frac{1}{N_m}\sum_{j=0}^{N_m}(\mathbf{x}_j + \mathbf{v}_j)$ is the predicted center of the m-th instance. Learning p_k^m can be considered a binary classification problem, which is supervised by a binary cross-entropy loss:

$$\mathcal{L}_{\text{cv}} = -\frac{1}{M}\sum_{m=1}^{M}\frac{1}{N}\sum_{k=1}^{N}[y_k \ln(p_k^m) + (1 - y_k)\ln(1 - p_k^m)], \tag{5.61}$$

where $y_k = 1$ indicates that voxel k belongs to instance m and $y_k = 0$ otherwise. N represents the total number of active voxels.

Occupancy Prediction. For each voxel in the input model, a positive scalar called the occupancy is predicted to estimate the size of its target instance. We denote the predicted occupancy for the k-th voxel as o_k. This term is supervised by the following occupancy loss:

$$\mathcal{L}_{\text{occupancy}} = \frac{1}{M}\sum_{m=1}^{M}\frac{1}{N_m}\sum_{k=1}^{N_m}||o_k - \ln(N_m)||, \tag{5.62}$$

where N_m represents the voxel number of the m-th instance. Note that instead of directly predicting the occupancy value, o_k is supervised to regress to the logarithm of the actual occupancy value to ensure numerical stability.

After the per-point occupancy o_k is predicted, the occupancy values of a proposed instance are averaged to predict the occupancy for the instance proposal.

Note that in practice, these three loss terms are assigned different scale factors before being summed as the total loss.

Clustering Stage

After the network stage, the semantic segmentation labels can be directly obtained based on the per-point semantic classification probabilities. However, instance segmentation requires a clustering stage. The aim of the clustering stage is to generate the final instance proposals based on the per-point features predicted using the network. Instead of using conventional approaches such as mean shift to cluster instances [73], we adopt a bottom-up graph merging approach, in which the predicted occupancy value is introduced into the clustering process as an additional constraint. We initialize the graph by applying an existing supervoxel segmentation algorithm [26] to the input point cloud. This preprocessing step helps to reduce the initial graph size, which accelerates the later iterative merging process. For each supervoxel i, we determine the average spatial center C_i and the average occupancy O_i with the following formulas:

$$\mathbf{C}_i = \frac{1}{|\Omega_i|}\sum_{k\in\Omega_i}(\mathbf{v}_k + \mathbf{x}_k), \tag{5.63}$$

$$\mathbf{O}_i = \frac{1}{|\Omega_i|}\sum_{k\in\Omega_i} o_k, \tag{5.64}$$

where Ω_i is the set of voxels inside the i-th supervoxel.

On the basis of the supervoxel segmentation result, we build an undirected graph representation as $G = (V, E, W)$, where each vertex v_i represents a supervoxel. an edge An edge $e_{i,j} \in E$ exists between each pair of vertices v_i, v_j with a weight $w_{i,j} \in W$. The weight $w_{i,j}$ indicates the probability that the connected nodes v_i, v_j belong to the same instance and can be calculated as

$$w_{i,j} = \frac{\exp\left(-\left(\frac{||\mathbf{S_i}-\mathbf{S_j}||}{\sigma_s}\right)^2 - \left(\frac{||\mathbf{C_i}-\mathbf{C_j}||}{\sigma_d}\right)^2\right)}{\max(r_{i,j}, r_0)}, \tag{5.65}$$

$$r_{i,j} = \frac{O_{i,j}}{|\Omega_{i,j}|}. \tag{5.66}$$

where r_0 is a threshold, which is set to 0.5. The numerator of $w_{i,j}$ represents the similarity measure based on the distance to the corresponding voting center. The denominator acts as an adaptive weight to adjust $w_{i,j}$ by evaluating whether the predicted occupancy is consistent with the actual occupancy. Specifically, we define $\Omega_{i,j} = \Omega_i \Cap \Omega_j$ as the resulting node after merging two supervoxels, and $O_{i,j}$ is calculated using Eq. (5.64). Then, the occupancy ratio $r_{i,j}$ is defined as the size of the merged node divided by the predicted occupancy, as shown in Eq. (5.66).

Fig. 5.39 Qualitative instance segmentation comparisons with an existing method [73] based on the ScanNetV2 [26] validation set. **a** GT. **b** Ours. **c** Lahoud [73]

If $r << 1$, the current node is smaller than expected, indicating oversegmentation. A smaller r value leads to a larger $w_{i,j}$, which encourages node merging so that the node size approaches the predicted value. Otherwise, if $r >> 1$, the weight $w_{i,j}$ is decreased, and merging is discouraged. With this approach, we generate plausible instance proposals and eliminate noisy predictions, as shown in Fig. 5.39.

We merge the graph in an iterative manner, as explained in Algorithm 1. The parameter T_0 serves as the stop condition and is set to 0.5. For each iteration, we merge the edges with the largest weights until no edges have weights larger than T_0. The final graph is output as the instance segmentation result.

Algorithm 1 Iterative occupancy guided instance clustering

Require: Current graph (G, E, W), threshold T_0.
procedure MERGE GRAPH
  $(i, j) \leftarrow argmax_{i,j} W_{i,j}$ ▷ Find the edge with the largest weight.
  while do$W_{i,j} > T_0$ Merge v_i, v_j. Update all weights.
    $(i, j) \leftarrow argmax_{i,j} W_{i,j}$

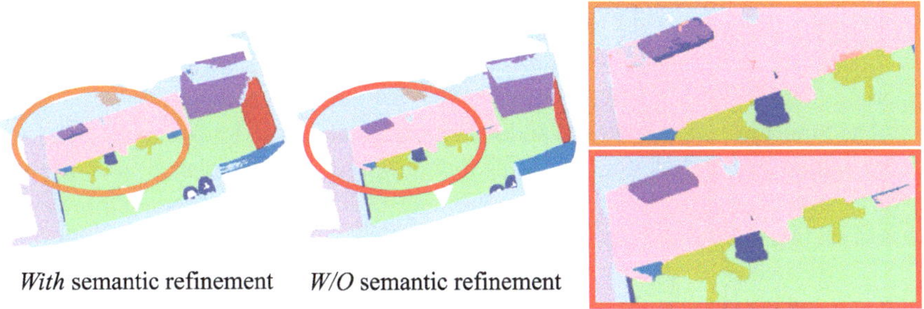

Fig. 5.40 The effect of semantic refinement on the semantic segmentation results

Instance-Guided Semantic Refinement. We use the instance segmentation results to refine the semantic prediction. Based on the merged graph G, for each node v_m, we let the semantic probability that each voxel belongs to the node be the average semantic probability within the node:

$$\mathbf{c}_i = \frac{1}{|\Omega_m|} \sum_{k \in \Omega_m} (\mathbf{c}_k), \text{ for each } i \in \Omega_m. \tag{5.67}$$

Since the instance clustering process explicitly encodes the spatial information, the semantic refinement process effectively reduces the prediction noise that often occurs at incomplete object boundaries, as shown in Fig. 5.40.

Experiment

In this section, we evaluate our 3D semantic and instance segmentation method based on public datasets. We first evaluate the offline 3D semantic instance segmentation accuracy of the occupancy-based semantic instance segmentation method. The online semantic and instance segmentation results are discussed in the next section. Then, an ablation study is performed and discussed.

Offline 3D Semantic and Instance Segmentation. In this subsection, we evaluate the offline segmentation results of the occupancy-based semantic instance segmentation pipeline. Several public datasets are used in the experiments, which we list as follows:

ScanNetv2 [26] consists of 1513 interior scenes with 3D semantic and instance labels for training and evaluation, as well as an additional test set of 100 scenes for benchmark evaluation. For the validation set results, we follow the same test/validation split as ScanNetv2 [26]. Quantitative comparisons of the semantic and instance segmentation results are shown in Tables 5.9 and 5.10, with our method denoted as **Ours (OFF)**. Note that **ON** and **OFF** represent online and offline, respectively. We separate the existing methods into two groups: online methods and offline methods. Note that our offline segmentation results are the best among the existing methods.

The qualitative instance segmentation results are shown in Fig. 5.41. Qualitative comparisons of the instance segmentation results are presented in Fig. 5.42. Our method obtains better qualitative results than the existing methods.

The Stanford Large-Scale 3D Indoor Space Dataset, also known as S3DIS [5], is a 3D dataset that includes 6 large-scale indoor scenes that cover an area of 6000 m^2. Thirteen object classes are included in this dataset. The instance segmentation results based on this dataset are summarized in Table 5.12.

The SceneNN [5] dataset includes 50 scenes in the training set and 26 scenes in the validation set. SceneNN is a small dataset, which is suitable for evaluating the generalizability of our method in scenarios with limited training data. The per-class instance segmentation results based on this dataset are reported in Table 5.11. Our method shows great generalizability for small datasets.

Ablation Study. We next perform an ablation study to analyze how occupancy prediction affects the instance segmentation accuracy. The effect of the proposed occupancy prediction is analyzed by comparing our method with the baseline method, which only uses the voting vector for instance clustering by setting the occupancy ratio r to a constant 1.0 in Eq. (5.65) in the clustering stage. We also ablate the semantic and spatial terms by separately removing them. We report the mAP, mAP@0.5 and mAP@0.25 values based on the ScanNetv2 validation set in Table 5.13. The results demonstrate that the occupancy prediction considerably improves the instance segmentation accuracy.

5.4.2 *Efficient and Incremental Sparse Convolutions for Online 3D Semantic Understanding*

Although voxel-based segmentation methods that use backbones such as SCN [44] achieve state-of-the-art performance on 3D semantic understanding problems [50], their computational complexity is too high for real-time applications on portable devices. In this chapter, an efficient sparse convolution approach is proposed to enable online inference with sparse convolution networks without reducing accuracy. Based on this approach, online 3D semantic and instance segmentation can be achieved.

We first introduce an adaptive chunk-based sparse convolution module. Then, we present an incremental sparse convolution inference framework and evaluate its real-time performance.

Table 5.9 Per-class semantic segmentation results based on the ScanNetV2 [26] test set in terms of the mIoU scores of 20 classes, using the m64 model

Method	**mIoU**	Bookshelf	Bed	Bath	Cabinet	Curtain	Counter	Chair	Desk	Other furniture	Floor	Door	Picture	Sink	Shower	Fridge	Sofa	Wall	Toilet	Table	Window
FA [152]	63.0	76.6	74.1	60.4	59.0	73.4	50.1	74.7	50.3	45.4	91.9	52.7	42.3	67.8	42.0	55.0	68.8	79.5	89.6	54.4	62.7
SV [57]	63.5	71.9	71.1	65.6	6[illegible].3	76.5	44.4	75.7	53.4	47.8	92.8	56.6	27.2	66.4	53.1	63.6	64.5	79.2	86.4	50.8	61.1
SC [44]	72.5	84.6	82.1	64.7	72.1	75.4	53.3	86.9	60.3	57.2	95.5	61.4	32.5	72.4	87.0	71.0	82.3	86.5	93.4	62.8	68.3
MK [23]	73.6	83.2	81.8	85.9	70.9	85.3	52.1	84.0	66.0	54.4	95.1	64.3	28.6	67.5	89.3	73.1	77.2	85.2	87.4	68.3	72.7
Ours (OFF)	76.4	83.9	79.6	75.8	74.6	85.0	56.2	90.7	68.0	61.0	97.8	67.2	33.5	84.7	81.9	77.7	83.0	88.5	97.2	69.1	72.7
Ours (ON)	71.7	81.2	75.9	75.1	70.4	84.2	53.7	86.8	60.9	53.4	95.3	60.8	29.3	71.9	86.4	61.6	79.3	84.5	93.3	64.0	66.3

Table 5.10 Per-class instance segmentation results based on the ScanNetV2 [26] test set in terms of the mAP@50 scores of 18 classes, using the m64 model

Method	**mAP**	Bookshelf	Bed	Bath	Cabinet	Curtain	Counter	Chair	Desk	Picture	Other furniture	Door	Fridge	Sofa	Sink	Shower	Window	Toilet	Table
PF [91] (ON)	47.8	59.5	71.2	66.7	25.9	61.3	0.0	55.0	17.5	43.7	43.4	25.0	41.1	59.1	48.5	85.7	35.9	94.4	26.7
BO [140] (OFF)	25.3	25.1	32.4	51.9	13.7	41.9	3.1	34.5	6.9	5.2	13.1	16.2	20.2	30.1	14.7	33.8	17.8	65.1	30.3
PG [61] (OFF)	63.6	62.4	76.5	100.0	50.5	69.6	11.6	79.7	38.4	47.6	55.9	44.1	59.6	75.6	66.6	100.0	51.3	99.7	55.6
MT [73] (OFF)	54.9	58.8	80.7	100.0	32.7	81.5	0.4	64.7	18.0	18.2	36.4	41.8	44.5	68.8	44.2	100.0	39.6	100.0	57.1
Ours (OFF)	67.2	68.2	75.8	100.0	57.6	50.4	47.7	84.2	52.4	45.1	58.5	56.7	55.7	79.7	75.1	100.0	46.7	100.0	56.3
Ours (ON)	65.7	66.7	76.0	100.0	58.1	65.5	32.3	86.3	47.7	43.2	54.9	47.3	65.0	73.8	65.5	100.0	47.2	94.4	58.5

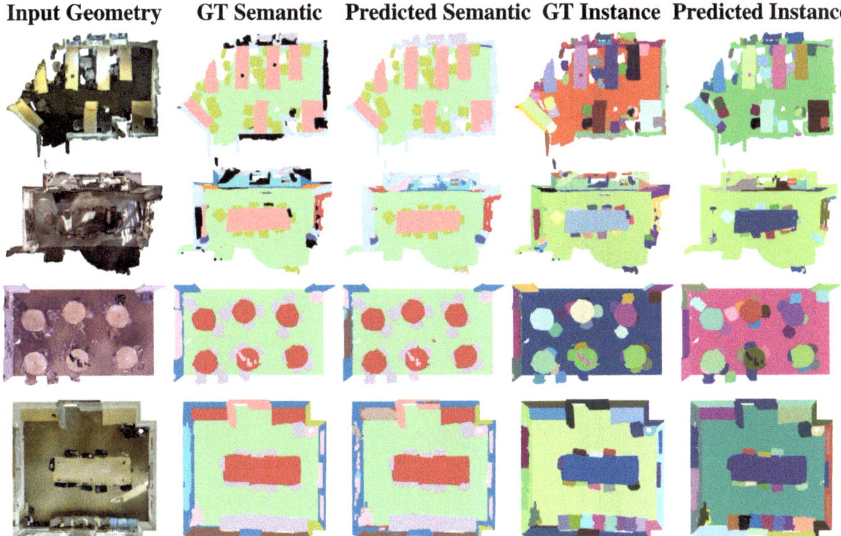

Fig. 5.41 Qualitative offline 3D instance segmentation results based on the ScanNetv2 and S3DIS validation sets

Review of Sparse Convolutions

We first briefly review the concept of sparse convolution. Conventional sparse convolutions disregard empty regions and store and perform convolutions based on only the nonempty input data locations. To prevent dilation of nonempty sites, submanifold sparse convolutions (SSCs) [44] calculate the output features only for active sites in the input. Formally, SSC operations are performed based on a K^D spatial neighborhood of each site u:

$$\mathbf{x}_u^{\text{out}} = \sum_{i \in K^D} W_i \mathbf{x}_{u+i}^{\text{in}} \text{ if } u \in A, \tag{5.68}$$

where K is a predefined kernel size, D indicates the spatial dimension (equal to 3 for 3D convolutions), and W_i is the weight matrix at location i for input features $\mathbf{x}_{u+i}^{in}$. A denotes the set of nonempty sites in the input tensor. More details can be found in [44].

In the implementation of the original SCN [44], the sparse convolution is calculated by Algorithm 2. When implementing sparse convolutions, a convolution rulebook is generated. Since the input is spatially sparse, the input tensor to the sparse convolution is usually stored as a 'COO' format sparse tensor, and the features are stored in a contiguous matrix with a size of $N \times C$, where N denotes the number of active sites, and C denotes the number of feature channels. To access a feature at a spatial location $\mathbf{x}$, the row index of its corresponding feature in the feature matrix

Fig. 5.42 Qualitative comparisons of offline 3D instance segmentation results based on ScanNetV2 [26]. **a** Input **b** SCN [44] **c** SGPN [130] **d** Lahoud [73] **e** Instance GT **f** Ours

Table 5.11 Offline instance segmentation results based on the SceneNN [55] dataset

Method	**mAP@0.5**	Bed	Floor	Cabinet	Wall	Chair	Sofa	Prop	Desk	TV	Table
MT-PNet [100]	8.5	15.0	27.3	0.0	13.1	21.2	0.0	2.0	0.0	6.0	0.7
MLS-CRF [100]	12.1	32.9	44.5	0.0	13.9	12.9	0.0	0.8	10.8	0.0	5.7
Ours	**47.1**	**66.7**	**93.8**	**5.7**	**39.0**	**91.3**	**8.7**	**7.14**	**31.6**	**76.9**	**50.0**

Table 5.12 Quantitative comparison of the offline 3D instance segmentation results based on the S3DIS [5] dataset

Method	mRec	mPrec
PartNet [88]	43.4	56.4
ASIS [131]	47.5	63.6
3D-BoNet [140]	47.6	65.6
Ours	**60.3**	**72.8**

Table 5.13 Ablation study of the occupancy-based instance segmentation method based on the ScanNetV2 validation set

Method	mAP@0.25	mAP@0.5	mAP
w/o semantic	62.6	51.8	36.7
w/o spatial	69.7	58.5	42.8
w/o occupancy	67.4	55.7	40.9
Ours	71.9	60.7	44.2

cannot be inferred directly, as in dense tensors, and is instead identified with a hash table. The hash table uses the 3D location $\mathbf{x}$ as input and outputs the row index of the corresponding feature. Instead of querying the memory indices when performing the convolution operations, the SCN [44] introduces a rulebook, which prepares the memory addresses before the convolutions are performed. The rulebook $\mathbf{R}$ is built based on the active sites A_{in}, A_{out} and the kernel size. The rulebook $\mathbf{R}$ contains K^D items $\mathbf{R}_i$, where i ranges from 0 to $K^D - 1$. Each item $\mathbf{R}_i$ is a list. Each entry is denoted as (I_j, O_j), which indicates that the I_j-th input feature should be multiplied by the i-th weight and added to the O_j-th output feature.

Algorithm 2 Original sparse convolution algorithm

Require: input feature X, output feature Y, convolution weights $\mathbf{W}$, kernel size K, convolution dimension D, convolution rulebook $\mathbf{R}$.

procedure RULEBOOK GENERTATION
  for all $O_i \in A_{\text{out}}$ **do**
    for all $I_i \in A_{\text{in}}$ and $I_i \in$ the receptive field of O_i **do**
      ind $\leftarrow$ the relative position of I_i in the receptive field of O_i.
      Insert (I_i, O_i) to $\mathbf{R}_{\text{ind}}$
procedure SPARSE CONVOLUTION
  $Y \leftarrow \mathbf{0}$
  for all $i \in \{1, \ldots, K^D\}$ **do**
    for all (I_j, O_j) in $\mathbf{R}_i$ **do**
      $Y_{O_j} \leftarrow Y_{O_j} + W_i * X_{I_j}$ // Add bias if applicable

Block Cache Sparse Convolution

As mentioned above, the input tensor for the sparse convolution is usually stored as a 'COO' format sparse tensor, and the features are stored in a $N \times C$ matrix. Each row in the matrix corresponds to an active site. However, the orders of the row indices are not guaranteed and could be modeled as random permutations. As a result, directly following Eq. (5.68) requires frequent random accesses to the feature matrix. GPU-accelerated computing models have severe performance issues because the feature matrix is stored in the GPU global memory, and random global memory access could lead to degraded memory bandwidth.

Thus, we were motivated to design a caching mechanism to exploit the locality of the sparse convolution to obtain optimal GPU memory efficiency. We propose to spatially divide the input tensor into several blocks. Within each block, we cache the input and output features inside the block in shared memory before performing the convolution operations.

Formally, we denote the block size as B. The sparse convolutions are performed for each block in parallel, and each block generates its own output features. The detailed sparse convolution procedure for a single block is shown in Algorithm 3.

Algorithm 3 Block cache sparse convolution

Require: input features X, output feature Y, convolution weights $\mathbf{W}$, kernel size K, convolution dimension D, convolution rulebook $\mathbf{R}$.

Generate Rulebook $\mathbf{R}$.
procedure SPARSE CONVOLUTION OF A SINGLE BLOCK
    $Y \leftarrow \mathbf{0}$
    Copy $\mathbf{W}$, X into shared memory
    for all $i \in \{0, 1, \ldots, K^D - 1\}$ **in parallel do**
        for all $(I_j, O_j) \in \mathbf{R}_i^{\text{block}}$ **in parallel do**
            $Y_{O_j} \leftarrow Y_{O_j} + W_i * X_{I_j}$

Note that if an output site is on the boundary of a block, its convolution output may depend on input features outside the block, as shown in Fig. 5.43. In the 2D example, the kernel size is $K = 3$, and the block size is $B = 4$. To calculate the output feature located at the red mark, we must consider the input features in the $K * K$ region around the output site. Therefore, it is necessary to cache not only the input features in the block but also those within a $(K - 1)/2$ margin.

Adaptive Block Division. In the above algorithm, we use a fixed block size. However, since point clouds usually have very uneven point density, the number of active sites contained in each block may also differ. If a block contains many active sites, the required cache size may exceed that available in the hardware. Thus, the cache utilization may be low and inefficient. Therefore, we propose an adaptive block division strategy. As shown in Fig. 5.44, we use 3 levels to achieve an even block division. We set the initial block size B and the desired number of points per block S. In each level, we evaluate which blocks contain more points than $N_{\max}$. These

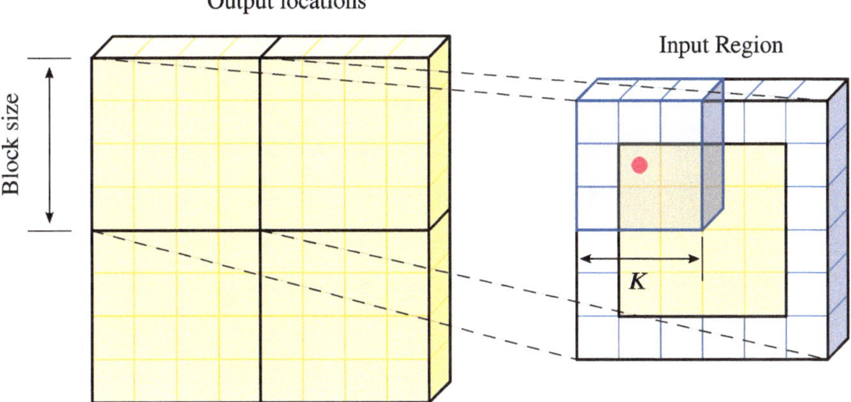

Fig. 5.43 Example of a block cache sparse convolution operation performed at the block boundary

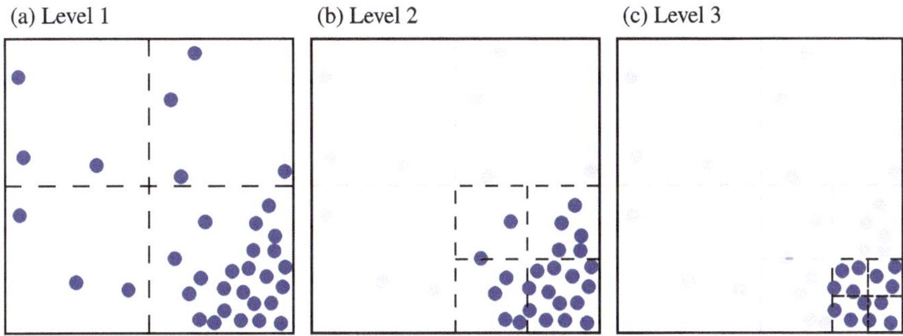

Fig. 5.44 Adaptive block division strategy

blocks are identified and divided further in the next level. In the second and third stages, the block size is set to $B/2$ and $B/4$, respectively, and the other operations are performed as in level 1.

Eventually, all blocks have varying sizes and contain similar numbers of points. With this approach, we achieve the optimal cache utilization.

We next explain the implementation details of the parallel structure. Figure 5.45 shows a simple example of a sparse convolution forward calculation, where $(N_{\text{in}}, N_{\text{out}})$ are the numbers of active input and output locations within the block region. C_{in} and C_{out} denote the number of input and output channels, respectively. K represents the kernel size, and D denotes the kernel dimension, which is 3 for a 3D convolution. B is a parameter that controls the parallel process, which is set to 64 in practice. Note that "$*$" does not indicate a matrix multiplication operation and instead represents a sparse convolution operation guided by the rulebook. Since N_{in}, N_{out} and C_{in}, C_{out} are usually large in practice, we divide these values by B. In our implementation, several C_{out}/B CUDA blocks are processed in parallel to calcu-

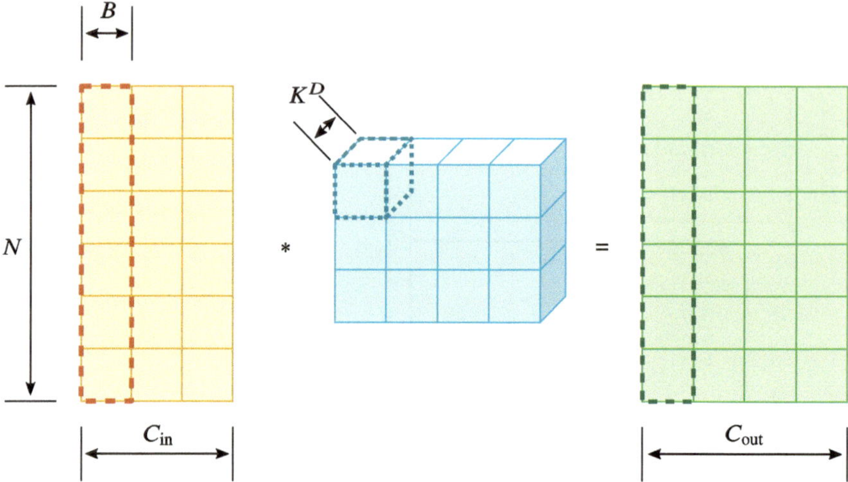

Fig. 5.45 Parallel structure for the block cache sparse convolution operation

late the sparse convolution for a single input block. Each CUDA block is responsible for calculating B output channels, which are marked in the green box in Fig. 5.45. Within each CUDA block, the threads are organized with dimension $(N_{\text{out}}/B, B, B)$. In each iteration, all the threads compute the B input channels in parallel (marked in the orange box in Fig. 5.45), and the results are added to the output location. Then, after iterating C_{in}/B times, all input channels are looped over, and the computation is complete. It is worth noting that within each iteration, some of the input/output features and weights within the marked boxes are shared across all threads in the CUDA block and are cached in shared memory. Therefore, the minimum shared memory per CUDA block is $2 * (N_{\text{in}} * B) + (B * B * K^D)$. We can adjust the parameter B based on the hardware.

Efficient Rulebook Generation. The algorithm for generating the SCN rulebook [44] is implemented via a CPU-based sequential approach, as shown in Algorithm 2, which leads to a significant efficiency bottleneck. Similar to the adaptive block cache sparse convolution approach, we generate the local rulebooks for each block in parallel, harnessing the power of the GPU. Instead of using the CPU-based hash table, which only supports sequential insertions or retrievals, we apply a GPU-based hash table CUDPP [2], which supports parallel insertions and retrievals. The block rulebook generation algorithm is shown in Algorithm 4.

Algorithm 4 Block rulebook generation algorithm

```
procedure BLOCK RULEBOOK GENERATION
    Divide the input/output space into B * B * B blocks
    for all blocks in parallel do
        for all active sites x ∈ A_out in parallel do
            for all sites y in the receptive field of x in parallel do
                Examine if the point is active by querying the hash tables in parallel.
                Assign a local memory index to the input point.
                Append the entry to the rulebook R.
```

To reduce the collision rate, we employ the following hash function for the 3D coordinates as keys:

$$key = ((((R * P) \text{ bitor } x) * P) \text{ bitor } y) * P \text{ bitor } z, \tag{5.69}$$

where bitor denotes the bitwise OR operation and $P = 216613626$ is a large prime number. (x, y, z) is the 3D coordinate of the query site. R is a random integer, which is used to build multiple hash functions for the CUDPP [2] algorithm.

Incremental Sparse Convolution Inference

In this part, incremental updates are used to accelerate the sparse convolution inference process. In real-time RGB-D reconstruction tasks, the scenes reconstructed at each time stamp form an incrementally expanding 3D geometric sequence, where the residuals between two continuous 3D frames are typically sparse. By utilizing an incremental inference process based on the residuals of continuous frames, a substantial amount of duplicate work can be avoided. We introduce a novel incremental sparse convolution inference framework, which we call *INS-Conv*. Based on INS-Conv, we adapt the semantic instance segmentation pipeline presented in Section **OccuSeg** into an incremental semantic instance segmentation pipeline, as shown in Fig. 5.46.

Insight. Let function f be a linear map that satisfies:

$$f(x + y) = f(x) + f(y), \ f(cx) = cf(x). \tag{5.70}$$

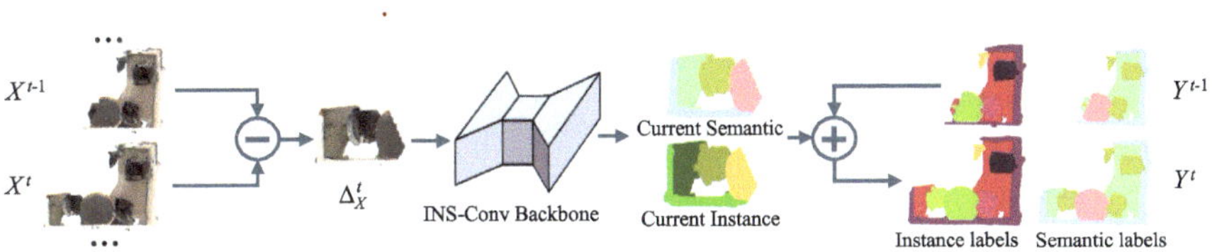

Fig. 5.46 Overview of the incremental 3D semantic and instance segmentation pipeline

In addition, let the combination of linear maps also be a linear map:

$$f(g(x+y)) = f(g(x)+g(y)) = f(g(x)) + f(g(y)). \tag{5.71}$$

Many modules in neural networks are linear maps, such as the convolution layer and linear layer. Some advanced modules, such as batch normalization layers and residual blocks, also satisfy the above equations by omitting the possible bias element for simplicity. In accordance with Eq. (5.71), neural networks composed of linear modules are also linear maps.

In our case, neural network inference is performed based on an incrementally reconstructed scene. We define x^t as the color features of all voxels that have been built at time t and Δ_x^t as the residual (difference) between x^t and x^{t-1}. The color features of the voxels are input into our neural network f, which outputs the labels of each voxel. At time t, the network can be divided into two parts:

$$f(x^t) = f(x^{t-1} + \Delta_x^t) = f(x^{t-1}) + f(\Delta_x^t), \tag{5.72}$$

where $f(x^{t-1})$ was previously computed. Thus, we can simply use the cached result and calculate $f(\Delta_x^t)$. The calculation of $f(\Delta_x^t)$ indicates that the network propagates residuals of features because for every linear map layer l, $l(\Delta_x) = l(x+\Delta_x) - l(x) = \Delta_y$, where x and y denote the input and output features of this layer.

In conclusion, we demonstrate that neural network inference based on successive inputs can be formulated as propagating the residuals of the input features, thus enabling incremental prediction. Based on this understanding, we propose INS-Conv, a fast and accurate incremental sparse convolutional network for online 3D segmentation, as detailed in the following subsections.

Layer Design of INS-Conv. We define an incremental submanifold sparse convolution (denoted as *INS-SSC*) layer that performs submanifold sparse convolutions based on the residuals. This layer takes residuals as input and approximates the conventional submanifold sparse convolution [44]. After the sparse convolution network is trained, we replace the submanifold convolution layers with our INS-SSC layer without modifying the network architecture to enable incremental inference.

Recall that the sparse convolution is performed based on the input sites with nonempty features. We denote this site set as A and also define an active residual site set B, which includes input sites that have nonempty residuals. Let the input and output features of the current layer at frame t be x^t and y^t, respectively. Then, the residuals of the input at frame t become $\Delta_x^t = x^t - x^{t-1}$, and our goal is to compute Δ_y^t. The propagation rule for the INS-SSC layer is defined as

$$\Delta_{y_u}^t = \begin{cases} \sum_i W_i \Delta_{x_{u+i}}^t & \text{if } u \in B^t \cap A^{t-1}, \\ \sum_i W_i (\Delta_{x_{u+i}}^t + x_{u+i}^{t-1}) & \text{if } u \in B^t \setminus A^{t-1}. \end{cases} \tag{5.73}$$

Figure 5.47 comprehensively demonstrates the INS-SSC layer employing 1-D sparse convolutions with a kernel size of 3. After propagating to the previous frame, as

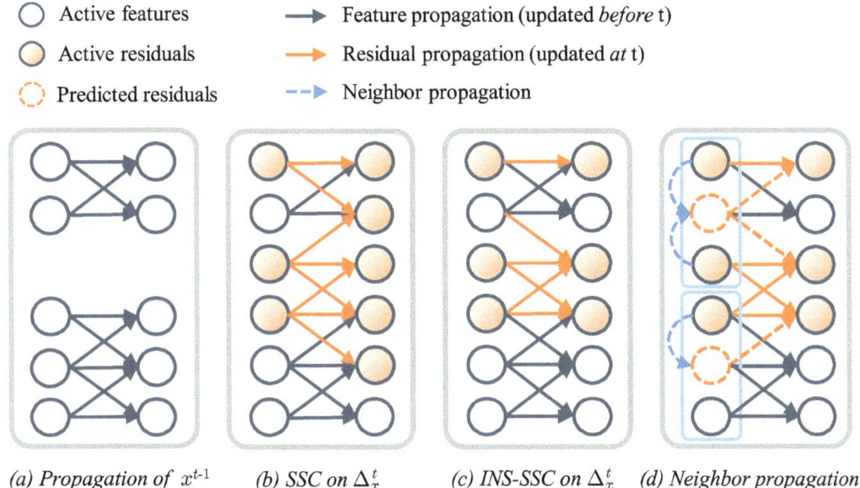

Fig. 5.47 Illustration of the INS-SSC layer with a kernel size of 3 for a 1D sparse convolution example

described in (a), the operation in the INS-SSC layer is performed based on the residuals of the current frame, as described in (c). (b) demonstrates that the typical SSC rule may lead to residual dilatation; thus, this rule is not appropriate for residual propagation. (d) demonstrates the INS-SSC layer with neighbor propagation, with the residuals of the unchanging sites estimated based on the residuals of their neighbors.

Compared to the SSC operation [44], the INS-SSC operation differs in that: (1) The INS-SSC layer takes the residual as input. (2) The INS-SSC layer processes the set of active residual sites B instead of the set of all active features A. Therefore, the INS-SSC operation is more effective because B has a considerably lower density than A. (3) The INS-SSC layer constrains the output active residual set to be identical to the input set, while the SSC operation 'dilates' the active residual set after each layer, as shown in Fig. 5.47b and c. In contrast to the SSC operation, the INS-SSC operation ensures that the output active residual set is identical to the input, as illustrated in Fig. 5.47b and c. (4) The INS-SSC operation uses a distinct set of convolution rules. For situations in which u is a newly active site that was previously inactive, the SSC criteria may produce inaccurate results. In particular, the prior feature y_u^{t-1} is set to zero under the sparse convolution rule that disregards inactive sites; however, this feature should be considered when u becomes active in the present frame, which we denote as $\hat{y}_u^{t-1}$.

In accordance with Eq. (5.73), this issue can be compensated for by adding $\hat{y}_u^{t-1}$ to the propagating residual.

The sparse residual propagation rule in the INS-SSC layer guarantees that the number of active residual sites does not increase, hence reducing the computational cost. Unfortunately, when the supposed residuals outside of B^t are discarded, the INS-SSC operation is no longer identical to the 'full' SSC propagation. As shown in Fig. 5.48,

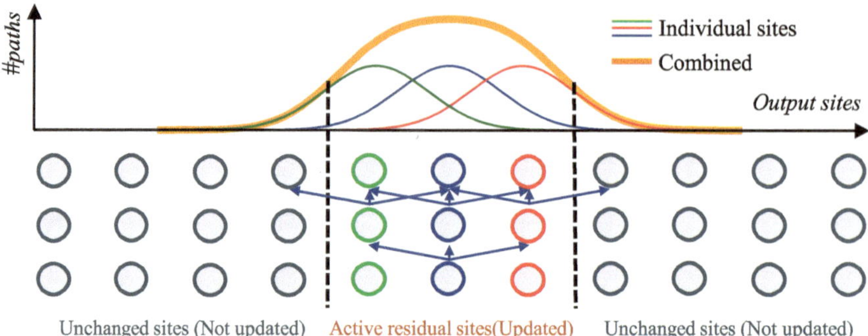

Fig. 5.48 Intuition underlying why the residual propagation rule of the INS-SSC operation leads to a low approximation error

Fig. 5.49 Visualization of the approximate error of INS-Conv. The error is calculated by the KL divergence of the output semantic probabilities between INS-Conv and the 'full' propagation

we used a 1D convolutional network to analyze the approximated error. According to Nakandala [90], in general, the distribution of the effect of a changed input feature on deeper layers resembles a Gaussian distribution centered on the changed input site, with the number of unique propagation paths serving as an approximate proxy. Due to the spatial proximity of the active residual sites in our scenario, the cumulative effect of all active residual sites approaches a Gaussian distribution (shown in orange in Fig. 5.48). Consequently, the effect of truncating the residual propagation outside of B^t is rather minor. Please note that we only perform the computations for the active residual sites; therefore, only errors at these sites are considered. As depicted in Fig. 5.49b, the error map indicates that the overall error is small and primarily located near the borders of the active residual sites as a result of truncating the surrounding residuals.

A neighbor propagation strategy is applied to further reduce the border approximation error. Our fundamental observation is that spatially adjacent sites should have similar characteristics and residuals. This elucidates why we can use the weighted average of the residuals of the surrounding active sites to approximate the residuals of the unmodified input sites $m \in A^t \setminus B^t$ that are not updated in the INS-SSC layer but have direct connections to the output active residual sites. The weighted average operation is defined as

$$\Delta^t_{x_m} = \sum_{n \in \mathcal{N}_m} w_{mn} \Delta^t_{x_n}, \tag{5.74}$$

where the weights w_{mn} are computed by similarities of features,

$$w_{mn} = \frac{\exp(s(x_m^{t-1}, x_n^{t-1}))}{\sum_{k \in \mathcal{N}_m} \exp(s(x_m^{t-1}, x_k^{t-1}))}. \tag{5.75}$$

Here, $s(x_m, x_n) = l(x_m - x_n)$ denotes the similarity of features x_m and x_n, and l represents a linear layer. $\mathcal{N}_m$ denotes the neighboring region of site m and is set to a kernel size of approximately m in the implementation.

The INS-SSC layer with neighbor propagation is depicted in Fig. 5.47d. As illustrated in Fig. 5.49c, the approximation error is efficiently reduced by utilizing the projected residuals at unaltered sites.

Feature maps are downsampled/upsampled using the sparse convolution and deconvolution layers. The propagation criteria of the INS convolution layer are identical to those of the INS-SSC layer, with the exception that active residual sites may undergo dilatation. The INS deconvolution layer is the opposite of the INS convolution layer.

In general, nonlinear layers are not linear maps and thus cannot directly propagate residuals. Formally, for a nonlinear function g, Δ^t_y is not equal to $g(\Delta^t_x)$. However, we can calculate the output residuals Δ^t_y using the definition of residual:

$$\Delta^t_{y_u} = g(\Delta^t_{x_u} + x_u^{t-1}) - y_u^{t-1}, \tag{5.76}$$

x_u^{t-1} and y_u^{t-1} are cached at the previous time step.

For sites that are not included in the active residual site set, the residuals are set to zero. Thus, these sites are disregarded across all layers. Based on the previously mentioned layer designs, INS-Conv requires only nonzero residual voxels as input. These input sites are included in the active residual site set in the first layer.

Network Architecture of INS-Conv. To ensure that the instance segmentation pipeline introduced in Section **OccuSeg** is compatible with INS-Conv, special designs are required for the incremental inference scenario.

As presented in Section **OccuSeg**, the backbone consists of a sparse convolutional network similar to a conventional UNet. This network functions identically to typical sparse convolutional networks during training. During the inference process, the layers are replaced with INS-Conv layers to perform incremental inference operations. In addition to the different features learned for semantic and instance segmentation, as described in Section **OccuSeg**, the uncertainty term and temporal consistency constraint are described in detail below.

Recall that in INS-Conv, we select only voxels with updated color features as input voxels. Although this procedure is simple, we can improve this process. INS-Conv only computes feature changes for active residual sites, which are specified by the input voxels according to the no dilation rule of INS-SSC. If we know which voxels will change significantly in future, we can include these voxels in the input even though they may not have color changes due to view limitations. Alternatively, if a voxel has already been accurately predicted, this voxel does not need to be input into the framework again.

This selection scheme can be implemented by assigning each voxel an uncertain probability. Uncertainty is defined as the inability to make a correct prediction based on a voxel due to having an incomplete scenario at the prediction time. A voxel's state is more likely to change in the future if the voxel has a larger uncertainty value. We propose to train the network to identify voxels with large uncertainty values. The analysis of each voxel is formulated as a binary classification task. If a voxel's semantic prediction in an incomplete scene differs from the semantic prediction in a complete scene or if the distance between the voxel's instance embedding and the complete scene's embedding is greater than δ_d, the voxel's uncertainty in the incomplete scene is defined as positive. Here, δ_d is set to 0.8.

To supervise the training process, we generate different complete versions of each scene and add each scene and its partial scenes to the same batch. The ground truth labels of the uncertainty term are generated using the entire scene prediction and partial scene prediction.

As described in Section **OccuSeg**, the discriminative loss function is used to ensure that the instance embeddings of voxels belonging to the same instance are located in nearby regions in feature space. In our online environment, we also include a temporal consistency loss that requires an instance's embeddings to remain close over time. This loss is important in the instance fusion stage to better match instances over time. The temporal consistency loss can be expressed using the following formula based on the training strategy that considers both complete and partial scenes:

$$\mathcal{L}_{con} = \frac{1}{K}\sum_{k=1}^{K}\frac{1}{C}\sum_{c=1}^{C}\frac{1}{N_c^k}\sum_{i=1}^{N_c^k}[||\mathbf{u_c} - e_i^k|| - \delta_v]_+^2. \tag{5.77}$$

Here, K denotes the number of partial scenes for a scene, C is the number of instances in the complete scene, N_c^k is the number of voxels of the c-th instance in partial scene k, $\mathbf{u}_c$ is the mean embedding of instance c in the complete scene, and e_i^k is the predicted embedding of the i-th voxel of instance c in partial scene k. δ_v is set to 0.1. This loss term requires the voxel embeddings in partial scenes of a scene to be statistically close to the mean embeddings of their respective instances in the full scene.

Online Semantic and Instance Segmentation. For each time step, we first calculate the difference between each voxel's color feature in the TSDF and the last frame. The nonzero sites in the residual volume have frustum shapes since the SLAM system updates the voxels in the frustum of the current view. We use the uncertainty

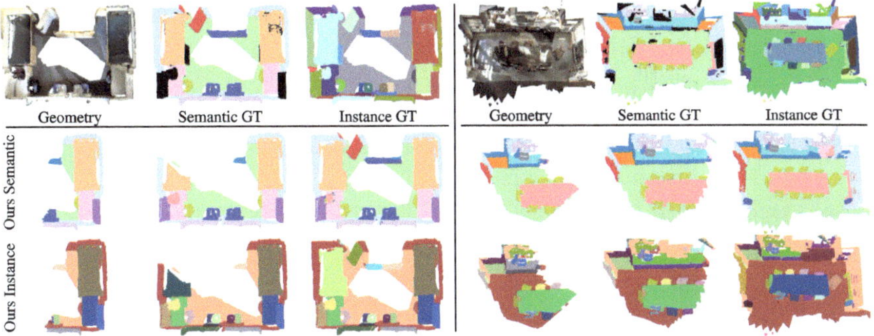

Fig. 5.50 Visualization of online semantic and instance results based on the validation set of ScanNetv2. We show the online segmentation results at 3 different time steps for each scene

value to select the input voxels. In particular, we remove voxels in the frustum with uncertainty values less than θ. In addition, the residuals for voxels near the frustum with uncertainty values greater than θ are treated as zeros. Here, $theta$ is set to 0.4. It should be noted that the input voxels are set as active residual sites regardless of whether their residual values are zero. Due to the no dilation criterion of INS-Conv, we obtain updated predictions only for the input voxels, which are then clustered and fused to produce the final semantic and instance segmentation results (Fig. 5.50).

We adapt a clustering-based instance segmentation scheme in Section **OccuSeg** for the incremental inference process and only perform instance clustering based on the updated voxels using the predicted embeddings to achieve high efficiency. The current instance set I_c is then fused with the global instance set I_g. For every instance $i \in I_c$, we compute the similarity $S(i, j)$ between i and every instance $j \in I_g$. In previous approaches [91], only the position overlap is used to match instances. However, these methods cannot compensate for errors and have difficulty calculating the overlap ratio during the matching process. For instance, if two instances were mistakenly merged into a single instance, they cannot be separated later. By comparing the mean instance embeddings, the matching relation can be computed; therefore, the robustness can be increased by taking advantage of the temporally consistent instance embeddings. In detail, we store the mean predicted embedding $\mathbf{u}_j$ for each instance $j \in I_g$. The distance d_{ij} between i and j can be computed as: $d_{ij} = \exp(-||\mathbf{u}_i - \mathbf{u}_j||^2)$.

The position overlap can also be used to evaluate the similarity between different instances. The overall similarity $S(i, j)$ is formulated as: $S(i, j) = (1 + \frac{O(i,j)}{2N_i})d_{ij}$, where $O(i, j)$ denotes the number of voxels of instance i that overlap with global instance j, and N_i is the total number of voxels of instance i. The maximum similarity $S_{max}(i)$ between instance i and global instance $\hat{j}$ is

$$S_{max}(i) = \max_{j \in I_g} S(i, j), \hat{j} = \underset{j \in I_g}{argmax}\, S(i, j) \tag{5.78}$$

If $S_{max}(i) > \alpha$, instance i matches global instance $\hat{j}$; otherwise, instance i is assigned a new instance label. Here, α is a hyperparameter that is set to 0.65 in our experiments.

To obtain a more spatially consistent semantic map, we force all points in the same instance $i \in Ic$ to have the same semantic label, i.e., the majority label of i, as opposed to simply utilizing the raw predicted semantic probabilities. Additionally, we employ the fusion approach proposed in [91] to fuse the most recent semantic results with the results of overall system to obtain more temporally consistent results.

Experiment

In this section, we evaluate the online 3D semantic and instance segmentation results of the proposed efficient sparse convolution method, as well as the runtime efficiency.

Online 3D Semantic and Instance Segmentation. The network is trained with the Adam optimizer. The learning rate is set to $1e-3$. We train the model from scratch for 320 epochs. For all the following experiments, we use the same model hyperparameters.

We first train the UNet-like sparse convolution network (Section **OccuSeg**) based on a public dataset. After model training is complete, the sparse convolution layers are replaced by the INS-Conv layers mentioned in ***Incremental Sparse Convolution Inference***. CPU-only and GPU versions of the efficient sparse convolution model are both implemented. The CPU-only version only uses INS-Conv layers for incremental inference, while the GPU version uses both block caching and INS-Conv operations.

The INS-Conv algorithm is executed for each input frame. The internal features of the network are updated every 100 frames based on the current full scene using the standard sparse convolution (introduced in ***Block Cache Sparse Convolution***) to reduce drifting errors. The network outputs from this step are not included in the following evaluation, which aims to demonstrate the impact of the INS-Conv layer. *m32* (smaller) and *m64* (larger) are two different-sized models that are evaluated to demonstrate the trade-off between accuracy and efficiency.

The performance is first evaluated based on the ScanNetv2 [26] dataset. In Tables 5.14 and 5.15, our model is denoted as **Ours (ON)**. In these tables, we present the mean intersection-over-union (mIoU) for the semantic segmentation results and the average precision at 0.5 IoU (mAP@50) for the instance segmentation results based on both the validation set and the test set. The test set results for the m64 model are reported here. For the online methods, the runtime is expressed as the number of frames per second (FPS). We present the paper results for methods without publicly available source codes; therefore, the FPS may not be exactly comparable due to the use of different hardware and experimental setups. Although both of our models are slightly slower than SVCNN [57], our models achieve the best mIoU scores among the online approaches for semantic segmentation. In addition, our models perform online segmentation substantially faster than the offline approaches and are comparable with state-of-the-art offline approaches.

Table 5.14 Online semantic segmentation results based on the ScanNetv2 dataset

Method	Type	mIoU		FPS	
		Validation	Test	GPU	CPU
Fusion-Aware [152]	Online	67.2	63.0	10	–
SVCNN [57]	Online	68.3	63.5	20	–
SCN [44]	Offline	69.3	72.5	–	–
MinkowskiNet [23]	Offline	72.2	73.6	–	–
Ours-m32	Online	71.5	–	15	10
Ours-m64	Online	72.4	71.7	10	8

Table 5.15 Online instance segmentation results based on the ScanNetv2 dataset

Method	Type	mAP@50		FPS	
		Validation	Test	GPU	CPU
PanopticFusion [91]	Online	–	47.8	4.3	–
PointGroup [61]	Offline	56.9	63.6	-	–
Ours (Offline)	Offline	60.7	67.2	–	–
Ours-m32	Online	57.4	–	15	10
Ours-m64	Online	61.4	65.7	10	8

The only existing online segmentation method that offers instance prediction for instance segmentation is PanopticFusion (PF) [91]. Compared to PF, the proposed method is quicker and obtains a substantially higher mAP@50 score (17.9%). Note that the PF method requires two GPUs, while our method only uses one GPU. Our method obtains similar mAP@50 scores when compared with offline techniques. It is important to note that our method does not use any postprocessing techniques, whereas offline methods frequently use postprocessing techniques to improve accuracy. The qualitative online semantic and instance segmentation results of our method are shown in Fig. 5.50.

Seventy-six indoor scenes with semantic and instance annotations are included in the SceneNN dataset [55]. To verify the generalizability of our semantic segmentation model, we train our model based on ScanNetv2 and evaluate the model based on SceneNN using the same parameters as SVCNN [57]. The average mean accuracy (mAcc) was used as the evaluation metric. Table 5.16a demonstrates that our method outperforms all existing online techniques.

The instance segmentation model was trained from scratch based on the SceneNN dataset. Fifty scenes were chosen for training, while 20 holdout scenes were selected as the test set, which is the same setting as SceneNN [55]. Table 5.16b shows comparisons with existing offline methods. The proposed method achieves a much better mAP@50 score than the existing methods due to the use of the proposed training strategy.

Table 5.16 Online instance segmentation results based on the SceneNN dataset using the m64 model: (**a**) Semantic mAcc (%), compared with other online methods; (**b**) Instance mAP@50(%), compared with offline methods

(a) Semantic mAcc based on SceneNN		**(b) Instance mAcc based on SceneNN**	
Method (Online)	mAcc	Method (Offline)	mAP@50
Fs-A [152]	71.5	MLS-CRF [100]	12.1
SVCNN [57]	76.5	MT-PNet [100]	8.5
Ours (Online)	79.5	Ours (Online)	57.6

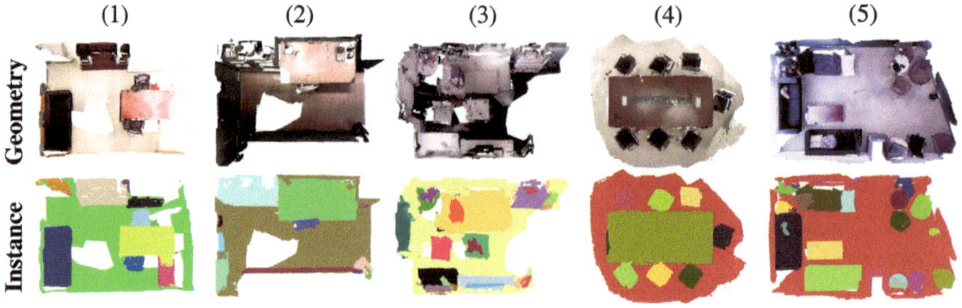

(1) (2) (3) (4) (5)

Geometry

Instance

Fig. 5.51 Online 3D instance segmentation results based on real-world scenes

Table 5.17 Runtime of each stage in our online semantic instance segmentation pipeline

Model	Network stage (ms)	Clustering stage (ms)	FPS (Hz)
m32	61	67	15
m64	99	67	10

We also test our online instance segmentation model based on real-world data, as shown in Fig. 5.51. The geometry models are reconstructed by FlashFusion [48]. Note that our model is trained based on only ScanNetv2, with no finetuning. The good results of our model prove the generalizability of the proposed methods.

Runtime Analysis. A representative large-scale scene from ScanNetv2, called *scene0645_01*, was used to evaluate the calculation time, following PanopticFusion [91]. Our 3D segmentation pipeline includes two stages, the network stage and the clustering stage, and we present the average runtime for each step in Table 5.17. The network operates in parallel with other stages on a GPU. The runtimes of the joint semantic and instance segmentation operations for the m32 and m64 models are 15 and 10 Hz, respectively. The runtimes for the CPU version are 10 and 8 Hz, respectively.

Ablation Study We next perform an ablation study to demonstrate the efficacy of the proposed method and validate our various design choices.

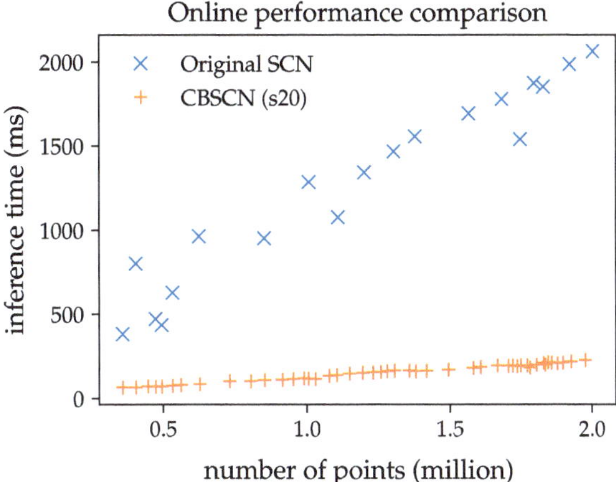

Fig. 5.52 Inference time comparison *with* and *w/o* the block cache sparse convolution

Block Cache Sparse Convolution. As shown in Fig. 5.52, we compare the inference speed of our block cache sparse convolution with that of the original SCN [44], i.e., *w/o* block caching. We record the time needed by the 3D sparse convolution network (*m32*) to perform a forward propagation operation based on various sized input point clouds. Note that the incremental inference process is not considered in this experiment. Figure 5.52 shows that the inference times of both methods increase linearly with the number of points in the input point cloud. However, the slope of our method is much flatter, suggesting that our method performs the inference step faster. On average, our method needs 242.1 ms to process a scene in the ScanNetv2 validation set, while the original SCN needs 1267 ms, suggesting that the block cache sparse convolution operation is 5× faster than the original SCN.

Incremental Sparse Convolution Inference. The effectiveness of different elements of the proposed method is discussed. The ScanNetv2 validation set is used for the ablation experiments. All models have the same size as the m64 model.

Recall that the incremental inference process with sparse convolutions accelerated the inference time. We evaluate the online 3D semantic instance segmentation performance based on *scene0645_01*. For our method, we perform incremental network inference using INS-Conv operations for each frame. For the baseline model (*w/o* INS-Conv), we predict the current full scene for every frame using the same model structure. On average, the baseline model needs 649ms for this process, while the INS-Conv model needs only 99ms, suggesting a 6.5x acceleration in the speed of network inference. With both block caching and INS-Conv, the model is more than 30× faster than the original SCN.

Since INS-Conv is technically not equivalent to the original SCN but is an approximation of the full SCN propagation process, we perform an ablation study to show how INS-Conv affects the segmentation accuracy. The results are shown in Table 5.18. The INS-Conv model achieves good approximation results. Compared with the full propagation process, the incremental inference process with INS-Conv does not

Table 5.18 Ablation study based on INS-Conv. The semantic and instance results show that this model obtains significantly higher accuracy than the other models. The comparison with the 'full' propagation demonstrates the approximation ability of the model

INS-Conv	Neighbor propagation	mIoU	mAP@50
—	—	58.0	30.5
✓	—	72.1	60.9
✓	✓	72.2	61.2
Full propagation		72.2	61.3

Table 5.19 Full propagation ($\times 10^{-3}$) approximation error, calculated by averaging the per-frame MSE of the final layer features

	MSE
w/o neighbor propagation	9.3
Ours	5.0

affect the semantic segmentation accuracy (mIoU) and only has a slight effect on the instance segmentation accuracy (mAP@50). The results of this experiment show that replacing the SCN layers with INS-Conv layers has minimal effect on the model prediction accuracy but can considerably accelerate the speed, demonstrating the feasibility of the proposed incremental inference operation.

The performance of the INS-Conv model is compared with that of the standard sparse convolution [44], which is performed based on the same input points, i.e., the points in the frustum of the current view, to demonstrate the effectiveness of the proposed model. The first and second rows of Table 5.18 display comparisons of the semantic and instance segmentation results. Without INS-Conv, the accuracy is drastically reduced. This is because INS-Conv can approximate the inference process for the entire scene, while the naive method can only "see" the current frustum. This demonstrates the importance of global information in instance and 3D semantic segmentation tasks.

Additionally, the neighbor propagation module is analyzed, which can reduce the approximation error of INS-Conv. Figure 5.49 displays the qualitative effects of the approximation error reduction. Tables 5.18 and 5.19 present the quantitative results. The neighbor propagation module reduces the approximation error of the final layer's features by approximately 50%, achieving roughly the same semantic and instance performance as the 'full' propagation while being significantly faster.

In the instance fusion stage, the temporally consistent embeddings help match the current instances with the global instances. The network was trained without the consistency loss term to demonstrate the significance of the temporal consistency constraint. The comparison results are displayed in Table 5.20. The results show the substantial decrease in the instance mAP score without this loss term. Without

Table 5.20 Ablation study on the temporal consistency constraint. The instance mAP@50 results are reported

	mAP@50
w/o consistency	59.6
Ours	61.2

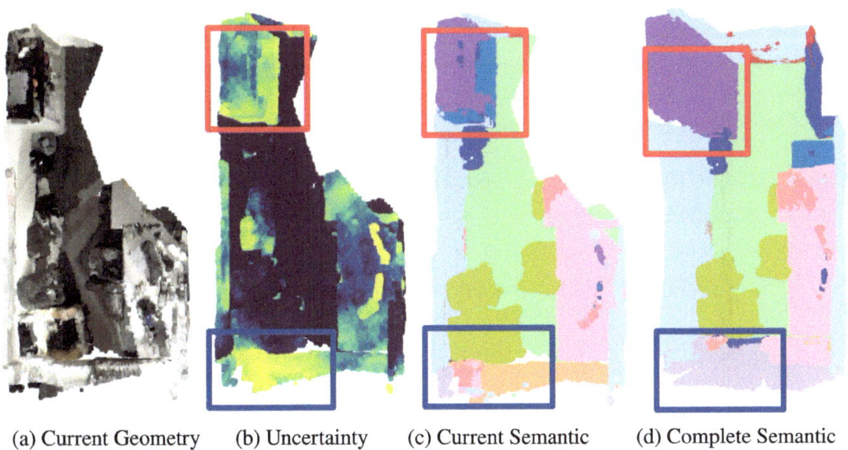

(a) Current Geometry (b) Uncertainty (c) Current Semantic (d) Complete Semantic

Fig. 5.53 Visualization of the predicted uncertainty map

temporally consistent embeddings, the current instances usually do not match the global instances. Thus, oversegmentation occurs as a result of reclassifying these instances.

When points are selected as input for the proposed INS-Conv model using the uncertainty probability, the system becomes more intelligent. The uncertainty map of an incomplete scene is depicted in Fig. 5.53. The uncertainty is larger in regions where the predictions are more difficult due to missing information in the immediate surroundings and lower in regions where the predictions are more reliable. The proposed uncertainty-guided input selection method is quantitatively compared with a naive approach, i.e., selecting points within the frustum current view. We use the m64 model and evaluate the precision and speed of the network running on a CPU and GPU. Table 5.21 shows the semantic and instance segmentation performance based on the validation set of ScanNetv2. For *scene0645_01*, the average execution time of the CPU-only and GPU models and the average number of input points per frame were evaluated. Table 5.21 shows that the use of the uncertainty not only reduces the computational time but also improves the precision. The average number of points processed for each frame is reduced by approximately 66%. Moreover, the time savings are the greatest when the model is run on the CPU because the CPU processes data sequentially. By exploiting the uncertainty to select the input, the

Table 5.21 Ablation study on the uncertainty-guided input selection method

	mIoU	mAP@50	Avg. CPU	Time GPU	#pts
w/o uncert.	72.2	61.2	328	122	19990
Ours	72.4	61.4	125	99	6820

amount of data that needs to be processed can be drastically reduced, which enables real-time 3D segmentation using only the CPU-based model.

5.4.3 BuildingFusion: Semantic-Aware Structural Reconstruction

The issue of scalable 3D reconstruction and comprehension is crucial but unfinished. There are three main challenges for large-scale perception. First, the complexity of pose optimization linearly increases with the number of input frames, and a large amount of memory is needed to reconstruct models of large-scale scenes. Therefore, real-time building-scale scene reconstruction is difficult to achieve on consumer computing devices because more frames and memory are required than for single-room scenes. Second, traditional loop closure detection (LCD) methods are based on local scene similarity and are not reliable when there are multiple rooms with similar appearances. Finally, most existing systems use single-camera scanning strategies, which are inefficient and have low fault tolerance rates for building-scale scene reconstruction (Fig. 5.54).

To address the above challenges, BuildingFusion [157] was developed; this method substantially expands the reconstruction scale of previous methods and

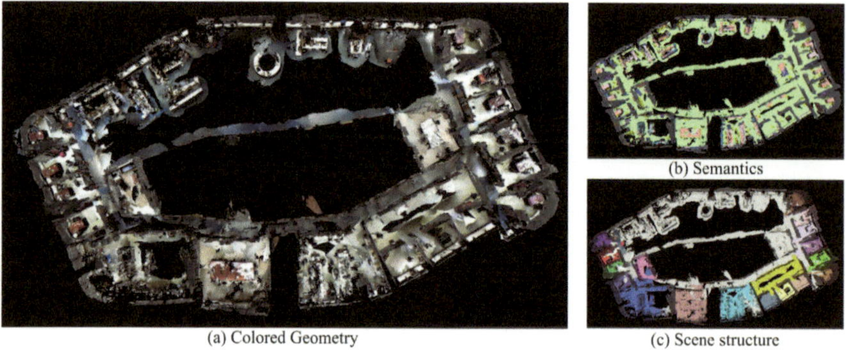

Fig. 5.54 The BuildingFusion results based on the Lab dataset. **a** shows the colored geometry; **b** shows the semantic labels; **c** shows the scene structure, where different colors represent different rooms. Gray indicates the place does not belong to any room

achieves semantic-aware building-scale reconstruction. Several techniques are employed to address the above three challenges. First, a submap strategy is adopted to efficiently use memory and achieve real-time performance, and the complexity of the global optimization process is reduced by adjusting the poses in the submaps instead of the number of frames. In addition, a new rendering strategy with different levels of details based on the submaps is developed to reduce the CPU and GPU memory requirements. Second, to address the issues with traditional LCD methods, a novel semantic-aware room-level LCD method is proposed. The key finding is that, despite the similarity in local views among different rooms, the object appearances and placements within each room typically differ, suggesting that the semantic and instance information in each room can be used to create distinct representations for place recognition. Finally, at the system level, a centralized and unstructured architecture is utilized to enable collaborative scanning. The system allows users to easily add or remove agents and does not stipulate the scanning route of the agents. When overlaps are discovered using the room-level LCD method, which is applied at the server level, each agent reconstructs a portion of the scene, and the results of each agent are combined. BuildingFusion [157] not only enables cooperative building-scale dense reconstruction but also rapidly obtains semantic and structural data. Thus, this method has considerable potential in indoor navigation and large-scale AR/VR applications.

An overview of BuildingFusion is presented in Fig. 5.55. This method uses raw RGBD observations as input and manages the input frames using keyframes based on conventional reconstruction algorithms [27, 133], where the relative poses between the local frames and keyframes are fixed. This keyframe organization approach ensures efficient pose graph optimization and LCD. However, when reconstructing large-scale environments, the number of keyframes increases rapidly as the scanned

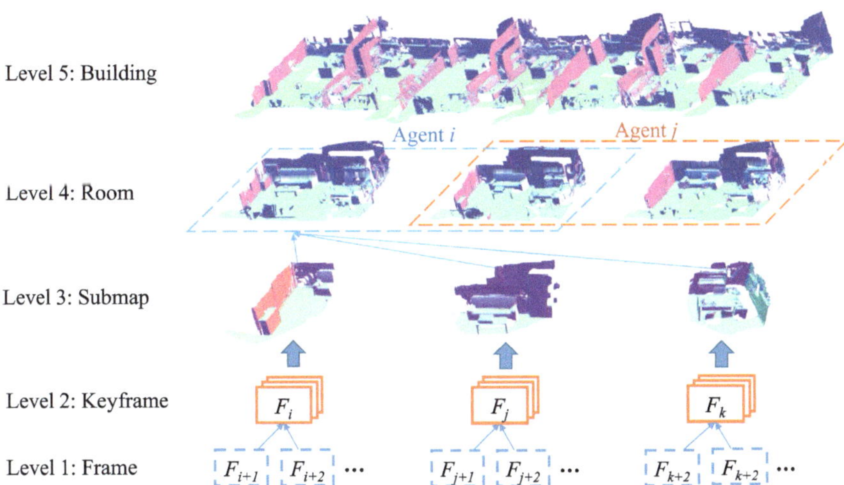

Fig. 5.55 Overview of the hierarchical data structure of our method

area increases. Thus, we group the keyframes as submaps every k times that the TSDF fusion operation is performed. Adjacent submaps show only limited overlap and thus can be reintegrated efficiently. Rooms that are detected across submaps contain abundant semantic and instance information and can thus be distinguished. We employ rooms as the basic elements in the global LCD approach; once multiple agents scan the same room, their reconstructed maps can be optimized based on the detected loop closure constraints. The overall collaborative reconstruction pipeline is shown in Fig. 5.73.

Data Structure and Notations

To explain the operation of our system, some basic data structures and notations need to be clarified. The RGBD frame with index i is denoted as F_i, where r_i and d_i represent the color and depth observations, respectively. Following previous works on globally consistent dense 3D reconstruction [48], ORB features $\mathbf{f_i} = f_i^k, k \in [0, \ldots, |\mathbf{f_i}|]$ are extracted from r_i for frame-to-frame registration and image-based LCD. The relative position of f_i^k is denoted as $\mathbf{p_i^k}$, which can be estimated based on the intrinsic camera parameters and the raw depth observations. The camera pose of f_i is denoted as T_i.

The frames are divided into keyframes and local frames based on the similarity of their color maps. Each keyframe is used as an anchor and has corresponding local frames, which are registered to the anchor following the conventional keyframe technique used in previous works [27, 133]. For pose optimization, only the poses of the keyframes are optimized, and the local frames are adjusted based on their corresponding keyframes.

Each keyframe belongs to a submap and has a relative pose within the submap $s(i)$, which indicates the submap index of the keyframe F_i. The global frame pose T_i is then determined based on the relative pose T_i^r and the submap pose $T_{s(i)}^S$:

$$T_i = T_{s(i)}^S T_i^r. \tag{5.79}$$

To estimate the globally consistent camera poses, our proposed pipeline employs a 2D image-based LCD approach for local consistency and a room-level LCD approach for global consistency.

For the local image-based LCD method, we employ MILD [47] to evaluate the appearance similarity between the current frame F_i and all the previous frames $F_j, j \in [0, \ldots, i)$. Successfully detected loop closures for frame F_i are denoted as Φ_i.

The rooms are detected by the global room-level LCD method. For the i-th room R_i, each instance is labeled as $I_i^k, k \in [0, \ldots, |I_i|)$. Here, T_i^R denotes the pose of room R_i. Then, the room poses are used to update the poses of the corresponding submaps, thereby maintaining global consistency among the multiple agents.

Finally, the globally consistent poses are estimated by minimizing the alignment error of all the observations:

$$E(\mathbf{T}) = E_f + E_r + E_n. \tag{5.80}$$

Here, $\mathbf{T}$ represents a collection of all submap poses, and E_f indicates the frame-to-frame constraints based on the frame tracking and local loop closure constraints. E_r represents the global room-level loop closure constraints, and E_n constrains the normal vectors of the detected floors so that the floors are perpendicular to the gravity direction.

Submap-Based Scalable Dense 3D Reconstruction

For dense 3D reconstruction, previous volumetric reconstruction methods [27] have obtained good results. However, since the computational complexity and memory requirements increase linearly with the size of the reconstructed region, the scalability of existing methods is limited. We adopt a submap strategy, which was inspired by [62], and develop a submap dispatching strategy to achieve scalable reconstruction. Each submap has a TSDF voxel hashing table, based on which high-resolution meshes can be extracted. Each submap is assigned a status, and the submaps can be activated or deactivated depending on the current location of the camera. Submaps are important for large-scale reconstruction tasks for several reasons:

(1) Fast pose optimization: The camera poses must be optimized to maintain global consistency due to cumulative drift and noise. The reconstruction of a building-scale scene, on the other hand, includes a large number of optimization variables, which slows the optimization process. By combining multiple keyframes in one submap, we can fuse the corresponding nodes in the pose graph into a larger node. Therefore, the original optimization process can be divided into local pose optimization processes within each submap and a global optimization process across all the submaps. This decreases the number of optimization variables, thereby reducing the complexity of the global optimization process.
(2) Seamless fusion: The integration of the depth and color information is a crucial step in creating a globally consistent model. However, when the device detects a loop closure over a long distance, the poses of most frames can shift dramatically, increasing the time of the reintegration process. The use of submaps can lead to a more efficient reintegration process. Since each submap has local consistency, only small overlapping regions need to be reintegrated when the pose changes.
(3) Memory efficiency: Although some previous methods efficiency used memory and expanded the reconstruction scale to large rooms or houses, the average linear increase in the memory consumption still limits the scalability of existing methods. In addition, simultaneously maintaining good interactivity while rendering large models is impractical due to limited computing resources. As a result, sparse local models are unavoidable. In particular, we utilize a submap

manager that dispatches submaps and renders them with various levels of details. Most of the required memory, including the frames, voxel grids, and dense models, is constrained, while the remaining memory is negligible and increases at a slow rate.

RGBD Odometry. Odometry is the first step toward SLAM and 3D reconstruction. Essentially, the input to an odometry system is two frames, and the output is a relative pose. The existing methods can be categorized as sparse feature-based and dense pixel-based methods. In our system, we combine these methods for robustness.

Sparse Matching. For each input frame F with RGB value r and depth d, we first extract the ORB features $\mathbf{f}$ from r. Then, a sparse matcher [49] is used for efficient feature matching with the last keyframe. Based on the feature correspondence $C_{i,j}^F$, we use the RANSAC algorithm to filter outliers, as shown in Fig. 5.56, and compute the relative transformation. The sparse matching process is fast, and the average frame rate can reach 50 Hz. However, this method could result in poor accuracy if not enough visual features are considered.

Dense Tracking. The goal of dense tracking is to maximize the photometric and depth consistency. The cost function is defined based on the reprojection errors in the color and depth terms:

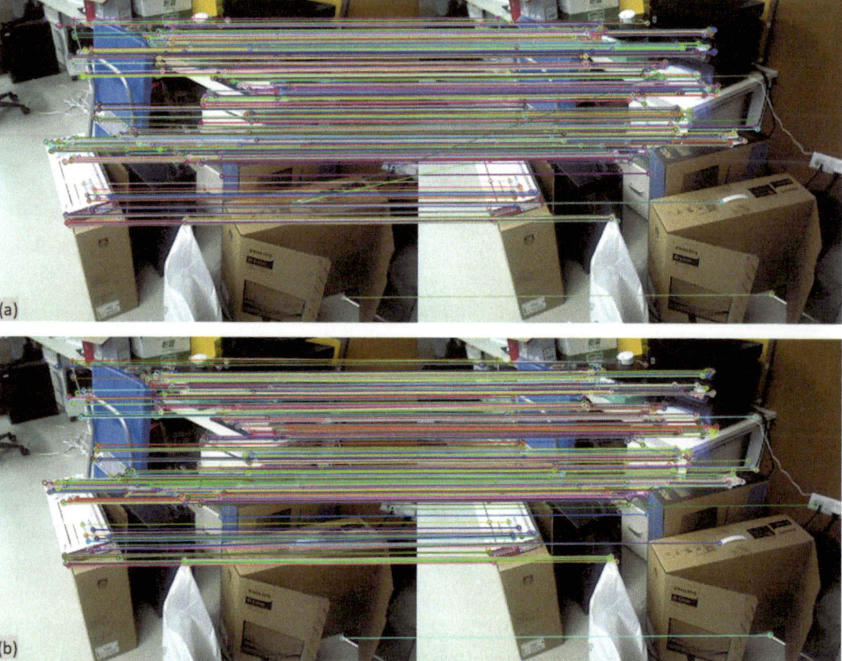

Fig. 5.56 Sparse matching. **a** Correspondences generated by a sparse matcher [49]. **b** Correspondences after outlier filtering

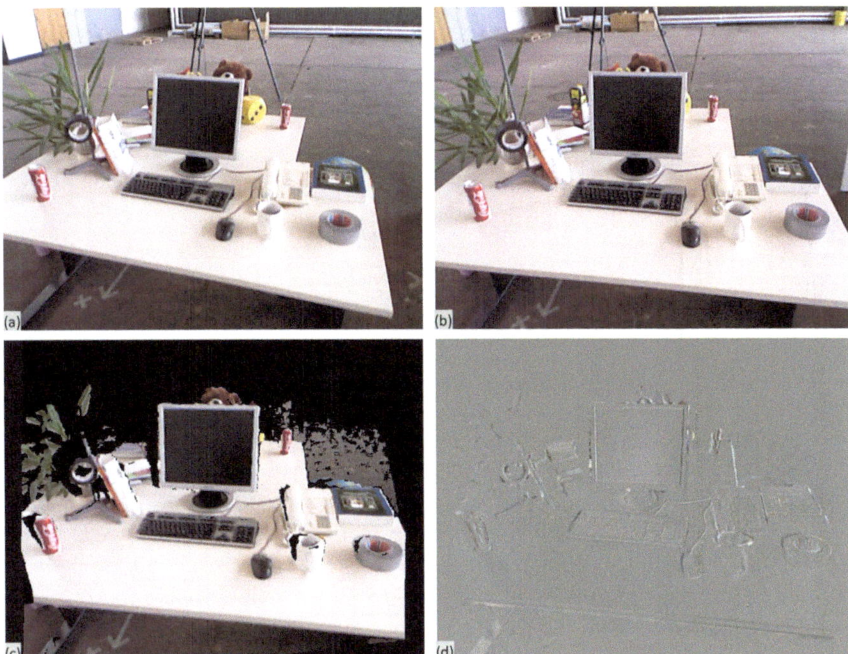

Fig. 5.57 Dense tracking. **a** Source RGB image; **b** Target RGB image; **c** Reprojected RGB image based on a relative transformation; **d** Differences between the source RGB and reprojected target RGB images

$$E_{\text{dense}} = \sum_k \alpha \cdot \left\| r_i\left(\mathcal{P}(\mathbf{p}_i^k)\right) - r_j\left(\mathcal{P}(T \cdot \mathbf{p}_i^k)\right)\right\|^2 + (1-\alpha) \cdot \left\| \left[T \cdot \mathbf{p}_i^k\right]_z - d_j\left(\mathcal{P}(T \cdot \mathbf{p}_i^k)\right)\right\|^2, \tag{5.81}$$

where α is the weight of the color term, $\mathcal{P}(\cdot)$ represents the operation of projecting a 3D point to a 2D pixel coordinate based on the pinhole camera model, and $[\cdot]_z$ is a function that returns the z-entry of a 3D point. Figure 5.57 visualizes the error of the color term. While the sparse matching process first finds the correspondence and then estimates the transformation, in the dense tracking process, we first initialize the transformation and then find the correspondences based on this transformation. The transformation is updated by minimizing Eq. 5.81. The dense tracking approach uses an iterative strategy that is similar to the point-to-point ICP [8]; however, it obtains a much better performance because the projection operation is easier than the closest neighbor search operation.

In our implementation, dense tracking is much slower (10 Hz) than sparse matching, but it can obtain a reasonable result even when the visual information is unreliable. Therefore, when sparse matching fails, dense tracking is applied to estimate the final pose.

Submap-Based Pose Estimation. For the reconstruction within a submap, we use a pipeline similar to FlashFusion [48], which performs local optimization for accurate local pose estimation and local consistency. The relative transformations of keyframes within a submap are fixed in the global optimization. When a new keyframe is registered, MILD [47] is used to match the new keyframe with previous keyframes. If the reference frame and the new frame do not belong to the same submap, the generated frame correspondences provide submap constraints E_f for the global optimization:

$$E_f = \sum_{(i,j)\in\Omega_S} \sum_{k=0}^{|C^S_{i,j}|-1} \|T^S_{s(i)} T^r_i \mathbf{p}^k_i - T^S_{s(j)} T^r_j \mathbf{p}^k_j\|^2, \tag{5.82}$$

where $s(i)$ is the submap index of frame F_i, $\Omega_S = \{(i, j)|s(i) \neq s(j)\}$ indicates all the keyframe pairs collected for the global optimization process, and $C^S_{i,j}$ is the set of corresponding feature points.

While MILD [47] can provide reliable visual constraints for room-scale reconstruction, the image appearance information is prone to errors in large-scale indoor environments, where repeated textures frequently occur in different places such as corridors and rooms with the same style, necessitating the rejection of false local LCD results. We reject correspondences that introduce large reprojection errors and those for which the source and target submaps are in different rooms. Note that the local LCD process is used only for keyframes for which the number of feature points is more than a certain threshold. This approach is reasonable because the visual information is less accurate and distinguishable in textureless areas.

Ground Normal Vector Constraints. The floor in the resulting model may not be flat due to accumulating drift errors, particularly when the scanning trajectory is very long. Gravity constraints are considered to address this issue based on the assumption that the floors in most indoor environments are perpendicular to the direction of gravity. For each submap and room, we extract the floor points based on the semantic labels and fit the ground plane using these points. The gravity direction is determined by the normal vector of the plane. Denoting the normal vector as N, the gravity alignment error is defined as

$$E_n = \sum_{i=0}^{|\Phi_S|} \|r(T^S_i)N^R_i - G\|^2 + \sum_{i=0}^{|\Phi_R|} \|r(T^R_i)N^R_i - G\|^2, \tag{5.83}$$

where G is the direction of gravity, Φ_S is the set of all the submaps, Φ_R indicates all the rooms, and $r(T^S_i)$, $r(T^R_i)$, N^S_i and N^R_i denote the rotation matrix and ground normal vector of submap i and room i, respectively. This process is visualized in Fig. 5.58.

Submap-Based Fusion. We create a comprehensive submap model based on the integration of the depth information of the input RGBD frames, following the hierarchical data structure shown in Fig. 5.55. Similar to [48], the volume is divided

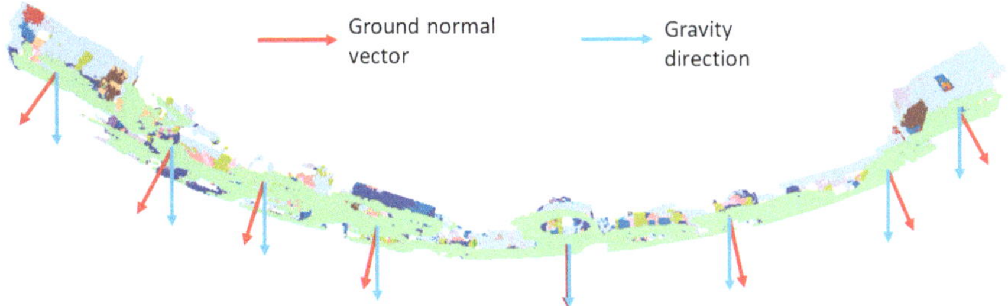

Fig. 5.58 Illustration of ground normal vector constraints

into small chunks, which are stored in a dynamic hash map. Each chunk has a unique hash entry based on the center coordinate and contains $8 \times 8 \times 8$ voxels. When a new keyframe is registered, we use sparse voxel sampling to select valid chunks and update the TSDF, color and weight values of the voxels in the visible chunks according to the following rule:

$$d = \frac{w_o d_o + w_i d_i}{w_o + w_i}, \mathbf{c} = \frac{w_o \mathbf{c}_o + w_i \mathbf{c}_i}{w_o + w_i}, w = w_o + w_i, \tag{5.84}$$

where d_o, w_o and $\mathbf{c}_o$ represent the original TSDF, weight and color values, respectively. d_i, w_i and $\mathbf{c}_i$ indicate the new TSDF, weight and color values, respectively. d, w, and $\mathbf{c}$ are the updated TSDF, weight and color values, respectively. To simplify the notation, we write this two-voxel operation as follows:

$$\{d, \mathbf{c}, w\} = \{d_o, \mathbf{c}_o, w_o\} \oplus \{d_i, \mathbf{c}_i, w_i\}. \tag{5.85}$$

Dense meshes are extracted based on the implicit representation using the accelerated marching cube algorithm.

An intuitive approach to achieve a unified model given the geometry models and submap poses is to directly transform all the submap models to the world frame. However, since different submaps use different keyframes, even for the same scene, color gaps between keyframes are unavoidable. As a result, before creating the final mesh, we suggest updating the TSDF by reintegrating different submaps based on their transformation. The function for extracting the TSDF of voxel $\mathbf{v}$ from submap S_s is denoted by $F(s, \mathbf{v})$ and defined as follows:

$$F(s, \mathbf{v}) = \{d_{s,\mathbf{v}}, \mathbf{c}_{s,\mathbf{v}}, w_{s,\mathbf{v}}\}, \tag{5.86}$$

where $d_{s,\mathbf{v}}$, $\mathbf{c}_{s,\mathbf{v}}$, and $w_{s,\mathbf{v}}$ are the TSDF, color and weight values, respectively. Then, the integration of the voxel v in the submap pair (S_r, S_n) can be formulated as

$$\hat{F}(\mathbf{v}) = F\left(n, (T_n^S)^{-1}\mathbf{v}\right) \oplus F\left(r, (T_r^S)^{-1}\mathbf{v}\right), \tag{5.87}$$

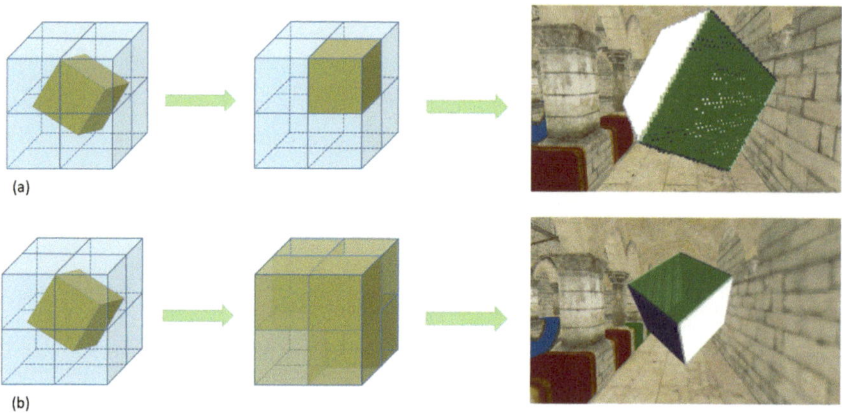

Fig. 5.59 Comparison of two voxel transformation methods: **a** Neighbor sampling and **b** eight-neighbor interpolation

where $\hat{F}(\mathbf{v})$ is the updated TSDF value of voxel $\mathbf{v}$. $\oplus$ is an operational symbol defined in Eq. (5.85).

The voxel coordinates are real numbers after the transformation, so the SDF value of the remapped voxel needs to be recomputed. Simply using the SDF value of the nearest voxel as the new SDF value can lead to information loss, as shown in Fig. 5.59. To perform subvoxel interpolation when updating the TSDF value, we must first find the hashing entries of the 8 nearest neighbors. As a result, transforming all of the chunks in a submap takes a long time. Due to the continuous motion of the camera, the overlap between submaps is usually minimal, and the other sections are not modified; thus, a transformation of the entire submap is unnecessary. In our system, when we detect a loop closure based on other submaps, we reproject the current frame to several submaps to identify the common chunks, which are then transformed and integrated using SIMD operations. Finally, the meshes are extracted based on the reintegrated chunks as follows:

$$M_g = \sum_{i=0}^{|\Phi_S|} T_i^S M_i, \tag{5.88}$$

where M_i is the local TSDF value of submap S_i, M_g is the global map, and Φ_S indicates the collection of all submaps.

Submap Dispatching. Since the reconstructed model consists of multiple submaps and only a few submaps need to be modified at a time, we assign each submap a status value, which shows whether the submap is active or inactive. A new rendering strategy with different levels of detail based on our dispatching scheme is developed to reduce the GPU burden. The submaps under the current camera view are displayed at a high resolution and updated, while the other submaps are moved to external storage and displayed using simplified models. We define two operations to dispatch submaps: *deactivation* and *activation*.

Deactivation occurs when the camera viewpoint moves away from the current submap, so there is no need to change the corresponding chunks. The status of a submap changes from active to inactive when the submap is deactivated. The chunks are sent to external storage in binary format. In addition, all keyframes for this submap, including the depth maps, color maps, and descriptors, are moved to external storage. Although the chunks can only be shared as a whole, individual keyframes can be loaded automatically in a short amount of time if the keyframe is required in the later frame registration process. We also utilize fast quadric edge collapse decimation to simplify the extracted meshes so that inactive submaps are rendered at a low level of detail.

Fast quadric edge collapse decimation is an improved implementation of quadric edge collapse decimation [40]. The intuition of the original algorithm is to collapse the edge that introduces the smallest quadric error. In a triangle mesh, each vertex can be considered an intersection of several planes. The distance between a point $\mathbf{v} = [x, y, z, 1]^\top$ and a plane $\mathbf{P} = [\mathbf{n}^\top, d]^\top$ is $\mathbf{v}^\top \mathbf{p}$. Therefore, the squared distance is

$$\begin{aligned} |D|^2 &= \left(\mathbf{v}^\top \mathbf{p}\right)\left(\mathbf{v}^\top \mathbf{p}\right) \\ &= \mathbf{v}^\top \mathbf{p}\mathbf{p}^\top \mathbf{v}. \end{aligned} \tag{5.89}$$

Thus, we can use a matrix $\mathbf{p}\mathbf{p}^T$ to compute the distance. Another advantage of this representation is that we can directly add the results if we want to compute the sum of squared distances:

$$\begin{aligned} \Delta(\mathbf{v}) &= \sum_{\mathbf{p} \in \text{ planes } (\mathbf{v})} \left(\mathbf{v}^\top \mathbf{p}\right)\left(\mathbf{p}^\top \mathbf{v}\right) \\ &= \sum_{\mathbf{p} \in (\text{ planes } (\mathbf{v})} \mathbf{v}^\top \left(\mathbf{p}\mathbf{p}^\top\right) \mathbf{v} \\ &= \mathbf{v}^\top \left(\sum_{\mathbf{p} \subset \text{ planes}} \mathbf{p}\mathbf{p}^\top \right) \mathbf{v} \\ &= \mathbf{v}^\top Q \mathbf{v}. \end{aligned} \tag{5.90}$$

We compute the Q matrix for each vertex in the mesh. For each edge $e = [\mathbf{v}_1, \mathbf{v}_2]$, if we want to collapse this edge and replace it with a new vertex $\mathbf{v}_n$, the edge error is

$$E_e = \mathbf{v}_n^\top (Q_1 + Q_2) v_n. \tag{5.91}$$

Therefore, the best v_n is to minimize the edge error is

$$\mathbf{v_n} = \arg\min_{\mathbf{v}} \mathbf{v}^\top Q \mathbf{v}. \tag{5.92}$$

If we extend $\mathbf{v}^\top Q\mathbf{v}$ as

$$\begin{aligned}\mathbf{v}^\top \mathbf{Q}\mathbf{v} = & q_{11}x^2 + 2q_{12}xy + 2q_{13}xz + 2q_{14}x + q_{22}y^2 \\ & + 2q_{23}yz + 2q_{24}y + q_{33}z^2 + 2q_{34}z + q_{44},\end{aligned} \tag{5.93}$$

after the derivation, we can easily obtain the direct solution by solving

$$\begin{bmatrix} q_{11} & q_{12} & q_{13} & q_{14} \\ q_{12} & q_{22} & q_{23} & q_{24} \\ q_{13} & q_{23} & q_{33} & q_{34} \\ 0 & 0 & 0 & 1 \end{bmatrix} \mathbf{v} = \begin{bmatrix} 0 \\ 0 \\ 0 \\ 1 \end{bmatrix}. \tag{5.94}$$

In the original algorithm, we compute the edge errors and collapse the edge with the minimal edge error. After updating the mesh, the Q values of certain vertices need to be computed, and the edges need to be resorted, which is time-consuming. In our implementation, we use a priority queue to store the edges and assign a *dirty* flag and *deleted* flag to each edge. We collapse all the edges with errors less than a threshold e_t, while edges with true *dirty* or *deleted* flags are skipped. We also prevent mesh flipping by checking two conditions to maintain the topology of the mesh, as shown in Fig. 5.60. Figure 5.61 shows the results of our implementation.

Activation could occur in two possible scenarios, in contrast to the deactivation operation. When a new submap is formed, it is marked as active. We initialize the pose as the global transformation of its first keyframe. In addition, if the chunks of an inactive submap fall into the frustum of the current camera view, which is identified by the local LCD, the submap is reactivated. The corresponding chunks are loaded into memory and used to create high-resolution meshes. Note that we do not need to manually load keyframes for the submap because of the automatic keyframe loading, as discussed for the deactivation operation.

In Fig. 5.62, the blue and red arrows indicate the geometry and texture details of the active and inactive submaps, respectively. The simplified model retains most of the geometric information but requires only 5% of the memory needed by the original model.

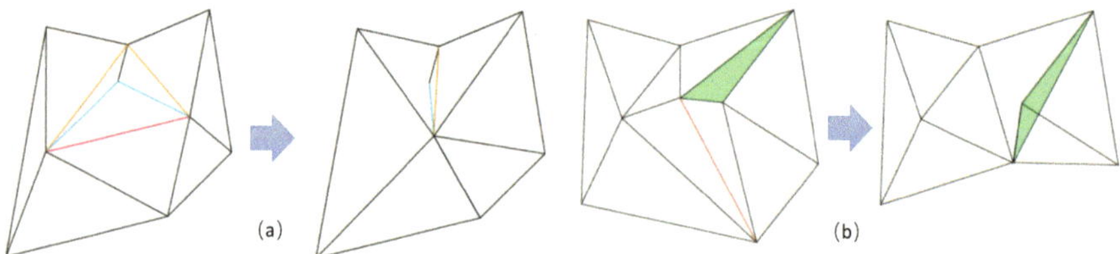

Fig. 5.60 Two types of mesh flipping. **a** Nonmanifold mesh; **b** Normal vector flipping

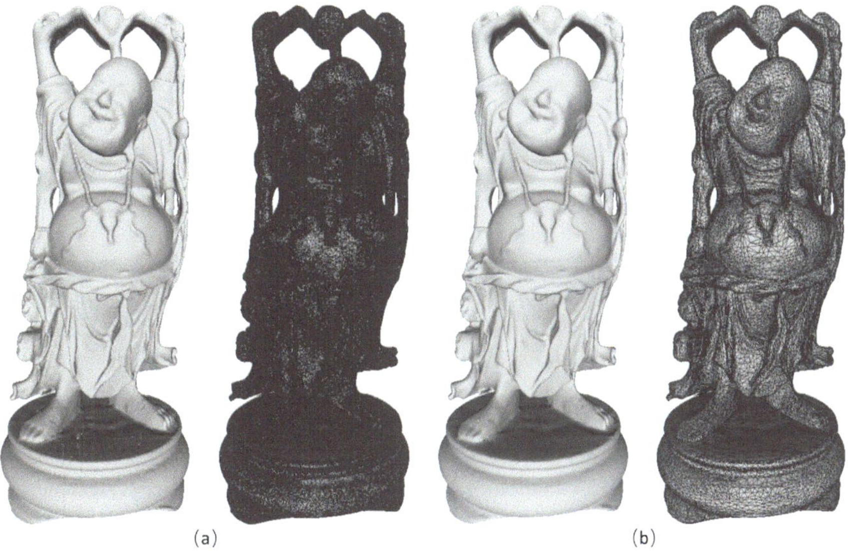

Fig. 5.61 Mesh decimation result. **a** Original model with 1087716 triangles and 543652 points; **b** Decimated model with 108770 triangles and 54179 points

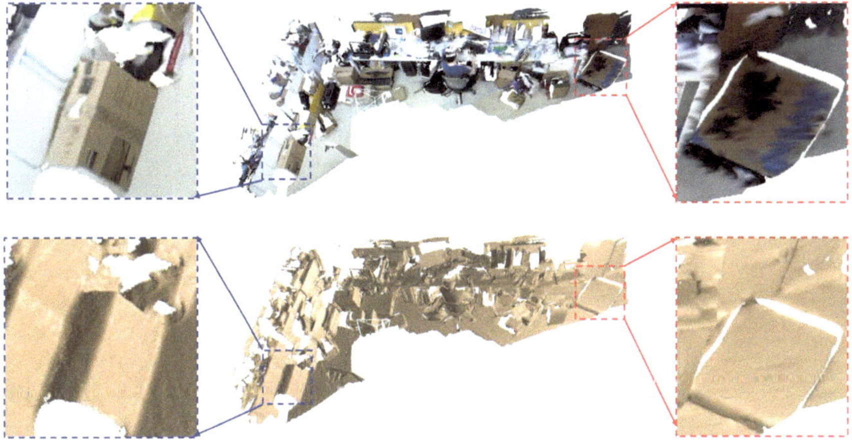

Fig. 5.62 Visualization of submap dispatching

Room-Level Loop Closure Detection

Although 2D image-based LCD methods perform admirably in the reconstruction of room-scale scenes, they have serious limitations when the scene is scaled up to the scale of a building. Three main challenges are listed as follows:

(1) *Similar rooms.* The rooms in the same building typically have similar appearances, such as offices in office buildings and bedrooms in hotels. Local 2D views struggle to distinguish similar rooms without higher-level knowledge, resulting in false loop closures.
(2) *Ambiguity of 2D views.* Lighting conditions, viewpoints, and other variables can distort the feature extraction and matching results based on 2D images. The recall rate is reduced for collaborative and large-scale reconstruction tasks because we cannot guarantee that the users are scanning at the same view angle or distance from the reconstructed surface.
(3) *Scalability of 2D LCD methods.* As the size of the image database increases, the execution time and recall rate increase. Furthermore, storing the 2D features of all images consumes a significant amount of memory, which is both inefficient and redundant.

Figure 5.63 shows an example of a false loop closure, in which an erroneous edge was added to the pose graph, reducing the quality of the reconstructed model.

Considering these fundamental issues with 2D LCD methods, we propose that loop closures can be detected at the room level because we could distinguish similar rooms using internal global object information such as the furniture placement. Furthermore, using 3D data automatically resolves the uncertainties associated with 2D data. In addition, room-level representations are much more compact than raw image sequences (Fig. 5.64).

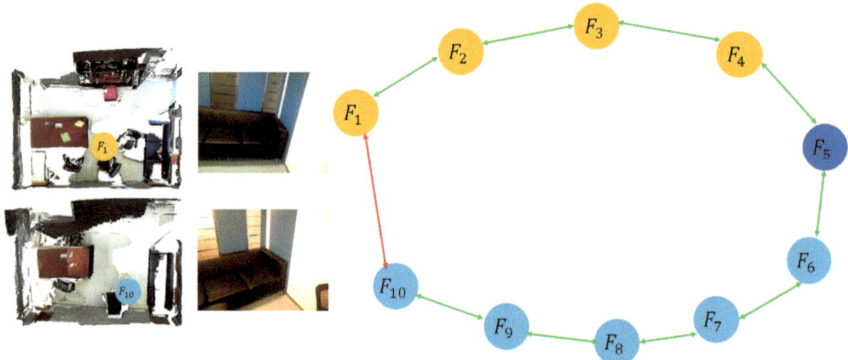

Fig. 5.63 False loop closure. The orange color indicates frames captured in *Room 1*, and the blue color indicates frames captured in *Room 2*. F_1 and F_{10} are from different rooms but were mistakenly aligned because of their similar appearances

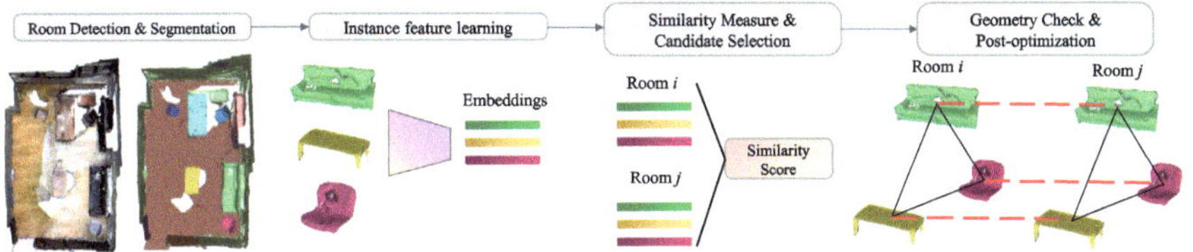

Fig. 5.64 The room-level LCD process includes room detection and segmentation, instance feature learning, similarity measurement and candidate selection, and geometry verification and post-optimization

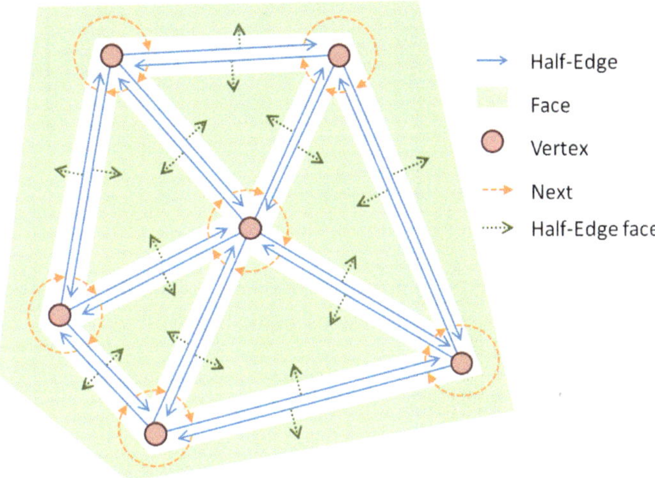

Fig. 5.65 Visualization of DCEL

Room Detection. Instead of using a complete model, as in [89], we detect the room during the reconstruction process. Our system uses the generated model as input to OccuSeg [50], a state-of-the-art joint 3D semantic/instance segmentation method. The output is the semantic and instance information. Then, we extract the points with semantic labels of 'wall' or 'window' as the room borders. A region growing process [14] is applied to extract the representative planes of the room borders, which are then projected onto the ground as lines to create complex 2D maps. To store these line arrangements, we use a doubly connected edge list (DCEL), which can be used to locate the cells of arbitrary points with minimal calculation. The data structure of DCEL is shown in Fig. 5.65.

Once the 2D cell complex is stored in the DCEL, we compute the weight of each edge in the DCEL, as shown in Fig. 5.66. Recall that the edges are generated by lines, and the lines are the projections of planes that are fitted by 3D points. Each edge corresponds to one line, and each line could correspond to several groups of points. We project the points into 2D space and create a tight 2D bounding box for each

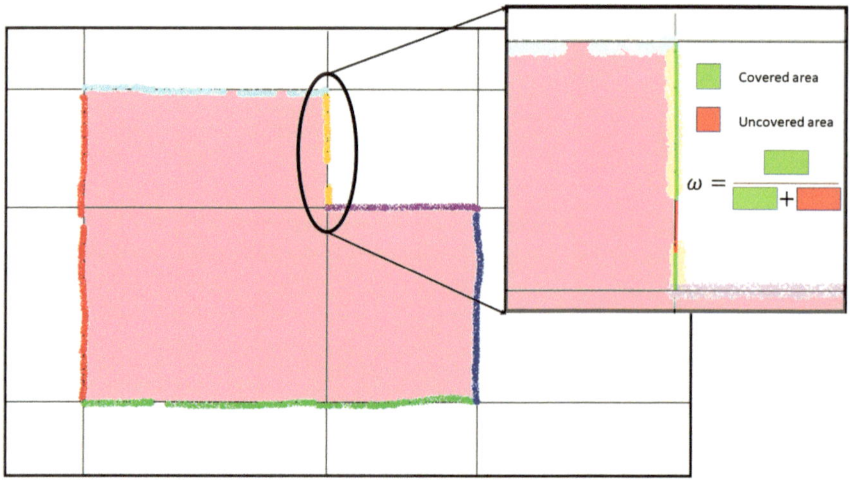

Fig. 5.66 Process of edge weight computing

group of points. Then, we consider the overlap between the edge and bounding box as the covered area l_c. The weight w is computed by the following equation:

$$w = \frac{\sum l_c}{l}, \tag{5.95}$$

where l is the length of the edge.

Furthermore, 2D room maps can be divided by simulating the process of heat diffusion in the cell complex. Based on the edge weights, we can construct an affinity matrix $\mathbf{L}$ for the cell complex, which is similar to a Laplacian in mesh analysis. The entry $\mathbf{L}_{ij}$ for $\mathbf{L}$ is set to

$$\mathbf{L}_{ij} = \begin{cases} e^{-w/\sigma}, & \text{if } i \neq j \text{ and } f_i, f_j \text{ are adjacent,} \\ 1, & \text{if } i = j, \\ 0 & \text{otherwise} \end{cases} \tag{5.96}$$

A Markov probability transition matrix $\mathbf{M}$ is computed for L as

$$\mathbf{M} = \mathbf{D}^{-1}\mathbf{L}, \tag{5.97}$$

where $\mathbf{D} = \text{diag}\left(\sum_{j=1}^{n} \mathbf{L}_{ij}\right)$. Each entry M_{ij} represents the local affinity value between faces f_i and f_j, which is determined by considering direct connectivity. We use diffusion maps [32] to spread these local affinities, which are considered to be noise-resistant and therefore well suited for our task. The faces are embedded in a high-dimensional Euclidean space by the diffusion map. The corresponding embedding space coordinate of a face f_i is

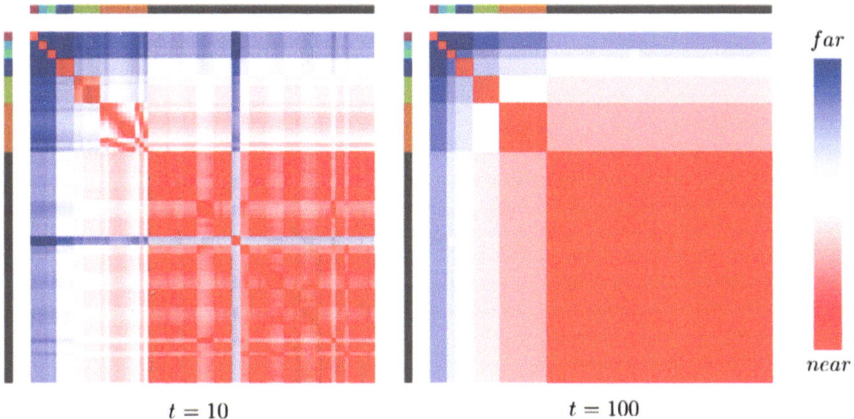

Fig. 5.67 Distance matrices for different diffusion times t

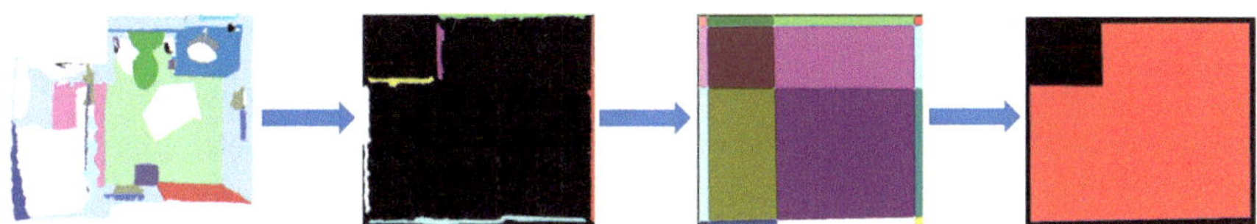

Fig. 5.68 Visualization of our room detection process

$$\Phi\left(f_i\right)=\left(\lambda_1^t \phi_1\left(f_i\right), \lambda_2^t \phi_2\left(f_i\right), \ldots, \lambda_m^t \phi_m\left(f_i\right)\right), \tag{5.98}$$

where λ_k and ϕ_k are the k-th eigenvalue and eigenvector of $\mathbf{M}$, respectively. t is the diffusion time, and m is the embedding dimension. A higher dimension m and longer time t are usually needed to distinguish more faces. Figure 5.67 shows the distance matrices of the computed embeddings for different diffusion times t. After the embeddings are obtained, the k-medoids algorithm [66] is used to cluster the faces into rooms. Figure 5.68 visualizes the 2D room map construction process.

The 2D map may not exactly fit the shape of the room model due to the incompleteness and noise in the created model. As a result, directly using a 2D map as a filter has a low fault tolerance rate and may lead to the creation of partial room models. A simple but efficient assumption is made to address this issue. The position of the camera always falls in the room map because we are usually inside the room during scanning. Therefore, we determine the average camera position for each submap and group the submaps for which the average camera position falls in the map as a room. Since the camera and the wall are usually separated by a certain distance, this solution has a higher fault tolerance rate and allows us to create a full room model. We also discard the room if the current camera position does not fall into the room map based on this assumption.

Room detection is performed only when we detect local loop closures among different submaps to improve the online performance. This is based on the fact that at least one closed loop will be detected before we finish scanning a room. Once the camera leaves the room space, we create a new submap, and the previous submap is considered part of the detected room. The room model with semantic and instance labels predicted by Occuseg [50] is then used for room-level LCD.

Instance Feature Learning. We use $\mathcal{P}$ to denote the vertices of the room model as a point cloud with predicted semantic and instance labels. The per-point semantic and instance results are denoted as S and I, respectively. Then, the point cloud of each predicted instance is extracted and masked with I,

$$\mathcal{P}_i = \{p_k \odot \delta(I_{p_k}, i)\}, k \in [1, |\mathcal{P}|], \tag{5.99}$$

where $\delta(\cdot, \cdot)$ equals 1 when the two variables are the same and 0 otherwise.

$\mathcal{P}_i$ is then centered at the origin and voxelized at a resolution of $r = 0.2$, resulting in a 256^3 voxel grid. During training, a random rotation around the z-axis is applied to $\mathcal{P}_i$, as we assume the gravity vector is perpendicular to the x-y plane in the input scene. Thus, the network and the resulting embedding are rotation invariant.

The network used to extract the instance-level embeddings is named *Embedding-Net*, and the architecture of this network is shown in Fig. 5.69.

An encoder architecture is built with the sparse convolutional network [43, 83], which takes advantage of the sparsity of the 3D point cloud. Therefore, the memory usage and computations are reduced, allowing the creation of larger and deeper networks. After the input layer, Embedding-Net contains a series of convolutional blocks which downsample the input 5 times, followed by a fully connected layer, which outputs the embedding. We denote Embedding-Net as $f(\cdot)$ and the resulting embedding as $\phi_i = f(\mathcal{P}_i), \phi_i \in \mathbb{R}^D$, where $D = 256$ is the embedding dimension.

A triplet loss is used to train Embedding-Net. Specifically, the triplet loss takes a tuple of embeddings as input and minimizes the distance between an anchor and a positive sample while maximizing the distance between the anchor and a negative

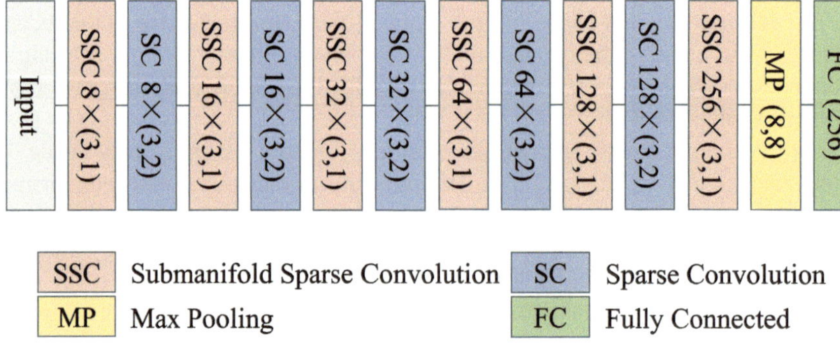

Fig. 5.69 The architecture of Embedding-Net

sample. In our training process, we use two instances $\mathcal{P}_i, \mathcal{P}_j$ as the anchor and negative sample and obtain the positive sample by applying a random rotation R around the z-axis to $\mathcal{P}_i$:

$$\mathcal{P}_i' = \left\{ R\boldsymbol{p}_k | \boldsymbol{p}_k \in \mathcal{P}_i \right\}. \tag{5.100}$$

The triplet loss then becomes

$$L_{triplet} = \left[d(f(\mathcal{P}_i), f(\mathcal{P}_i)) + d(f(\mathcal{P}_i), f(\mathcal{P}_j)) + m \right]_+, \tag{5.101}$$

where d is the Euclidean distance function and m is a margin. Note that $\mathcal{P}_i$ and $\mathcal{P}_j$ are randomly selected from the training set and are not necessarily from the same room. Following Hermans et al. [150], we use an online hardest negative mining strategy, as illustrated in Fig. 5.70, with a batch size of 64. In practice, we train Embedding-Net based on the ScanNet [26] dataset.

Similarity Measure and Candidate Selection. The similarity of different rooms can be computed by comparing their corresponding instance embeddings, as mentioned above. The shape information is encoded by the embedding learned for each instance, so similar instances appear close in the high-dimensional space. Therefore, two rooms are similar if their instances have close embeddings.

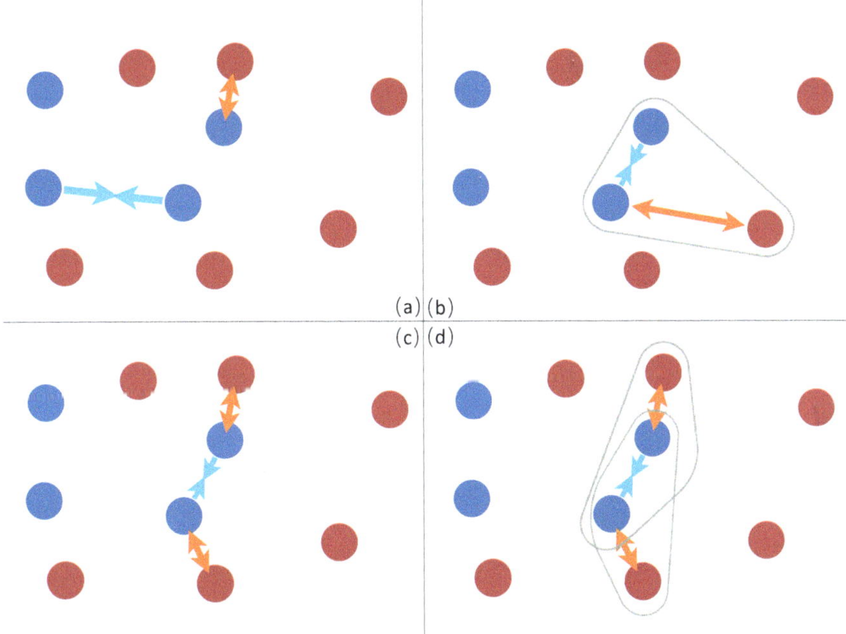

Fig. 5.70 Visualization of the triplet loss and hardest negative mining strategy. **a** Contrastive loss; **b** Triplet loss; **c** Hardest contrastive mining strategy; **d** Hardest triplet mining strategy

For a pair of room scans $(\mathcal{P}, \mathcal{Q})$, the similarity between the rooms is defined as

$$D(\mathcal{P}, \mathcal{Q}) = \frac{\sum_i \sum_j \mathcal{F}(\phi_i^p, \phi_j^q)}{|\Phi_p||\Phi_q|}, \tag{5.102}$$

where $\Phi = \{\phi_1, \phi_2 \ldots, \phi_k\}$, and k is the instance number of the given room. $\mathcal{F}(\cdot, \cdot)$ measures the similarity of two embeddings as

$$\mathcal{F}(\phi_i, \phi_j) = \begin{cases} \exp(-d^2/\sigma^2), & d \leq d_0 \\ 0 & d > d_0. \end{cases} \tag{5.103}$$

Here, d is the Euclidean distance between ϕ_i and ϕ_j. σ is a weighting factor. d_0 is a distance threshold. We sort the database based on the similarity scores and choose the rooms with the highest N scores as candidates based on the similarity measure.

Geometry Verification and Post-optimization Process. The aforementioned similarity measurement does not consider the relative geometric relationships between instances; thus, this measure is useful only for obtaining a rough candidate collection. Therefore, a geometry verification module is used to ensure that the identified candidates are true loop closures. Following Finman et al. [34], we use graph matching to verify the geometry results. Since multiple candidate objects of the same shape may have been identified, a graph-based approach is applied instead of a direct matching approach using object embeddings. As a result, the geometric structure must be utilized in this process.

For each room, the instances within the room form a graph $\mathcal{G} = (V, E)$, where each vertex includes the center of the i-th instance, $\mathbf{c}_i \in \mathbb{R}^3$, and its semantic label s_i:

$$v_i \triangleq \{\mathbf{c}_i, s_i\}. \tag{5.104}$$

The edge $e_{i,j}$ is determined based on the distance metric between instance centers, i.e., $||\mathbf{c}_i - \mathbf{c}_j||$. The goal is to find correspondences between two graphs $\mathcal{G}_1$ and $\mathcal{G}_2$. The correspondences are shown in the form of an assignment matrix X of size $|V_1| \times |V_2|$.

$$X_{i,j} = \begin{cases} 1, & \text{when } v_i^{\mathcal{G}_1} \text{ is matched to } v_j^{\mathcal{G}_2} \\ 0, & \text{otherwise.} \end{cases} \tag{5.105}$$

According to [30], the problem can then be formulated as minimizing the score function:

$$\hat{X} = \arg\max_X \ \text{score}(X), \tag{5.106}$$

$$\text{score}(X) = \sum_{i_1, i_2, j_1, j_2} H_{i_1,i_2,j_1,j_2} X_{i_1,j_1} X_{i_2,j_2}, \tag{5.107}$$

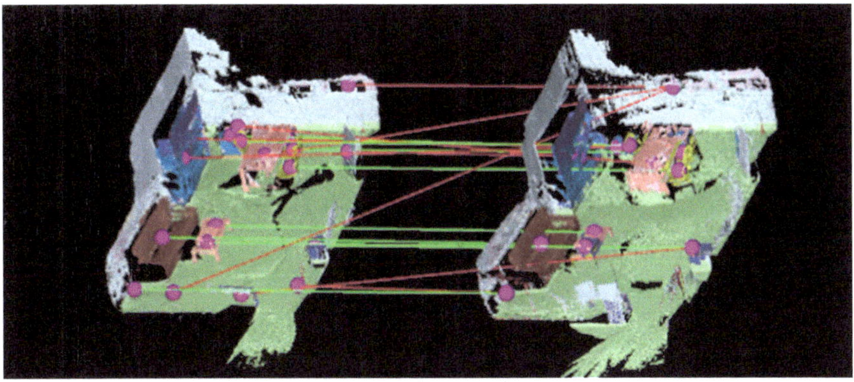

Fig. 5.71 Visualization of the geometry verification results

where H is the second-order potential corresponding vertex pairs of (v_{i_1}, v_{i_2}) based on $\mathcal{G}_1$ and (v_{j_1}, v_{j_2}) based on $\mathcal{G}_j$. H is computed as

$$H_{i_1,i_2,j_1,j_2} = h(e_{i_1,i_2}, e_{j_1,j_2})\delta(s_{i_1}, s_{j_1})\delta(s_{i_2}, s_{j_2}), \tag{5.108}$$

$$h(e_{i_1,i_2}, e_{j_1,j_2}) = \exp(-\gamma||e_{i_1,i_2} - e_{j_1,j_2}||). \tag{5.109}$$

Then, $\hat{X}$ is solved using power iteration, as described in [30].

RANSAC is used to obtain a robust transformation matrix between the two graphs based on the predicted correspondences. If the number of correspondences reaches a minimum value n, we consider the proposed candidate as a correct loop closure. Figure 5.71 shows an example of this process. In this figure, the purple spheres represent graph nodes. The graph matching results are shown by the lines; the green lines indicate RANSAC-identified correspondences, while the red lines indicate incorrect correspondences caused by incomplete segmentation outcomes. Thus, the transformation estimate may be resistant to erroneous geometric shapes or incorrect semantic/instance segmentation results.

Since we cannot guarantee the density and completeness of the sampling during scanning, the relative locations of the instance centers may vary slightly between scans, resulting in errors in the transformation matrix between two rooms. Furthermore, due to the scarcity of the instance centers, rejecting all the false loop closures is difficult. To reject all false results and reduce the alignment error in the correct room correspondence results, we refine the transformation and propose a scoring technique for each room correspondence, which gives a low score when different rooms have the same layout and similar room models and a high score when two models of the same room have different appearances due to incomplete scans.

For the source point cloud $\mathcal{P}_i$ and the target point cloud $\mathcal{P}_i$ in a room correspondence (i, j), ICP is performed with the initial transformation matrix and the RANSAC matrix. We use $\mathcal{P}^k$ to denote a set of points in point cloud $\mathcal{P}$ that belong to the kth instance and $\mathcal{P}^{inlier}$ to denote a collection of inlier points in $\mathcal{P}$ registered by the ICP algorithm. The inlier ratio of the kth instance in point cloud $\mathcal{P}$, denoted as r_k, is defined as

$$r_k = \frac{|\mathcal{P}^{inlier} \cap \mathcal{P}^k|}{|\mathcal{P}^k|}. \tag{5.110}$$

Then, we calculate the score of the source point cloud $\mathcal{S}_{\mathcal{P}_i}$ as

$$\mathcal{S}_{\mathcal{P}_i} = \sum_{k}^{|I_i|} f_s(r_k), \tag{5.111}$$

where I_i is a set of instance labels for the source point cloud $\mathcal{P}_i$. The score function of the instances is defined as

$$f_s(r) = \begin{cases} r, & \text{if } r > r_{min} \\ r - 1, & \text{otherwise} \end{cases} \tag{5.112}$$

We apply a penalty to instances with inlier ratios close to zero by introducing a negative score. The score of the target point cloud $\mathcal{S}_{\mathcal{P}_j}$ is calculated in the same way, and we use the lower value as the final score:

$$\mathcal{S}_{(i,j)} = \min\left(\mathcal{S}_{\mathcal{P}_i}, \mathcal{S}_{\mathcal{P}_j}\right). \tag{5.113}$$

Correspondences with scores greater than a threshold $\mathcal{S}_{min}$ are represented as correct room-level loop closures.

Room Constraints. The room-level LCD method provides room constraints to maintain global consistency across multiple agents throughout the global pose optimization, as shown in Eq. 5.80. We define the room error as

$$E_r = \sum_{(i,j)\in\Omega_R} \sum_{k=0}^{|C_{i,j}^R|-1} \|T_i^R P_i^k - T_j^R P_j^k\|^2, \tag{5.114}$$

where $C_{i,j}^R$ is the corresponding point for the room pair (i, j) and Ω_R is a collection of all the room correspondences.

Collaborative 3D Reconstruction System

The above sections described the methods used in the different modules. In this section, we connect all these parts by introducing our collaborative system. Furthermore, we present an optional offline optimization approach to improve the quality of the details in each room model (Figs. 5.72 and 5.73).

Fig. 5.72 Different parts of the *Lab* dataset that were collaboratively reconstructed by different agents

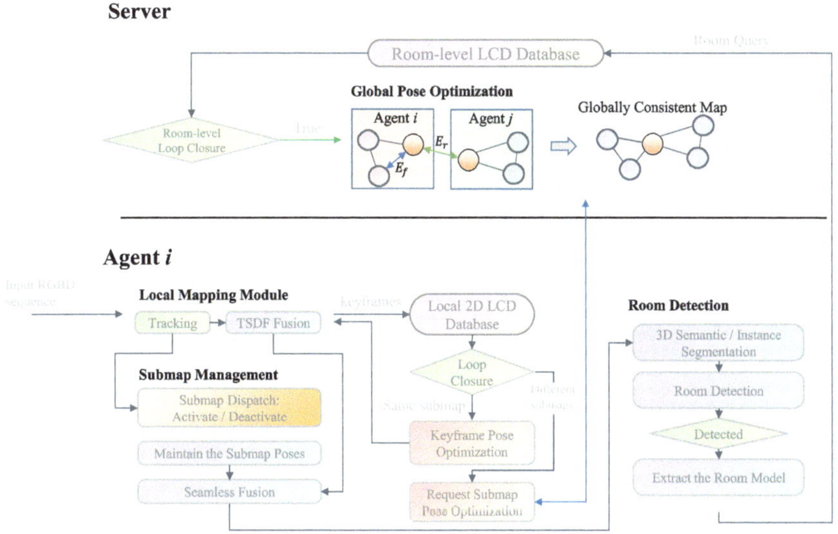

Fig. 5.73 System pipeline of our collaborative reconstruction architecture

Centralized Collaborative Reconstruction. For collaborative dense 3D reconstruction, a centralized system with a server and multiple agents is built. The ROS [118] protocol is used for communication among the agents and the server. The server collects the results from all agents and maintains a globally consistent sparse map, while each agent is responsible for producing a consistent model based on its own scans. The global poses for each submap and room are optimized in the server using an efficient method, while the reconstruction results based on each agent have the same complexity as FlashFusion [48]. Once room-based loop closures are detected across multiple agents, the server combines all of these agents into a single node, allowing their poses to be optimized simultaneously. Ideally, only one node should remain at the end of the process, resulting in a globally consistent model that is composed of maps created by all agents.

Each agent in our system produces a room-scale consistent dense reconstruction result. After capturing data, robust RGBD odometry is used for local pose estimation. We first use sparse binary feature matching for efficiency; if sparse odometry fails, dense tracking [119] based on the last frame is used for robustness to jointly minimize the photometric and geometric errors. If the current frame is registered as a keyframe, the keyframe database is queried to find all the loop closure candidates through MILD [47]. If inter-submap loop closures are identified and verified by our frame matcher, the agent transmits the correspondences to the server for submap pose optimization, as well as keyframe information such as descriptors and 3D keypoints. Once the agents receive a response from the server, the keyframe is integrated into the voxel-hashing representation and becomes a part of the generated dense model, which is then transmitted to OccuSeg [50] to obtain accurate semantic and instance labels. Figure 5.74 shows the details of the reconstruction process for each agent.

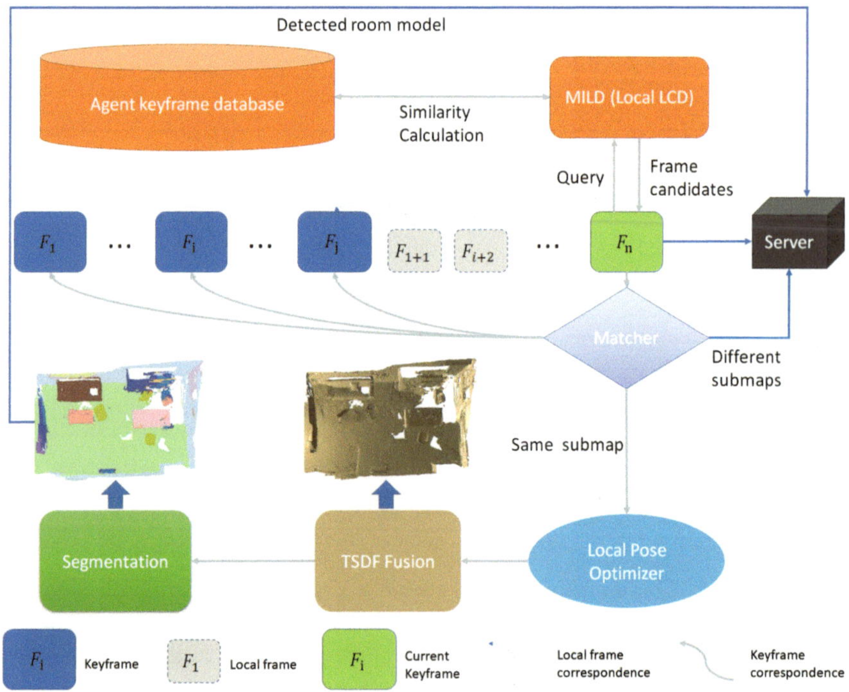

Fig. 5.74 Pipeline of the local map reconstruction process for each agent

Note that each agent also manages its own submaps and clusters them into rooms when cross-submap loop closures are detected.

In this system, a server is used for global pose optimization and global model visualization. The server performs global room-level LCD for room correspondences. Initially, all the agents are run separately, and we consider each agent as a node until we first detect room correspondences between different agents. Then, the corresponding nodes are merged, which is essentially a union-find set problem. When an agent detects a room, the room model is transmitted to the server, and the room is registered through our room-level LCD method. Then, the global poses are optimized across all the agents in the same node. For pose optimization in one node, the server jointly minimizes the errors of the submap, gravity and room constraints. We call a submap a *room submap* if it belongs to one particular room or a *single submap* if it does not belong to any room. A factor graph is used to represent the relationships among the submaps, rooms and agents in the server. Three factors correspond to the three constraints. The submap factors are the basic edges in the factor graph because they connect all the submap vertices for the same agent. Note that the submap factors are fixed if the corresponding submaps belong to the same room. The room factors are used to connect rooms across all the agents. Finally, each room and single submap have a gravity factor, which is used to align the ground normal vector with the gravity direction. A visualization of the factor graph is shown in Fig. 5.75.

Offline Postprocessing. We use frame reprojections to identify common chunks among multiple submaps. Although this approach is highly efficient, it does not guarantee a complete reintegration of all overlapping parts among the submaps, which can lead to incomplete local information. Intuitively, the room quality is particularly important for large-scale indoor scene reconstruction. As a result, we suggest a postprocess to enhance the local room details. We use our hierarchical scene structure, as shown in Fig. 5.55, to find all the frames of each room. Because there is no need to use the submaps and reject frame correspondences of textureless areas in the room-scale reconstruction task, we apply an exhaustive local LCD method [47] to find all the possible frame-to-frame correspondences among these frames and opti-

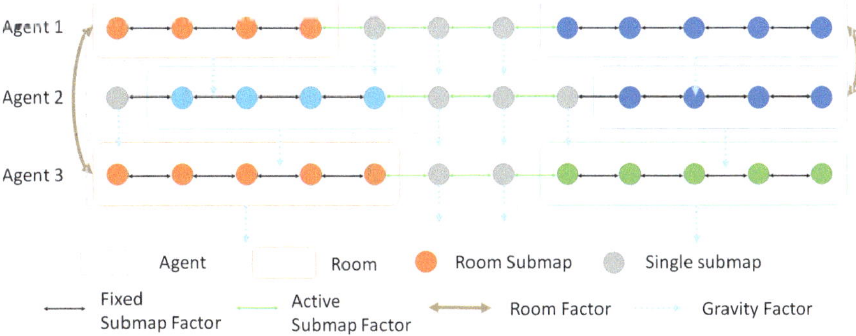

Fig. 5.75 Factor graph on the server

mize their poses based on the global poses calculated during online reconstruction. Finally, these frames are integrated into a single voxel hashing table, based on which we extract a more detailed dense model for each room that directly replaces the corresponding parts in the global model.

Experiments

In this section, room-level LCD evaluations and building-scale structural reconstruction results are presented.

Room-Level LCD. We evaluate our Embedding-Net training and room-level LCD method based on two datasets, both of which are generated from the ScanNetv2 [26] dataset. ScanNetv2 is an RGBD video dataset containing 1513 scans acquired in 707 different locations, as well as reconstructed 3D models and ground truth labels for 3D semantic/instance segmentation. This dataset is commonly used for 3D semantic and instance segmentation tasks [50].

We need a series of repeated scans of the same room to simulate the loop closure events that occur in building-scale 3D reconstruction to evaluate our model. The following dataset organization strategy is adopted in our experiments:

(1) *Instance segmentation.* To train the instance segmentation network, i.e., Occuseg [50], we split the training/test data based on the original ScanNet instance segmentation results. There are 1201 scenes in the training set and 312 scenes in the test set.
(2) *Embedding learning.* Using the ground truth instance segmentation labels, we create an object dataset using the same split as above. We extract point clouds for all the objects based on the restored 3D meshes except for those with ‘floor,’ ‘wall,’ or ‘ceiling’ semantic labels because these objects are not distinguishable enough in different scenes. In total, 15266 objects are included in the training set, and 4245 objects are included in the test set.
(3) *Room-level LCD.* Based on the ScanNet test set, we evaluate the room-level LCD results. The test set includes 312 scans of 142 distinct scenes, with 100 scenes having multiple scans. The query list[1] consists of 100 scans of distinct scenes, while the database consists of the remaining scans. A result is identified as a correct detection if the query result and the query object are scans of the same scene. Each query has at least one correct candidate in the dataset with this configuration.

Comparison with Instance-Based Methods. In this section, we compare the performance of our Embedding-Net model with that of several baseline methods that also use instance segmentation results as input.

[1] Query scans with the format `scene_xxxx_01` are selected.

The evaluation includes the following phases:

(1) *Instance segmentation.* We generate instance labels for each scan in the ScanNet test set in this step. We use the state-of-the-art joint semantic-instance segmentation network Occuseg [50] as the upstream network in this and all subsequent experiments. Objects with few points are ignored.
(2) *Instance embedding.* For each instance created in the previous stage, Embedding-Net is used to predict the embedding. Before the instances are input into the network, each proposed instance is preprocessed and voxelized, and the corresponding point cloud is centered at the origin.
(3) *Similarity measurement.* We evaluate the similarity of each room pair by computing the similarity score described in Eq. 5.102. We compare each query room with other rooms in the databases. The average recall rate for the N nearest query candidates is computed as recall@N.

We compare Embedding-Net with two other baseline models to demonstrate the viability of the proposed network.

The first baseline is a histogram of semantic labels. We obtain the semantic category of each object after instance segmentation by determining the mode of its corresponding semantic labels. Next, the similarity of two scans is determined using the L2 norm based on the semantic label histograms. Higher similarity values have smaller histogram distances.

PointNet [101] is another baseline model. To create the object embeddings, we use the original PointNet model pretrained based on the ModelNet40 dataset [136] to assign the object categories. We aim to demonstrate the generalizability of the PointNet architecture and determine whether this model can be applied to room-level LCD tasks.

To create a 256-dimensional embedding, we discard the last softmax layer in the PointNet model. These embeddings are then used for similarity measurements in the same manner as in our model.

We train Embedding-Net using the training set of the object dataset that is generated based on the ScanNetv2 [26] dataset. Our model is trained using the online hardest triplet loss with a 64 batch size. The Adam optimizer is used for optimization, and the learning rate is set to $1e-3$. The dimension of the embedding is set to 256. The length of the voxel is set to 0.2. In our implementation, we set $\sigma = 0.71$, $d = 1.3$. The model was trained for 150 epochs, and the room-level LCD results using Embedding-Net are shown in Table 5.23. Our method outperforms the PointNet baseline and semantic label histogram by a large margin in terms of Recall@1. Furthermore, we show the recall rates of 1–25 query candidates in Fig. 5.77. In general, the semantic label histogram and our method outperform PointNet, demonstrating that PointNet fails to generalize well for the ScanNet object database and does not produce strong shape embeddings. The good Recall@5 performance of the semantic label histogram method indicates that the semantic/instance segmentation results contain considerable scene knowledge. Since the semantic label histogram method cannot discriminate between different objects in the same category, such as two chairs with vastly

Table 5.22 Overview of the datasets used for instance segmentation, embedding learning and room-level LCD evaluation

	Instance seg	Embedding	Room-level LCD	
	#Scans	#Objects	#Database	#Queries
Train	1201	15266	–	–
Test	312	4245	212	100

Table 5.23 Comparison of the results of our room-level LCD method with different baseline models (%)

Method	Type	Recall@1	Recall@5
PVLAD [4]	3D Non-instance	36	66
FPFH [108]	3D Non-instance	29	55
BoBW [39]	2D Image-based	12.1	–
PointNet [101]	3D Instance-based	15	31
Histogram	3D Instance-based	73	96
Ours	3D Instance-based	**86**	**98**

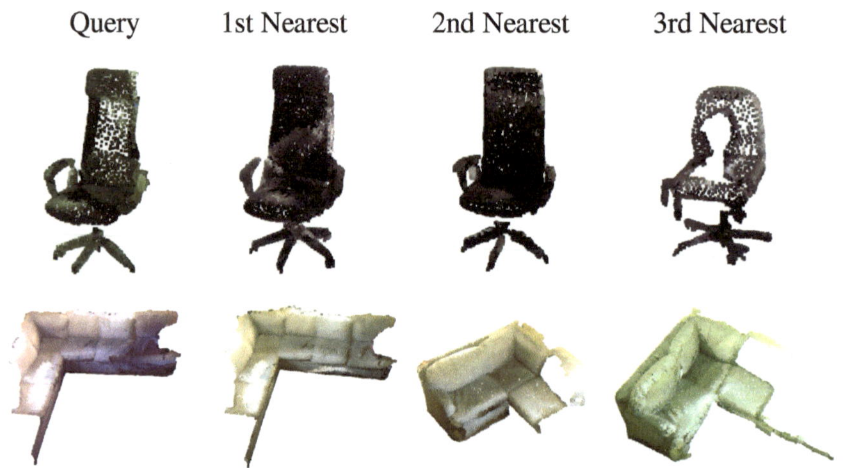

Fig. 5.76 Nearest neighbor search of embeddings based on the object test set

different shapes, Embedding-Net aims to obtain more comprehensive descriptions for each object to enhance the room-level LCD accuracy. The embedding quality is qualitatively assessed in Fig. 5.76. The results are obtained by identifying the nearest neighbor embeddings in the test set. The results show that geometrically similar objects are successfully grouped using the embeddings (Table 5.22).

We also investigate how the embedding dimension affects the room-level LCD performance. We train three Embedding-Net models with output dimensions of

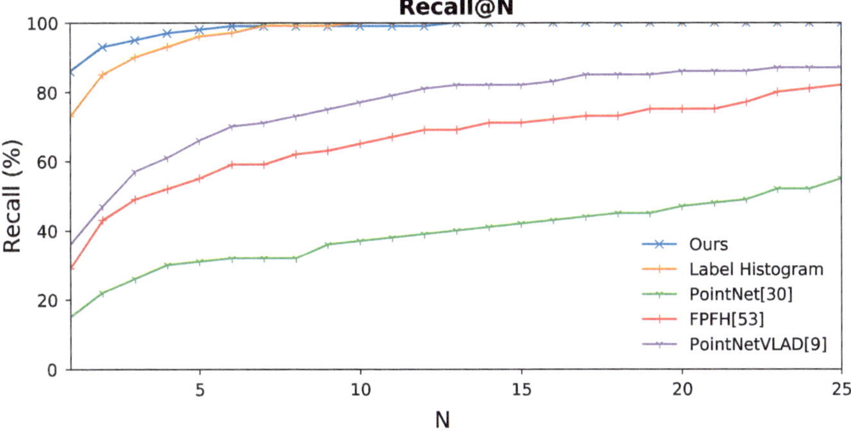

Fig. 5.77 Recall@N comparisons of our room-level LCD method with the baseline models

Table 5.24 Comparison of the results of our Embedding-Net trained under different output dimension settings

	Recall@1 (%)		Recall@5 (%)	
	GT	PRED	GT	PRED
D-128	88	83	98	95
D-256	**89**	**86**	97	**98**
D-512	85	86	**99**	92

128, 256, and 512. The results are shown in Table 5.24. Note that 'PRED' and 'GT' represent using the instance labels predicted by Occuseg [50] or the ground truth labels as the input to the room-level LCD model. We choose $D = 256$ because this model produces the best results with the predicted instance labels (Fig. 5.78).

Comparison with Non-Instance-Based Methods. In the above evaluation, we compare our model with instance-based room-level LCD methods, which use instance segmentation information. To justify the use of the instance-based approach, we also compare our method with non-instance-based methods, which do not use instance segmentation information.

We train a PointNetVLAD [4] network based on the same training set used to train our model. PointNetVLAD [4] is a state-of-the-art 3D LCD method that learns global descriptors for point clouds. We use the best pretrained PointNetVLAD model [4] and finetune the model based on our dataset to improve the performance (Figs. 5.79 and 5.80).

In addition, we compare our model with another baseline: hand-crafted local 3D features, or fast point feature histograms (FPFHs) [108].

We first uniformly downsample the input query scene by voxelization, with the voxel size set to 0.05 m. Then, we compute the pointwise FPFH descriptors. Since

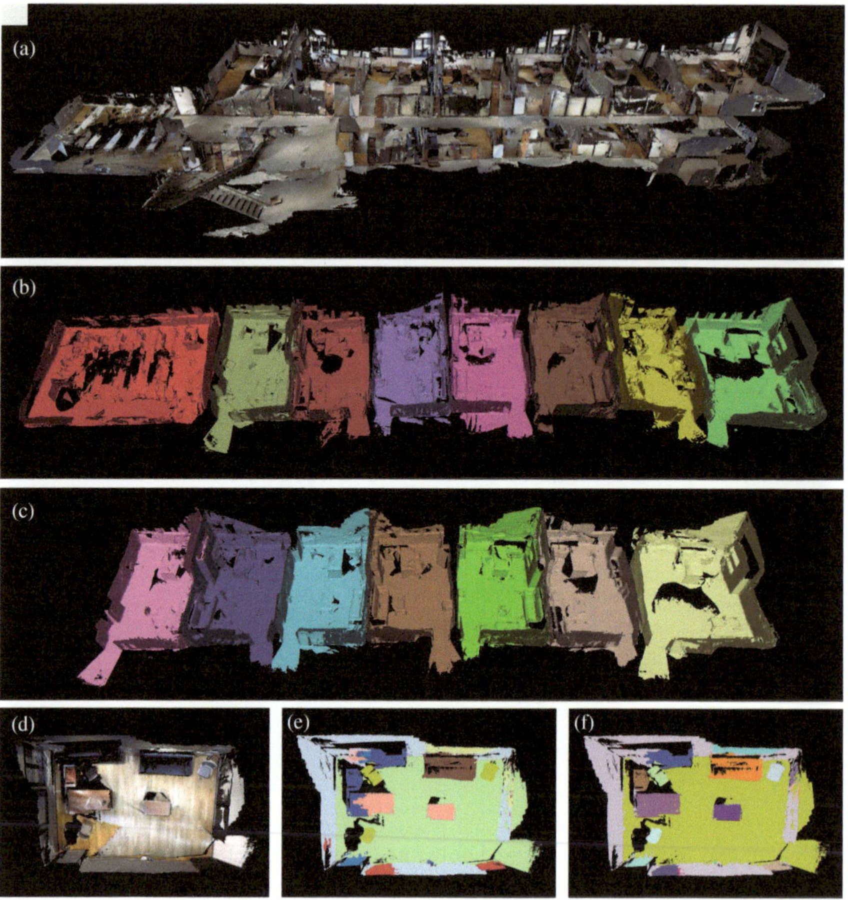

Fig. 5.78 Results based on the *Office* dataset. **a** Colored geometry; **b** scene structure (the first floor); **c** scene structure (the second floor); **d**, **e**, **f** detailed color, semantic and instance labels for *room_11*. At the bottom, we present the semantic label colors

the number of FPFH features for each scene is too large to apply the direct similarity measure shown in Eq. 5.102, we firstrandomly sample 600 features for each scene. Then, a bag-of-words model is used to aggregate the features, with the cluster number set to 50, resulting in a global histogram of dimension 50. The similarity of two scans is then determined by calculating the L2 distance of the histograms.

Figure 5.77 and Table 5.23 show the Recall@N results of our method and the two baselines. Our method outperforms both baselines, showing that our method generates more accurate descriptions of the scenes (Fig. 5.81).

Comparison with 2D LCD Methods. We also compare our method with the state-of-the-art 2D LCD method BoBW [39]. We also evaluate BoBW based on the ScanNetv2 validation set. The query and database scenes are the same as those in the

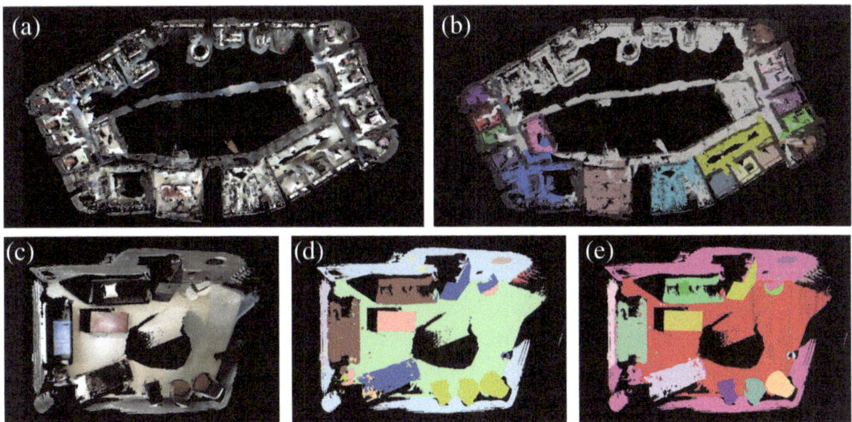

Fig. 5.79 Results based on the *Lab* dataset. **a** Colored geometry; **b** scene structure; **c**, **d**, **e** detailed color, semantic and instance labels for *room_6*

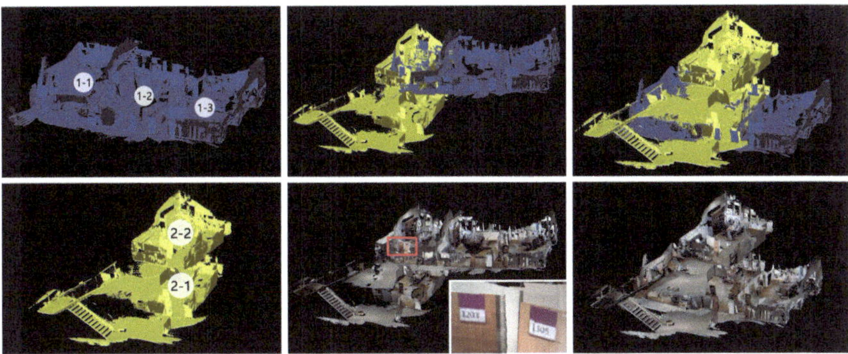

Fig. 5.80 Comparison of the results with and without room-level LCD

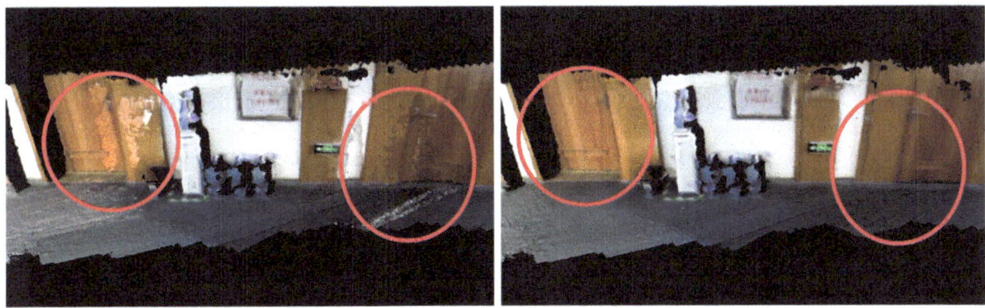

Fig. 5.81 Comparison of the results with and without submap reintegration

room-level LCD dataset shown in Table 5.22. We use the RGB scan sequences of the corresponding scenes as the input to the BoBW model. We use the predefined parameters and pretrained vocabulary for the BoBW model. We start by adding all the images from the database scenes to the BoBW database. Then, we perform the LCD query using t((he images from the query scenes without inserting these images into the database. The resulting database contains 366462 images. Since we cannot directly generate frame-to-frame correspondences using the frame camera poses (different scans have different world coordinates), we instead consider all the candidates that pass the geometric verification in the BoBW model [39] to be correct loop closures, as this model is reported to have very high accuracy [39]. Considering possible false-positives, the actual recall of the BoBW model [39] might be slightly lower than the calculated result shown in Table 5.23. Additionally, the BoBW model [39] uses different candidate selection strategies rather than purely comparing similarity scores. Therefore, the Recall@5 metric is not directly applicable in Table 5.23.

Nevertheless, the results show the advantages of the room-level LCD method over traditional 2D methods when the frame number increases to the building-scale level. The relatively low recall of the BoBW model [39] may be due to two reasons: (1) The large database may cause the hit rate to decrease when performing fast vocabulary tree searches; (2) The BoBW model [39] is not good at relocalizing for novel poses. These problems do not occur with room-level LCD methods because the number of rooms increases much slower than the number of frames, and 3D models are not affected by 2D camera poses.

Building-Scale Reconstruction. In this section, we first compare our method with other large-scale RGBD reconstruction methods. Then, we present ablation studies based on the room-level LCD method, ground normal vector constraints and submap reintegration process. Finally, we show our results based on two building-scale datasets that we captured.

We captured two building-scale reconstruction sequences, *lab* and *office*, to evaluate our system. All data were acquired using a Surface Book equipped with an ASUS Xtion depth camera. We captured the RGB and depth images at a resolution of 640×480 with a frame rate of 30 Hz. We covered the whole scene with several different sequences to enable multiagent collaborative reconstruction. Table 5.25 presents more details about the two datasets.

In this experiment, we run the server and agents on PCs, each equipped with a CPU, i7-8700, and a Nvidia GTX1070 GPU with 8 GB memory. In both datasets, we set the voxel resolution to 0.00625 m. The threshold of the 3D reprojection error for the visual odometry system is set to 0.01, and the weights of the normal constraints

Table 5.25 Details of our building-scale reconstruction datasets

Name	#Sequences	#Rooms	#Frames	Area (m^2)
Lab	7	19	99126	957.6
Office	9	15	87952	708.8

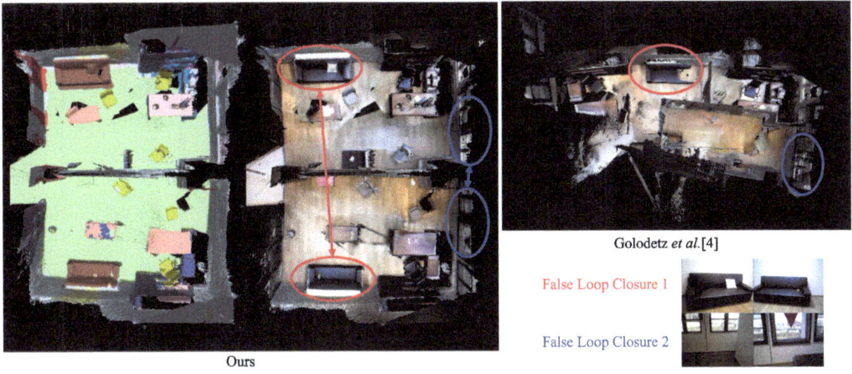

Fig. 5.82 Comparison of the reconstruction results with those of the method by [42]. The similar frames that caused LCD failures for the method by [42] are shown on the bottom right

and room constraints are set to 2.0 and 10.0, respectively, while the weight of the frame constraints is set to 1.0.

A qualitative comparison of the reconstruction results is shown in Fig. 5.82. Golodetz et al. [42] developed a state-of-the-art large-scale reconstruction method. However, due to frequent tracking losses and loop closure errors, Golodetz et al. [42] could not generate reasonable results for challenging large-scale building scenes.

Taking the partial sequence in our *office* dataset with two rooms as an example, the method of Golodetz et al. [42] failed to faithfully reconstruct the geometry of the two rooms because two false loop closure events occurred. As highlighted by the red circles in Fig. 5.82, the two couches have similar appearances; thus, the relocalization strategy in the method by [42], which uses a 2D image-based regression tree LCD method [12], falsely relocalized the couch in the second room based on the couch in the first room, causing the second room to overlap with the first room. In addition, the similar windows in the two rooms caused erroneous loop closures, as highlighted by the blue circles. As a result, the geometric models of the two rooms incorrectly overlap.

In contrast to the method developed by Golodetz et al. [12], our system is more robust and benefits from the semantic/instance segmentation and structure understanding modules. Note that our local loop closure detection operation is only performed within rooms, and the submap correspondences across different rooms are discarded; thus, the reconstruction of the next room is not affected by similar 2D frames in the previously scanned rooms.

Additionally, the localization accuracy is evaluated by comparing our method with BundleFusion [27], ElasticFusion [133], and FlashFusion [48] based on the synthetic ICL-NUIM dataset [51] (with noise) and the real-world TUM RGBD dataset [121]. As shown in Tables 5.26 and 5.27, BundleFusion (offline) is competitive among the comparison methods, and our system achieves comparable tracking accuracy. In terms of system performance, we show a comparison of the average frame rates of the

Table 5.26 Comparison of the localization error based on the TUM RGBD dataset [121] (cm)

Method	*fr1/desk*	*fr2/xyz*	*fr3/office*	*fr3/nst*
RGBD SLAM [33]	2.3	**0.8**	3.2	1.7
ElasticFusion [133]	2.0	1.1	**1.7**	1.6
BundleFusion (online) [27]	1.7	1.4	2.9	1.6
BundleFusion (offline) [27]	**1.6**	1.1	2.2	**1.2**
FlashFusion [48]	1.9	1.3	2.5	1.8
Ours	2.0	1.4	2.5	1.8

Table 5.27 Comparison of the localization error based on the ICL-NUIM dataset [51] (cm)

Method	*kt0*	*kt1*	*kt2*	*kt3*
RGBD SLAM [33]	2.6	0.8	1.8	43.3
ElasticFusion [133]	0.9	0.9	1.4	10.6
BundleFusion (online) [27]	0.8	0.5	1.1	1.2
BundleFusion (offline) [27]	**0.6**	**0.4**	**0.6**	**1.1**
FlashFusion [48]	0.7	0.8	1.1	1.4
Ours	0.9	0.9	1.2	1.4

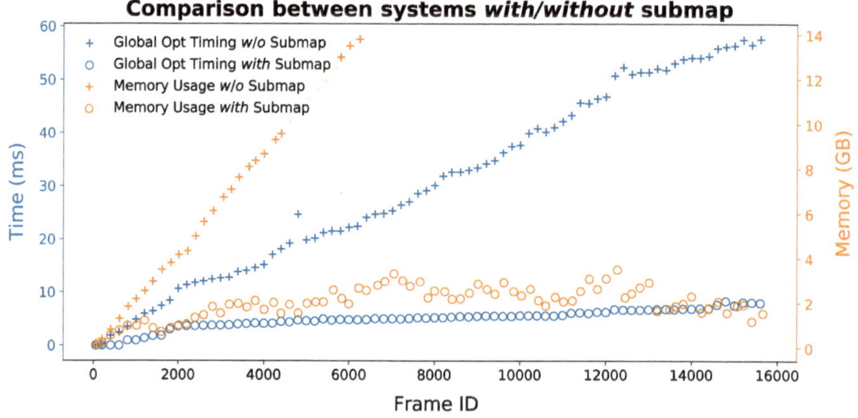

Fig. 5.83 Comparison of the global optimization run time and the memory usage of the reconstructed model between the keyframe and submap strategies

different methods in Table 5.28. We compare our system with BundleFusion [27] and FlashFusion [48] based on the public dataset provided by BundleFusion [27]. Our system outperforms FlashFusion [48], showing the advantage of the submap strategy. While BundleFusion [27] has a higher frame rate than our method, our method is more scalable and provides online semantic information without affecting the real-time performance (Fig. 5.83).

Table 5.28 Comparison of average frame rate based on the BundleFusion dataset (fps)

	BundleFusion [27]	FlashFusion [48]	Ours
Apt 0	**34.8**	–	17.3
Apt 1	**35.7**	–	16.6
Copy room	**36.8**	17.2	17.9
Office 0	**37.9**	14.2	16.8
Office 1	**36.0**	14.4	16.8
Office 2	**37.1**	–	17.0
Office 3	**36.0**	–	16.0

Fig. 5.84 Comparison of the results with and without the ground normal vector constraints

Ablation Study. We evaluate the importance of the room-level LCD strategy by comparing our results *with* and *without* the room-level LCD method. For the model *without* the room-level LCD strategy, we use a local 2D LCD module [47] to detect loop closures across agents. Different agents use a shared 2D LCD database.

The input sequences are taken from the *office* dataset. The results are shown in Fig. 5.80. For convenience, we label the five rooms in the two input sequences, as shown in the first column. Room *1–2* and Room *2–1* are the same room. According to the results of the method *w/o* the room-level LCD strategy, Room *1–3* and Room *2–1* are falsely aligned because the 2D LCD module generates false loop closures. These errors are inevitable because the rooms have similar appearances. The misalignment can be clearly seen in the zoomed-in view shown in the red box, where the aligned rooms have different doorplates. With the room-level LCD method, we can correctly align the two agents because the room-level LCD method matches rooms using the room-level geometry and semantic information instead of 2D frames.

We also consider ground normal vector constraints in the global optimization. Figure 5.84 shows the results when the ground normal vector constraints are ignored. The red model was generated without considering the normal vector constraint, and the blue model shows our results. The green line represents the ground truth model based on the floor points, which is constructed by projecting the floor points onto the ground plane with normal vector $[0, 0, 1]^\top$. The results show that our ground normal vector constraints significantly reduce the cumulative drift in the gravity direction.

As shown in Fig. 5.83, as the number of input frames increases, the keyframe-based optimization process slows due to the increasing number of optimization variables, while the submap-based optimization strategy is still highly efficiency. Furthermore, the keyframe-based reconstruction approach suffers from high memory requirements, and this program tends to crash when the number of input frames reaches approximately 6000, while the proposed submap-based scheme successfully processes all frames. Tables 5.26 and 5.27 present detailed comparisons of the localization accuracy between the keyframe (FlashFusion [48]) and submap-based (ours) methods with public datasets, showing that the submap-based scheme achieves comparable localization accuracy with the keyframe method.

Submap reintegration is important to generate seamless reconstruction results because there may be color inconsistencies among the submaps. Thus, directly combining the geometries of two neighboring submaps may lead to color gaps. We evaluate the advantages of the submap reintegration scheme, and the online reconstruction results are shown in Fig. 5.81. The results *with* the submap reintegration scheme show smoother colors than the results *without* the submap reintegration scheme.

Evaluation Based on Real-World Large-Scale Datasets. We first show our results based on the *Lab* dataset in Fig. 5.79. We show the color geometry and scene structure and the detailed geometry, semantic, and instance labels for a room. Note that in Fig. 5.79b, we show the structure by color-coding the detected rooms. Different rooms are assigned different colors, and gray indicates that the place is not detected as a room. We also assign a number to label each room for visual guidance and convenience. Figure 5.79d, e, and f show zoomed-in results for *room_6*. Specifically, in all of our experiments, we use the same color scheme as ScanNetv2 [26] to visualize the semantic labels. The color scheme is shown at the bottom of Fig. 5.79. In the color scheme for the instance labels, each instance is assigned a random color, and different colors indicate different instances.

We also show how the scene is collaboratively reconstructed in Fig. 5.72, where we use different colors to represent the parts that are reconstructed by different agents.

Similarly, we show the results based on the *Office* dataset in Fig. 5.78. Note that in Fig. 5.78b and c, we remove the hallway and stairs and separate the two floors to provide a clearer view, where (b) shows the lower floor and (a) shows the upper floor. Different rooms are assigned different colors and numbers to show the structure of the scene. The collaborative parts are shown in Fig. 5.85 (Fig. 5.86).

We analyze the complexity of our method to demonstrate its scalability. The execution time of the agent-level restoration process is shown in Fig. 5.87. We can infer that the execution times of agent-level modules do not substantially increase over time based on the statistics of the tracking, TSDF fusion, mesh extraction, instance/semantic segmentation and room detection results.

This is because the TSDF fusion and meshing process are performed within submaps. The instance/semantic segmentation and room detection times are also bounded because these processes are only performed for active submaps. The semantic information of previously detected rooms does not need to be updated, and only new submaps need updated semantic information.

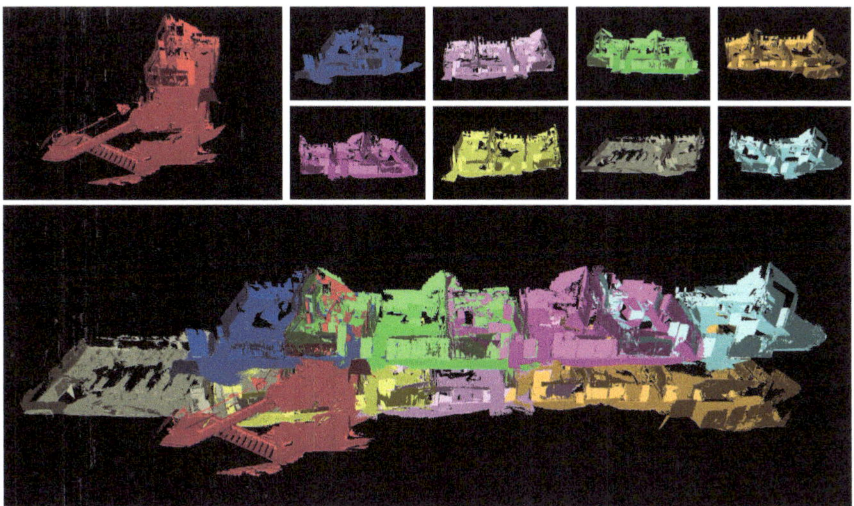

Fig. 5.85 Different parts in the *Office* dataset that are collaboratively reconstructed by different agents

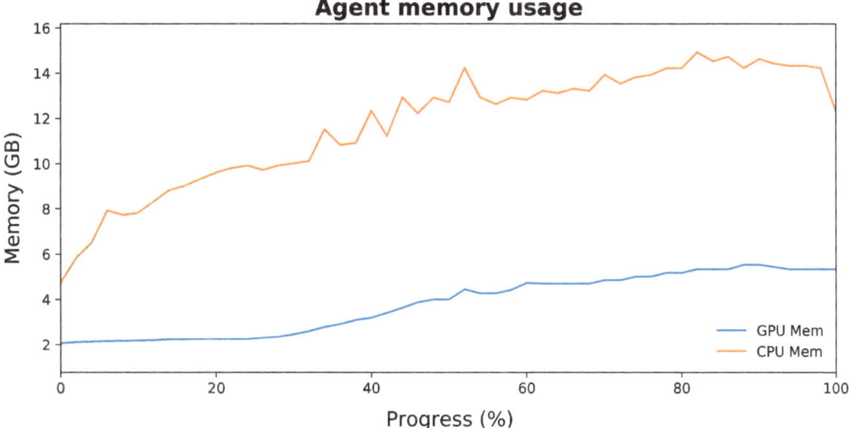

Fig. 5.86 Agent memory usage based on a sequence in the *Lab* dataset

Note that room detection is performed sparsely because it is triggered onlywhen local 2D loop closures are detected for efficiency.

We also show the agent-level memory usage in Fig. 5.86. The only limitation for the agent-level reconstruction process is the CPU memory, which increases slowly because dense models are regularly offloaded from memory. Only simplified models are stored in memory for the purpose of visualization. In terms of GPU memory, since we only use the GPU for segmentation and instance embedding, the GPU memory usage is bounded because only active submaps are involved in these processes.

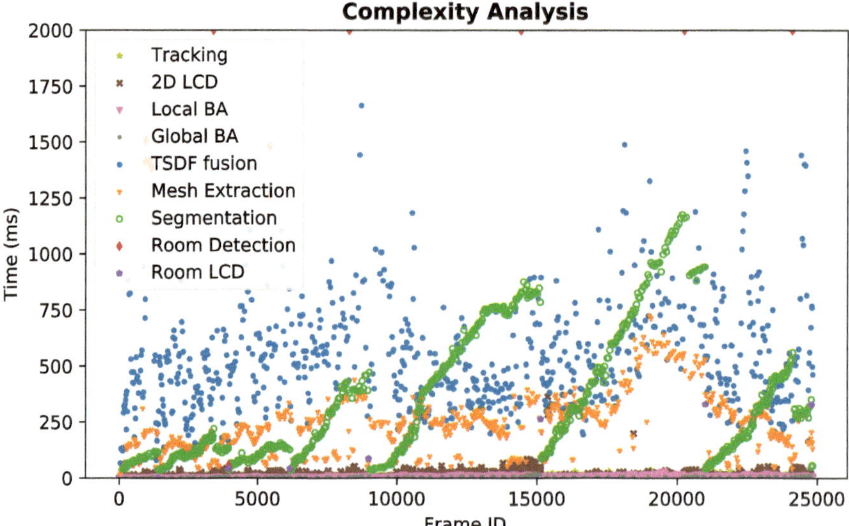

Fig. 5.87 Complexity analysis based on a sequence in the *Lab* dataset

Table 5.29 Server execution time based on the *Lab* dataset (ms)

		Min	Avg	Max
Submap pose optimization		1.70	20.74	69.55
Room level LCD	Embedding	114.58	179.66	328.24
	Similarity	1.47	10.74	30.93
	Geometry check	82.20	181.24	840.33
	Post-optimization	0.0	2832.22	12186.00
	Total (LCD)	260.32	3205.28	12507.48

The server-level execution times are shown in Table 5.29. We provide the room-level LCD query time and the submap pose optimization time when performing a full reconstruction based on the *Lab* dataset. The submap optimization time is consistently low. The room-level LCD query time is acceptable because the room-level LCD queries occur sparsely and are performed only when a room is detected. Note that the minimum execution time of the post-optimization process is zero because if no candidates pass the geometry verification, the post-optimization process is not necessary. The server only uses GPU memory for visualization and stores the simplified global model only in CPU memory. In the experiments performed with both datasets, the CPU memory usage is consistently below 6 GB. Based on the above analysis, our system can handle larger scenes and more agents than existing approaches.

References

1. Henrik Aanæs, Rasmus Ramsbøl Jensen, George Vogiatzis, Engin Tola, and Anders Bjorholm Dahl. Large-scale data for multiple-view stereopsis. *International Journal of Computer Vision*, 120(2):153–168, 2016.
2. Dan A. Alcantara, Andrei Sharf, Fatemeh Abbasinejad, Shubhabrata Sengupta, Michael Mitzenmacher, John Douglas Owens, and Nina Amenta. Real-time parallel hashing on the gpu. In *SIGGRAPH 2009*, 2009.
3. Kara-Ali Aliev, Artem Sevastopolsky, Maria Kolos, Dmitry Ulyanov, and Victor Lempitsky. Neural point-based graphics. In *Computer Vision–ECCV 2020: 16th European Conference, Glasgow, UK, August 23–28, 2020, Proceedings, Part XXII 16*, pages 696–712. Springer, 2020.
4. Mikaela Angelina Uy and Gim Hee Lee. Pointnetvlad: Deep point cloud based retrieval for large-scale place recognition. In *Proceedings of the IEEE Conference on Computer Vision and Pattern Recognition*, pages 4470–4479, 2018.
5. Iro Armeni, Ozan Sener, Amir R Zamir, Helen Jiang, Ioannis Brilakis, Martin Fischer, and Silvio Savarese. 3d semantic parsing of large-scale indoor spaces. In *Proceedings of the IEEE Conference on Computer Vision and Pattern Recognition*, pages 1534–1543, 2016.
6. Jonathan T Barron, Ben Mildenhall, Matthew Tancik, Peter Hedman, Ricardo Martin-Brualla, and Pratul P Srinivasan. Mip-nerf: A multiscale representation for anti-aliasing neural radiance fields. In *Proceedings of the IEEE/CVF International Conference on Computer Vision*, pages 5855–5864, 2021.
7. F. Bernardini, J. Mittleman, H. Rushmeier, C. Silva, and G. Taubin. The ball-pivoting algorithm for surface reconstruction. *IEEE Transactions on Visualization and Computer Graphics*, 5(4):349–359, 1999. 10.1109/2945.817351.
8. P.J. Besl and Neil D. McKay. A method for registration of 3-d shapes. *IEEE Transactions on Pattern Analysis and Machine Intelligence*, 1992.
9. Tom E Bishop, Sara Zanetti, and Paolo Favaro. Light field superresolution. In *2009 IEEE International Conference on Computational Photography (ICCP)*, pages 1–9. IEEE, 2009.
10. Mark Boss, Varun Jampani, Raphael Braun, Ce Liu, Jonathan T. Barron, and Hendrik P. A. Lensch. Neural-pil: Neural pre-integrated lighting for reflectance decomposition. In *NeurIPS*, 2021.
11. Matthew Brown and David G Lowe. Automatic panoramic image stitching using invariant features. *International journal of computer vision*, 74(1):59–73, 2007.
12. Tommaso Cavallari, Stuart Golodetz, Nicholas A Lord, Julien Valentin, Luigi Di Stefano, and Philip HS Torr. On-the-fly adaptation of regression forests for online camera relocalisation. In *Proceedings of the IEEE conference on computer vision and pattern recognition*, 2017.
13. Eric Chan, Marco Monteiro, Petr Kellnhofer, Jiajun Wu, and Gordon Wetzstein. pi-gan: Periodic implicit generative adversarial networks for 3d-aware image synthesis. *2021 IEEE/CVF Conference on Computer Vision and Pattern Recognition (CVPR)*, pages 5795–5805, 2021.
14. Anne-Laure Chauve, Patrick Labatut, and Jean-Philippe Pons. Robust piecewise-planar 3d reconstruction and completion from large-scale unstructured point data. In *2010 IEEE Computer Society Conference on Computer Vision and Pattern Recognition*, 2010.
15. Qimin Chen, Vincent Nguyen, Feng Han, Raimondas Kiveris, and Zhuowen Tu. Topology-aware single-image 3d shape reconstruction. *2020 IEEE/CVF Conference on Computer Vision and Pattern Recognition Workshops (CVPRW)*, pages 1089–1097, 2020a.
16. Rui Chen, Songfang Han, Jing Xu, and Hao Su. Point-based multi-view stereo network. In *The IEEE International Conference on Computer Vision (ICCV)*, pages 1538–1547, 2019.
17. Rui Chen, Songfang Han, Jing Xu, et al. Visibility-aware point-based multi-view stereo network. *IEEE Transactions on Pattern Analysis and Machine Intelligence*, 2020b.
18. Shenchang Eric Chen and Lance R. Williams. View interpolation for image synthesis. *Proceedings of the 20th annual conference on Computer graphics and interactive techniques*, 1993.

19. Weifeng Chen, Zhao Fu, Dawei Yang, and Jia Deng. Single-image depth perception in the wild. In *Advances in neural information processing systems*, pages 730–738, 2016.
20. Yinbo Chen, Sifei Liu, and Xiaolong Wang. Learning continuous image representation with local implicit image function. In *Proceedings of the IEEE/CVF Conference on Computer Vision and Pattern Recognition*, pages 8628–8638, 2021.
21. Shuo Cheng, Zexiang Xu, Shilin Zhu, Zhuwen Li, Li Erran Li, Ravi Ramamoorthi, and Hao Su. Deep stereo using adaptive thin volume representation with uncertainty awareness. In *Proceedings of the IEEE/CVF Conference on Computer Vision and Pattern Recognition*, pages 2524–2534, 2020.
22. Sungjoon Choi, Qian Yi Zhou, and Vladlen Koltun. Robust reconstruction of indoor scenes. In *IEEE Conference on Computer Vision and Pattern Recognition*, 2015.
23. Christopher Bongsoo Choy, JunYoung Gwak, and Silvio Savarese. 4d spatio-temporal convnets: Minkowski convolutional neural networks. *2019 IEEE/CVF Conference on Computer Vision and Pattern Recognition (CVPR)*, pages 3070–3079, 2019.
24. Jaebum Chung, Gloria W Martinez, Karen C Lencioni, Srinivas R Sadda, and Changhuei Yang. Computational aberration compensation by coded-aperture-based correction of aberration obtained from optical fourier coding and blur estimation. *Optica*, 6(5):647–661, 2019.
25. Özgün Çiçek, Ahmed Abdulkadir, Soeren S. Lienkamp, Thomas Brox, and Olaf Ronneberger. 3d u-net: Learning dense volumetric segmentation from sparse annotation. In Sebastien Ourselin, Leo Joskowicz, Mert R. Sabuncu, Gozde Unal, and William Wells, editors, *Medical Image Computing and Computer-Assisted Intervention – MICCAI 2016*, pages 424–432, Cham, 2016. Springer International Publishing. ISBN 978-3-319-46723-8.
26. Angela Dai, Angel X. Chang, Manolis Savva, Maciej Halber, Thomas A. Funkhouser, and Matthias Nießner. Scannet: Richly-annotated 3d reconstructions of indoor scenes. *2017 IEEE Conference on Computer Vision and Pattern Recognition (CVPR)*, pages 2432–2443, 2017a.
27. Angela Dai, Matthias Nießner, Michael Zollhöfer, Shahram Izadi, and Christian Theobalt. Bundlefusion: Real-time globally consistent 3d reconstruction using on-the-fly surface reintegration. *ACM Transactions on Graphics (ToG)*, 36(4):1, 2017b.
28. Yuchao Dai, Zhidong Zhu, Zhibo Rao, and Bo Li. Mvs2: Deep unsupervised multi-view stereo with multi-view symmetry. In *2019 International Conference on 3D Vision (3DV)*, pages 1–8. IEEE, 2019.
29. Liuyun Duan and Florent Lafarge. Towards large-scale city reconstruction from satellites. In *European Conference on Computer Vision*, pages 89–104. Springer, 2016.
30. Olivier Duchenne, Francis Bach, In-So Kweon, and Jean Ponce. A tensor-based algorithm for high-order graph matching. *IEEE transactions on pattern analysis and machine intelligence*, 33(12):2383–2395, 2011.
31. Vincent Dumoulin, Ethan Perez, Nathan Schucher, Florian Strub, Harm de Vries, Aaron C. Courville, and Yoshua Bengio. Feature-wise transformations. *Distill*, 2018.
32. Martin Ehler. Applied and computational harmonic analysis. 2008.
33. Felix Endres, Jürgen Hess, Nikolas Engelhard, Jürgen Sturm, Daniel Cremers, and Wolfram Burgard. An evaluation of the rgb-d slam system. In *Robotics and Automation (ICRA), 2012 IEEE International Conference on*, pages 1691–1696. IEEE, 2012.
34. Ross Finman, Liam Paull, and John J Leonard. Toward object-based place recognition in dense rgb-d maps. In *ICRA Workshop Visual Place Recognition in Changing Environments, Seattle, WA*, 2015.
35. Yasutaka Furukawa and Jean Ponce. Accurate, dense, and robust multiview stereopsis. *IEEE Transactions on Pattern Analysis and Machine Intelligence*, 32(8):1362–1376, 2010.
36. Yasutaka Furukawa, Brian Curless, Steven M Seitz, and Richard Szeliski. Manhattan-world stereo. In *2009 IEEE Conference on Computer Vision and Pattern Recognition*, pages 1422–1429. IEEE, 2009.
37. Silvano Galliani, Katrin Lasinger, and Konrad Schindler. Massively parallel multiview stereopsis by surface normal diffusion. In *IEEE International Conference on Computer Vision*, pages 873–881, 2015.

38. David Gallup, Jan-Michael Frahm, and Marc Pollefeys. Piecewise planar and non-planar stereo for urban scene reconstruction. In *2010 IEEE computer society conference on computer vision and pattern recognition*, pages 1418–1425. IEEE, 2010.
39. Dorian Gálvez-López and Juan D Tardos. Bags of binary words for fast place recognition in image sequences. *IEEE Transactions on Robotics*, 28(5):1188–1197, 2012.
40. Michael Garland and Paul S. Heckbert. Surface simplification using quadric error metrics. *Proceedings of the 24th annual conference on Computer graphics and interactive techniques*, 1997.
41. Clément Godard, Oisin Mac Aodha, and Gabriel J Brostow. Unsupervised monocular depth estimation with left-right consistency. In *Proceedings of the IEEE Conference on Computer Vision and Pattern Recognition*, pages 270–279, 2017.
42. Stuart Golodetz, Tommaso Cavallari, Nicholas A Lord, Victor A Prisacariu, David W Murray, and Philip HS Torr. Collaborative large-scale dense 3d reconstruction with online inter-agent pose optimisation. *IEEE transactions on visualization and computer graphics*, 24(11):2895–2905, 2018.
43. Benjamin Graham and Laurens van der Maaten. Submanifold sparse convolutional networks. *arXiv preprint*arXiv:1706.01307, 2017.
44. Benjamin Graham, Martin Engelcke, and Laurens van der Maaten. 3d semantic segmentation with submanifold sparse convolutional networks. *2018 IEEE/CVF Conference on Computer Vision and Pattern Recognition*, pages 9224–9232, 2018.
45. Thibault Groueix, Matthew Fisher, Vladimir G Kim, Bryan C Russell, and Mathieu Aubry. A papier-mâché approach to learning 3d surface generation. In *Proceedings of the IEEE conference on computer vision and pattern recognition*, pages 216–224, 2018.
46. Xiaodong Gu, Zhiwen Fan, Siyu Zhu, Zuozhuo Dai, Feitong Tan, and Ping Tan. Cascade cost volume for high-resolution multi-view stereo and stereo matching. In *Proceedings of the IEEE/CVF Conference on Computer Vision and Pattern Recognition*, pages 2495–2504, 2020.
47. Lei Han and Lu Fang. Mild: Multi-index hashing for appearance based loop closure detection. In *2017 IEEE International Conference on Multimedia and Expo (ICME)*, pages 139–144. IEEE, 2017.
48. Lei Han and Lu Fang. Flashfusion: Real-time globally consistent dense 3d reconstruction using cpu computing. In *Robotics: Science and Systems*, volume 1, page 7, 2018.
49. Lei Han, Guyue Zhou, Lan Xu, and Lu Fang. Beyond sift using binary features for loop closure detection. *arXiv preprint*arXiv:1709.05833, 2017.
50. Lei Han, Tian Zheng, Lan Xu, and Lu Fang. Occuseg: Occupancy-aware 3d instance segmentation. *2020 IEEE/CVF Conference on Computer Vision and Pattern Recognition (CVPR)*, pages 2937–2946, 2020.
51. A. Handa, T. Whelan, J. McDonald, and A. J. Davison. A benchmark for rgb-d visual odometry, 3d reconstruction and slam. In *2014 IEEE International Conference on Robotics and Automation (ICRA)*, pages 1524–1531, May 2014. 10.1109/ICRA.2014.6907054.
52. Ji Hou, Angela Dai, and Matthias Nießner. 3d-sis: 3d semantic instance segmentation of rgb-d scans. *2019 IEEE/CVF Conference on Computer Vision and Pattern Recognition (CVPR)*, pages 4416–4425, 2019.
53. Qingyong Hu, Bo Yang, Linhai Xie, Stefano Rosa, Yulan Guo, Zhihua Wang, Agathoniki Trigoni, and Andrew Markham. Randla-net: Efficient semantic segmentation of large-scale point clouds. *2020 IEEE/CVF Conference on Computer Vision and Pattern Recognition (CVPR)*, pages 11105–11114, 2020.
54. Wenbo Hu, Hengshuang Zhao, Li Jiang, Jiaya Jia, and Tien-Tsin Wong. Bidirectional projection network for cross dimension scene understanding. *2021 IEEE/CVF Conference on Computer Vision and Pattern Recognition (CVPR)*, pages 14368–14377, 2021.
55. Binh-Son Hua, Quang-Hieu Pham, Duc Thanh Nguyen, Minh-Khoi Tran, Lap-Fai Yu, and Sai-Kit Yeung. Scenenn: A scene meshes dataset with annotations. *2016 Fourth International Conference on 3D Vision (3DV)*, pages 92–101, 2016.

56. Baichuan Huang, Hongwei Yi, Can Huang, Yijia He, Jingbin Liu, and Xiao Liu. M^3vsnet: Unsupervised multi-metric multi-view stereo network. 2020.
57. Shi-Sheng Huang, Ze-Yu Ma, Tai-Jiang Mu, Hongbo Fu, and Shi-Min Hu. Supervoxel convolution for online 3d semantic segmentation. *ACM Trans. Graph.*, 40:34:1–34:15, 2021.
58. Zeng Huang, Tianye Li, Weikai Chen, Yajie Zhao, Jun Xing, Chloe LeGendre, Linjie Luo, Chongyang Ma, and Hao Li. Deep volumetric video from very sparse multi-view performance capture. In *Proceedings of the European Conference on Computer Vision (ECCV)*, pages 336–354, 2018.
59. Mengqi Ji, Juergen Gall, Haitian Zheng, Yebin Liu, and Lu Fang. Surfacenet: An end-to-end 3d neural network for multiview stereopsis. In *Proceedings of the IEEE International Conference on Computer Vision*, pages 2307–2315, 2017.
60. Mengqi Ji, Jinzhi Zhang, Qionghai Dai, and Lu Fang. Surfacenet+: An end-to-end 3d neural network for very sparse multi-view stereopsis. *IEEE Transactions on Pattern Analysis and Machine Intelligence*, 43(11):4078–4093, 2020.
61. Li Jiang, Hengshuang Zhao, Shaoshuai Shi, Shu Liu, Chi-Wing Fu, and Jiaya Jia. Pointgroup: Dual-set point grouping for 3d instance segmentation. *2020 IEEE/CVF Conference on Computer Vision and Pattern Recognition (CVPR)*, pages 4866–4875, 2020.
62. Olaf Kähler, Victor Adrian Prisacariu, and David W. Murray. Real-time large-scale dense 3d reconstruction with loop closure. In *Computer Vision - ECCV 2016 - 14th European Conference, Amsterdam, The Netherlands, October 11-14, 2016, Proceedings, Part VIII*, pages 500–516, 2016.
63. James T Kajiya. The rendering equation. In *Proceedings of the 13th annual conference on Computer graphics and interactive techniques*, pages 143–150, 1986.
64. Abhishek Kar, Christian Häne, and Jitendra Malik. Learning a multi-view stereo machine. In *Advances in Neural Information Processing Systems*, pages 365–376, 2017.
65. Brian Karis. Real shading in unreal engine 4 by. 2013.
66. Leonard Kaufman and Peter J. Rousseeuw. Clustering by means of medoids, 1987.
67. Jan Kautz, Pere-Pau Vázquez, Wolfgang Heidrich, and Hans-Peter Seidel. Unified approach to prefiltered environment maps. In *Rendering Techniques*, 2000.
68. Tejas Khot, Shubham Agrawal, Shubham Tulsiani, Christoph Mertz, Simon Lucey, and Martial Hebert. Learning unsupervised multi-view stereopsis via robust photometric consistency. *arXiv preprint*arXiv:1905.02706, 2019.
69. Arno Knapitsch, Jaesik Park, Qian-Yi Zhou, and Vladlen Koltun. Tanks and temples: Benchmarking large-scale scene reconstruction. *ACM Transactions on Graphics*, 36(4), 2017a.
70. Arno Knapitsch, Jaesik Park, Qian-Yi Zhou, and Vladlen Koltun. Tanks and temples: Benchmarking large-scale scene reconstruction. *ACM Transactions on Graphics (ToG)*, 36(4):1–13, 2017b.
71. Johannes Kopf, Matt Uyttendaele, Oliver Deussen, and Michael F Cohen. Capturing and viewing gigapixel images. *aCm Transactions on Graphics (TOG)*, 26(3):93–es, 2007.
72. Abhijit Kundu, Yin Li, and James M. Rehg. 3d-rcnn: Instance-level 3d object reconstruction via render-and-compare. *2018 IEEE/CVF Conference on Computer Vision and Pattern Recognition*, pages 3559–3568, 2018.
73. Jean Lahoud, Bernard Ghanem, Marc Pollefeys, and Martin R. Oswald. 3d instance segmentation via multi-task metric learning. *2019 IEEE/CVF International Conference on Computer Vision (ICCV)*, pages 9255–9265, 2019.
74. Marc Levoy and Pat Hanrahan. Light field rendering. *Proceedings of the 23rd annual conference on Computer graphics and interactive techniques*, 1996.
75. Jie Liao, Yanping Fu, Qingan Yan, and Chunxia Xiao. Pyramid multi-view stereo with local consistency. In *Computer Graphics Forum*, volume 38, pages 335–346. Wiley Online Library, 2019.
76. Chen-Hsuan Lin, Oliver Wang, Bryan C Russell, Eli Shechtman, Vladimir G Kim, Matthew Fisher, and Simon Lucey. Photometric mesh optimization for video-aligned 3d object reconstruction. In *IEEE Conference on Computer Vision and Pattern Recognition (CVPR)*, 2019.

77. Chen Liu, Jimei Yang, Duygu Ceylan, Ersin Yumer, and Yasutaka Furukawa. Planenet: Piece-wise planar reconstruction from a single rgb image. In *Proceedings of the IEEE Conference on Computer Vision and Pattern Recognition*, pages 2579–2588, 2018.
78. Chen Liu, Kihwan Kim, Jinwei Gu, Yasutaka Furukawa, and Jan Kautz. Planercnn: 3d plane detection and reconstruction from a single image. In *Proceedings of the IEEE/CVF Conference on Computer Vision and Pattern Recognition*, pages 4450–4459, 2019a.
79. Fayao Liu, Chunhua Shen, and Guosheng Lin. Deep convolutional neural fields for depth estimation from a single image. In *Proceedings of the IEEE Conference on Computer Vision and Pattern Recognition*, pages 5162–5170, 2015.
80. Lingjie Liu, Jiatao Gu, Kyaw Zaw Lin, Tat-Seng Chua, and Christian Theobalt. Neural sparse voxel fields. *Advances in Neural Information Processing Systems*, 33:15651–15663, 2020.
81. Shichen Liu, Tianye Li, Weikai Chen, and Hao Li. Soft rasterizer: A differentiable renderer for image-based 3d reasoning. In *Proceedings of the IEEE International Conference on Computer Vision*, pages 7708–7717, 2019b.
82. Xiaoxiao Long, Lingjie Liu, Christian Theobalt, and Wenping Wang. Occlusion-aware depth estimation with adaptive normal constraints. In *European Conference on Computer Vision*, pages 640–657. Springer, 2020.
83. Benjamin luvi, Martin Engelcke, and Laurens van der Maaten. 3d semantic segmentation with submanifold sparse convolutional networks. In *Proc. Computer Vision and Pattern Recognition (CVPR), IEEE*, 2018.
84. Ricardo Martin-Brualla, Noha Radwan, Mehdi SM Sajjadi, Jonathan T Barron, Alexey Dosovitskiy, and Daniel Duckworth. Nerf in the wild: Neural radiance fields for unconstrained photo collections. *arXiv preprint*arXiv:2008.02268, pages 7210–7219, 2020.
85. Paul Merrell, Philippos Mordohai, Jan Michael Frahm, and Marc Pollefeys. Evaluation of large scale scene reconstruction. In *IEEE International Conference on Computer Vision*, 2007.
86. Lars Mescheder, Michael Oechsle, Michael Niemeyer, Sebastian Nowozin, and Andreas Geiger. Occupancy networks: Learning 3d reconstruction in function space. In *Proceedings of the IEEE Conference on Computer Vision and Pattern Recognition*, pages 4460–4470, 2019.
87. Ben Mildenhall, Pratul P. Srinivasan, Matthew Tancik, Jonathan T. Barron, Ravi Ramamoorthi, and Ren Ng. Nerf: Representing scenes as neural radiance fields for view synthesis. In *ECCV*, pages 405–421. Springer, 2020.
88. Kaichun Mo, Shilin Zhu, Angel X Chang, Li Yi, Subarna Tripathi, Leonidas J Guibas, and Hao Su. Partnet: A large-scale benchmark for fine-grained and hierarchical part-level 3d object understanding. In *Proceedings of the IEEE Conference on Computer Vision and Pattern Recognition*, pages 909–918, 2019.
89. Claudio Mura, Oliver Mattausch, Alberto Jaspe Villanueva, Enrico Gobbetti, and Renato Pajarola. Automatic room detection and reconstruction in cluttered indoor environments with complex room layouts. *Computers & Graphics*, 2014.
90. Supun Nakandala, Kabir Nagrecha, Arun Kumar, and Yannis Papakonstantinou. Incremental and approximate computations for accelerating deep cnn inference. *ACM Transactions on Database Systems (TODS)*, 45(4):1–42, 2020.
91. Gaku Narita, Takashi Seno, Tomoya Ishikawa, and Yohsuke Kaji. Panopticfusion: Online volumetric semantic mapping at the level of stuff and things. *2019 IEEE/RSJ International Conference on Intelligent Robots and Systems (IROS)*, pages 4205–4212, 2019.
92. Peer Neubert and Peter Protzel. Compact watershed and preemptive slic: On improving trade-offs of superpixel segmentation algorithms. In *2014 22nd international conference on pattern recognition*, pages 996–1001. IEEE, 2014.
93. Michael Niemeyer, Lars Mescheder, Michael Oechsle, and Andreas Geiger. Differentiable volumetric rendering: Learning implicit 3d representations without 3d supervision. In *Proceedings of the IEEE/CVF Conference on Computer Vision and Pattern Recognition*, pages 3504–3515, 2020.
94. Michael Niemeyer, Jonathan T Barron, Ben Mildenhall, Mehdi SM Sajjadi, Andreas Geiger, and Noha Radwan. Regnerf: Regularizing neural radiance fields for view synthesis from sparse inputs. In *Proceedings of the IEEE/CVF Conference on Computer Vision and Pattern Recognition*, pages 5480–5490, 2022.

95. David Novotny, Samuel Albanie, Diane Larlus, and Andrea Vedaldi. Semi-convolutional operators for instance segmentation. In *Proceedings of the European Conference on Computer Vision (ECCV)*, pages 86–102, 2018.
96. Michael Oechsle, Lars Mescheder, Michael Niemeyer, Thilo Strauss, and Andreas Geiger. Texture fields: Learning texture representations in function space. In *Proceedings of the IEEE International Conference on Computer Vision*, pages 4531–4540, 2019.
97. Jeong Joon Park, Peter Florence, Julian Straub, Richard Newcombe, and Steven Lovegrove. Deepsdf: Learning continuous signed distance functions for shape representation. In *Proceedings of the IEEE Conference on Computer Vision and Pattern Recognition*, pages 165–174, 2019.
98. Despoina Paschalidou, Osman Ulusoy, Carolin Schmitt, Luc Van Gool, and Andreas Geiger. Raynet: Learning volumetric 3d reconstruction with ray potentials. In *Proceedings of the IEEE Conference on Computer Vision and Pattern Recognition*, pages 3897–3906, 2018.
99. Ethan Perez, Florian Strub, Harm de Vries, Vincent Dumoulin, and Aaron C. Courville. Film: Visual reasoning with a general conditioning layer. In *AAAI*, 2018.
100. Quang-Hieu Pham, Thanh Nguyen, Binh-Son Hua, Gemma Roig, and Sai-Kit Yeung. Jsis3d: Joint semantic-instance segmentation of 3d point clouds with multi-task pointwise networks and multi-value conditional random fields. In *Proceedings of the IEEE Conference on Computer Vision and Pattern Recognition*, pages 8827–8836, 2019.
101. C. Qi, Hao Su, Kaichun Mo, and Leonidas J. Guibas. Pointnet: Deep learning on point sets for 3d classification and segmentation. *2017 IEEE Conference on Computer Vision and Pattern Recognition (CVPR)*, pages 77–85, 2017a.
102. C. Qi, Or Litany, Kaiming He, and Leonidas J. Guibas. Deep hough voting for 3d object detection in point clouds. *2019 IEEE/CVF International Conference on Computer Vision (ICCV)*, pages 9276–9285, 2019.
103. Charles R. Qi, Li Yi, Hao Su, and Leonidas J. Guibas. Pointnet++: Deep hierarchical feature learning on point sets in a metric space. In *Proceedings of the 31st International Conference on Neural Information Processing Systems*, NIPS'17, pages 5105–5114, Red Hook, NY, USA, 2017b. Curran Associates Inc. ISBN 9781510860964.
104. Gernot Riegler and Vladlen Koltun. Free view synthesis. In *European Conference on Computer Vision*, pages 623–640. Springer, 2020.
105. Gernot Riegler and Vladlen Koltun. Stable view synthesis. In *Proceedings of the IEEE/CVF Conference on Computer Vision and Pattern Recognition*, pages 12216–12225, 2021.
106. Gernot Riegler, Ali Osman Ulusoy, Horst Bischof, and Andreas Geiger. Octnetfusion: Learning depth fusion from data. In *2017 International Conference on 3D Vision (3DV)*, pages 57–66. IEEE, 2017.
107. Darius Rückert, Linus Franke, and Marc Stamminger. Adop: Approximate differentiable one-pixel point rendering. *arXiv preprint*arXiv:2110.06635, 2021.
108. Radu Bogdan Rusu, Nico Blodow, and Michael Beetz. Fast point feature histograms (fpfh) for 3d registration. In *2009 IEEE international conference on robotics and automation*, pages 3212–3217. IEEE, 2009.
109. Ruwen Schnabel, Roland Wahl, and Reinhard Klein. Efficient ransac for point-cloud shape detection. In *Computer graphics forum*, volume 26, pages 214–226. Wiley Online Library, 2007.
110. Johannes L Schonberger and Jan-Michael Frahm. Structure-from-motion revisited. In *Proceedings of the IEEE conference on computer vision and pattern recognition*, pages 4104–4113, 2016.
111. Johannes Lutz Schönberger, Enliang Zheng, Marc Pollefeys, and Jan-Michael Frahm. Pixelwise view selection for unstructured multi-view stereo. In *European Conference on Computer Vision (ECCV)*, pages 501–518. Springer, 2016.
112. Thomas Schops, Johannes L Schonberger, Silvano Galliani, Torsten Sattler, Konrad Schindler, Marc Pollefeys, and Andreas Geiger. A multi-view stereo benchmark with high-resolution images and multi-camera videos. In *Proceedings of the IEEE Conference on Computer Vision and Pattern Recognition*, pages 3260–3269, 2017.

113. Steven M Seitz, Brian Curless, James Diebel, Daniel Scharstein, and Richard Szeliski. A comparison and evaluation of multi-view stereo reconstruction algorithms. In *2006 IEEE computer society conference on computer vision and pattern recognition (CVPR'06)*, volume 1, pages 519–528. IEEE, 2006.
114. Jonathan Shade, Steven J. Gortler, Li wei He, and Richard Szeliski. Layered depth images. *Proceedings of the 25th annual conference on Computer graphics and interactive techniques*, 1998.
115. Harry Shum and Sing Bing Kang. Review of image-based rendering techniques. In *Visual Communications and Image Processing*, 2000.
116. Sudipta Sinha, Drew Steedly, and Rick Szeliski. Piecewise planar stereo for image-based rendering. In *2009 International Conference on Computer Vision*, pages 1881–1888, 2009.
117. Vincent Sitzmann, Michael Zollhöfer, and Gordon Wetzstein. Scene representation networks: Continuous 3d-structure-aware neural scene representations. In *Advances in Neural Information Processing Systems*, pages 1121–1132, 2019.
118. Stanford Artificial Intelligence Laboratory et al. Robotic operating system.
119. Frank Steinbrücker, Jürgen Sturm, and Daniel Cremers. Real-time visual odometry from dense rgb-d images. In *2011 IEEE international conference on computer vision workshops (ICCV Workshops)*, 2011.
120. Christoph Strecha, Wolfgang Von Hansen, Luc Van Gool, Pascal Fua, and Ulrich Thoennessen. On benchmarking camera calibration and multi-view stereo for high resolution imagery. In *2008 IEEE Conference on Computer Vision and Pattern Recognition*, pages 1–8. Ieee, 2008.
121. Jürgen Sturm, Nikolas Engelhard, Felix Endres, Wolfram Burgard, and Daniel Cremers. A benchmark for the evaluation of rgb-d slam systems. In *Intelligent Robots and Systems (IROS), 2012 IEEE/RSJ International Conference on*, pages 573–580. IEEE, 2012.
122. Towaki Takikawa, Joey Litalien, Kangxue Yin, Karsten Kreis, Charles Loop, Derek Nowrouzezahrai, Alec Jacobson, Morgan McGuire, and Sanja Fidler. Neural geometric level of detail: Real-time rendering with implicit 3d shapes. In *Proceedings of the IEEE/CVF Conference on Computer Vision and Pattern Recognition*, pages 11358–11367, 2021.
123. Maxim Tatarchenko, Alexey Dosovitskiy, and Thomas Brox. Octree generating networks: Efficient convolutional architectures for high-resolution 3d outputs. In *Proceedings of the IEEE International Conference on Computer Vision*, pages 2088–2096, 2017.
124. Justus Thies, Michael Zollhöfer, and Matthias Nießner. Deferred neural rendering: Image synthesis using neural textures. *ACM Transactions on Graphics (TOG)*, 38(4):1–12, 2019.
125. Hugues Thomas, C. Qi, Jean-Emmanuel Deschaud, Beatriz Marcotegui, François Goulette, and Leonidas J. Guibas. Kpconv: Flexible and deformable convolution for point clouds. *2019 IEEE/CVF International Conference on Computer Vision (ICCV)*, pages 6410–6419, 2019.
126. Engin Tola, Christoph Strecha, and Pascal Fua. Efficient large-scale multi-view stereo for ultra high-resolution image sets. *Machine Vision and Applications*, pages 1–18, 2012.
127. Dor Verbin, Peter Hedman, Ben Mildenhall, Todd E. Zickler, Jonathan T. Barron, and Pratul P. Srinivasan. Ref-nerf: Structured view-dependent appearance for neural radiance fields. *ArXiv*, abs/2112.03907, 2021.
128. Michael Waechter, Mate Beljan, Simon Fuhrmann, Nils Moehrle, Johannes Kopf, and Michael Goesele. Virtual rephotography: Novel view prediction error for 3d reconstruction. *ACM Transactions on Graphics (TOG)*, 36(1):1–11, 2017.
129. Qianqian Wang, Zhicheng Wang, Kyle Genova, Pratul P Srinivasan, Howard Zhou, Jonathan T Barron, Ricardo Martin-Brualla, Noah Snavely, and Thomas Funkhouser. Ibrnet: Learning multi-view image-based rendering. In *Proceedings of the IEEE/CVF Conference on Computer Vision and Pattern Recognition*, pages 4690–4699, 2021.
130. Weiyue Wang, Ronald Yu, Qiangui Huang, and Ulrich Neumann. Sgpn: Similarity group proposal network for 3d point cloud instance segmentation. In *Proceedings of the IEEE Conference on Computer Vision and Pattern Recognition*, pages 2569–2578, 2018.
131. Xinlong Wang, Shu Liu, Xiaoyong Shen, Chunhua Shen, and Jiaya Jia. Associatively segmenting instances and semantics in point clouds. In *Proceedings of the IEEE Conference on Computer Vision and Pattern Recognition*, pages 4096–4105, 2019.

132. Xueyang Wang, Xiya Zhang, Yinheng Zhu, Yuchen Guo, Xiaoyun Yuan, Liuyu Xiang, Zerun Wang, Guiguang Ding, David Brady, Qionghai Dai, et al. Panda: A gigapixel-level human-centric video dataset. In *Proceedings of the IEEE/CVF conference on computer vision and pattern recognition*, pages 3268–3278, 2020.
133. Thomas Whelan, Renato F Salas-Moreno, Ben Glocker, Andrew J Davison, and Stefan Leutenegger. Elasticfusion: Real-time dense slam and light source estimation. *The International Journal of Robotics Research*, 35(14):1697–1716, 2016.
134. Gaochang Wu, Yebin Liu, Lu Fang, Qionghai Dai, and Tianyou Chai. Light field reconstruction using convolutional network on epi and extended applications. *IEEE transactions on pattern analysis and machine intelligence*, 41(7):1681–1694, 2018.
135. Wenxuan Wu, Zhongang Qi, and Fuxin Li. Pointconv: Deep convolutional networks on 3d point clouds. *2019 IEEE/CVF Conference on Computer Vision and Pattern Recognition (CVPR)*, pages 9613–9622, 2019.
136. Zhirong Wu, Shuran Song, Aditya Khosla, Fisher Yu, Linguang Zhang, Xiaoou Tang, and Jianxiong Xiao. 3d shapenets: A deep representation for volumetric shapes. In *Proceedings of the IEEE conference on computer vision and pattern recognition*, 2015.
137. Qingshan Xu and Wenbing Tao. Multi-scale geometric consistency guided multi-view stereo. In *Proceedings of the IEEE Conference on Computer Vision and Pattern Recognition*, pages 5483–5492, 2019.
138. Qingshan Xu and Wenbing Tao. Planar prior assisted patchmatch multi-view stereo. In *AAAI*, 2020.
139. Jianfeng Yan, Zizhuang Wei, Hongwei Yi, Mingyu Ding, Runze Zhang, Yisong Chen, Guoping Wang, and Yu-Wing Tai. Dense hybrid recurrent multi-view stereo net with dynamic consistency checking. In *European Conference on Computer Vision*, pages 674–689. Springer, 2020.
140. Bo Yang, Jianan Wang, Ronald Clark, Qingyong Hu, Sen Wang, Andrew Markham, and Niki Trigoni. Learning object bounding boxes for 3d instance segmentation on point clouds. In H. Wallach, H. Larochelle, A. Beygelzimer, F. d'Alché-Buc, E. Fox, and R. Garnett, editors, *Advances in Neural Information Processing Systems*, volume 32. Curran Associates, Inc., 2019a.
141. Gengshan Yang, Joshua Manela, Michael Happold, and Deva Ramanan. Hierarchical deep stereo matching on high-resolution images. In *The IEEE Conference on Computer Vision and Pattern Recognition (CVPR)*, pages 5515–5524, June 2019b.
142. Yao Yao, Zixin Luo, Shiwei Li, Tian Fang, and Long Quan. Mvsnet: Depth inference for unstructured multi-view stereo. In *Proceedings of the European Conference on Computer Vision (ECCV)*, pages 767–783, 2018.
143. Yao Yao, Zixin Luo, Shiwei Li, Tianwei Shen, Tian Fang, and Long Quan. Recurrent mvsnet for high-resolution multi-view stereo depth inference. In *Proceedings of the IEEE Conference on Computer Vision and Pattern Recognition*, pages 5525–5534, 2019.
144. Yao Yao, Zixin Luo, Shiwei Li, Jingyang Zhang, Yufan Ren, Lei Zhou, Tian Fang, and Long Quan. Blendedmvs: A large-scale dataset for generalized multi-view stereo networks. In *Proceedings of the IEEE/CVF Conference on Computer Vision and Pattern Recognition*, pages 1790–1799, 2020.
145. Lior Yariv, Yoni Kasten, Dror Moran, Meirav Galun, Matan Atzmon, Basri Ronen, and Yaron Lipman. Multiview neural surface reconstruction by disentangling geometry and appearance. *Advances in Neural Information Processing Systems*, 33, 2020.
146. Hongwei Yi, Zizhuang Wei, Mingyu Ding, Runze Zhang, Yisong Chen, Guoping Wang, and Yu-Wing Tai. Pyramid multi-view stereo net with self-adaptive view aggregation. *arXiv preprint*arXiv:1912.03001, 2019.
147. Haiyang Ying, Jinzhi Zhang, Yuzhe Chen, Zheng Cao, Jing Xiao, Ruqi Huang, and Lu Fang. Parsemvs: Learning primitive-aware surface representations for sparse multi-view stereopsis. In *Proceedings of the 30th ACM International Conference on Multimedia*, 2022.
148. Xiaoyun Yuan, Mengqi Ji, Jiamin Wu, David J Brady, Qionghai Dai, and Lu Fang. A modular hierarchical array camera. *Light: Science & Applications*, 10(1):1–9, 2021.

149. Kaan Yucer, Changil Kim, Alexander Sorkine-Hornung, and Olga Sorkine-Hornung. Depth from gradients in dense light fields for object reconstruction. In *2016 Fourth International Conference on 3D Vision (3DV)*, pages 249–257. IEEE, 2016.
150. Yao Zhai, Xun Guo, Yan Lu, and Houqiang Li. In defense of the classification loss for person re-identification. In *2019 IEEE/CVF Conference on Computer Vision and Pattern Recognition Workshops (CVPRW)*, pages 1526–1535, 2019.
151. Jianing Zhang, Jinzhi Zhang, Shi Mao, Mengqi Ji, Guangyu Wang, Zequn Chen, Tian Zhang, Xiaoyun Yuan, Qionghai Dai, and Lu Fang. Gigamvs: A benchmark for ultra-large-scale gigapixel-level 3d reconstruction. *IEEE Transactions on Pattern Analysis & Machine Intelligence*, (01):1–1, 2021a.
152. Jiazhao Zhang, Chenyang Zhu, Lin tao Zheng, and Kai Xu. Fusion-aware point convolution for online semantic 3d scene segmentation. *2020 IEEE/CVF Conference on Computer Vision and Pattern Recognition (CVPR)*, pages 4533–4542, 2020a.
153. Jingyang Zhang, Yao Yao, Shiwei Li, Zixin Luo, and Tian Fang. Visibility-aware multi-view stereo network. *arXiv preprint*arXiv:2008.07928, 2020b.
154. Jinzhi Zhang, Mengqi Ji, Guangyu Wang, Xue Zhiwei, Shengjin Wang, and Lu Fang. Surrf: Unsupervised multi-view stereopsis by learning surface radiance field. *IEEE Transactions on Pattern Analysis and Machine Intelligence*, 2021b.
155. Kai Zhang, Gernot Riegler, Noah Snavely, and Vladlen Koltun. Nerf++: Analyzing and improving neural radiance fields. *arXiv preprint* arXiv:2010.07492, 2020c.
156. Richard Zhang, Phillip Isola, Alexei A Efros, Eli Shechtman, and Oliver Wang. The unreasonable effectiveness of deep features as a perceptual metric. In *IEEE/CVF Conference on Computer Vision and Pattern Recognition (CVPR)*, pages 586–595, 2018.
157. Tian Zheng, Guoqing Zhang, Lei Han, Lan Xu, and Lu Fang. Building fusion: semantic-aware structural building-scale 3d reconstruction. *IEEE Transactions on Pattern Analysis and Machine Intelligence*, 2020.
158. Yichao Zhou, Haozhi Qi, Yuexiang Zhai, Qi Sun, Zhili Chen, Li-Yi Wei, and Yi Ma. Learning to reconstruct 3d manhattan wireframes from a single image. In *Proceedings of the IEEE/CVF International Conference on Computer Vision*, pages 7698–7707, 2019.
159. Jacek Zienkiewicz, Akis Tsiotsios, Andrew Davison, and Stefan Leutenegger. Monocular, real-time surface reconstruction using dynamic level of detail. In *2016 Fourth International Conference on 3D Vision (3DV)*, pages 37–46. IEEE, 2016.

Chapter 6
GigaVision: When Computer Vision Meets Gigapixel Videography

In previous chapters, we have explored advanced plenoptic imaging and reconstruction techniques, enabling images and videos to reach gigapixel-level resolution. This breakthrough unlocks new possibilities for a wide range of applications and industries. However, traditional computer vision methods, tailored for megapixel-level data, are ill-equipped to handle the complexities of gigapixel-level data, which often feature large-scale scenes with hundreds of objects and intricate interactions. As a result, these methods face significant limitations in both precision and efficiency.

In this chapter, we examine the performance of leading computer vision algorithms in the realm of gigapixel videography and introduce specialized approaches for processing gigapixel-level images and videos. Specifically, Sect. 6.1 introduces the groundbreaking gigapixel-level human-centric dataset, PANDA. Sections 6.2 through 6.5 focus on state-of-the-art methods designed for specific computer vision tasks in the context of gigapixel videography, such as object/group detection, multiple object tracking, pedestrian intent understanding, and crowd reconstruction.

6.1 PANDA: Gigapixel-Level Human-Centric Video Dataset

It has been widely recognized that the recent success of computer vision techniques, especially deep learning-based methods, relies heavily on large-scale and well-annotated datasets. For example, ImageNet [1] and CIFAR-10/100 [2] are important datasets for studying deep convolutional neural networks [3, 4], Pascal VOC [5] and MS COCO [6] are important datasets for studying object detection and segmentation tasks, LFW [7] is an important dataset for face recognition tasks, and Caltech Pedestrians [8] and the MOT benchmark [9] are important datasets for person detection and tracking tasks. Among these tasks, human-centric visual analysis tasks are critical yet challenging. These tasks have many subtasks, e.g., pedestrian detection,

L. Fang, *Plenoptic Imaging and Processing*, Advances in Computer Vision and Pattern Recognition, https://doi.org/10.1007/978-981-97-6915-5_6

Fig. 6.1 Image examples from the MS COCO dataset

tracking, action recognition, anomaly detection, and attribute recognition, and have attracted considerable interest in the last decade [10–17]. While significant progress has been made, there is a lack of long-term analysis of crowd activities over large spatiotemporal ranges with clear local details (Fig. 6.1).

One potential reason for this is that existing datasets [6, 8, 9, 18–20] suffer an inherent tradeoff between a wide FoV and high resolution. For example, consider a football game; a wide-angle camera may cover the entire panoramic scene, but each player shows significant scale variations, leading to very low spatial resolution. In contrast, while a telephoto lens camera can be used to capture the local details of a particular player, the scope of the content is highly restricted to a small FoV. Although multiple surveillance camera setups may acquire more information, the prerequisite of identification based on scattered video clips highly affects continuous analyses of real-world crowd behavior. Overall, existing human-centric datasets remain constrained by the limited spatial and temporal information provided. The problems of low spatial resolution [18, 21, 22], lack of video information [17, 23–25], unnatural human appearances and actions [26–28], and limited scope of activities with short-term annotations [20, 29–31] inevitably influence our understanding of the complicated behaviors and interactions in crowds.

Fig. 6.2 A representative video of a marathon in the PANDA dataset. The characteristics of a wide field-of-view and high spatial resolution enable large-scale, long-term, and multiobject visual analysis

To address the aforementioned problems, a novel gigaPixel-level humAN-centric viDeo dAtaset (PANDA)[1] was proposed. The videos in the PANDA dataset were captured by a gigacamera [32, 33], which can cover a large-scale area with many high-resolution details. A representative video example of a marathon is presented in Fig. 6.2. Due to the rich information in this dataset, PANDA is a competitive dataset with multiscale features, including (1) a globally wide field-of-view, where the visible area exceeds 1km^2; (2) local high-resolution details with gigapixel-level spatial resolution; (3) temporally long-term crowd activities, with 43.7 k frames in total; and (4) real-world scenes with diverse human attributes, behavioral patterns, scales, densities, occlusions, and interactions. PANDA also includes rich and hierarchical ground truth annotations, with 15, 974.6 k bounding boxes, 111.8 k fine-grained attribute labels, 12.7 k trajectories, 2.2 k groups, and 2.9 k interactions in total.

Data Collection

It is known that single camera-based imaging must balance a wide FoV with high spatial resolution. The recently developed array camera-based gigapixel videography techniques demonstrate the feasibility of high-performance imaging systems [32, 33]. Advanced computational algorithms are designed, and many microcameras work simultaneously to generate seamless gigapixel-level videos in real time. As a result, the tradeoff between the field-of-view and spatial resolution can be eliminated. The most recently developed gigacamera [33] was adopted to collect the data for the PANDA dataset, where the horizontal FoV is approximately 70 °C, and the video resolution reaches 25×14 k at 30 Hz. A representative video of a marathon in the dataset is shown in Fig. 6.2, demonstrating the uniqueness of the PANDA dataset, with the data having both a globally wide FoV and local high-resolution details.

Currently, PANDA includes 21 real-world outdoor scenes, with diverse features, high pedestrian density, many trajectory distributions, and different group activities. In each scene, approximately 2 h of 30 Hz video was collected as the raw data. Afterward, approximately 3600 frames (approximately two-minute-long segments)

[1] PANDA Website: http://www.panda-dataset.com.

were extracted. Approximately 30 representative frames per video, 600 in total, were selected as images to be annotated, covering different crowd distributions and activities.

Data Annotation

Full-image annotation of the images and videos in the PANDA dataset is difficult due to the gigapixel-level resolution. Herein, following the idea of divide-and-merge, the full image is partitioned into 4 to 16 subimages by considering the pedestrian density and size. After the subimages are separately annotated, the annotation results are mapped back to the full image. The objects cut by block borders are labeled with a special status and relabeled after merging all the blocks together. All the labels were provided by a well-trained professional annotation team.

Figure 6.3 presents a typical large-scale real-world scene of the OCT Harbor in the PANDA dataset, where the crowd shows significant diversity in terms of scale, location, occlusion, activity, and interactions. In addition to the fine-grained bounding boxes shown in Fig. 6.3b, each pedestrian is assigned a fine-grained label showing the detailed attributes, as presented in Fig. 6.3c. Five categories are assigned based on posture: walking, standing, sitting, riding, and holding in arms (for children). Pedestrians with key parts occluded are labeled as "unsure". The "riding" label is further subdivided into bicycle riders, tricycle riders, and motorcycle riders. Another detailed attribute is "child" or "adult", which are distinguished based on appearance and behavior, as shown in Fig. 6.3a.

Fig. 6.3 Visualization of annotations in the PANDA dataset. **a** The scale variation of the pedestrians in a large-scale scene. **b** Three fine-grained bounding boxes on the human body. **c** Five categories for human body postures. **d** Group information and the intragroup interactions (TK = Talking, PC = Physical contact), where the circle and short line denote pedestrians and their face orientation, respectively

Table 6.1 Pedestrian datasets comparison (statistics of CityPersons only contain public available training set). # Image is the total number of images, # Person is the total number of persons, Density denotes person density (average number of person per image) and PANDA-C is the PANDA-Crowd subset

	Caltech	CityPersons	PANDA	PANDA-C
Resolution	480P	2048 × 1024	> **25** k ×**4** k	> **25** k ×**4** k
# Images	**249.9** k	5 k	555	45
# Person	**289.9** k	35 k	111.8 k	122.1 k
Density	1.16	7	**201.4**	**2, 713.8**

Image Annotation

The PANDA dataset includes 600 annotated images captured from 21 diverse scenes for the multiobject detection task. Among them, the PANDA-Crowd subset is composed of 45 images labeled with human head bounding boxes, which are selected from 3 extremely crowded scenes with many pedestrians. The remaining 555 images from 18 real-world daily scenes include a total of 111 k pedestrians, labeled with head points, head bounding boxes, visible body bounding boxes, and estimated full body bounding boxes close to the border of the pedestrian. For crowds that are too far or too dense to distinguish individuals, people reflected in glass and people with more than 80% occluded areas are marked as "ignore" and not included in the benchmark.

Quantitative comparisons between the PANDA dataset and the representative Caltech [8] and CityPersons image datasets [34] are provided (Table 6.1). Table 6.1 shows that each image in the PANDA dataset has gigapixel-level resolution, which is approximately 100 times that of the images in the existing datasets. Although the number of images in the PANDA dataset is much smaller than the number of images in the other datasets, due to the high resolution and wide FoV, the PANDA dataset has much higher pedestrian density per image than the other datasets, especially the extremely crowded PANDA-Crowd subset, and the total number of pedestrians in PANDA is comparable to that in the Caltech image dataset.

Video Annotation

The video annotation process was focused on labels related to activities/interactions. In addition to the bounding box of each person, the face orientation (quantified into eight bins) and the occlusion ratio (without, partial and heavy) were labeled for the data in the PANDA dataset. Pedestrians who were completely occluded for a short time were labeled with a virtual body bounding box and marked as "disappearing". Multiobject tracking (MOT) annotations are available for all the videos in the PANDA dataset except those in the PANDA-Crowd subset.

Table 6.2 Comparison of multiobject tracking datasets (statistics of KITTI only contain public available training set). Resolution means video resolution. # Clips, # Frames, # Tracks and # Boxs denote the number of video clips, video frames, tracks and bounding boxes, respectively. Density means average number of person per frame

	KITTI-T	MOAT16	MOT19	PANDA
Resolution	1392 × 215	1080P	1080P	> **25** k × **4** k
# Clips	**20**	14	8	15
# Frames	19.1 k	11.2 k	13.4 k	**43.7** k
# Tracks	204	1.3 k	3.9 k	**12.7** k
# Boxs	13.4 k	292.7 k	2,259.2 k	**15, 480.5** k
Density	0.7	26.1 k	168.6	**354.6**

Quantitative comparisons between the PANDA dataset and the KITTI-T [35] and MOT [36] video datasets are provided (Table 6.2). PANDA is competitive, with the largest number of frames, tracks, and bounding boxes.[2] Moreover, the tracking duration in PANDA is much longer than that in KITTI-T and MOT because the data in PANDA have wider FoVs. This property makes PANDA an excellent dataset for large-scale and long-term tracking. Moreover, more tracks in PANDA have partial or heavy occlusions in terms of both absolute number and relative portion, increasing the difficulty of the tracking task.

The advantages of PANDA, with wide FoV global information, high-resolution local details, and temporal activities, enable more reliable annotations for group detection tasks. In contrast to existing group-based datasets, which focus on either the similarity of global trajectories [22] or the stability of local spatial structures [23], we utilized social signal processing [37] to label the group attributes at the interaction level.

More specifically, the annotated bounding boxes were first used to label the group members based on the scene characteristics and social signals such as the interpersonal distance [38] and interactions [37]. Then, each group was assigned a category label denoting the relationship, such as acquaintance, family, or business, as shown in Fig. 2d. To enrich the features for group identification, the interactions between members within each group were labeled, including the interaction category (including physical contact, body language, facial expression, eye contact, and talking, using a multilabel annotation) and the start/end time. The mean duration of each interaction was 518 frames (17.3 s). To reduce the number of overly subjective or ambiguous cases, three rounds of cross verification were performed.

Because the PANDA dataset includes comprehensive and multiscale information, PANDA can be used for a variety of fundamental yet challenging tasks in image/video-based human-centric analysis. However, accuracy and efficiency issues

[2] Since the movement speed of the pedestrians is relatively slow and stable and the postures of the pedestrians rarely change rapidly or dramatically, the objects were sparsely labeled in every k-th frame (k = 6 to 15 based on the scene content) to reduce the labeling cost.

need to be addressed for detection methods based on the PANDA dataset. The accuracy is influenced by the significant scale variations and complex occlusions, while the efficiency is highly affected by the gigapixel resolution. Thus, the task of tracking was considered as a benchmark. With the large-scale, long-term, and multiobject properties of our dataset, the tracking task is challenging due to the complex occlusions as well as large-scale and long-term activities in the real-world scenes. Moreover, the PANDA dataset enables the distinct task of identifying the group relationships for the crowd shown in the video, which is termed interaction-aware group detection. In this task, a novel global-to-local zoom-in framework known as GigaGroup is proposed to reveal the mutual effects between global trajectories and local interactions. Note that these three tasks are inherently correlated. Although the detection task may be biased due to local high-resolution details and the tracking task may focus on global trajectories, the former significantly improves the latter. Moreover, the spatial-temporal trajectories deduced from the detection and tracking tasks are used in the group analysis. In summary, PANDA is a standardized dataset for the field that can be used to investigate new algorithms to understand complicated social behaviors of crowds in large-scale real-world scenarios.

6.2 Object and Group Detection in Gigapixel Image

6.2.1 Classical Methods for Object and Group Detection

Object Detection

Object detection is a basic, common, but challenging task aimed at locating and recognizing objects in images or videos. For most artificial intelligence applications, such as surveillance video analysis and person reidentification, the performance significantly depends on the output of the preset object detection method. In some special scenarios, such as crowd flow analysis and anomaly detection in surveillance videos, detection efficiency is a key issue, and (near) real-time execution is needed. In recent years, object detectors based on deep learning have significantly improved, achieving precise and efficient detection performance with common images. For example, for popular benchmark datasets such as Pascal VOC [5] and MS COCO [6], two-stage detectors, including Faster R-CNN [10], RetinaNet [12], and R-FCN [39], have achieved satisfactory performance. Several single-stage detectors have shown faster speed, achieving speeds of more than 25 FPS on GPUs with megapixel input images, including the YOLO series [40–42], SSD [43], and SqueezeDet [44]. In addition, some anchor-free methods, such as FCOS [45], CenterNet [46], and CornerNet [47], have been proposed to address the issues introduced by preset anchors.

Although state-of-the-art object detectors have achieved great success, remaining shortcomings must be addressed for newly emerging scenarios such as gigapixel-level videos. Gigapixel-level videos have been obtained with the development of photography techniques. They are usually captured by gigacameras in outdoor scenes, and the FoV often covers a fairly wide range. Gigapixel-level surveillance videos have recently become important data for analyzing group behavior in public spaces [48]. Most existing object detectors are designed to process images containing only dozens of objects with resolutions ranging from 640 px to 2000 px [49], and these object detectors cannot be directly used with the large images included in gigapixel videography data.

There have been some preliminary attempts toward object detection with gigapixel-level images and videos. Since it is impossible to load the whole image as input for deep learning models due to GPU memory limitations, sliding windows are used to scan the whole image in a block-by-block manner, and the results of each grid are combined to obtain the final detection results. Existing methods have achieved unsatisfactory performance compared to conventional detectors. For example, when Faster R-CNN [10] was used as a basic detector, only 75.5% $AP^{IoU=.50}$ for a large visible body and 0.10 FPS with the gigapixel-level PANDA dataset [50] were achieved, which is far lower than the performance on benchmarks such as Pascal VOC [5]. Even when using a single-stage detector such as YOLO [40], the inference process takes several seconds since scanning all grids requires considerable time. Selecting an appropriate grid size is also a challenge, which may result in cases in which small objects are found in large grids or out-of-range objects occupy entire small grids. In summary, conventional detection methods are limited in terms of both accuracy and efficiency when they are applied with gigapixel-level images. As a result, new detection methods for gigapixel-level images should be investigated.

Group Detection in Gigapixel Image

Group detection aims at identifying groups of people in crowds. Existing datasets focus on either the similarity of global trajectories [22] or the stability of local spatial structures [23]. The PANDA dataset, with its wide FoV global information, high-resolution local details, and temporal activities, includes rich information for group detection tasks.

As indicated by the recent advances in trajectory embedding [51, 52], trajectory prediction [53, 54], and interaction modeling in video recognition tasks [28, 55, 56], these tasks are strongly correlated with the group detection task. For example, modeling group interactions can help improve trajectory prediction performance [53, 54, 57], while learning good trajectory embeddings is beneficial for video action recognition [51, 52, 58]. However, no previous studies have investigated how multimodal information can be incorporated into the group detection task. Hence, an interaction-aware group detection task is proposed, in which video data and multimodal annotations (spatial-temporal human trajectories, face orientations, and human interactions) are provided as input for group detection.

6.2.2 *GigaDet: Object Detection with Gigapixel Images*

GigaDet Framework

To address the difficulties of existing object detectors in balancing accuracy and efficiency, a novel progressive strategy called GigaDet is proposed for efficient and accurate detection based on gigapixel videos. The patch generation network (PGN) and decorated detector (DecDet) are two important components of GigaDet. The PGN quickly processes the thumbnail of the video frame to obtain proper patches with the highest probability of containing valid objects. DecDet is designed to perform precise detection based on the patches in parallel. The detection results of each patch are then remapped to the source video frame according to each patch's offset information, and the final detection results are obtained. Our idea was inspired by the human visual system [59], which first obtains a rough look and then performs detailed inspections if necessary when scanning scenes. The PGN module is designed to execute the scanning action, which effectively reduces the possibility of processing large regions that do not contain objects. This progressive mechanism contributes to the acceleration in the detection speed while maintaining good accuracy. The overall framework of GigaDet is shown in Fig. 6.4.

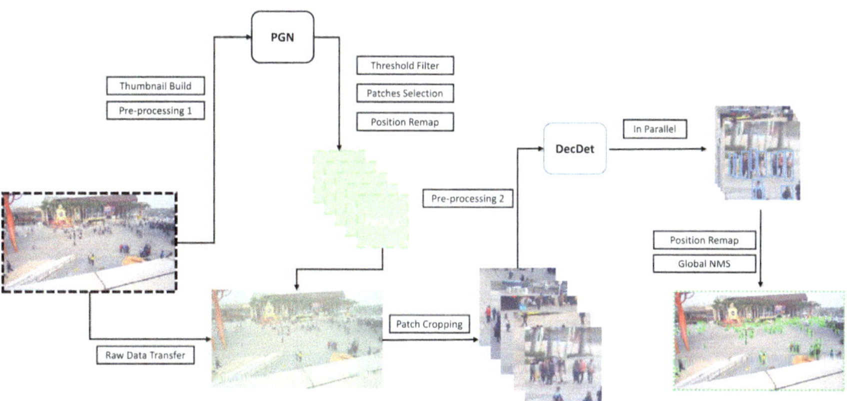

Fig. 6.4 Framework of GigaDet. The original gigapixel-level image is loaded into memory and then fed into two branches. The image is first processed as a thumbnail and fed into the PGN module to obtain patches of interest. Then, the patches and the original raw data are used to obtain cropped images, which are fed into the DecDet module. Parallel detection tasks are executed, and multiple outputs are remapped to the global coordinates. Postprocessing steps such as nonmaximum suppression are applied, and the detection results for the whole gigapixel-level image are produced. GigaDet accelerates the inference speed since the PGN module only processes the thumbnail of the resized gigapixel-level image, which does not slow down the inference task, and the DecDet module receives a batch of images that do not depend on each other with considerably fewer pixels than the original image. This progressive mechanism contributes to the acceleration of the inference task for gigapixel-level images while maintaining acceptable accuracy

Patch Generation

The patch generation network (PGN) extracts patches that may contain valid objects. The PGN assigns patch candidates for the entire input video frame and estimates the score (reflecting the probability of that patch containing objects) for each patch candidate, which serves as the input for the DecDet module. The term *Patch* represents the subregions in an image. Suppose we have a source frame *GI* with width W and height H; then, a *Patch* is defined as

$$Patch_i = \text{Crop}\,(GI, l_i, t_i, r_i, b_i)\,. \tag{6.1}$$

Here, Crop(·) denotes the cropping operation for the source video frame. The parameters l_i, t_i, r_i, and b_i represent the coordinates of $Patch_i$. In particular, the width of the patch is $w_i = r_i - l_i$, and the height of the patch is $h_i = b_i - t_i$. The above values satisfy the basic geometric constraint:

$$0 \le l_i < r_i < W \text{ and } 0 \le t_i < b_i < H. \tag{6.2}$$

Based on this definition, any Patch state of the art within image *GI* can be obtained by operation Crop (GI, l_i, t_i, r_i, b_i). Furthermore, according to the object annotation information for the image *GI*, there may be several objects located entirely within the $Patch_i$. We define O_i as the set of objects that are located entirely within $Patch_i$:

$$O_i = \left\{ \bigcup obj_c \mid obj_c \in Patch_i \right\}, \tag{6.3}$$

where $\in$ indicates that object obj_c is completely found in patch *Patch*, which means that no part of the object is outside the range of this subregion. Equivalently, in this case, *Patch* contains obj_c. More specifically, we define O_i' as the subset of O_i for which the objects are all valid. An object is valid when its size satisfies a certain range:

$$O_i' = \left\{ \bigcup obj_c \mid obj_c \in Patch_i \text{ and } obj_c \sim Range_i \right\}. \tag{6.4}$$

The $Range_i$ is determined by the size of $Patch_i$, and the symbol $\sim$ denotes that the size of obj_c satisfies the condition of $Range_i$. We define $CountO_i$ as the size of set O_i, which also represents the number of objects that $Patch_i$ contains

$$CountO_i = \|O_i\|. \tag{6.5}$$

The PGN obtains multiple patch candidates with different scales in the image, and the network is trained to estimate the property $CountO_i$ for each patch candidate i. The estimations are based on a feature learning procedure, where the input is determined by the features of the input image, and the corresponding label is calculated through the aggregation of the annotated object information. Multiscale patch candidates are used to ensure that objects of different sizes can be contained by at least

Fig. 6.5 Illustration of the patch candidates in an image. A patch candidate is the output of the cropping operation performed with specific parameters on a large image. Three different sizes of patches are visualized in the image. "4×" indicates that the side–length ratio is four times that of the unit ratio, which means that the patch candidate is four times larger than the patch candidate with the "1×" notation. Objects may have different sizes in distant/close views, so it is critical to select a patch candidate with an appropriate size that contains many valid objects. The PGN presets different sizes of patch candidates that are uniformly distributed throughout the image, which serve as the candidates for the final generated patches. As shown in the figure, a "4×" patch candidate in the close area contains several people of appropriate size, and the "1×" patch in the distant area also contains many people. A larger patch candidate does not always contain more valid people than a smaller patch candidate, as this property depends on the number of people in that region. The number of people who appear in the "4×" patch in the distant area may be too small

one patch. An object can be located in one or multiple patches. The restriction for valid objects ensures that the objects are neither too small nor too large relative to the patches to which they belong. In the patch selection procedure, the $CountO_i$ property can be used as the assessment criterion for each patch candidate. An illustration of the patch candidates in a gigapixel-level image is shown in Fig. 6.5

For a gigapixel-level image, we suggest that different regions have distinct semantic information, and a minority of regions may contain a large proportion of the objects in the image. The proposed PGN is designed to generate patches to cover these important regions. In consideration of the computational efficiency for subsequent tasks, the number of patches generated by the PGN should be limited. We use $S_P = \{\bigcup_{k=1}^{K} Patch_k\}$ to represent a set of patches that is a subset of all preset patch candidates, where K is much smaller than the total number of patches.

Based on the estimated patch candidates, the PGN uses a conditional strategy to select the most valuable patches. All patch candidates are first sorted by their $CountO$ value. Since the patch candidates have overlapping regions, a conditional removal operation is applied to reduce redundancy in the patch selection process. Finally, the PGN retrieves the top K patches and organizes them as the output.

DecDet

The decorated detector (DecDet) is used to meticulously detect objects in each patch. An important property of the DecDet module is its parallel execution ability. Since there are no dependencies among the patches generated by the PGN, the detection results for each patch can be obtained in parallel. This design saves considerable time compared to serial calculation.

The location information of the patch and the bounding boxes of objects detected in the patch can be remapped to the original gigapixel image. After performing global postprocessing operations such as nonmaximum suppression (NMS) [60], the detection result for each patch is processed as the final detection result for the whole video frame. In theory, the DecDet module could include any existing object detector. Here, YOLO [40] was adopted because of its agile and simple nature.

In the detector reasoning process, all patches are adjusted to the same size in the preprocessing stage, and the data are normalized. After obtaining the detection results for each patch, the NMS postprocessing operation is used to merge the results. The size of the input image can be dynamically set according to the size of the patch. The module simply sets the input size of all patches to 320 × 320. The confidence threshold is set to 0.001, and the $IOU_{threshold}$ of the NMS operation is set to 0.6. After the PGN and DecDet modules were trained, the whole reasoning process was organized into an end-to-end process. First, a gigapixel video frame is loaded and preprocessed to build a thumbnail image, which is used by the PGN module. The PGN module then generates a list of patches, which are used to crop the input image from the original gigapixel video frame for use in the for DecDet module. Finally, the detection process is executed in parallel for each patch, and some postprocessing operations are performed. The detection results for all patches are merged by the global NMS operation.

Object Detection Performance in Gigapixel Image

Evaluation Metrics

Despite the difficulty of processing gigapixel-level images, the process can be viewed as a common detection task. The popular evaluation metrics $AP@0.5$ and AR are used as the primary performance indicators. $AP@50$ represents the average precision at $IoU = 0.50$, and AR is the average recall, with the IoU score ranging from $[0.5, 0.95]$ with a stride of 0.05. The precision and recall are defined as follows:

$$\text{precision} = \frac{\text{TP}}{\text{TP} + \text{FP}} \tag{6.6}$$

$$\text{recall} = \frac{\text{TP}}{\text{TP} + \text{FN}}, \tag{6.7}$$

where TP, FP, and FN are the numbers of true positives, false positives, and false negatives, respectively. The intersection over union (IoU) between two bounding boxes is defined as follows:

$$\text{IoU} = \frac{|A \cap B|}{|A \cup B|}, \tag{6.8}$$

where A and B are the pixel areas of the predicted and ground truth bounding boxes, respectively.

The inference speed is also taken into account as an important assessment criterion for our proposed method, which aims at improving the efficiency of the detection task to achieve real-time performance.

Baseline

Faster R-CNN [10], Cascade R-CNN [11], and RetinaNet [12] are chosen as the baseline detectors with a ResNet101 [3] backbone. To train our network with the gigapixel images, the original images are resized at multiple scales, and the images are partitioned into blocks with appropriate sizes as input to the neural network. The objects cut by block borders are retained if the preserved area is over 50%. Similarly, for evaluation, the original images are resized at multiple scales, and the sliding window approach is used to generate appropriate size blocks for the detector. To better analyze the detector performance and limitations, the test results are split into subsets according to the object size.

Inference Speedup

The person annotations in the PANDA dataset are chosen as our detection target. GigaDet is evaluated based on the annotated labels, including the visible body, full body, and head labels. The proposed framework improves the inference speed to ~ 5 FPS, which is at least 50× faster than the baseline method, while maintaining approximately the same precision. The $AP@50$ and AR results are shown in Table 6.3.

Many hyperparameter settings in the evaluation procedure can affect the tradeoff between precision and efficiency. The hyperparameter values are set as follows. The scales of the patch candidates are set to $[1, 2, 4, 8]$. The K value for patch generation is set to 45. The IoU threshold for patch generation is set to 0.2. The input size for DecDet is set to 320. The confidence threshold for DecDet is set to 0.001. The IoU threshold for the NMS operation is set to 0.6.

GigaDet maintains approximately the same $AP@50$ performance while achieving an approximately $5FPS$ increase in the inference speed for the detection task based on gigapixel-level images, which is approximately 50× faster than the original baseline method. Through the coarse-to-fine progressive detection process, many invalid operations with empty regions are eliminated, reducing the time cost. Thus, a potential real-time object detection framework was developed, and this framework can be optimized to achieve real-time analysis for gigapixel-level videos in the future.

Table 6.3 Performance of detection methods on PANDA. Subset means subset of different object sizes, where Small, Middle, and Large indicate object size being $< 96 \times 96, 96 \times 96 - 288 \times 288$, and $> 288 \times 288$

	Sub.	Visible body		Full body		Head		Speed
		AP	AR	AP	AR	AP	AR	FPS
Faster R-CNN [10]	S	0.201	0.137	0.190	0.128	0.031	0.023	~ 0.10
	M	0.560	0.381	0.552	**0.376**	0.157	0.088	
	L	0.755	**0.523**	**0.744**	0.512	0.202	0.105	
Cascade R-CNN [11]	S	0.204	**0.140**	0.227	**0.160**	0.028	0.018	~ 0.07
	M	0.561	**0.388**	0.579	0.384	0.168	0.091	
	L	0.747	0.532	0.765	**0.518**	0.241	0.116	
RetinaNet [12]	S	0.171	0.121	0.221	0.150	0.023	0.018	~ 0.13
	M	0.547	0.370	0.561	0.360	0.143	0.081	
	L	0.725	0.482	0.740	0.479	0.259	0.149	
GigaDet [61]	S	**0.210**	0.093	**0.236**	0.100	**0.575**	**0.289**	**~ 5**
	M	**0.599**	0.339	**0.605**	0.326	**0.769**	**0.443**	
	L	**0.762**	0.467	0.728	0.418	**0.596**	**0.462**	

Necessity of Patch Generation

The key to accelerating the detection of objects in gigapixel images is selecting the most important patches. This section explores the necessity of patch generation and verifies that the PGN module can help to select effective information.

The aim of patch generation is to extract high-value information from gigapixel images. During the patch generation process, all patch candidates with $CountO$ values less than the threshold $\xi_C = 1$ are filtered out, and the remaining candidates are sorted by their $CountO$ value. In principle, the larger the threshold ξ_C is, the fewer patches remain in the following step. Because patches with lower $CountO$ values are ranked lower in the sorting step, the threshold is set as 1 so that patches that contain at least one object are taken into account in the sorting operation. This approach reserves the information to the maximum extent, and the time cost of sorting is negligible.

The hyperparameter K can be adjusted based on the application scenario to remove some of the patches in the sorted patch list. More patches improve the AP metric but increase the computing time. The experimental results in Table 6.4 show the effects on the K value on the patch generation results.

Patch candidates with fixed scales of [1, 2, 4, 8] and an input size as 320 were used, without the option of direct detection. The results show that when more top K patches are selected, the $AP@50$ value and inference speed both increase. When K increases from 64 to 128, the increase in K has very little effect on the results, which suggests that the patch addition is saturated at this point.

For each K setting, another experiment was conducted to verify the necessity of patch generation. When the image is directly divided into K grids, the results

Table 6.4 Study on the K value in patch generation procedure. The larger the K value is, the more patches are generated and fed into subsequent decorated detection process, which means the increment of expenditure of time. Specific setting of the value K can be decided by the practical scenario

K^*	Speed	$AP.50$ for visible body			
		S	M	L	Total
8	0.69×	0.028	0.343	0.364	0.346
16	0.83×	0.058	0.448	0.559	0.498
32	0.91×	0.139	0.556	0.715	0.633
45	1.00×	0.210	0.599	0.762	0.684
64	1.38×	0.275	0.607	0.784	0.707
128	2.47×	0.281	0.607	0.785	0.708

Table 6.5 Results of split image into K^* grids directly. Instead of using PGN module to generate patches, we did control experiment about directly splitting the image into K^* grids. For example, for $K^* = 32$, we split the image into 8×4 grids. Obviously this rough method cannot obtain satisfactory result

K^*	Speed	$AP.50$ for visible body			
		S	M	L	Total
8	0.25×	0.000	0.011	0.424	0.231
16	0.28×	0.000	0.039	0.574	0.331
32	0.55×	0.005	0.305	0.692	0.511
45	0.76×	0.004	0.427	0.700	0.561
64	0.79×	0.003	0.360	0.554	0.461
128	1.61×	0.007	0.550	0.638	0.589

show that the performance seriously deteriorates, which verifies the necessity of the proposed PGN module. The results are shown in Table 6.5 for comparison.

Scales of Patch Candidates

Patch generation is critical for obtaining good performance. State-of-the-art PGN module uses preset patch candidates for patch selection. Experiments were conducted to determine the optimal scale settings. The generated patches are used as inputs for the decorated detector module; thus, because the patches are resized and cropped to squares during that step, there is no need to set various irregular shapes for the preset candidates. Thus, the aspect ratio is fixed to 1. Different sizes are necessary because patches of different sizes should contain objects of corresponding sizes. The multiscale settings for the patch candidates are critical in capturing objects from near or distant views.

Many settings were explored, as shown in Table 6.6. Different scales of patch candidates contain objects of different scales. For example, the setting [1, 2, 4, 8] means that four sizes of patch candidates are preset that are 1, 2, 4, and 8 times the unit size.

Table 6.6 Study on PGN anchors settings. The array of digits in Anchors column means the sizes of anchor bases. For example, [1, 2, 4] means three sizes of anchor base are preset, and they are 1, 2, and 4 times the unit size, respectively. Anchors of larger size are supposed to capture valid objects with larger scale and vice versa

Anchor setting	$AP.50$ for visible body			
	S	M	L	Total
[1, 2, 4]	0.137	0.453	0.534	0.434
[1, 2, 4, 8]	0.210	0.599	0.762	0.684
[1, 2, 4, 8, 16]	0.197	0.566	0.765	0.645

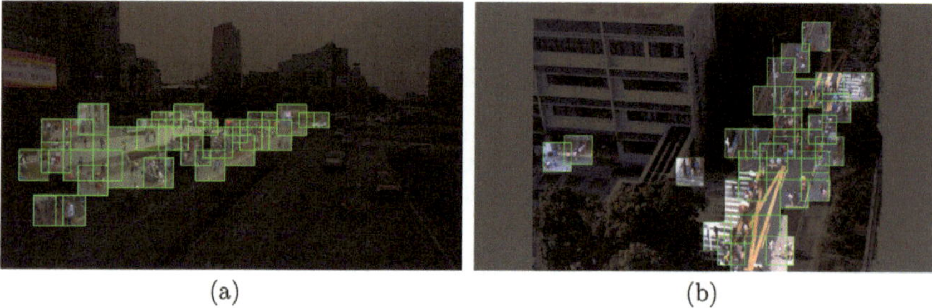

(a) (b)

Fig. 6.6 Example of anchor setting and patches selection. Different sizes of anchors are placed over the whole image. Each anchor is responsible for one value *CountO*, which represents the number of valid objects contained in that patch. The visualized schematic diagram shows that, anchors in a large area of the road (which is not marked due to its low *CountO* value) cannot be selected since no person is at that location. For the areas near streets and buildings where crowds gather, the anchors have high *CountO* values. Note that in the distant view, anchors with larger sizes do not obtain higher scores since objects of extremely small sizes are not valid objects

The unit size is determined by the downsampling ratio of the PGN model, and in our case, it is $\frac{1}{16}$ of the shorter size of the input image. Patch candidates with larger sizes should include valid objects with larger scales, and vice versa. The diverse scales enable us to capture more objects in the images; however, they also introduce a difficult selection dilemma during the patch generation process. The logic of the patch selection process is visualized in Fig. 6.6.

6.2.3 GigaGroup: Group Detection with Gigapixel Image

Framework of GigaGroup

To detect different groups in gigapixel images, a global-to-local zoom-in framework known as GigaGroup is designed, as shown in Fig. 6.7, to validate the incremental effectiveness of local visual clues to global trajectories. More specifically, human

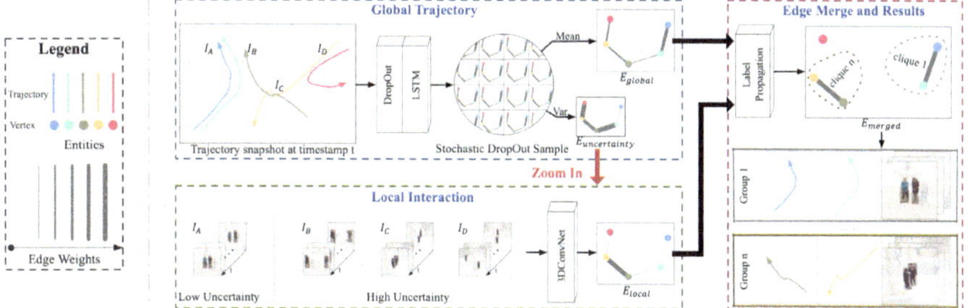

Fig. 6.7 GigaGroup framework for interaction-aware group detection. The global trajectory, local interaction, zoom-in, and edge merging modules are considered. The different color vertices and trajectories represent different human entities. The line thickness represents the edge weights in the graph. (1) Global trajectory module: The trajectories are first fed into the LSTM encoder with a dropout layer to obtain the embedding vectors; then, a graph in which the edge weights are the ℓ_2 distances between embedding vectors is constructed. (2) Zoom-in module: By repeating inference with dropout activated as stochastic sampling [66], E_{global} and $E_{uncertainty}$ are obtained from the sample mean and variance, respectively. (3) Local interaction module: The local interaction videos corresponding to high uncertainty edges ($I_B \sim I_D$) are investigated using a video interaction classifier (3DConvNet [68]). (4) Edge merging and results: The edges are merged using a label propagation scheme [69], and the objects remaining in the graph are the group detection results

entities and their relationships are represented as vertices and edges, respectively, in graph $G = (V, E)$. Features from multiple scales and modalities, such as global trajectories, face orientation vectors, and local interaction videos are used to generate edge sets E_{global} and E_{local}. Following a global-to-local strategy [62–65], E_{global} is first obtained by calculating the ℓ_2 distance in feature space for each trajectory embedding vector, which are determined by an LSTM encoder, similar to common practice [51]. Then, uncertainty-based [66, 67] and random selection policies are adopted to determine the subset of edges that need to be verified further using visual clues. Then, video interaction scores among the different entities are estimated by a spatial-temporal ConvNet [68]. The combinations of the obtained edge sets, e.g., $E_{global} \cup E_{local}$ or E_{global}, are merged using a label propagation strategy [69], and the remaining objects in the graph are the group detection results. Finally, the incremental effectiveness is estimated based on the performance metrics specified in [23] for different edge set combinations.

Zoom-in Policy

The zoom-in module is used to select a subset of edges $E_{uncertainty}$ to calculate the local interaction scores given E_{global}. Each edge $e \in E_{uncertainty}$ is fed into the local interaction module, and E_{local} and w_{local} are obtained as described above. Here, two methods are compared: random selection and uncertainty-based methods. For the random selection method, $E_{uncertainty}$ consists of η samples that are randomly selected from E_{global} and predicted to be positive. For the uncertainty-based method, the top η positive predicted uncertain edges are selected. To estimate the uncertainty, stochastic dropout sampling [66] is adopted. More specifically, with the dropout layer activated, we perform an inference process τ times for each input. Thus, for each edge score, there are τ estimations, and we can use the variance among the estimations as the desired uncertainty.

Edge Merging Strategy

Given E_{global} and E_{local} and the w_{global} and w_{local} values defined based on them, a label propagation strategy [69] is adopted to delete or merge edges with an adaptive threshold in an iterative manner. The edges are gradually deleted, and the graph is divided into several disconnected components, which are used as the group detection results.

Trajectory Source in Group Detection

We encourage users to explore an integrated solution that considers the MOT result trajectory as the group detection input. However, in our experiment, even the state-of-the-art MOT method cannot address the serious ID switch and trajectory fragmentation issues. Thus, we separate the MOT task and group detection task in the first benchmark and the previous incremental effectiveness experiment. The released dataset includes sufficient annotations, and we encourage users to explore more robust MOT methods or an integrated solution based on the two tasks. Because we use trajectory annotations, the training/testing set split differs from that used in the previous task.

Group Detection Performance with Gigapixel Image

Evaluation Metrics

The half metric refers to a detected group prediction that is positive if the detected group contains at least half of the elements of the ground truth group (and vice versa). Then, we can calculate the precision, recall, and F1-score based on the positive and negative samples. More specifically, each detected group (Grp_{pd}) and the ground truth group (Grp_{gt}) are a set of group members:

$$Grp_{*} = \{v \mid v \in V \text{ and } v \text{ belong to the group}\}. \tag{6.9}$$

One detected group is regarded as correct under the half metric if and only if it satisfies the following criterion:

$$\frac{Grp_{pd} \cap Grp_{gt}}{\max\left(\left|Grp_{pd}\right|, \left|Grp_{gt}\right|\right)} > 0.5. \tag{6.10}$$

Results

The performance of the G2L framework for group detection is shown in Table 6.7. The metrics [23] include the precision, recall, and F1-score, where a group member $IoU = 0.5$ is used for evaluation. The performance is significantly improved

Table 6.7 Incremental effectiveness (half metric [23]). The random zoom-in policy randomly selects several local videos to estimate interaction score while the uncertainty-based one selects local videos depending on the uncertainty estimation from stochastic dropout sample [66]

Edge sets	Zoom in	Precision	Recall	F1
E_{global}	/	0.237	0.120	0.160
$E_{global} \cup E_{local}$	Random	0.244	0.133	0.172
$E_{global} \cup E_{local}$	Uncertainty	0.293	0.160	0.207

by leveraging E_{local} and the uncertainty estimation approach, which validates the effectiveness of the local visual clues provided by the PANDA dataset.

6.3 Tracking and Trajectory Prediction with Gigapixel-Level Video

The pedestrian tracking task aims to associate pedestrians at different spatial positions and temporal frames. The superior properties of the PANDA dataset make it naturally suitable for long-term tracking tasks. However, complex scenes with many pedestrians in crowds also lead to various challenges.

6.3.1 Conventional Multiobject Tracking Methods

Multiobject tracking (MOT) plays an essential role in video understanding. MOT aims to detect and track all specific classes of objects in a frame-by-frame manner. In this section, we review several conventional multiobject tracking methods.

Separate and Joint Trackers

MOT methods can be classified as separate and joint trackers. Separate trackers [13, 13, 14, 70–74] apply the tracking-by-detection paradigm; these methods first localize targets and then associate them with appearance, motion, or other information. With the rapid development of object detection methods [10, 42, 45, 75, 76], separate object trackers have mainly been applied in MOT tasks. Recently, several joint trackers [77–83] have been proposed to jointly train detection networks and other components, e.g., motion, embedding, and association models. The main benefit of these trackers is their low computational cost and comparable performance. However, joint trackers face two major problems: competition between different components and limited data for jointly training the various components. These two problems limit the tracking accuracy. Therefore, the tracking-by-detection paradigm still achieves the optimal tracking performance.

Global Link in MOT

To exploit rich global information, several methods refine the tracking results with a global link model [84–88]. These models tend to generate accurate but incomplete tracks by using spatiotemporal and/or appearance information. Then, these tracks are linked by exploring global information in an offline manner. TNT [87] is a multiscale TrackletNet that was designed to measure the connectivity between two tracks. This model encodes motion and appearance information in a unified network by using multiscale convolution kernels. TPM [85] applies a track-plane matching process to move easily confused tracks into different track planes, which helps reduce errors in the track matching step. ReMOT [88] is an improved version of ReMOTS [89]. Given any tracking results, ReMOT splits imperfect trajectories into tracks and then merges the different tracks based on their appearance features. GIAOTracker [84] uses a complex global link algorithm that encodes track appearance features by using an improved ResNet50-TP model [90] and associates different tracks with low spatial and temporal costs. Although these methods have led to notable improvements, they all rely on appearance features, which increase computational costs.

Tracking Performance with Gigapixel Video

Evaluation Metrics

To evaluate the performance of multiperson tracking algorithms, the metrics used in the MOTChallenge are adopted [9, 91], including the multiobject tracking accuracy (MOTA), multiobject tracking precision (MOTP), ID F1-score (IDF1), false alarm rate (FAR), mostly tracked targets (MT), and processing speed. The multiobject tracking accuracy (MOTA) computes the accuracy considering three error sources: false positives, false negatives/missed targets, and identity switches. The multiobject tracking precision (MOTP) considers the misalignment between the ground truth and the predicted bounding boxes. The ID F1-score (IDF1) measures the ratio of the correctly identified detections to the average number of ground truth and computed detections. The false alarm rate (FAR) measures the average number of false alarms per frame. The mostly tracked target (MT) metric measures the ratio of ground truth trajectories that are covered by predicted tracks for at least 80% of their respective lifespan. The processing speed of the algorithm is measured in units of Hz.

Baselines

Three representative algorithms DeepSORT [13], DAN [15], and MOTDT [14] are evaluated. All of them follow the tracking-by-detection strategy. In our experiments, the bounding boxes are generated from three detection algorithms [10–12] in the previous subsection. For the sake of fairness, it uses the same pretrained weights on the COCO dataset and detection threshold scores (0.7) for them. Default model parameters provided by the authors are used for evaluating three trackers.

Overall Performance

Table 6.8 shows the results of DeepSORT [13], MOTDT [14], and DAN [15] based on the PANDA dataset. The time costs to process a single frame are 18.36, 19.13, and 8.29 s for DeepSORT, MOTDT, and DAN, respectively. MOTDT shows better bounding box alignment than the other models according to the MOTP and FAR metrics. DAN shows better IDF1 and MT values, implying its stronger capability to establish correspondences between the detected objects in different frames. The experimental results also demonstrate the challenge of the PANDA dataset. The best MOTA values for DeepSORT, DAN, and MOTDT based on MOT16 are 61.4, 52.42, and 47.6, respectively, and these values decrease by more than half with the PANDA dataset. With regard to the object detectors, Faster R-CNN performs the best, and Cascade R-CNN shows similar performance. However, the performance of RetinaNet is relatively poor except for the MOTP and FAR metrics because RetinaNet has low recall under a confidence threshold of 0.7 for the detection results.

This section further analyzes the influence of different pedestrian properties, including the (a) tracking duration; (b) tracking distance; (c) movement speed; (d) scale (height); (e) scale variation (standard deviation of the height); and (f) occlusions. For each property, we divided the pedestrian targets into three subsets, ranging from easy to hard. In addition, to eliminate the influence of the detector performance, we used the ground truth bounding boxes as input. Figure 5b and c shows that the tracking distance and movement speed have the largest effects on the trackers' performance. In Fig. 5a, the impact of the tracking duration on the tracker performance is not obvious because the scene contains many stationary or slow-moving people (Fig. 6.8).

Table 6.8 Performance of multiple object tracking methods on PANDA. ↑ denotes higher is better and vice versa

Tracker	Detector	MOTA ↑	MOTP ↑	IDF1 ↑	FAR ↓	MT ↑
DeepSORT [13]	FR [10]	**25.53**	76.67	21.14	20.45	762
	CR [11]	24.35	76.31	21.39	15.59	661
	RN [12]	16.36	78.0	15.16	4.32	259
DAN [15]	FR [10]	25.06	74.81	**21.85**	25.95	**826**
	CR [11]	24.24	78.55	20.13	12.42	602
	RN [12]	15.57	79.90	13.43	3.33	227
MOTDT [14]	FR [10]	13.51	78.82	14.92	6.52	257
	CR [11]	13.54	80.25	14.89	4.41	255
	RN [12]	10.77	**80.62**	11.86	**1.90**	162

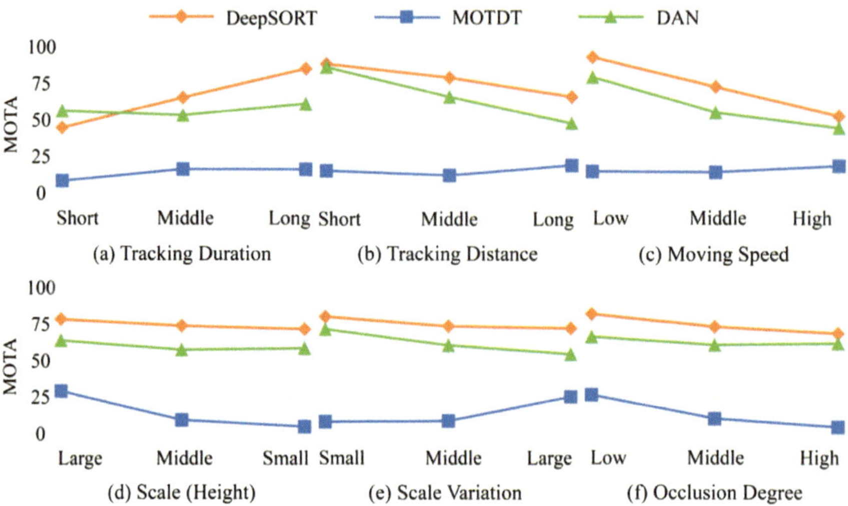

Fig. 6.8 Influence of target properties on tracker's MOTA. The pedestrian targets are divided into three subsets from easy to hard for each property

6.3.2 *Conventional Trajectory Prediction Methods*

Predicting pedestrian trajectories is a long-term problem in studies on human behavior [92, 93] and unmanned systems [94, 95]. In recent years, with the development of well-annotated video datasets, including ETH [22]/UCY [21], Stanford Drone Dataset (SDD) [96], and Intersection Drone Dataset (InD) [97], learning-based approaches [98–103] have shown great promise. Although the state-of-the-art methods show good trajectory prediction performance with small-scale scenes, long-term predictions in large-scale open scenes ($\sim$1km^2) are necessary for advanced industries, including smart cities. However, existing datasets cannot be used to achieve these goals since existing imaging technologies are restricted to small FoVs to distinguish trajectories. Fortunately, the PANDA dataset, which includes scenes with both a wide FoV and high resolution, can potentially support long-term trajectory predictions in large-scale scenes. A comparison of the most popular datasets is shown in Fig. 6.9. Therefore, for the first time, we present a new task, namely, long-term trajectory predictions in plaza-scale open scenes, considering global scenes, sequential local details, historic trajectories, and interactions as optional inputs.

Mathematical Methods

Existing mathematical methods typically assume some pre-established structures to address the problem. Social force-based, velocity obstacle-based, and energy-based formulations are usually considered in the process of modeling human move-

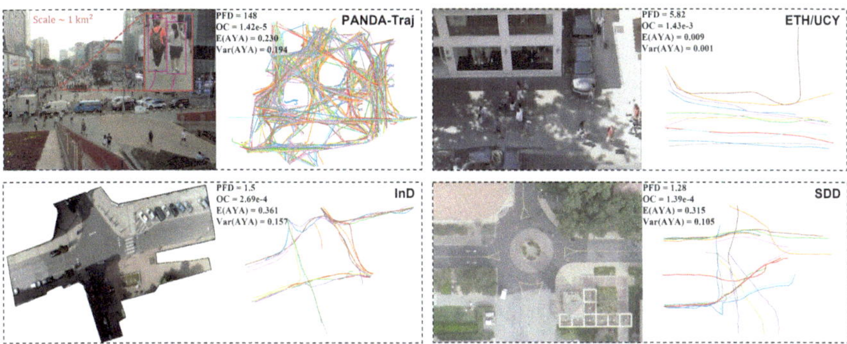

Fig. 6.9 Four scenes from the PANDA-Traj, ETH/UCY, InD, and SDD datasets. The trajectory groups and three indicators, i.e., the pedestrian frame density (PFD), orientation consistency (OC), and average yaw angle (AYA), are shown in the figures

ment [104–107]. For example, Social Forces proposes modeling interactions as attractive and repulsive forces and future trajectories as deterministic paths evolving on the basis of these forces. Many approaches have been applied to the problem of trajectory forecasting, formulating the problem as a time-series regression problem and applying methods such as Gaussian process regression (GPR) [108–110] and inverse reinforcement learning (IRL) [111] to achieve good effects.

Deep Learning Methods

Deep learning-based methods use a large amount of data to learn agent behaviors without making assumptions about the underlying motivations. Alahi et al. [98] proposed the Social-LSTM model, which applies a pooling mechanism to extract social features among pedestrians and uses an LSTM network to model their motion patterns. Based on Social-LSTM, Bartoli et al. [112] defined a context-aware pooling layer and considered the objects around it. Sun et al. [113] used an LSTM model to learn context-specific and time-specific human activity patterns in a target environment over long time periods. Huynh et al. [114] combined LSTM-based trajectory prediction with local transition patterns in specific scenarios. Fernando et al. also proposed an LSTM-based structure, which extends the encoding–decoding framework in conventional sequence-to-sequence modeling and introduces soft and hard attention mechanisms to map the current peripheral position information to the pedestrian's future position.

Generative approaches have also been developed as trajectory forecasting methods. Generative methods are usually GAN based or CVAE based, which can predict multiple feasible trajectories and optimize the prediction accuracy through game theory. Gupta et al. [99] introduced Social-GAN, which uses a generative framework to recursively infer future trajectories. Guided by the discriminator, this network can generate various acceptable future trajectories. SoPhie [115] is an interpretable framework based on a GAN that captures historical path and scene context information in the image and introduces social and physical attention to generate more authentic samples. Amirian et al. [116] used Info-GAN to resolve the problem of potential mode collapsing and mode dropping in traditional GANs and achieved good training effects for multimodal pedestrian trajectory prediction. DESIRE is a CVAE-based model that uses a latent variable to account for the uncertainty in the future paths, and a sampling model is trained to generate multiple potential future trajectories based on the given observations. Many CVAE-based models [117] were inspired by this approach to generate a variety of future predictions.

In graph neural network (GNN)-based works, pedestrians are regarded as nodes in a graph, and edge weights are used to represent their relations, which accelerates the modeling of social interactions. Social-STGCNN [100] utilizes a graph convolutional network [118] to carry out trajectory prediction; the social interaction between pedestrians is embedded in an adjacency matrix, and the spatial information of distances among pedestrians is aggregated. SimAug [119] uses graph attention networks [120], assigning higher weights to edges with higher social affinity to infer feasible trajectories. Huang et al. [101] used a graph attention mechanism to capture spatial social behaviors at each time step. Zhou et al. [102] extended the Social-STGCNN model by adding an attention module to improve the network's ability to understand pedestrians' social behavior. Trajectron++ [121] integrated trajectories in a graph-structured recurrent network, including the relationships between agents and environmental information.

More recently, Liu et al. [122] modeled actual negative examples such as collisions using contrastive learning. MemoNet [103] obtains trajectory prediction results by querying similar scene information in memory and combining this information with the current scene status. Mangalam et al. [123] proposed a factorized Y-net model including epistemic uncertainty and aleatoric uncertainty. Yue et al. [124] designed a novel method named the neural differential equation model, which used an explicit physics model with learnable parameters.

Trajectory Prediction Performance with Gigapixel Video

Evaluation Metrics

We consider two standard evaluation metrics, including the **average displacement error (ADE)** and **final displacement error (FDE)**. The ADE indicates the average Euclidean distance between the predicted future steps and the ground truth steps over the entire trajectory. The FDE indicates the Euclidean distance between the ground truth and the final position. In the case of multiple future predictions, we report minADE, which is the minimum of the average displacement error with respect to the ground truth over all predicted trajectories. Note that the latter metric is more important for the proposed long-term prediction task since we are concerned with whether the final predicted location is consistent with the agent's destination.

Following previous methods, we generate 20 predicted samples for each input trajectory. To train the model, we can use pixel coordinates or world coordinates. To obtain the predictions, we adopt world coordinates to ensure fair comparisons.

Baselines

Two state-of-the-art algorithms, NSP-SFM [124] and Y-Net [123], are set as baselines. These algorithms show the best performance with well-known datasets, including the SDD and ETH/UCY datasets. Y-Net has been applied to study long-term prediction tasks based on the SDD and InD datasets. NSP-SFM shows the best short-term prediction performance based on the SDD and ETH/UCY datasets. We apply the NSP-SFM-yaw and NSP-SFM-world models and combine the results, which is denoted as NSP-SFM-world-yaw.

Overall Performance

We consider two tasks: a 35 s prediction task and a 45 s prediction tasks. The long-term prediction results in large-scale free space with multiple agents are shown in Table 6.9. We note the following observations. (i) Overall, the error increases with the prediction duration, which suggests that the difficult of the problem increases over time. This is consistent with our intuitive perception. (ii) Not surprisingly, NSP-SFM outperforms Y-Net for both tasks. (iii) NSP-SFM-yaw achieves better performance than NSP-SFM. Remarkably, we achieve this improvement with simple loss reweighting approach. (iv) The NSP-SFM-world and NSP-SFM-world-yaw models both outperform the NSP-SFM model. Considering a non-top-down view of the data, the prediction results are better when directly using physical coordinates and discarding inappropriate environmental repulsive forces. (v) Compared with NSP-SFM-world, NSP-SFM-world-yaw achieves better FDE results and similar ADE results. This demonstrates that the utilization of the yaw metric is very effective for improving performance.

Table 6.9 Performance of classical methods for trajectory prediction in PANDA

Models	Task 1: 35 s		Task 2: 45 s	
	minADE	minFDE	minADE	minFDE
Y-Net [123]	3.373	5.313	4.457	6.391
NSP-SFM [124]	2.069	4.935	2.956	6.320
NSP-SFM-yaw	1.970 (–0.099)	4.809 (–0.126)	2.786 (–0.170)	6.189 (–0.131)
NSP-SFM-world	**0.339 (–1.730)**	1.540 (–3.395)	**0.365 (–2.591)**	2.423 (–3.897)
NSP-SFM-world-yaw	0.352 (–1.717)	**1.351 (–3.584)**	0.389 (–2.567)	**2.177 (–4.143)**

6.4 Pedestrian Intention Understanding with Gigapixel Video

6.4.1 Classical Methods for Pedestrian Intention Understanding

Understanding pedestrian dynamics is critical to a variety of real-world tasks, e.g., autonomous driving [94, 95], robot navigation [125, 126], pedestrian flow analysis [127, 128], and crowd evacuation [129, 130]. Interestingly, humans have an instinctive ability to anticipate the future actions of other people while navigating in crowded spaces and interacting with other pedestrians [131–135], e.g., avoiding head-on collisions and maintaining pace with peers at comfortable distances. Analogically, as shown in Fig. 6.10a, this ability to comprehend and anticipate the actions of pedestrians could improve the performance of unmanned systems in urban environments.

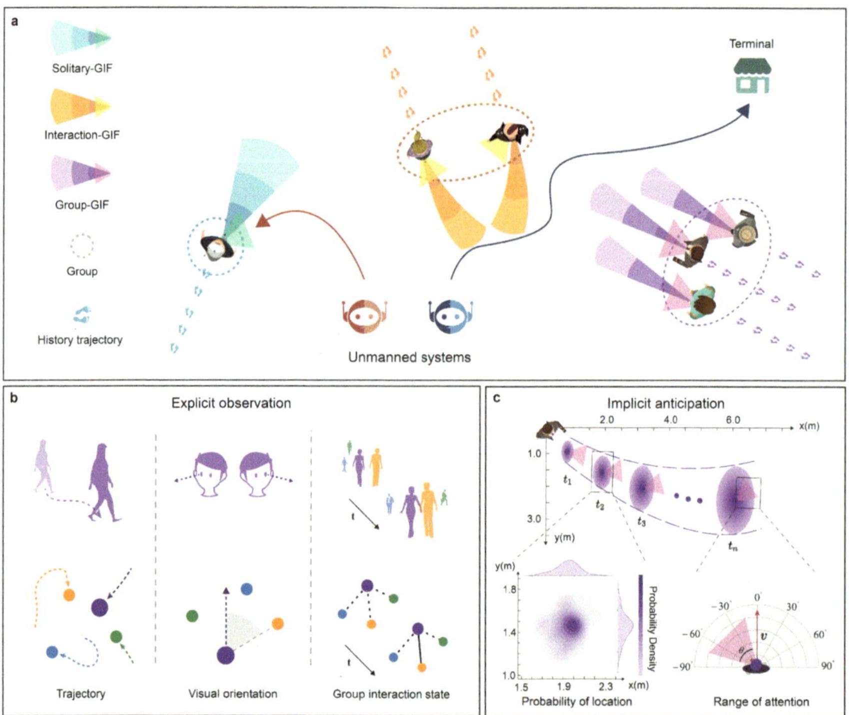

Fig. 6.10 Group interaction field (GIF). **a** The GIF, consisting of a proxemics field and an attention field, estimated from explicit observations, implicitly represents anticipated pedestrian behaviors. **b** Explicit observations include the trajectory, visual orientation, and observable group interaction state. **c** The proxemics field and the attention field are represented by a sequence of 2D probabilistic distribution maps and a sequence of angular ranges, respectively. Two representative applications are demonstrated in a; the GIF can help unmanned systems avoid disturbing pedestrians or attract pedestrians' attention

In recent decades, pedestrian anticipation was modeled by bidirectional flow [135, 136], cellular automaton [132, 137], or time-to-collision [107, 134, 138] models to simulate collective behaviors. Recently, machine learning technology has been utilized, and the future states of each pedestrian can be forecasted [100, 115, 121, 139–142]. Essentially, the above methods mainly model individual behaviors in collision avoidance tasks and do not consider group-related social behaviors. However, humans are naturally social beings that socially interact and form social groups [143], e.g., up to 70% of observed pedestrians on a street are in groups [144]. Pedestrians conform to expected social norms in groups and act accordingly under the influence of nearby neighbors [122], and intra/intergroup interactions have critical effects on pedestrians' social cognition [145] and behavior patterns [92, 93]. To model the group or interaction information, graph neural networks (GNNs) have been utilized for pedestrian/agent dynamics understanding [99–101, 122, 146–148], which are state-of-the-art methods. However, the influence of group behaviors on pedestrians is not only diverse but also subtle, and different group relationships or interaction states have varied impacts on future pedestrian states. For example, a family group (e.g., mother and daughter) and a tour group usually show different behaviors in similar scenarios, as the attention of children is less focused. These subtle differences cannot be modeled well by simple relationship or interaction graphs. Existing methods cannot distinguish different group relationships among pedestrians and thus cannot accurately predict the differences in anticipated pedestrian behaviors influenced by group behaviors [99–101, 121, 122, 135, 146, 149, 150].

6.4.2 *GIFNet: Understand Pedestrian Intention in Large-Scale Scenes*

In complex scenes, it is important but difficult to understand and label group relationships. Thus, the group interaction field (GIF), a novel group-aware representation, was developed to implicitly quantify anticipated pedestrian behaviors. In particular, the GIF consists of a proxemics field and an attention field, which represent anticipated pedestrian behaviors based on the probability fields of pedestrians' future locations and attention orientations, respectively. Moreover, GIFNet was developed to estimate the GIF based on explicit multidimensional observations, including the trajectory, visual orientation, and group interaction state. GIFNet can quantify the diverse and subtle influences of group behaviors by formulating group interaction graphs with propagation and attention mechanisms based on group size and dynamic interaction states.

Group Interaction Field

Humans are naturally social beings that interact socially and form social groups [143]. Previous social investigations have shown that up to 70% of observed pedestrians in a street are in groups [144]. Therefore, to accurately predict pedestrian behaviors, individual trajectories and group relationships should be considered. However, in real complex scenes, it is difficult to accurately understand and label the group relationships. Thus, the group interaction field (GIF), a group-aware representation, was developed to implicitly quantify anticipated pedestrian behaviors. As an implicit estimation of anticipated pedestrian behaviors, the GIF consists of a proxemics field and an attention field. The proxemics field is a sequence of 2D probabilistic distribution maps denoting the future locations of the pedestrian of interest (Fig. 6.10c), with time span T and temporal resolution R. Similarly, the attention field is a sequence of angular ranges θ, representing the possible orientation and range of visual attention of a given pedestrian. Formally, given a timestamp in an observation sequence $t \in [1, \ldots, T]$ of the pedestrian of interest i, the GIF is defined as $GIF_i^T = \left[P_i^T, A_i^T\right]$, with proxemics field P_i^T and attention field A_i^T.

As shown in Fig. 6.10a, the GIFs of solitary pedestrians (cyan), grouped pedestrians without interactions (purple), and grouped pedestrians with interactions (orange) have apparent differences. The single pedestrian has a long and wide proxemics field, while the grouped pedestrians without interactions have shorter and narrower fields. Moreover, the grouped pedestrians with interactions tend to approach each other closely. The GIF can be used to predict pedestrians' future locations and attention orientations and thus has great potential in unmanned system applications. Figure 6.10a shows two representative applications of the GIF. The proxemics field can help the unmanned system (blue) plan the path to avoid disturbing pedestrians, while the attention field can guide the unmanned system (red) to approach a pedestrian and attract their attention.

GIFNet

To accurately estimate the GIF, GIFNet was developed, as illustrated in Fig. 6.11a. GIFNet takes three explicit observations as inputs, including the trajectory of the pedestrian of interest T_p, the visual orientation of the pedestrian of interest F_p, and neighbor trajectories T_n in an interaction graph I_t, and outputs the GIF of the pedestrian of interest. Given the pedestrian of interest (purple), the remaining pedestrians in the same group (other colors) are denoted as its neighbors. Specifically, the group interaction graph I_t is a graph sequence that organizes the group interaction state; the edges represent whether the pedestrian of interest interacts with the corresponding neighbors at each time step.

GIFNet consists of four modules: (i) a trajectory encoder that models the historical trajectories of the pedestrians of interest; (ii) an attention encoder that models the pedestrian of interest's visual orientation information; (iii) GIF-GAT, which models the interaction information between the pedestrian of interest and its neighbors; and

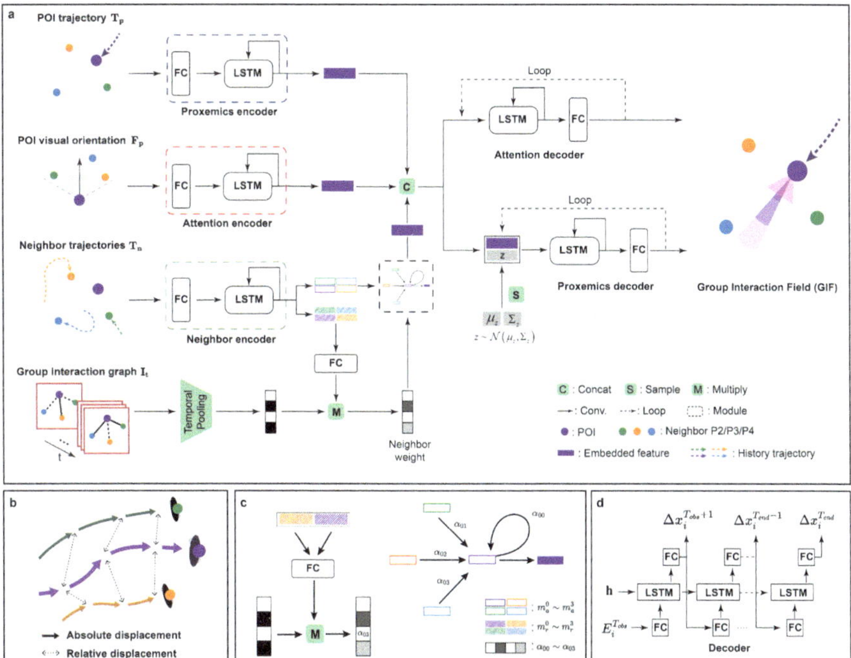

Fig. 6.11 GIFNet. **a** Network structure of GIFNet, including three encoders (trajectory, visual orientation, and neighbor), GIF-GAT, and two decoders (proxemics and attention). POI, pedestrian of interest. **b** Illustration of the relative displacement (the neighbor location relative to the pedestrian of interest) and the absolute displacement (neighbors' self-displacement). **c** Network structure of GIF-GAT. GIF-GAT takes the neighbors' relative embeddings (filled rectangles, m_r^i), the neighbors' absolute embeddings (hollow rectangles, m_a^i), and the dynamic group interaction graph as inputs and outputs a fixed-length embedding vector representing the influence of all group neighbors on the pedestrian of interest. **d**) Loop structure in the proxemics decoder and attention decoder for predicting the sequence of future states

(iv) a visual orientation decoder and proxemics decoder that estimate the proxemics field and attention field of the pedestrian of interest, respectively. In GIFNet, three encoders composed of a fully connected (FC) layer and a long short-term memory (LSTM) unit are used to extract features from T_p, F_p, and T_n. For the neighbor trajectories T_n, the encoder produces two embedding vectors (m_a^i and m_r^i) for the i-th neighbor, encoding the neighbor's absolute displacement and their displacement relative to the pedestrian of interest, respectively (Fig. 6.11b). The group interaction graph I_t and the features of the neighbor trajectories T_n are processed by a graph attention module (GIF-GAT, Fig. 6.11c). At each time step t, an FC layer is used to calculate the weights of the neighbors based on the relative displacement of the neighbors. The weights are multiplied by the group interaction graph I_t to obtain the final weight α_i for the i-th neighbor. The absolute displacement features of the neighbors (m_a^i) are then summed with the weights, using α_i as the final neighbor embedding vector. With this approach, GIFNet propagates the influence of neighbors and the group interactions through the graph to learn an embedding feature vector. Finally, the embedding feature vectors of the four kinds of explicit observations are input to the decoders to estimate the proxemics field and attention field (Fig. 6.11d). For the proxemics decoder, a Gaussian sampling module is used to learn the uncertainty of the proxemics field and produce a sequence of probability distribution maps representing the pedestrian of interest's future location.

Trajectory Encoder

The purpose of the trajectory encoder is to encode the historical trajectory information and generate a trajectory embedding. The trajectory encoder consists of $LSTM_x$. It first calculates the relative location of each pedestrian i with respect to the location at the previous timestep, i.e., $\Delta x_i^t = x_i^t - x_i^{t-1}$. Therefore, the past trajectory information of pedestrian i is $X_i = \{\Delta x_i^1, \ldots, \Delta x_i^{T_{obs}}\}$.

For time steps $t = \{1, \ldots, T_{obs}\}$, the following update operation is performed in the FC layer in Fig. 6.11a to embed the relative locations into a fixed-length vector:

$$e_i^t = \phi(\Delta x_i^t; W_{ee}). \tag{6.11}$$

Then, the embedding vectors are used as inputs to the LSTM unit as follows:

$$m_x^t(i) = LSTM_x(m_x^{t-1}(i), e_i^t; W_x), \tag{6.12}$$

where the function ϕ denotes the embedding of the past trajectory information of pedestrian i in the FC layer, W_{ee} is the embedding weight, $m_x^t(i)$ is the hidden state of $LSTM_x$ at timestep t, and W_x is the weight of the $LSTM_x$ cell. These parameters are shared among all the pedestrians in the scene.

Visual Orientation Encoder

The purpose of the visual orientation encoder is to encode the historical visual orientation information and generate a visual orientation embedding. The past visual orientation information of pedestrian i is represented by an ordered set of their orientation values in a unit vector form as follows:

$$\begin{aligned} A_i &= \{a_i^1, \ldots, a_i^{T_{obs}}\} \\ a_i^t &= (\cos\theta_1, \sin\theta_1), \end{aligned} \tag{6.13}$$

where θ_1 is the inner angle of visual orientation concerning the forward orientation, as illustrated in Fig. 6.13b. Similar to the trajectory encoder module, the visual attention sequence A_i with the hidden state $m_o^t(i)$ is fed into the attention encoder $LSTM_o$ for pedestrian i. The operation is as follows:

$$\begin{aligned} e_i^t &= \phi(\Delta a_i^t; W_{ee}) \\ m_o^t(i) &= LSTM_o(m_o^{t-1}(i), e_i^t; W_o), \end{aligned} \tag{6.14}$$

where W_o is the weight of the $LSTM_o$ cell. For simplicity, we use ϕ and W_{ee} to represent the embedding function and the embedding weight in the hidden state, respectively. The final vector $m_o^{T_{obs}}(i)$ is an ensemble of the visual orientation information for pedestrian i.

GIF-GAT

For efficiency and simplicity, GIF-GAT adopts a similar mechanism as the trajectory encoder to encode the neighbor trajectories. For the pedestrian of interest i, the relative location of each neighbor j with respect to the location at the previous timestep is calculated as $\Delta x_j^t = x_j^t - x_j^{t-1}$. Then, the relative location of each neighbor j with respect to the location of the pedestrian of interest i is calculated at each timestep as $\Delta x_{ij}^t = x_i^t - x_j^t$.

The neighbor locations are denoted in the forms of Δx_j^t and Δx_{ij}^t, which are referred to as the absolute displacement and relative displacement, respectively. The operations are as follows:

$$\begin{aligned} e_i^t &= \phi(\Delta x_j^t; W_{ee}) \\ e_{ij}^t &= \phi(\Delta x_{ij}^t; W_{ee}). \end{aligned} \tag{6.15}$$

We obtain two distance-sensitive context embeddings, including the neighbor's relative embedding $m_r^t(j)$ and the neighbor's absolute embedding $m_a^t(j)$, by feeding the corresponding vectors into the neighbor encoder. The operations are as follows:

$$\begin{aligned} m_a^t(i) &= LSTM_e(m_a^{t-1}(i), e_i^t; W_a) \\ m_r^t(i) &= LSTM_e(m_r^{t-1}(i), e_{ij}^t; W_a). \end{aligned} \tag{6.16}$$

The GAT is used as a sharing mechanism to aggregate the interaction information between the pedestrian of interest and its neighbors. As shown in Fig. 6.11, the pedestrians in a scene are considered as nodes, and the edges in the graph are used to represent the human–human interaction information. The GAT is constructed by stacking graph attention layers. The previous group interaction graph for the pedestrian of interest i is represented by a sequence of dummy variables as

$$\beta = \{\beta_{ij}^t | j \in \{1, \ldots, D\}, t \in \{1, \ldots, T_{obs}\}\}, \tag{6.17}$$

where β_{ij}^t is a dummy variable indicating the existence of an interaction between the pedestrian of interest i and neighbor j. Temporal pooling for β_{ij}^t is adopted to generate a pooled context vector C_{ij}, which is composed of the interaction information across the observation period, i.e., $C_{ij} = \frac{1}{T} \sum_i^T \beta_{ij}^t$.

Let $m_r^{T_{obs}}(j)$ denote the final relative embedding and $m_a^{T_{obs}}(j)$ denote the final absolute embedding of neighbor j. In the observation period, $m_a^{T_{obs}}(j)$ is fed into the graph attention layer. The coefficients in the attention mechanism of the node pair (i, j) can be computed by multiplying $m_r^{T_{obs}}(j)$ and C_{ij} as

$$\alpha_{ij} = m_r^{T_{obs}}(i) \cdot C_{ij}. \tag{6.18}$$

The output of one graph attention layer for node i (pedestrian of interest i) is given by

$$\hat{m}_a^{T_{obs}}(i) = \sigma(\sum_{j \in N_i} \alpha_{ij} W m_a^{T_{obs}}(j)), \tag{6.19}$$

where σ is a nonlinear function and N_i represents the neighbors of node i. $W \in R^{F' \times F}$ is the parameter matrix of a shared linear projection that is applied to each neighbor separately (F is the dimension of the input, and F' is the dimension of the output). In addition, $\hat{m}_a^{T_{obs}}$ is a fixed-length embedding for the pedestrian of interest i at the observation time, representing the influence of all neighbors on the pedestrian of interest.

Visual Orientation and Proxemics Decoder

Decoders are used to generate the proxemics field and attention field conditioned on $E_t(i) = [E_p, E_v, E_n]$:

$$\begin{aligned} E_p &= \phi(m_x^{T_{obs}}(i); W_{ee}) \\ E_v &= \phi(m_o^{T_{obs}}(i); W_{ee}) \\ E_n &= \phi(m_a^{T_{obs}}(i); W_{ee}), \end{aligned} \tag{6.20}$$

where E_p, E_v, and E_n are the trajectory, visual orientation, and neighbor influence embeddings, respectively. Then, the attention field of the pedestrian of interest i is updated using

$$\begin{aligned} h_a^{T_{obs}+1}(i) &= LSTM_{dec}^p(h_a^{T_{obs}}(i), [E_t(i), z]; W_{dec}^a) \\ a_i^{T_{obs}+1} &= \phi(h_a^{T_{obs}+1}(i); W_{ee}). \end{aligned} \tag{6.21}$$

This method directly concatenates a noise vector z sampled from a Gaussian distribution and the context embeddings $E_t(i)$ as the input to the proxemics decoder.

$$\begin{aligned} h_p^{T_{obs}+1}(i) &= LSTM_{dec}^p(h_p^{T_{obs}}(i), [E_t(i), z]; W_{dec}^p) \\ \Delta x_i^{T_{obs}+1} &= \phi(h_p^{T_{obs}+1}(i); W_{ee}), \end{aligned} \tag{6.22}$$

where $\Delta x_i^{T_{obs}+1}$ and $a_i^{T_{obs}+1}$ are the location and visual orientation of the pedestrian of interest i at $T_{obs} + 1$, respectively. The notations $h_a^t(i)$ and $h_p^t(i)$ are used to represent the hidden states of the visual orientation decoder and proxemics decoder, respectively. W_{dec}^p and W_{dec}^a are used to represent the embedding weights of the proxemics field decoder and the attention field decoder, respectively.

Loss Function

For training the proxemics field decoder, the variety loss is used:

$$\mathcal{L}_p = \min_k \|Y_i^p(t) - \hat{Y}_i^p(t)^k\|_2, \tag{6.23}$$

where $\hat{Y}_i^p(t)^k$ is the k-th predicted location of pedestrian i and $Y_i^p(t)$ is the ground truth location. In addition, k is a hyperparameter for calculating the variety loss, where $k = 20$.

The ℓ_2 loss based on the attention field is applied to measure the difference between the prediction and the ground truth:

$$\mathcal{L}_a = \|Y_i^a(t) - \hat{Y}_i^a(t)\|_2^2, \tag{6.24}$$

where $\hat{Y}_i^a(t)$ is the predicted attention orientation of pedestrian i and $Y_i^a(t)$ is the ground truth attention orientation.

Performance of Pedestrian Intention Understanding

Dataset

The performance of our models is evaluated based on the PANDA dataset [50]. The videos in the PANDA dataset were captured by gigapixel cameras, and each video frame contains hundreds to thousands of pedestrians, with rich group interaction information. Our method only requires the trajectories, visual orientations, and group interaction information. Thus, we extract this information from the PANDA labels and form a new dataset with 21, 677 trajectories. We divide the dataset into training, testing, and validation sets with 15, 511, 3, 052, and 3, 141 trajectories, respectively. Then, we compute a homography matrix to map the images to the top view to obtain the locations of the pedestrians in world coordinates.

In contrast to existing group-based trajectory prediction datasets [141], in our dataset, each group is assigned several category labels, denoting the kinds of group relationships (acquaintance and family) and interaction states (no interactions, non-physical interactions, and physical interactions). Eye contact, body language such as waving hands, and talking are types of nonphysical interactions, while holding hands is a type of physical interaction. The group relationship information is identified through the interactions and characteristics of the members, such as the appearance, gender, age, and exchanges.

Evaluation Metrics

During the tests, a Gaussian model was applied to fit the predicted locations for all k samples, and the point with the highest probability was used as the optimal predicted location $\hat{Y}_i^p(t)$, calculated as

$$\hat{Y}_i^p(t) = \max_{\Delta x_i^t} P(\Delta x_i^t), \tag{6.25}$$

where $P(\Delta x_i^t)$ is the fitted Gaussian model for all predicted locations of pedestrian i at time t. The average displacement error (ADE) [100] and the final displacement error (FDE) [151] were used to evaluate the predicted trajectories as follows:

$$\begin{aligned} ADE &= \frac{1}{N}\sum_{t=1}^{N} |Y_i^p(t) - \hat{Y}_i^p(t)| \\ FDE &= |Y_i^p(T_{obs}) - \hat{Y}_i^p(T_{obs})|, \end{aligned} \tag{6.26}$$

where N is the number of predicted timesteps.

Similarly, we used the average angular error (AAE) and the final angular error (FAE) to evaluate the predicted visual orientations:

$$\begin{aligned} AAE &= \frac{1}{N}\sum_{t=1}^{N} |Y_i^a(t) - \hat{Y}_i^a(t)| \\ FAE &= |Y_i^a(T_{obs}) - \hat{Y}_i^a(T_{obs})|. \end{aligned} \tag{6.27}$$

Predicting the Proxemics Field

As the proxemics field represents the future location distribution of the pedestrian of interest, GIFNet was evaluated using the accuracy of the predicted locations based on the given dataset. Recent studies on crowd forecasting have indicated that the short-term motion of pedestrians is highly predictable [152]. We adopt similar settings as used in previous works, with a time span $T = 3$ s and temporal resolution $R = 1/3$ s. As shown in Fig. 6.12a, the mean displacement error (ADE) and final displacement error (FDE, displacement error at the endpoint, shown by the stars) of the predicted locations are used as the evaluation metrics. For each time step, the predicted location with the highest probability is used to calculate the ADE and FDE. In the experiments, GIFNet outperforms the state-of-the-art machine learning-based trajectory prediction methods (Sophie [115], STGAT [101], SGAN [99], Social-STGCNN [100], and SGCN [153]) and the baseline method "Linear". Among these methods, only our GIFNet encodes all four kinds of features: the trajectories, visual orientations, neighbor trajectories, and group interaction states. Sophie, STGAT, SGAN, Social-STGCNN, and SGCN encode only the trajectory and integrate the information of the surrounding neighbors with a relative-distance-dependent method. The baseline method Linear is a linear regressor that takes only the past trajectory as input. The results of GIFNet without neighbor information and group interaction information are also presented, showing their importance in future location prediction. Section 6.4.2 reports a more detailed ablative analysis.

For a more in-depth analysis of the neighbor and group interaction information, we divide the pedestrians into several categories (solitary pedestrians, members of the acquaintance group, members of the family group, group members without interactions, group members with nonphysical interactions, and group members with physical interactions) and plot the statistical analysis results in Fig. 6.12b–e. Figure 6.12b and c illustrates the distribution of the ground truth versus the estimated forward (movement direction at the current time step) and lateral (movement orthogonal to the forward direction) speeds. The prediction of GIFNet is highly consistent with the ground truth. The solitary pedestrians move faster than the grouped pedestrians in both directions ($p < 0.001$, $N = 12245$), and the pedestrians in the

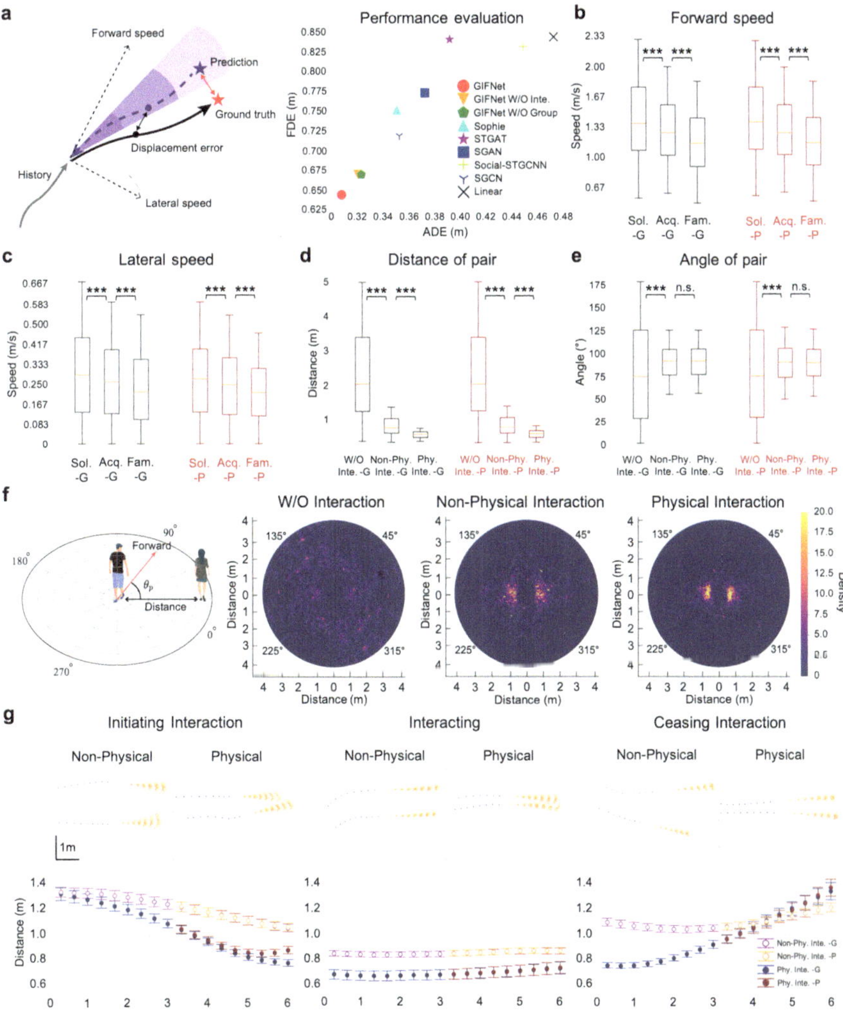

◄**Fig. 6.12 Evaluation of the proxemics field prediction.** **a** Left, illustration of the average displacement error (ADE) and final displacement error (FDE). Right, performance of GIFNet and other state-of-the-art methods. W/O Group, without group neighbor information input. W/O Inte., with group neighbor information input but no group interaction graph input. **bc** Boxplots of the forward and lateral speeds of solitary pedestrians and pedestrians in different groups (family and acquaintance). **de** Boxplots of the pairwise distance and angle (θ_p in **f**) of grouped pedestrians without interactions, grouped pedestrians with nonphysical interactions, and grouped pedestrians with physical interactions. **f** Top-view distribution maps of neighbors' locations. From left to right: pedestrian pairs without interactions, pedestrian pairs with nonphysical interactions, and pedestrian pairs with physical interactions. **g** Change in the proxemics fields of pedestrian pairs when the group interaction state changes. The top row shows representative estimated proxemics fields for examples with nonphysical or physical interactions; the blue points are the input observations, and the orange probability distribution maps are the predicted proxemics fields. The bottom row shows the changes in the pairwise distance. The gray background indicates cases with no interactions, and the white background indicates that the interaction is in progress. The error bars represent the standard error of the mean (s.e.m.). ***, $p < 0.001$; n.s., no significance, single-sided Mann–Whitney U-test. Sol., solitary. Fam., family. Acq., acquaintance. Inte., interaction. Phy., Physical. W/O, Without. G, ground truth. P, prediction

acquaintance group move faster than those in the family group in both directions ($p < 0.001$, $N = 12245$). However, grouped pedestrians with and without interactions show no significant difference, meaning that group interactions do not affect pedestrians' walking speed. In addition, the proxemics fields of the solitary pedestrians are more dispersed than those of the grouped pedestrians, i.e., the walking direction of solitary pedestrians has higher uncertainty. The above results indicate that groups directly affect pedestrian speed and walking direction.

The spatial organization of the walking pedestrian groups can be measured by the angle (the inner angle between the neighbor and forward orientation, θ_p in Fig. 6.12f) and distance between pedestrians [144]. Figure 6.12d and e illustrates the influence of interactions on the pairwise distance and angle of pedestrian pairs in groups. The distances between pedestrian pairs with physical, nonphysical, and no interactions increase significantly (both $p < 0.001$, $N = 3668$). The angles of pedestrian pairs with physical and nonphysical interactions are clustered at 90°, while the angles of pedestrian pairs without interactions are smaller ($p < 0.001$, $N = 3668$) and more uncertain. The distributions of the pairwise distance and angle are presented in Fig. 6.12f. A total of 559 pairs of pedestrians with no interactions, 137 pairs of pedestrians with nonphysical interactions, and 199 pairs of pedestrians with physical interactions are plotted on three 2D histograms: the pedestrian of interest is located in the center, with a 90-degree forward direction, and the neighbors are plotted based on the corresponding distance and inner angle. The pedestrian pairs with interactions are more concentrated than those without interactions, and pedestrians in this case tend to walk in parallel (angles clustered at 90°).

Figure 6.12g illustrates the changes in the proxemics field and pairwise distance at the time of initiating, performing, and ceasing physical and nonphysical interactions. We randomly sampled 400 pairs for each state and plotted the time–distance curve (bottom part of Fig. 6.12g), and the proxemics field of a representative pair is plotted in the top part of Fig. 6.12g for each stage. Pedestrian pairs with physical and nonphysical interactions show similar changes. When the pedestrians initiate interactions, they move closer; during the interaction, the distance remains stable; and when the interaction stops, they tend to separate. Compared with the case of nonphysical interactions, pedestrian pairs with physical interactions have smaller pairwise distances. The high correlation between the predicted curves and the ground truth shows that GIFNet can effectively capture the changes in the group interaction state and predict accurate future locations for all states.

Predicting the Attention Field

As shown in Fig. 6.10c, the attention field is an angular range denoting the visual attention of the pedestrian of interest. Here, we fix the angular range to 30°, corresponding to the aperture of the cone of visual attention [154], and predict the central orientation. The ground truth attention fields are calculated based on the annotated visual orientations in the dataset. Similarly, we set the time span $T = 3$ s and temporal resolution $R = 1/3$ s and evaluate GIFNet using the average angular error (AAE) and final angular error (FAE). Since there are no visual orientation prediction methods, we modify the state-of-the-art trajectory prediction methods for visual orientation prediction, denoted as Sophie-A, STGAT-A, SGAN-A, Social-STGCNN-A, and SGCN-A. "Linear" is the baseline linear regression method. Figure 6.13a shows that our GIFNet model (red circle) achieves the best AAE and FAE results among all the methods. Similar to the proxemics field prediction results, the group neighbor and interaction information have notable impacts on the attention field prediction (yellow triangle and green pentagon).

For a more in-depth analysis of the influence of such information on pedestrian anticipation, we evaluate the forward-attention angle (θ_1 in Fig. 6.13b), cross-attention angle (θ_2 in Fig. 6.13b), and neighbor-attention angle (θ_3 in Fig. 6.13b). The forward-attention angle measures the consistency between the pedestrians' attention orientation and the forward direction; the cross-attention angle measures the consistency between the attention orientations of pedestrian pairs, and the neighbor-attention angle reflects whether pedestrians' attention is attracted by their neighbors. As illustrated in Fig. 6.13c–f, all three angles predicted by our GIFNet model (red) show good consistency with the ground truth (black). As shown in Fig. 6.13c–e, the forward-attention angles of pedestrians with physical interactions, pedestrians without interactions, and pedestrians with nonphysical interactions increase significantly (both $p < 0.001$, $N = 893$); the cross-attention angles of pedestrian pairs with nonphysical interactions are significantly smaller than those of pedestrian pairs without interactions ($p < 0.001$, $N = 2986$); and the neighbor-attention angles of pedestrian pairs with physical interactions, pedestrian pairs with no interactions, and pedestrian pairs with nonphysical interactions decrease significantly (both $p < 0.001$, $N = 2986$). The above results indicate that grouped pedestrians with physical interactions tend to focus on the forward direction (cross-attention angles close to 0° and neighbor-attention angles close to 90°), while pedestrians with nonphysical interactions are more likely to look at each other. This is because nonphysical interactions mainly include verbal communication and

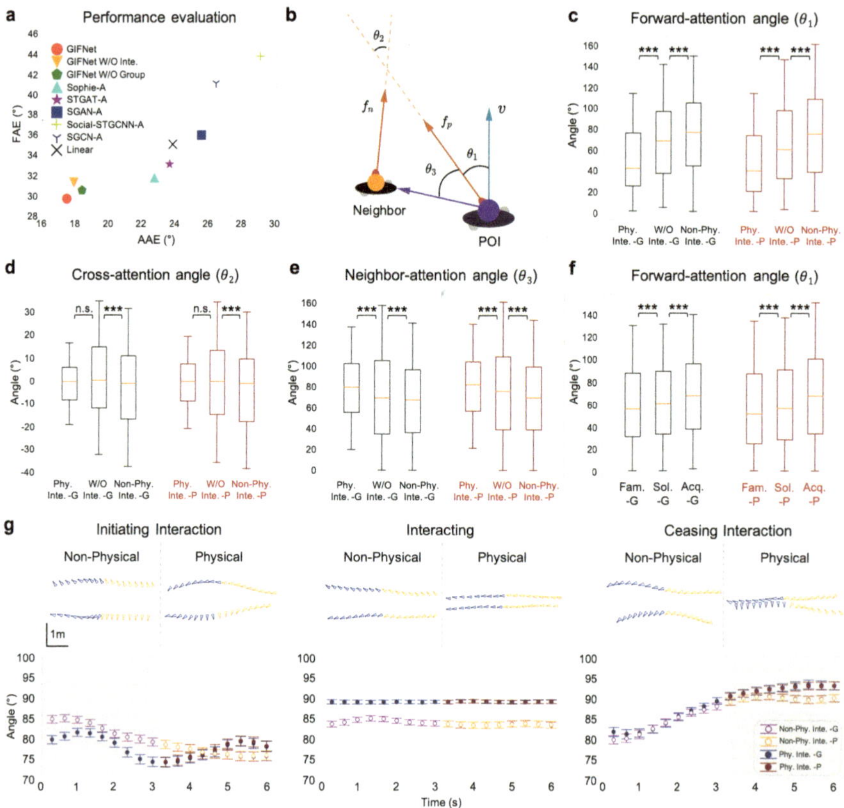

Fig. 6.13 Evaluation of the attention field prediction. a) Comparison of the average angular error (AAE) and final angular error (FAE) between GIFNet and other methods. W/O Group, without group neighbor information input. W/O Inte., with group neighbor information input but no group interaction graph input. **b**) Illustration of the forward-attention angle (θ_1, the inner visual orientation angle in the forward orientation), cross-attention angle (θ_2, the inner angle between the visual orientations of the pedestrian of interest f_p and neighbor f_n; the angle is negative when the two orientations converge inward), and neighbor-attention angle (θ_3, the inner visual orientation angle based on the orientation of the neighbor). **c, f**) Distributions of the ground truth and predicted forward-attention angles (θ_1 in b). **d**) Distribution of the ground truth and predicted cross-attention angles (θ_2 in b). **e**) Distribution of the ground truth and predicted neighbor-attention angles (θ_3 in b). **g**) Change in the attention fields of pedestrian pairs when the group interaction state changes. The top row shows representative estimated attention fields with nonphysical or physical interactions. The blue fans represent the input observations; the center is the location of the pedestrian, and the direction is oriented toward the attention direction. The orange fans are predictions for future time steps. The bottom row shows the changes in the neighbor-attention angles. The gray background indicates cases with no interactions, and the white background indicates that the interaction is in progress. The error bars represent the standard error of the mean (s.e.m.). ***, $p < 0.001$; n.s., no significance, single-sided Mann–Whitney U-test. Inte., interaction. Phy., physical. W/O, Without. Sol., solitary. Fam., family. Acq., acquaintance. G, ground truth. P, prediction

eye-to-eye behaviors that require visual attention, while pedestrians with physical interactions can focus more on walking because they can effectively perceive the location of their partner through touch instead of sight. Similar to Fig. 6.13c, f shows that the forward-attention angles of pedestrians in the family group, solitary pedestrians, and pedestrians in the acquaintance group increase significantly (both $p < 0.001, N = 4641$). This is because family members are more likely to interact with each other physically, while acquaintances are equally likely to demonstrate physical and nonphysical interactions. Hence, group interactions have a significant and diverse effect on pedestrians' visual attention.

Our work analyzes the significant influence of group interactions on pedestrian behavior and contributes to explaining the changes in anticipated pedestrian behavior. For example, pedestrians participating in nonphysical interactions are anticipated to look at the interacting peer, whereas pedestrians participating in physical interactions are anticipated to focus on the road ahead (Fig. 6.13c–e). We believe that our explanations of the changes in anticipated pedestrian behavior contribute to a more comprehensive understanding of the social nature of pedestrians, thereby improving the accuracy of pedestrian dynamics modeling.

Furthermore, our results demonstrate the changes in the attention field at the time of initiating, performing, and ceasing physical and nonphysical interactions. Similar to Fig. 6.12g, the top row of Fig. 6.13g shows the representative predicted attention fields for pedestrian pairs, and the bottom row shows the changes in the neighbor-attention angle. When the pairs start to interact, they tend to look at each other; during the interaction, both participants look forward, and the participants sometimes look at each other (more often during nonphysical interactions); and when the interaction stops, the participants look at each other again and then turn to the forward orientation. Similar to Fig. 6.12g, the curve in Fig. 6.13g shows the high correlation between the predicted results and ground truth, proving that GIFNet can capture the influence of group interactions on the attention field.

Ablation Study

In this section, we conduct a careful ablation study to show the effectiveness of GIFNet. In particular, the two best-performing baseline methods, Sophie and SGCN, are compared with our model. The ADE and FDE are adopted as evaluation metrics for the proxemics field. The AAE and FAE are adopted as evaluation metrics for the attention field. As shown in Fig. 6.14, for the proxemics field prediction, the visual orientation information input and the group interaction information input improve the performance at every time step. For attention field prediction, the trajectory information input and the group interaction information input improve the performance in the first 6 time steps.

Technically, the effective encoding of the group and group interaction information contributes to the superior accuracy of our method. In existing machine learning methods, although pooling-like operations [98, 155] and graph neural networks [101, 141, 156–158] have been used to encode the influence of different pedestrian information, only the relative spatial distance between pedestrians is used, while the group and group interactions are not considered. In addition to encoding physical

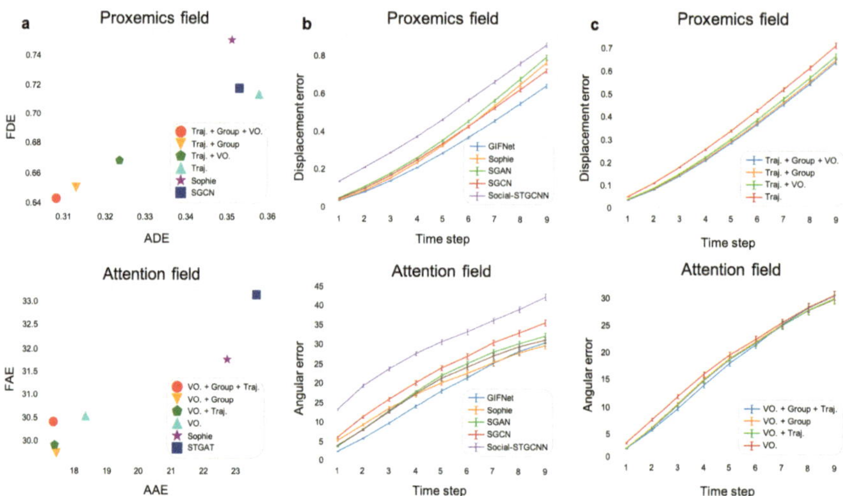

Fig. 6.14 Ablation analysis of the network. **a** Top: Comparison of the prediction of the proxemics field by GIFNet with models with no visual orientation information, no group interaction information, and neither type of information. Bottom: Comparison of the prediction of the attention field by GIFNet with models with no trajectory information, no group interaction information, and neither type of information. **b** Comparison of the prediction errors of GIFNet and other baseline methods at each time step (time span $T = 3$ s and temporal resolution $R = 1/3$ s, 9 time steps in total). For proxemics field prediction, GIFNet performs best at every time step. For the attention field prediction, GIFNet performs best in the first 6 time steps and was slightly worse than Sophie in the last 3 time steps. **c** Ablative comparison of the prediction results of GIFNet at each time step (time span $T = 3$ s and temporal resolution $R = 1/3$ s, 9 time steps in total). Top: Comparison of the proxemics field prediction by GIFNet with models with no visual orientation information, no group interaction information, and neither type of information. Bottom: Comparison of GIFNet with models with no trajectory information, no group interaction information, and no attention field prediction. The error bars represent the s.e.m. Traj., trajectory. VO., visual orientation

features such as the spatial distance, GIFNet introduces the group interaction graph with a graph attention module to propagate group neighbor information. With this approach, GIFNet explicitly determines the importance of each group neighbor based on the relative displacement and dynamic interaction states, enabling better quantification of the influence of group behaviors. In addition, anthropologists recognize that visual orientation is highly related to pedestrians' walking paths [159], and visual perception is conducive to forming group cohesion [160, 161]. However, most of the existing methods analyze pedestrian trajectories and visual orientation information separately. In contrast, GIFNet simultaneously encodes both types of information, which improves the prediction accuracies of the proxemics and attention fields.

GIF for Crowd-Aware Robot Navigation

With the development of unmanned systems (e.g., autonomous driving and service robots), their applications are expected to expand from isolated environments to social spaces shared with humans. Unmanned systems are expected not only to perform powerful functions but also to exhibit smart interactions, with comfort, naturalness, and sociability [163]. Our group-aware pedestrian anticipation method may lead to innovative unmanned systems that can operate in a human-like manner and comply with social norms. To validate our method, we propose a new robot navigation method based on the GIF (Fig. 6.15a). Existing robot navigation approaches usually regard pedestrians as simple circular obstacles and avoid pedestrians based on their current locations [162], which makes it difficult to prevent pedestrian disturbances while maintaining navigation efficiency. In contrast, by imparting the robot with human-like anticipation capabilities, our method can adaptively plan robots' paths according to pedestrians' proxemics field and attention field, which reflect pedestrians' behavioral intentions in a fine-grained manner. Thus, our method can effectively prevent pedestrian disturbances while maintaining the robot's driving efficiency (Fig. 6.15e–g). We anticipate that the GIF can promote harmonious human–machine relationships in broader applications.

6.5 Crowd Reconstruction with Gigapixel Image

Monocular multiperson reconstruction in large scenes, especially for hundreds of people, is beneficial for scene understanding and crowd analysis. The reconstruction of poses, shapes, and 3D locations for the people in surveillance scenes can help with recognizing actions/activities, including interactions between people based on their locations and poses, modeling crowd behavior for simulations and security monitoring, and improving individual tracking over time. However, existing human body regression methods cannot deal with large scenes containing many people. The PANDA dataset provides an opportunity to reconstruct hundreds of people with accurate global 3D locations in large scenes based on single RGB images.

To reconstruct hundreds of people in a large scene, an intuitive idea is to use single-person reconstruction methods [164–167] for cropped images containing one person. However, these methods can only recover 3D poses and shapes of people without 3D spatial location information. Some multiperson reconstruction methods [168–170] have achieved coherent reconstruction results; however, these methods use relative depth information that does not reflect the real positions. SMAP [171] can recover multiperson 3D poses with absolute depth information in small scenes. However, none of the above multiperson methods can be directly applied to large scenes due to memory limitations and the high-resolution requirements. One solution is image cropping; however, image cropping can lead to inconsistent reconstructions in large scenes due to the independent inference processes performed during the cropping operations. In general, there are two challenges in reconstructing hundreds of people coherently from a single large-scene image. Firstly, in a large scene, there are large amounts of people with a wide range of scales. Secondly, due to the depth ambiguity from a single view, it is difficult to estimate absolute 3D positions of people in the large scene.

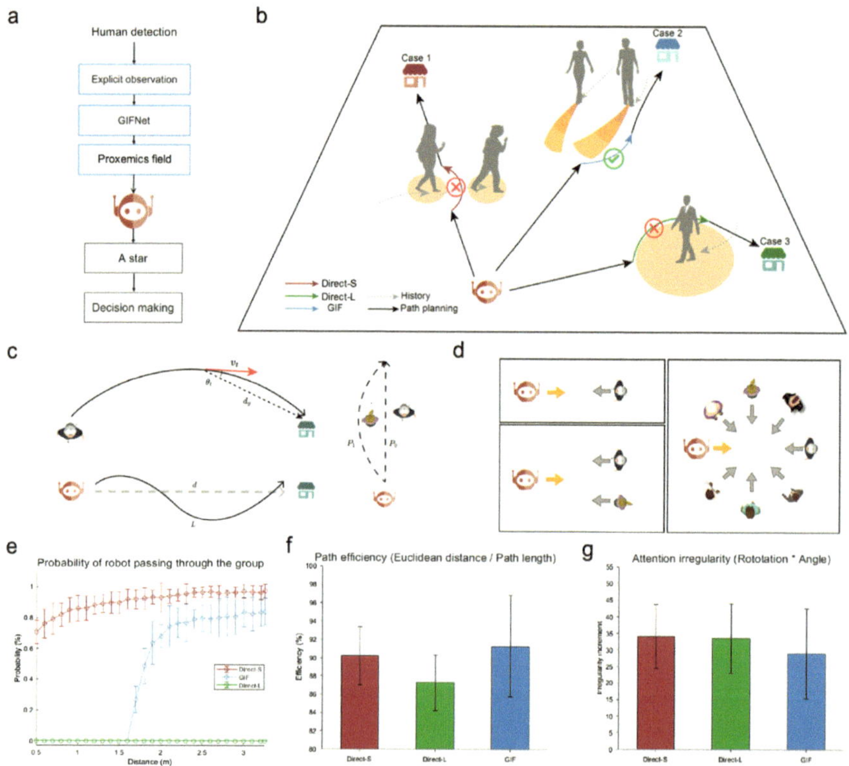

Fig. 6.15 **a**) Illustration of the pipeline of the GIF-based robot navigation method. Our method uses the GIF to guide robot path planning. **b**) Illustration of three different path planning methods in our experiments. The "Direct-S" and "Direct-L" methods are widely used robot navigation approaches that treat pedestrians as simple circular obstacles [162]. The "Direct-S" method treats pedestrians as small circular obstacles and often causes robots to disturb pedestrians. The "Direct-L" method treats pedestrians as large circular obstacles, which reduces the navigation efficiency and leads to longer paths. The "GIF" method treats the range of the predicted proxemics field as an obstacle. To ensure a fair comparison, we adopt the conventional A-star path planning method to evaluate the performance of the three methods. **c**) Three evaluation metrics are used. The robot path efficiency is the ratio between the robot's Euclidean distance d to the goal and the actual traveled distance L, which measures the robot's navigation efficiency. The attention irregularity is the sum of the pedestrians' unnecessary rotation angles (θ_t, the inner angle between the direction of the destination d_g and the direction of current speed v_t) caused by the robot, which measures the robot's disturbance to pedestrians. The probability of the robot passing through group P_0 is used to measure the disturbance of the robot to the pedestrian group. **d**) Illustration of the layout of our experiments. We evaluate the three methods in three different scenarios: one robot with one person, one robot with two people, and one robot with multiple people. **e**) The probability of the robot passing through a two-person group versus the distance between the two-person group. Our group-aware method can accurately identify groups (whose inner distance is usually less than 1.5 m) and ensure that robots do not pass through groups. **f–g**) A comparison of the path efficiency and attention irregularity of the different methods. Our GIF-based method has the highest path efficiency and the lowest attention irregularity, which suggests that our method effectively ensures that robots do not disturb pedestrians while maintaining the robot's driving efficiency. The error bars represent the standard deviation

In general, there are two challenges in reconstructing hundreds of people coherently based on a single large-scene image. First, in a large scene, there are many people with different scales. Second, due to the depth ambiguity in a single view, it is difficult to estimate the absolute 3D positions of people in large scenes.

Therefore, we propose Crowd3D, a coherent reconstruction framework for crowds in large scenes based on single images. To solve the problem of the large number of people and various human scales, we propose an adaptive human-centric cropping scheme to obtain more consistent scales for people in different cropped images by leveraging the pyramid-like changes in the scales of people in large-scene images. To obtain globally consistent spatial locations for a large number of people, we propose a progressive ground-guided position transform to convert the complex 3D crowd spatial localization information into 2D pixel localization information. Specifically, we assume that the scene has a dominant ground plane, where the people are on the ground plane and some people are standing, and estimate the ground plane equation and the parameters of the large-scene camera based on the human pose priors. With the estimated parameters, we propose a progressive ground-guided reconstruction network *Crowd3DNet* to transform the 2D local instances into global 3D spatial positions. In addition, we predict a 3D offset to refine the positions of the people. Our Crowd3D model reconstructs the 3D poses and shapes of people in crowds with global consistency based on adaptively cropped images.

6.5.1 Classical Methods for Crowd Reconstruction

There are many methods to address multiperson 3D pose estimation, which can be divided into top-down [172–175] and bottom-up [171, 176–180] paradigms. The top-down methods first detect the people and then estimate the 3D pose of each person separately. Moon et al. [172] estimated the root location and root-relative pose separately after detecting people. They regarded the area of the 2D bounding box as a prior and adopted a neural network to learn a correction factor. HMOR [175] divides human relations into three levels and formulates pairwise ordinal relations at each level. In contrast to the top-down paradigm, bottom-up methods directly detect all joints and group them. However, most methods optimize the translation via postprocessing [176] or ignore root localization. Inspired by monocular depth estimation methods, SMAP [171] directly utilizes a deep convolutional neural network (CNN) [3, 181] to estimate the root depth map and part relative depth maps. A depth-aware part association algorithm was proposed to associate detected joints with corresponding individuals using the estimated depth maps.

However, the above methods only estimate the 3D poses in the form of skeletons while lacking shape information, which is important in many applications, such as interpretation reasoning to prevent impossible poses, person reidentification, and crowd analysis.

Parametric human body models, e.g., SMPL [182], have been widely adopted to represent the 3D poses and shapes of people. Single-person 3D pose and shape estimation have been achieved with good results [164–166, 183–189], while multiperson 3D pose and shape estimation still face many challenges.

Some methods adopt two-stage frameworks by utilizing single-person reconstruction methods for each detected person. 3DCrowdNet [167] leverages 2D poses to distinguish different people and uses a joint-based regressor to estimate human model parameters. This kind of approach focuses on the accuracy of the pose and shape but ignores the 3D spatial locations of the people, which are important for obtaining a holistic understanding of the scene. To obtain coherent reconstruction results, Jiang et al. [168] proposed CRMH, an R-CNN-based architecture [10], to detect all the people in an image and estimate their SMPL parameters by using interpenetration loss and depth ordering-aware loss during training. This method calculates human depths based on the assumption that people have consistent heights, leading to incorrect depths for short individuals such as children. To solve the inherent body size and depth ambiguity problem, Ugrinovic et al. [190] proposed a multistage optimization-based method to optimize the 3D translations and scales of body meshes estimated by CRMH [168]. Different from multistage methods with redundant computations, BMP [169] is a single-stage multiperson mesh regression method that correlates the depth of a person with the features at different scales. In ROMP [191], the mesh and location information are obtained in combination with the camera map and SMPL map according to the center map. However, this method is based on the assumption of weak perspective projections and can only be used to determine the 2D locations of people in the image plane. Moreover, it uses an approximation method to obtain the depth ordering. To address this issue, BEV [170] uses bird's-eye-view representations to simultaneously determine the body centers in the image and the depth. However, all the above methods obtain relative depths rather than absolute positions, and they cannot be applied directly to large scenes.

6.5.2 *Crowd3D: Toward Hundreds of People Reconstruction from a Single Image*

Crowd3D aims to obtain a globally coherent reconstruction of hundreds of people based on a single large-scene image, as shown in Fig. 6.16.

Given a large-scene image, instead of using a uniform cropping approach, which cannot handle people of various sizes, we adopt an adaptive human-centric cropping scheme to crop the image into patches with hierarchical sizes. This approach ensures that the height ratios between people in the cropped images are as consistent as possible. Then, we estimate the intrinsic camera parameters and ground plane equation of the scene based on the human pose priors to perform the subsequent inference step. Taking the cropped images, ground plane equation, and camera parameters as inputs, we directly estimate the human meshes in the large scene in the camera coordinate system via Crowd3DNet, a progressive ground-guided reconstruction network. Finally, we remove duplicate people in overlapping adjacent patches with

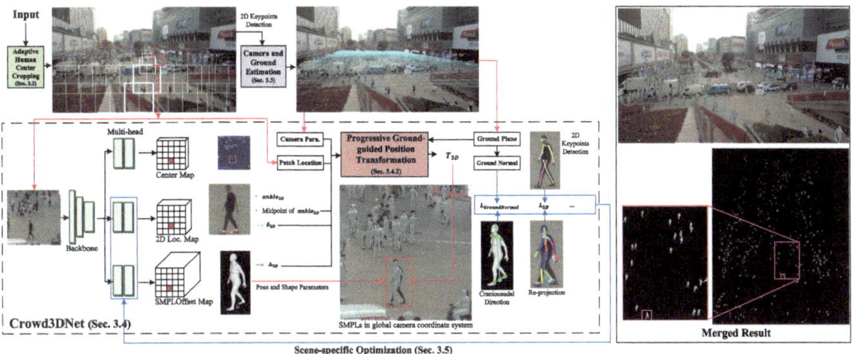

Fig. 6.16 Overview of Crowd3D framework

a merging operation. The merging operation retains the people farther away from the boundary of the overlapping region; this approach favors more complete people, thereby preventing truncation.

Adaptive Human-Centric Cropping

Unlike sliding window strategies [192, 193] that use uniform cropping schemes without considering the characteristics of people in the large-scene image, we propose an adaptive human-centric cropping strategy to ensure that people in different cropped images have appropriate scales. Our cropping method is based on the observation that human heights vary hierarchically following a pyramid-like structure in large-scene images. Our method is greedy and relies on the following assumptions: i) the vanishing line of the ground plane is horizontal and ii) the ratios between any adjacent nonoverlapping cropped images in the y-direction are approximately the same. We define the heights of the smallest and largest people and the upper and lower bounds of the processing area as $h_{\min}$, $h_{\max}$, s, and e, respectively. We use a square window and represent the size of the i-th window in the y-direction as c_i. We set the height of the people in the cropped image as half the height of the people in the cropped image; hence, we have $c_1 = 2 \times h_{\min}$, $c_n = c_1 \times q^{n-1}$ and $\sum_{i=1}^{n} c_i = e - s$, where q is the proportionality coefficient. This cropping problem is formulated as follows:

$$\arg\min_{n,q} |c_n - 2 \times h_{\max}|. \tag{6.28}$$

To ensure that each person appears completely in the cropped image, we create overlapping blocks between adjacent cropped images by combining the respective halves of the two cropped images. The cropping parameters $h_{\min}$, $h_{\max}$, s, and e are set manually or automatically.

Camera and Ground Plane Estimation

We use a ground plane as guidance, which has the following benefits: 1) the ground plane provides the global information to the local cropped images and 2) constraining the people to the ground plane allows us to better infer the positions of the different people. Crowd3D estimates a scene camera and a ground plane in the camera coordinate system based on the 2D keypoints detected by RMPE [194]. Specifically, we use a pinhole camera model with a focal length of f ($f = f_x = f_y$), where the principal point of the camera is the image center. The ground equation is represented as $N^T P_G + D = 0$, where $N \in \mathbb{R}^3$ is the ground normal with $\|N\|_2 = 1$, $P_G \in \mathbb{R}^3$ is a point on the ground plane, and D is a constant term. Please note that we only use a few standing people selected by the human pose priors to estimate the ground plane. Similar to [195], we define the midpoint of two ankle keypoints of a person as $X_b \in \mathbb{R}^3$ and the midpoint of two shoulder keypoints as $X_t \in \mathbb{R}^3$. We assume that X_b is on the ground plane, and the line from X_b to X_t is parallel to the ground normal. We also set a fixed height prior h. Different from [195], we use an optimization-based method because the large-scene image contains sufficient samples. We define the projections of X_b and X_t as $x_b = (u_b, v_b)$ and $x_t = (u_t, v_t)$, respectively. In addition, we have the image formation equation $z_b \times \bar{x}_b = K X_b$, where $\bar{x} = (u, v, 1)^T$ represents the homogeneous coordinate of $x = (u, v)$, K denotes the intrinsic parameter matrix of the scene camera, and z_b is the depth of X_b. Because X_b is on the ground plane, we have $N^T X_b + D = 0$; thus, we have

$$z_b = -\frac{D}{N^T K^{-1} \bar{x}_b}. \tag{6.29}$$

We compute the projection of the midpoint of two shoulder keypoints x_{t-p} by

$$z_t \times \bar{x}_{t-p} = z_t \times \begin{bmatrix} u_{t-p} \\ v_{t-p} \\ 1 \end{bmatrix} = K(z_b \times K^{-1}\bar{x}_b + h \times N), \tag{6.30}$$

where z_t is the depth of X_t. For the i-th person, we build the following loss:

$$L_{p_i} = \lambda_{\text{angle}} L_{\cos}(x_{t-p} - x_b, x_t - x_b) + \lambda_{\text{mod}} \frac{\left| \|x_{t-p} - x_b\|_2 - \|x_t - x_b\|_2 \right|}{\|x_t - x_b\|_2}, \tag{6.31}$$

where $L_{\cos}$ is the cosine distance and λ_{angle} and λ_{mod} are the weights of the corresponding loss terms. Finally, we translate 0.1 meters along the ground normal direction to obtain the real ground plane rather than the plane where the ankle keypoints are located.

Crowd3DNet

Our Crowd3DNet model is a progressive ground-guided reconstruction network that is designed to infer globally consistent human body meshes in large scenes based on cropped images. As shown in Fig. 6.16, Crowd3DNet is a one-stage multihead network. Given a cropped RGB image, Crowd3DNet simultaneously outputs a body center heatmap, a 2D localization map, and a SMPLOffset map. The body center heatmap is used to find people and predicts the probability that each location is the center of a human body. If the body center heatmap has a positive value, we sample relevant parameters from other maps at the corresponding center location to obtain the 2D localization information, SMPL parameters, and 3D offset of the person. With the progressive ground-guided position transform approach, Crowd3DNet combines the sampled parameters, ground plane equation, and scene camera parameters to infer the global 3D positions of people and place the body meshes in the large-scene camera coordinate system.

Progressive Ground-Guided Position Transform

To illustrate the progressive ground-guided position transform approach, we first define a unique 3D landing point for each person, which is the projection point from the center of the torso to the ground. We represent the landing point of the person in the large-scene camera coordinate system as $P = (P_x, P_y, P_z)^T \in \mathbb{R}^3$. The 2D projection points of P in the large-scene image and the local cropped image are represented as $p \in \mathbb{R}^2$ and $p_{\text{local}} \in \mathbb{R}^2$, respectively. For convenience, we denote p_{local} as the 2D localization information. Because P is on the ground plane in the large scene, based on the estimated camera parameters and ground equation, according to Eq. (6.29), we have

$$P = K^{-1}\bar{p} \times P_z, \quad \text{where } P_z = -\frac{-D}{N^T K^{-1}\bar{p}} \text{ and } p = p_{\text{local}} + t_p, \tag{6.32}$$

where $t_p \in \mathbb{R}^2$ is the pixel position coordinate of the upper left corner of the cropped image obtained in the cropping step. We use Eq. (6.32) to establish a direct mapping between the local 2D localization information of a person and their landing point in the large-scene camera coordinate system. The landing point provides an initial global position for the estimated body mesh. To refine this value, we introduce a local 3D offset Δ_{3D}, which represents the offset between the midpoint of the two ankle keypoints of the person and their landing point. Therefore, we can move the body mesh M in SMPL canonical space to the large-scene camera coordinate system with the following transform:

$$M_{\text{cam}} = M + T_{3D}, \quad \text{where } T_{3D} = -mean(J_{\text{ankles}}) + \Delta_{3D} + P, \tag{6.33}$$

where M_{cam} is the body mesh in the scene camera space, T_{3D} represents the global translation, and J_{ankles} denotes the 3D ankle joints of M. The progressive ground-guided position transform strategy converts the local 2D localization information into global 3D position information in a coarse-to-fine manner, establishing the connection between the local cropped image and the global scene.

Representations

We design our network using point-based representations [191], which have been shown to be robust to occlusions. In addition, we build a 2D localization map to estimate p_{local}, the 2D projection of the human landing point in the locally cropped image. We add the 3D offset Δ_{3D} estimation in the SMPL map to generate the SMPLOffset map.

Body Center Heatmap. The body center heatmap $\boldsymbol{C_m}$ represents the body center likelihood with a Gaussian kernel combining different body scales, where $C_m \in \mathbb{R}^{1\times H\times W}$ and $H = W = 64$.

2D Localization Map. The goal of the 2D localization map $\boldsymbol{F_m} \in \mathbb{R}^{6\times H\times W}$ is to estimate the 2D localization information of the person. Considering that the 2D localization map does not contain semantic information, we decompose the estimation into estimating semantic 2D ankle joints j_{ankle} and a 2D offset δ_{2D}. We obtain the 2D localization for a person as $p_{\text{local}} = mean(J_{\text{ankles}}) + \delta_{\text{2D}}$.

SMPLOffset Map. The SMPLOffset map $SO_m \in \mathbb{R}^{145\times H\times W}$ consists of the SMPL parameters and the 3D offset for a person. The SMPL parametric model can represent various shapes and poses with only a few parameters. The pose parameters θ and shape parameters β as used as inputs, and the model outputs a body mesh $M \in \mathbb{R}^{6890\times 3}$. Similar to ROMP, we adopt the 6D rotation representation [196] and remove the last two hand joints. Our 3D offset is the deviation between the centers of the two ankle keypoints of a person and their landing point, which is closely related to the human pose; hence, we combine this information with the SMPL parameters.

Loss Function

Crowd3DNet is supervised by the weighted sum of multiple loss items as follows:

$$\begin{aligned} L = &\lambda_{\text{center}} L_{\text{center}} + \lambda_{\text{mesh}} L_{\text{mesh}} + \lambda_{J_{\text{ankles}}} L_{J_{\text{ankles}}} + \lambda_{\delta_{\text{2D}}} L_{\delta_{\text{2D}}} \\ &+ \lambda_{\Delta_{\text{3D}}} L_{\Delta_{\text{3D}}} + \lambda_{\text{root}} L_{\text{root}} + \lambda_{\text{out}} L_{\text{out}}. \end{aligned} \tag{6.34}$$

Following previous works [170, 191], L_{center} and L_{mesh} are the 2D focal loss [12] and the standard mesh loss, respectively. $L_{J_{\text{ankles}}}$, $L_{\delta_{\text{2D}}}$, $L_{\Delta_{\text{3D}}}$, and L_{root} are all L_2 losses, which are used to supervise the 2D ankle keypoints, 2D offset, 3D local translation, and absolute root position, respectively. We also propose an out-of-bound loss to prevent people in large scenes from penetrating into the ground. Furthermore, we use the L_1 loss to punish the point with the most serious penetration into the ground plane, and our out-of-bound loss is defined as

$$L_{\text{out}} = |min(\{\bar{v}_i \cdot G \mid \bar{v}_i \cdot G < 0\})|, \quad \text{where } v_i \in M_{\text{cam}}, \; G = [N^T, D]^T. \tag{6.35}$$

Scene-Specific Optimization

To generalize to various camera and ground plane parameters, we design a scene-specific optimization approach, which allows Crowd3DNet to implicitly learn the mechanism for guiding the ground plane based on the craniocaudal directions of people in new scenes during the test. Please note that the scene-specific optimization is performed only once for a fixed camera scene, i.e., only one image of the scene is needed. Specifically, given a new scene during the test, we optimize a small set of weights in the head layer of Crowd3DNet based on the normal of the ground plane and the 2D poses estimated in the camera and ground plane estimation stages. The optimization loss L_{opt} is defined as

$$L_{\text{opt}} = \lambda_{\text{GN}} L_{\text{GN}} + \lambda_{J_{\text{ankles}}} L'_{J_{\text{ankles}}} + \lambda_{\text{mesh}} L'_{\text{mesh}} + \lambda_{\text{out}} L_{\text{out}}, \tag{6.36}$$

where $L_{\text{GN}} = L_{\cos}(X_t - X_b, N_G)$ is the ground plane normal loss, in which $N_G = (x_{N_G}, y_{N_G}, z_{N_G}) \in \mathbb{R}^3$ is the ground plane normal, with $y_{N_G} > 0$, and $X_t - X_b$ is the craniocaudal direction of the human. Different from Eq. (6.34), $L'_{J_{\text{ankles}}}$ and L'_{mesh} are supervised by the predicted 2D poses rather than the ground truth data.

Crowd Reconstruction Performance with Gigapixel Image

Evaluation Metrics

In addition to the mean per joint position error (MPJPE) metric used for 3D pose evaluation, we use the root error (RtError) [171], percentage of the correct ordinal depth (PCOD) [171], and pairwise percentage distance estimation error (PPDError) [197] to evaluate the accuracy of the spatial locations. RtError is used to evaluate the Euclidean distance between the predicted and ground truth root locations. PCOD is used to evaluate the ordinal depth relations between all pairs of people in the image. PPDError is the percentage error of the physical distance between all pairs of people in the scene. For the datasets that lack 3D pose annotations, we use the object keypoint similarity (OKS) metric to evaluate the accuracy of the projected 2D poses.

Comparison of Large Scene

We evaluate the performance of the proposed method based on the large-scene dataset *Crowd-Location* and compare the results with those of four state-of-the-art methods: SMAP [171], CRMH [168], ROMP [191], and BEV [170]. Note that none of these methods can directly handle large-scene images, and hence we use our adaptive human-centric cropping strategy to obtain the hierarchically cropped images to use as inputs. To obtain the reconstruction results of these methods based on the large-scene camera system, we first restore their 2D projections based on the large-scene image by using the position coordinates of the cropped images and provide the scene camera parameters estimated by our method. Following BEV [170], we directly

Table 6.10 Comparisons to the state-of-the-art methods on *Crowd-Location*

Method	S1_scene				S2_scene			
	Matched	PCOD	PPDError	OKS	Matched	PCOD	PPDError	OKS
SMAP [171]	96.2	93.4	728.6	95.4	68.0	91.7	807.8	**90.5**
CRMH [168]	98.9	91.1	29.8	94.0	97.1	84.5	49.4	83.7
ROMP [191]	**99.6**	93.2	22.5	**97.5**	97.2	89.3	39.4	**90.5**
BEV [170]	99.5	92.3	27.7	96.8	97.3	88.0	39.5	88.1
Ours	99.4	**98.1**	**4.5**	94.7	**97.4**	**98.6**	**7.7**	83.8

compute the 3D positions of the results of ROMP [191] by solving the PnP algorithm (RANSAC [198]). For CRMH [168], we use its bounding box to depth transform to infer the predicted depths in large scenes. SMAP [171] and BEV [170] both infer the depths through perspective camera models; however, they use only the cropped images. Following a previous method [199], we scale the depths of their results according to the focal length of our scene camera. We fine-tuned the comparison methods with *PANDA-Pose* and selected the best results (before or after fine-tuning) for each method. Table 6.10 shows the quantitative results. Our method outperforms all the state-of-the-art methods based on the PCOD and PPDError metrics, which demonstrates that our method not only obtains better depths but also better estimates the precise physical distances between people in large scenes. Figure 6.17 shows qualitative comparisons with CRMH [168] and BEV [170]. The complete bird's-eye view on the right shows that our predicted crowd distribution is consistent with the upper right scene image, while the compared methods are not consistent with this image. Taking persons with numbers as an example, only our method recovers correct relative positions. For the BEV method [170], the person in the orange bounding box is estimated to be larger than the ground truth, which is associated with a smaller depth (as shown in the middle column). CRMH [168] predicts reasonable relative positions for the persons indicated with different colors; this method is enhanced by our global camera predictions.

Comparison of Small Scene

We also evaluate our method based on the small-scene dataset *Panoptic* and compare the results with those of state-of-the-art methods. We do not test our method based on the Mafia dataset because the 3D ground truth data could not be obtained from the official website. For the SMAP model [171], we use the model provided by the authors that was not trained based on the *Panoptic* dataset for a fair comparison. We use small-scene images directly as inputs to our Crowd3DNet model and compare the results with those of the state-of-the-art methods. Because of the small number of people in the small scene, our method cannot predict the ground plane well based on only a single image. Fortunately, *Panoptic* is a video dataset, and the data contain

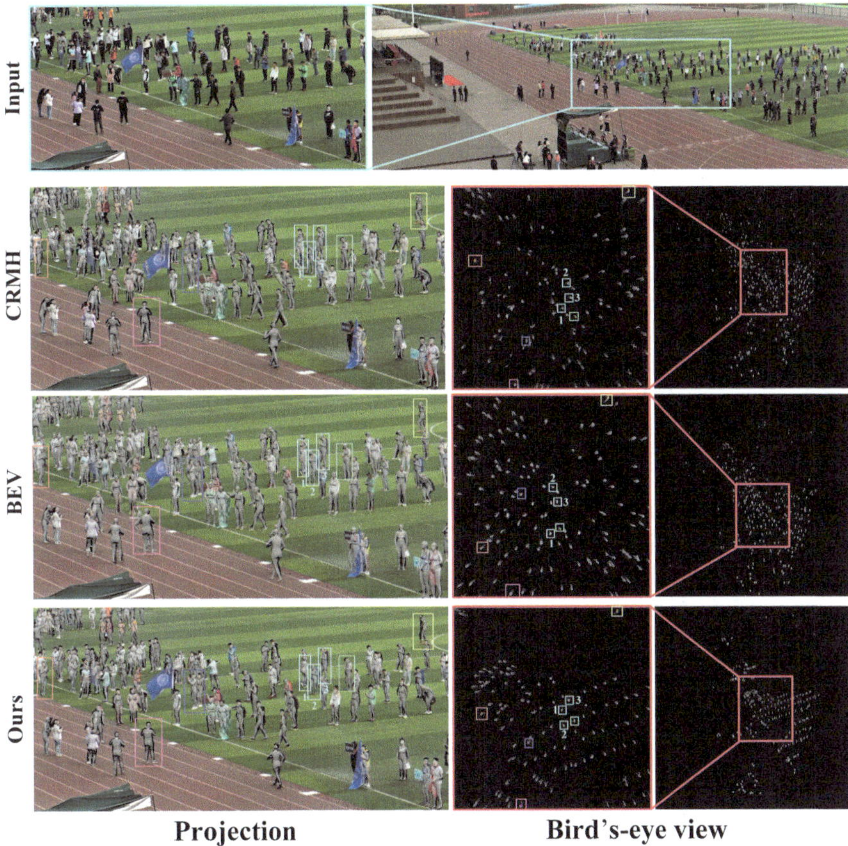

Fig. 6.17 Qualitative results based on the *Crowd-Location* dataset. The bounding boxes with the same color or number indicate corresponding persons

many standing people. Hence, we consider the people in the same scene at different times to predict the ground plane, and we also provide the results predicted using the ground truth ground plane. As shown in Table 6.11, our method achieves the first or second best results for both position inference and pose estimation, validating the effectiveness of our method for small scenes.

Table 6.11 Comparisons to the state-of-the-art methods on *Panoptic*

	Method	Haggling	Mafia	Ultim.	Pizza	Mean
MPJPE ↓	SMAP [171]	128.5	—	141.2	236.4	168.7
	CRMH [168]	129.6	133.5	153.0	156.7	143.2
	BMP [169]	120.4	132.7	140.9	147.5	135.4
	ROMP [191]	111.8	**129.0**	148.5	149.1	134.6
	3DCrowdNet [167]	109.6	135.9	129.8	135.6	127.6
	BEV [170]	**78.6**	—	**97.1**	116.2	**97.3**
	Our(Pre-ground)	107.2	—	114.1	**112.9**	111.4
	Our(GT-ground)	98.7	—	113.0	113.3	108.3
RtError ↓	SMAP [171]	432.8	—	529.8	1297.6	753.4
	CRMH [168]	2384.5	—	2301.0	2418.7	2368.1
	BEV [170]	791.2	—	672.6	847.8	770.5
	Our(Pre-ground)	275.3	—	429.6	351.6	352.2
	Our(GT-ground)	**166.9**	—	**330.1**	**240.1**	**245.7**
PCOD ↑	SMAP [171]	83.5	—	93.3	76.0	84.3
	CRMH [168]	**89.5**	—	93.2	74.8	85.8
	BEV [170]	88.4	—	**98.3**	**92.5**	**93.1**
	Our(Pre-ground)	86.9	—	92.5	88.8	89.4
	Our(GT-ground)	87.6	—	93.8	88.9	90.1
PPDError ↓	SMAP [171]	33.3	—	51.1	92.5	59.0
	CRMH [168]	89.0	—	43.2	84.2	72.1
	BEV [170]	27.7	—	**10.0**	32.7	23.5
	Our(Pre-ground)	18.4	—	18.2	27.3	21.3
	Our(GT-ground)	**16.3**	—	16.8	**24.5**	**19.2**

References

1. Olga Russakovsky, Jia Deng, Hao Su, Jonathan Krause, Sanjeev Satheesh, Sean Ma, Zhiheng Huang, Andrej Karpathy, Aditya Khosla, Michael Bernstein, et al. Imagenet large scale visual recognition challenge. *International journal of computer vision*, 115(3):211–252, 2015.
2. Antonio Torralba, Rob Fergus, and William T Freeman. 80 million tiny images: A large data set for nonparametric object and scene recognition. *IEEE transactions on pattern analysis and machine intelligence*, 30(11):1958–1970, 2008.
3. Kaiming He, Xiangyu Zhang, Shaoqing Ren, and Jian Sun. Deep residual learning for image recognition. In *Proceedings of the IEEE conference on computer vision and pattern recognition*, pages 770–778, 2016.
4. Alex Krizhevsky, Ilya Sutskever, and Geoffrey E Hinton. Imagenet classification with deep convolutional neural networks. *Communications of the ACM*, 60(6):84–90, 2017.
5. Mark Everingham, Luc Van Gool, Christopher KI Williams, John Winn, and Andrew Zisserman. The pascal visual object classes (voc) challenge. *International journal of computer vision*, 88(2):303–338, 2010.
6. Tsung-Yi Lin, Michael Maire, Serge Belongie, James Hays, Pietro Perona, Deva Ramanan, Piotr Dollár, and C Lawrence Zitnick. Microsoft coco: Common objects in context. In *European conference on computer vision*, pages 740–755. Springer, 2014.

7. Gary B Huang, Marwan Mattar, Tamara Berg, and Eric Learned-Miller. Labeled faces in the wild: A database forstudying face recognition in unconstrained environments. In *Workshop on faces in'Real-Life'Images: detection, alignment, and recognition*, 2008.
8. Piotr Dollar, Christian Wojek, Bernt Schiele, and Pietro Perona. Pedestrian detection: An evaluation of the state of the art. *IEEE transactions on pattern analysis and machine intelligence*, 34(4):743–761, 2011.
9. Anton Milan, Laura Leal-Taixé, Ian Reid, Stefan Roth, and Konrad Schindler. Mot16: A benchmark for multi-object tracking. *arXiv preprint*arXiv:1603.00831, 2016.
10. Shaoqing Ren, Kaiming He, Ross Girshick, and Jian Sun. Faster r-cnn: Towards real-time object detection with region proposal networks. *Advances in neural information processing systems*, 28, 2015.
11. Zhaowei Cai and Nuno Vasconcelos. Cascade r-cnn: Delving into high quality object detection. In *Proceedings of the IEEE conference on computer vision and pattern recognition*, pages 6154–6162, 2018.
12. Tsung-Yi Lin, Priya Goyal, Ross Girshick, Kaiming He, and Piotr Dollár. Focal loss for dense object detection. In *Proceedings of the IEEE international conference on computer vision*, pages 2980–2988, 2017.
13. Nicolai Wojke, Alex Bewley, and Dietrich Paulus. Simple online and realtime tracking with a deep association metric. In *2017 IEEE international conference on image processing (ICIP)*, pages 3645–3649. IEEE, 2017.
14. Long Chen, Haizhou Ai, Zijie Zhuang, and Chong Shang. Real-time multiple people tracking with deeply learned candidate selection and person re-identification. In *2018 IEEE international conference on multimedia and expo (ICME)*, pages 1–6. IEEE, 2018.
15. ShiJie Sun, Naveed Akhtar, HuanSheng Song, Ajmal Mian, and Mubarak Shah. Deep affinity network for multiple object tracking. *IEEE transactions on pattern analysis and machine intelligence*, 43(1):104–119, 2019a.
16. Francesco Zanlungo, Dražen Brščić, and Takayuki Kanda. Pedestrian group behaviour analysis under different density conditions. *Transportation Research Procedia*, 2:149–158, 2014.
17. Marco Cristani, Loris Bazzani, Giulia Paggetti, Andrea Fossati, Diego Tosato, Alessio Del Bue, Gloria Menegaz, and Vittorio Murino. Social interaction discovery by statistical analysis of f-formations. In *BMVC*, volume 2, pages 10–5244. Citeseer, 2011.
18. James Ferryman and Ali Shahrokni. Pets2009: Dataset and challenge. In *2009 Twelfth IEEE international workshop on performance evaluation of tracking and surveillance*, pages 1–6. IEEE, 2009.
19. Ergys Ristani, Francesco Solera, Roger Zou, Rita Cucchiara, and Carlo Tomasi. Performance measures and a data set for multi-target, multi-camera tracking. In *European conference on computer vision*, pages 17–35. Springer, 2016.
20. Scott Blunsden and RB Fisher. The behave video dataset: ground truthed video for multi-person behavior classification. *Annals of the BMVA*, 4(1-12):4, 2010.
21. Alon Lerner, Yiorgos Chrysanthou, and Dani Lischinski. Crowds by example. In *Computer graphics forum*, volume 26, pages 655–664. Wiley Online Library, 2007.
22. Stefano Pellegrini, Andreas Ess, Konrad Schindler, and Luc Van Gool. You'll never walk alone: Modeling social behavior for multi-target tracking. In *2009 IEEE 12th international conference on computer vision*, pages 261–268. IEEE, 2009.
23. Wongun Choi, Yu-Wei Chao, Caroline Pantofaru, and Silvio Savarese. Discovering groups of people in images. In *European conference on computer vision*, pages 417–433. Springer, 2014.
24. Loris Bazzani, Marco Cristani, Diego Tosato, Michela Farenzena, Giulia Paggetti, Gloria Menegaz, and Vittorio Murino. Social interactions by visual focus of attention in a three-dimensional environment. *Expert Systems*, 30(2):115–127, 2013.
25. Gloria Zen, Bruno Lepri, Elisa Ricci, and Oswald Lanz. Space speaks: towards socially and personality aware visual surveillance. In *Proceedings of the 1st ACM international workshop on Multimodal pervasive video analysis*, pages 37–42, 2010.

26. Xavier Alameda-Pineda, Jacopo Staiano, Ramanathan Subramanian, Ligia Batrinca, Elisa Ricci, Bruno Lepri, Oswald Lanz, and Nicu Sebe. Salsa: A novel dataset for multimodal group behavior analysis. *IEEE transactions on pattern analysis and machine intelligence*, 38(8):1707–1720, 2015.
27. Hanbyul Joo, Hao Liu, Lei Tan, Lin Gui, Bart Nabbe, Iain Matthews, Takeo Kanade, Shohei Nobuhara, and Yaser Sheikh. Panoptic studio: A massively multiview system for social motion capture. In *Proceedings of the IEEE International Conference on Computer Vision*, pages 3334–3342, 2015.
28. Mostafa S Ibrahim, Srikanth Muralidharan, Zhiwei Deng, Arash Vahdat, and Greg Mori. A hierarchical deep temporal model for group activity recognition. In *Proceedings of the IEEE conference on computer vision and pattern recognition*, pages 1971–1980, 2016.
29. Wongun Choi, Khuram Shahid, and Silvio Savarese. What are they doing?: Collective activity classification using spatio-temporal relationship among people. In *2009 IEEE 12th international conference on computer vision workshops, ICCV Workshops*, pages 1282–1289. IEEE, 2009.
30. Thor List, Jos Bins, Jose Vazquez, and Robert B Fisher. Performance evaluating the evaluator. In *2005 IEEE International Workshop on Visual Surveillance and Performance Evaluation of Tracking and Surveillance*, pages 129–136. IEEE, 2005.
31. Sangmin Oh, Anthony Hoogs, Amitha Perera, Naresh Cuntoor, Chia-Chih Chen, Jong Taek Lee, Saurajit Mukherjee, JK Aggarwal, Hyungtae Lee, Larry Davis, et al. A large-scale benchmark dataset for event recognition in surveillance video. In *CVPR 2011*, pages 3153–3160. IEEE, 2011.
32. David J Brady, Michael E Gehm, Ronald A Stack, Daniel L Marks, David S Kittle, Dathon R Golish, EM Vera, and Steven D Feller. Multiscale gigapixel photography. *Nature*, 486(7403):386–389, 2012.
33. Xiaoyun Yuan, Lu Fang, Qionghai Dai, David J Brady, and Yebin Liu. Multiscale gigapixel video: A cross resolution image matching and warping approach. In *2017 IEEE International Conference on Computational Photography (ICCP)*, pages 1–9. IEEE, 2017.
34. Shanshan Zhang, Rodrigo Benenson, and Bernt Schiele. Citypersons: A diverse dataset for pedestrian detection. In *Proceedings of the IEEE conference on computer vision and pattern recognition*, pages 3213–3221, 2017.
35. Andreas Geiger, Philip Lenz, and Raquel Urtasun. Are we ready for autonomous driving? the kitti vision benchmark suite. In *2012 IEEE conference on computer vision and pattern recognition*, pages 3354–3361. IEEE, 2012.
36. Patrick Dendorfer, Hamid Rezatofighi, Anton Milan, Javen Shi, Daniel Cremers, Ian Reid, Stefan Roth, Konrad Schindler, and Laura Leal-Taixe. Cvpr19 tracking and detection challenge: How crowded can it get? *arXiv preprint*arXiv:1906.04567, 2019.
37. Alessandro Vinciarelli, Maja Pantic, and Hervé Bourlard. Social signal processing: Survey of an emerging domain. *Image and vision computing*, 27(12):1743–1759, 2009.
38. Marshall P Duke and Stephen Nowicki. A new measure and social-learning model for interpersonal distance. *Journal of Experimental Research in Personality*, 1972.
39. Jifeng Dai, Yi Li, Kaiming He, and Jian Sun. R-fcn: Object detection via region-based fully convolutional networks. *Advances in neural information processing systems*, 29, 2016.
40. Joseph Redmon, Santosh Divvala, Ross Girshick, and Ali Farhadi. You only look once: Unified, real-time object detection. In *Proceedings of the IEEE conference on computer vision and pattern recognition*, pages 779–788, 2016.
41. Alexey Bochkovskiy, Chien-Yao Wang, and Hong-Yuan Mark Liao. Yolov4: Optimal speed and accuracy of object detection. *arXiv preprint*arXiv:2004.10934, 2020.
42. Zheng Ge, Songtao Liu, Feng Wang, Zeming Li, and Jian Sun. Yolox: Exceeding yolo series in 2021. *arXiv preprint*arXiv:2107.08430, 2021.
43. Wei Liu, Dragomir Anguelov, Dumitru Erhan, Christian Szegedy, Scott Reed, Cheng-Yang Fu, and Alexander C Berg. Ssd: Single shot multibox detector. In *European conference on computer vision*, pages 21–37. Springer, 2016.

44. Bichen Wu, Forrest Iandola, Peter H Jin, and Kurt Keutzer. Squeezedet: Unified, small, low power fully convolutional neural networks for real-time object detection for autonomous driving. In *Proceedings of the IEEE conference on computer vision and pattern recognition workshops*, pages 129–137, 2017.
45. Zhi Tian, Chunhua Shen, Hao Chen, and Tong He. Fcos: Fully convolutional one-stage object detection. In *Proceedings of the IEEE/CVF international conference on computer vision*, pages 9627–9636, 2019.
46. Kaiwen Duan, Song Bai, Lingxi Xie, Honggang Qi, Qingming Huang, and Qi Tian. Centernet: Keypoint triplets for object detection. In *Proceedings of the IEEE/CVF international conference on computer vision*, pages 6569–6578, 2019.
47. Hei Law and Jia Deng. Cornernet: Detecting objects as paired keypoints. In *Proceedings of the European conference on computer vision (ECCV)*, pages 734–750, 2018.
48. Marco Cristani, Ramachandra Raghavendra, Alessio Del Bue, and Vittorio Murino. Human behavior analysis in video surveillance: A social signal processing perspective. *Neurocomputing*, 100:86–97, 2013.
49. Yi Liu, Dingwen Zhang, Qiang Zhang, and Jungong Han. Part-object relational visual saliency. *IEEE Transactions on Pattern Analysis and Machine Intelligence*, 2021a.
50. Xueyang Wang, Xiya Zhang, Yinheng Zhu, Yuchen Guo, Xiaoyun Yuan, Liuyu Xiang, Zerun Wang, Guiguang Ding, David Brady, Qionghai Dai, et al. Panda: A gigapixel-level human-centric video dataset. In *Proceedings of the IEEE/CVF conference on computer vision and pattern recognition*, pages 3268–3278, 2020a.
51. Qiang Gao, Fan Zhou, Kunpeng Zhang, Goce Trajcevski, Xucheng Luo, and Fengli Zhang. Identifying human mobility via trajectory embeddings. In *IJCAI*, volume 17, pages 1689–1695, 2017.
52. John Co-Reyes, YuXuan Liu, Abhishek Gupta, Benjamin Eysenbach, Pieter Abbeel, and Sergey Levine. Self-consistent trajectory autoencoder: Hierarchical reinforcement learning with trajectory embeddings. In *International conference on machine learning*, pages 1009–1018. PMLR, 2018.
53. Rohan Chandra, Uttaran Bhattacharya, Aniket Bera, and Dinesh Manocha. Traphic: Trajectory prediction in dense and heterogeneous traffic using weighted interactions. In *Proceedings of the IEEE/CVF Conference on Computer Vision and Pattern Recognition*, pages 8483–8492, 2019.
54. Chiho Choi and Behzad Dariush. Learning to infer relations for future trajectory forecast. In *Proceedings of the IEEE/CVF Conference on Computer Vision and Pattern Recognition Workshops*, pages 0–0, 2019.
55. Tianmin Shu, Sinisa Todorovic, and Song-Chun Zhu. Cern: confidence-energy recurrent network for group activity recognition. In *Proceedings of the IEEE conference on computer vision and pattern recognition*, pages 5523–5531, 2017.
56. Christoph Feichtenhofer, Haoqi Fan, Jitendra Malik, and Kaiming He. Slowfast networks for video recognition. In *Proceedings of the IEEE/CVF international conference on computer vision*, pages 6202–6211, 2019.
57. Yanyu Xu, Zhixin Piao, and Shenghua Gao. Encoding crowd interaction with deep neural network for pedestrian trajectory prediction. In *Proceedings of the IEEE conference on computer vision and pattern recognition*, pages 5275–5284, 2018.
58. Limin Wang, Yu Qiao, and Xiaoou Tang. Action recognition with trajectory-pooled deep-convolutional descriptors. In *Proceedings of the IEEE conference on computer vision and pattern recognition*, pages 4305–4314, 2015.
59. Dirk Brockmann and Theo Geisel. The ecology of gaze shifts. *Neurocomputing*, 32:643–650, 2000.
60. Alexander Neubeck and Luc Van Gool. Efficient non-maximum suppression. In *18th International Conference on Pattern Recognition (ICPR'06)*, volume 3, pages 850–855. IEEE, 2006.
61. Kai Chen, Zerun Wang, Xueyang Wang, Dahan Gong, Longlong Yu, Yuchen Guo, and Guiguang Ding. Towards real-time object detection in gigapixel-level video. *Neurocomputing*, 477:14–24, 2022.

62. Mahyar Najibi, Bharat Singh, and Larry S Davis. Autofocus: Efficient multi-scale inference. In *Proceedings of the IEEE/CVF international conference on computer vision*, pages 9745–9755, 2019.
63. Mingfei Gao, Ruichi Yu, Ang Li, Vlad I Morariu, and Larry S Davis. Dynamic zoom-in network for fast object detection in large images. In *Proceedings of the IEEE conference on computer vision and pattern recognition*, pages 6926–6935, 2018.
64. Hongyang Li, Yu Liu, Wanli Ouyang, and Xiaogang Wang. Zoom out-and-in network with map attention decision for region proposal and object detection. *International Journal of Computer Vision*, 127(3):225–238, 2019.
65. Wuyang Chen, Ziyu Jiang, Zhangyang Wang, Kexin Cui, and Xiaoning Qian. Collaborative global-local networks for memory-efficient segmentation of ultra-high resolution images. In *Proceedings of the IEEE/CVF Conference on Computer Vision and Pattern Recognition*, pages 8924–8933, 2019.
66. Alex Kendall, Vijay Badrinarayanan, and Roberto Cipolla. Bayesian segnet: Model uncertainty in deep convolutional encoder-decoder architectures for scene understanding. *arXiv preprint*arXiv:1511.02680, 2015.
67. Alex Kendall and Yarin Gal. What uncertainties do we need in bayesian deep learning for computer vision? *Advances in neural information processing systems*, 30, 2017.
68. Kensho Hara, Hirokatsu Kataoka, and Yutaka Satoh. Can spatiotemporal 3d cnns retrace the history of 2d cnns and imagenet? In *Proceedings of the IEEE conference on Computer Vision and Pattern Recognition*, pages 6546–6555, 2018.
69. Xiaohang Zhan, Ziwei Liu, Junjie Yan, Dahua Lin, and Chen Change Loy. Consensus-driven propagation in massive unlabeled data for face recognition. In *proceedings of the European Conference on Computer Vision (ECCV)*, pages 568–583, 2018.
70. Erik Bochinski, Volker Eiselein, and Thomas Sikora. High-speed tracking-by-detection without using image information. In *2017 14th IEEE international conference on advanced video and signal based surveillance (AVSS)*, pages 1–6. IEEE, 2017.
71. Peng Chu and Haibin Ling. Famnet: Joint learning of feature, affinity and multi-dimensional assignment for online multiple object tracking. In *Proceedings of the IEEE/CVF International Conference on Computer Vision*, pages 6172–6181, 2019.
72. Weitao Feng, Zhihao Hu, Wei Wu, Junjie Yan, and Wanli Ouyang. Multi-object tracking with multiple cues and switcher-aware classification. *arXiv preprint*arXiv:1901.06129, 2019.
73. Mohamed A Naiel, M Omair Ahmad, MNS Swamy, Jongwoo Lim, and Ming-Hsuan Yang. Online multi-object tracking via robust collaborative model and sample selection. *Computer Vision and Image Understanding*, 154:94–107, 2017.
74. Fengwei Yu, Wenbo Li, Quanquan Li, Yu Liu, Xiaohua Shi, and Junjie Yan. Poi: Multiple object tracking with high performance detection and appearance feature. In *European Conference on Computer Vision*, pages 36–42. Springer, 2016.
75. Joseph Redmon and Ali Farhadi. Yolov3: An incremental improvement. *arXiv preprint*arXiv:1804.02767, 2018.
76. Mingxing Tan, Ruoming Pang, and Quoc V Le. Efficientdet: Scalable and efficient object detection. In *Proceedings of the IEEE/CVF conference on computer vision and pattern recognition*, pages 10781–10790, 2020.
77. Chao Liang, Zhipeng Zhang, Xue Zhou, Bing Li, Shuyuan Zhu, and Weiming Hu. Rethinking the competition between detection and reid in multiobject tracking. *IEEE Transactions on Image Processing*, 31:3182–3196, 2022.
78. Zhichao Lu, Vivek Rathod, Ronny Votel, and Jonathan Huang. Retinatrack: Online single stage joint detection and tracking. In *Proceedings of the IEEE/CVF conference on computer vision and pattern recognition*, pages 14668–14678, 2020.
79. Jinlong Peng, Changan Wang, Fangbin Wan, Yang Wu, Yabiao Wang, Ying Tai, Chengjie Wang, Jilin Li, Feiyue Huang, and Yanwei Fu. Chained-tracker: Chaining paired attentive regression results for end-to-end joint multiple-object detection and tracking. In *European conference on computer vision*, pages 145–161. Springer, 2020a.

80. Qiang Wang, Yun Zheng, Pan Pan, and Yinghui Xu. Multiple object tracking with correlation learning. In *Proceedings of the IEEE/CVF Conference on Computer Vision and Pattern Recognition*, pages 3876–3886, 2021a.
81. Yongxin Wang, Kris Kitani, and Xinshuo Weng. Joint object detection and multi-object tracking with graph neural networks. In *2021 IEEE International Conference on Robotics and Automation (ICRA)*, pages 13708–13715. IEEE, 2021b.
82. Yihong Xu, Aljosa Osep, Yutong Ban, Radu Horaud, Laura Leal-Taixé, and Xavier Alameda-Pineda. How to train your deep multi-object tracker. In *Proceedings of the IEEE/CVF Conference on Computer Vision and Pattern Recognition*, pages 6787–6796, 2020.
83. En Yu, Zhuoling Li, Shoudong Han, and Hongwei Wang. Relationtrack: Relation-aware multiple object tracking with decoupled representation. *IEEE Transactions on Multimedia*, 2022.
84. Yunhao Du, Junfeng Wan, Yanyun Zhao, Binyu Zhang, Zhihang Tong, and Junhao Dong. Giaotracker: A comprehensive framework for mcmot with global information and optimizing strategies in visdrone 2021. In *Proceedings of the IEEE/CVF International Conference on Computer Vision*, pages 2809–2819, 2021.
85. Jinlong Peng, Tao Wang, Weiyao Lin, Jian Wang, John See, Shilei Wen, and Erui Ding. Tpm: Multiple object tracking with tracklet-plane matching. *Pattern Recognition*, 107:107480, 2020b.
86. Bing Wang, Gang Wang, Kap Luk Chan, and Li Wang. Tracklet association by online target-specific metric learning and coherent dynamics estimation. *IEEE transactions on pattern analysis and machine intelligence*, 39(3):589–602, 2016.
87. Gaoang Wang, Yizhou Wang, Haotian Zhang, Renshu Gu, and Jenq-Neng Hwang. Exploit the connectivity: Multi-object tracking with trackletnet. In *Proceedings of the 27th ACM International Conference on Multimedia*, pages 482–490, 2019.
88. Fan Yang, Xin Chang, Sakriani Sakti, Yang Wu, and Satoshi Nakamura. Remot: A model-agnostic refinement for multiple object tracking. *Image and Vision Computing*, 106:104091, 2021.
89. Fan Yang, Xin Chang, Chenyu Dang, Ziqiang Zheng, Sakriani Sakti, Satoshi Nakamura, and Yang Wu. Remots: Self-supervised refining multi-object tracking and segmentation. *arXiv preprint*arXiv:2007.03200, 2020.
90. Jiyang Gao and Ram Nevatia. Revisiting temporal modeling for video-based person reid. *arXiv preprint*arXiv:1805.02104, 2018.
91. Laura Leal-Taixé, Anton Milan, Ian Reid, Stefan Roth, and Konrad Schindler. Motchallenge 2015: Towards a benchmark for multi-target tracking. *arXiv preprint*arXiv:1504.01942, 2015.
92. Lin Cheng, Ragamayi Yarlagadda, Clinton Fookes, and Prasad Yarlagadda. A review of pedestrian group dynamics and methodologies in modelling pedestrian group behaviours. *World journal of mechanical engineering*, 1(1):1–13, 2014.
93. Zeynep Yücel, Francesco Zanlungo, and Masahiro Shiomi. Modeling the impact of interaction on pedestrian group motion. *Advanced Robotics*, 32(3):137–147, 2018.
94. Amir Rasouli, Iuliia Kotseruba, and John K Tsotsos. Pedestrian action anticipation using contextual feature fusion in stacked rnns. *arXiv preprint*arXiv:2005.06582, 2020.
95. Yuanfu Luo, Panpan Cai, Aniket Bera, David Hsu, Wee Sun Lee, and Dinesh Manocha. Porca: Modeling and planning for autonomous driving among many pedestrians. *IEEE Robotics and Automation Letters*, 3(4):3418–3425, 2018.
96. Alexandre Robicquet, Amir Sadeghian, Alexandre Alahi, and Silvio Savarese. Learning social etiquette: Human trajectory understanding in crowded scenes. In *European conference on computer vision*, pages 549–565. Springer, 2016.
97. Julian Bock, Robert Krajewski, Tobias Moers, Steffen Runde, Lennart Vater, and Lutz Eckstein. The ind dataset: A drone dataset of naturalistic road user trajectories at german intersections. In *2020 IEEE Intelligent Vehicles Symposium (IV)*, pages 1929–1934. IEEE, 2020.
98. Alexandre Alahi, Kratarth Goel, Vignesh Ramanathan, Alexandre Robicquet, Li Fei-Fei, and Silvio Savarese. Social lstm: Human trajectory prediction in crowded spaces. In *Proceedings of the IEEE conference on computer vision and pattern recognition*, pages 961–971, 2016.

99. Agrim Gupta, Justin Johnson, Li Fei-Fei, Silvio Savarese, and Alexandre Alahi. Social gan: Socially acceptable trajectories with generative adversarial networks. In *Proceedings of the IEEE conference on computer vision and pattern recognition*, pages 2255–2264, 2018.
100. Abduallah Mohamed, Kun Qian, Mohamed Elhoseiny, and Christian Claudel. Social-stgcnn: A social spatio-temporal graph convolutional neural network for human trajectory prediction. In *Proceedings of the IEEE/CVF Conference on Computer Vision and Pattern Recognition*, pages 14424–14432, 2020.
101. Yingfan Huang, Huikun Bi, Zhaoxin Li, Tianlu Mao, and Zhaoqi Wang. Stgat: Modeling spatial-temporal interactions for human trajectory prediction. In *Proceedings of the IEEE/CVF International Conference on Computer Vision*, pages 6272–6281, 2019.
102. Hao Zhou, Dongchun Ren, Huaxia Xia, Mingyu Fan, Xu Yang, and Hai Huang. Ast-gnn: An attention-based spatio-temporal graph neural network for interaction-aware pedestrian trajectory prediction. *Neurocomputing*, 445:298–308, 2021a.
103. Chenxin Xu, Weibo Mao, Wenjun Zhang, and Siheng Chen. Remember intentions: Retrospective-memory-based trajectory prediction. In *Proceedings of the IEEE/CVF Conference on Computer Vision and Pattern Recognition*, pages 6488–6497, 2022.
104. Dirk Helbing and Peter Molnar. Social force model for pedestrian dynamics. *Physical review E*, 51(5):4282, 1995.
105. Jur van den Berg, Stephen J Guy, Ming Lin, and Dinesh Manocha. Reciprocal n-body collision avoidance. In *Robotics research*, pages 3–19. Springer, 2011.
106. Kota Yamaguchi, Alexander C Berg, Luis E Ortiz, and Tamara L Berg. Who are you with and where are you going? In *CVPR 2011*, pages 1345–1352. IEEE, 2011.
107. Ioannis Karamouzas, Brian Skinner, and Stephen J Guy. Universal power law governing pedestrian interactions. *Physical review letters*, 113(23):238701, 2014.
108. Matthias Seeger. Gaussian processes for machine learning. *International journal of neural systems*, 14(02):69–106, 2004.
109. Jack M Wang, David J Fleet, and Aaron Hertzmann. Gaussian process dynamical models for human motion. *IEEE transactions on pattern analysis and machine intelligence*, 30(2):283–298, 2007.
110. Peter Trautman and Andreas Krause. Unfreezing the robot: Navigation in dense, interacting crowds. In *2010 IEEE/RSJ International Conference on Intelligent Robots and Systems*, pages 797–803. IEEE, 2010.
111. Namhoon Lee and Kris M Kitani. Predicting wide receiver trajectories in american football. In *2016 IEEE Winter Conference on Applications of Computer Vision (WACV)*, pages 1–9. IEEE, 2016.
112. Federico Bartoli, Giuseppe Lisanti, Lamberto Ballan, and Alberto Del Bimbo. Context-aware trajectory prediction. In *2018 24th International Conference on Pattern Recognition (ICPR)*, pages 1941–1946. IEEE, 2018.
113. Li Sun, Zhi Yan, Sergi Molina Mellado, Marc Hanheide, and Tom Duckett. 3dof pedestrian trajectory prediction learned from long-term autonomous mobile robot deployment data. In *2018 IEEE International Conference on Robotics and Automation (ICRA)*, pages 5942–5948. IEEE, 2018.
114. Manh Huynh and Gita Alaghband. Trajectory prediction by coupling scene-lstm with human movement lstm. In *International Symposium on Visual Computing*, pages 244–259. Springer, 2019.
115. Amir Sadeghian, Vineet Kosaraju, Ali Sadeghian, Noriaki Hirose, Hamid Rezatofighi, and Silvio Savarese. Sophie: An attentive gan for predicting paths compliant to social and physical constraints. In *Proceedings of the IEEE/CVF conference on computer vision and pattern recognition*, pages 1349–1358, 2019.
116. Javad Amirian, Jean-Bernard Hayet, and Julien Pettré. Social ways: Learning multi-modal distributions of pedestrian trajectories with gans. In *Proceedings of the IEEE/CVF Conference on Computer Vision and Pattern Recognition Workshops*, pages 0–0, 2019.
117. Boris Ivanovic and Marco Pavone. The trajectron: Probabilistic multi-agent trajectory modeling with dynamic spatiotemporal graphs. In *Proceedings of the IEEE/CVF International Conference on Computer Vision*, pages 2375–2384, 2019.

118. Thomas N Kipf and Max Welling. Semi-supervised classification with graph convolutional networks. *arXiv preprint*arXiv:1609.02907, 2016.
119. Junwei Liang, Lu Jiang, and Alexander Hauptmann. Simaug: Learning robust representations from simulation for trajectory prediction. In *European Conference on Computer Vision*, pages 275–292. Springer, 2020.
120. Petar Veličković, Guillem Cucurull, Arantxa Casanova, Adriana Romero, Pietro Lio, and Yoshua Bengio. Graph attention networks. *arXiv preprint*arXiv:1710.10903, 2017.
121. Tim Salzmann, Boris Ivanovic, Punarjay Chakravarty, and Marco Pavone. Trajectron++: Dynamically-feasible trajectory forecasting with heterogeneous data. In *European Conference on Computer Vision*, pages 683–700. Springer, 2020.
122. Yuejiang Liu, Qi Yan, and Alexandre Alahi. Social nce: Contrastive learning of socially-aware motion representations. In *Proceedings of the IEEE/CVF International Conference on Computer Vision*, pages 15118–15129, 2021b.
123. Karttikeya Mangalam, Yang An, Harshayu Girase, and Jitendra Malik. From goals, waypoints & paths to long term human trajectory forecasting. In *Proceedings of the IEEE/CVF International Conference on Computer Vision*, pages 15233–15242, 2021.
124. Jiangbei Yue, Dinesh Manocha, and He Wang. Human trajectory prediction via neural social physics. In *European Conference on Computer Vision*, pages 376–394. Springer, 2022.
125. Peter Trautman, Jeremy Ma, Richard M Murray, and Andreas Krause. Robot navigation in dense human crowds: the case for cooperation. In *2013 IEEE international conference on robotics and automation*, pages 2153–2160. IEEE, 2013.
126. Xinjie Yao, Ji Zhang, and Jean Oh. Following social groups: Socially compliant autonomous navigation in dense crowds. *arXiv preprint*arXiv:1911.12063, 2019.
127. Jie Zhou and Zhong-Ke Shi. A new lattice hydrodynamic model for bidirectional pedestrian flow with the consideration of pedestrian's anticipation effect. *Nonlinear Dynamics*, 81(3):1247–1262, 2015.
128. Serge Hoogendoorn and Piet HL Bovy. Simulation of pedestrian flows by optimal control and differential games. *Optimal control applications and methods*, 24(3):153–172, 2003.
129. Xiaoping Zheng and Yuan Cheng. Conflict game in evacuation process: A study combining cellular automata model. *Physica A: Statistical Mechanics and its Applications*, 390(6):1042–1050, 2011.
130. S Bouzat and MN Kuperman. Game theory in models of pedestrian room evacuation. *Physical Review E*, 89(3):032806, 2014.
131. Qiancheng Xu, Mohcine Chraibi, and Armin Seyfried. Anticipation in a velocity-based model for pedestrian dynamics. *Transportation research part C: emerging technologies*, 133:103464, 2021.
132. Yushi Suma, Daichi Yanagisawa, and Katsuhiro Nishinari. Anticipation effect in pedestrian dynamics: Modeling and experiments. *Physica A: Statistical Mechanics and its Applications*, 391(1-2):248–263, 2012.
133. Stefan Nowak and Andreas Schadschneider. Quantitative analysis of pedestrian counterflow in a cellular automaton model. *Physical review E*, 85(6):066128, 2012.
134. Rafael Bailo, José A Carrillo, and Pierre Degond. Pedestrian models based on rational behaviour. In *Crowd Dynamics, Volume 1*, pages 259–292. Springer, 2018.
135. Hisashi Murakami, Claudio Feliciani, Yuta Nishiyama, and Katsuhiro Nishinari. Mutual anticipation can contribute to self-organization in human crowds. *Science Advances*, 7(12):eabe7758, 2021.
136. Hisashi Murakami, Claudio Feliciani, and Katsuhiro Nishinari. Lévy walk process in self-organization of pedestrian crowds. *Journal of the Royal Society Interface*, 16(153):20180939, 2019.
137. Robert M Roe, Jermone R Busemeyer, and James T Townsend. Multialternative decision field theory: A dynamic connectionst model of decision making. *Psychological review*, 108(2):370, 2001.
138. Francesco Zanlungo, Tetsushi Ikeda, and Takayuki Kanda. Social force model with explicit collision prediction. *EPL (Europhysics Letters)*, 93(6):68005, 2011.

139. Vineet Kosaraju, Amir Sadeghian, Roberto Martín-Martín, Ian Reid, Hamid Rezatofighi, and Silvio Savarese. Social-bigat: Multimodal trajectory forecasting using bicycle-gan and graph attention networks. *Advances in Neural Information Processing Systems*, 32, 2019.
140. Andrey Rudenko, Luigi Palmieri, Achim J Lilienthal, and Kai O Arras. Human motion prediction under social grouping constraints. In *2018 IEEE/RSJ International Conference on Intelligent Robots and Systems (IROS)*, pages 3358–3364. IEEE, 2018.
141. Jianhua Sun, Qinhong Jiang, and Cewu Lu. Recursive social behavior graph for trajectory prediction. In *Proceedings of the IEEE/CVF Conference on Computer Vision and Pattern Recognition*, pages 660–669, 2020.
142. Karttikeya Mangalam, Harshayu Girase, Shreyas Agarwal, Kuan-Hui Lee, Ehsan Adeli, Jitendra Malik, and Adrien Gaidon. It is not the journey but the destination: Endpoint conditioned trajectory prediction. In *European conference on computer vision*, pages 759–776. Springer, 2020.
143. Chen Zhou, Ming Han, Qi Liang, Yi-Fei Hu, and Shu-Guang Kuai. A social interaction field model accurately identifies static and dynamic social groupings. *Nature human behaviour*, 3(8):847–855, 2019a.
144. Mehdi Moussaïd, Niriaska Perozo, Simon Garnier, Dirk Helbing, and Guy Theraulaz. The walking behaviour of pedestrian social groups and its impact on crowd dynamics. *PloS one*, 5(4):e10047, 2010.
145. Hanne De Jaegher, Ezequiel Di Paolo, and Shaun Gallagher. Can social interaction constitute social cognition? *Trends in cognitive sciences*, 14(10):441–447, 2010.
146. Rui Zhou, Hongyu Zhou, Masayoshi Tomizuka, Jiachen Li, and Zhuo Xu. Grouptron: Dynamic multi-scale graph convolutional networks for group-aware dense crowd trajectory forecasting. *arXiv preprint*arXiv:2109.14128, 2021b.
147. Sergio Casas, Cole Gulino, Renjie Liao, and Raquel Urtasun. Spagnn: Spatially-aware graph neural networks for relational behavior forecasting from sensor data. In *2020 IEEE International Conference on Robotics and Automation (ICRA)*, pages 9491–9497. IEEE, 2020.
148. Harshayu Girase, Haiming Gang, Srikanth Malla, Jiachen Li, Akira Kanehara, Karttikeya Mangalam, and Chiho Choi. Loki: Long term and key intentions for trajectory prediction. In *Proceedings of the IEEE/CVF International Conference on Computer Vision*, pages 9803–9812, 2021.
149. Bosi Zhang, Wenyan Chen, Xian Ma, Ping Qiu, and Fupeng Liu. Experimental study on pedestrian behavior in a mixed crowd of individuals and groups. *Physica A: Statistical Mechanics and its Applications*, 556:124814, 2020.
150. Andrew C Gallup, Joseph J Hale, David JT Sumpter, Simon Garnier, Alex Kacelnik, John R Krebs, and Iain D Couzin. Visual attention and the acquisition of information in human crowds. *Proceedings of the National Academy of Sciences*, 109(19):7245–7250, 2012.
151. Pongsathorn Raksincharoensak, Takahiro Hasegawa, and Masao Nagai. Motion planning and control of autonomous driving intelligence system based on risk potential optimization framework. *International Journal of Automotive Engineering*, 7(AVEC14):53–60, 2016.
152. Alexandre Alahi, Vignesh Ramanathan, and Li Fei-Fei. Socially-aware large-scale crowd forecasting. In *Proceedings of the IEEE Conference on Computer Vision and Pattern Recognition*, pages 2203–2210, 2014.
153. Liushuai Shi, Le Wang, Chengjiang Long, Sanping Zhou, Mo Zhou, Zhenxing Niu, and Gang Hua. Sgcn: Sparse graph convolution network for pedestrian trajectory prediction. In *Proceedings of the IEEE/CVF Conference on Computer Vision and Pattern Recognition*, pages 8994–9003, 2021.
154. James Intriligator and Patrick Cavanagh. The spatial resolution of visual attention. *Cognitive psychology*, 43(3):171–216, 2001.
155. Parth Kothari, Sven Kreiss, and Alexandre Alahi. Human trajectory forecasting in crowds: A deep learning perspective. *IEEE Transactions on Intelligent Transportation Systems*, 2021.
156. Cunjun Yu, Xiao Ma, Jiawei Ren, Haiyu Zhao, and Shuai Yi. Spatio-temporal graph transformer networks for pedestrian trajectory prediction. In *European Conference on Computer Vision*, pages 507–523. Springer, 2020.

157. Jiezhong Qiu, Jian Tang, Hao Ma, Yuxiao Dong, Kuansan Wang, and Jie Tang. Deepinf: Social influence prediction with deep learning. In *Proceedings of the 24th ACM SIGKDD international conference on knowledge discovery & data mining*, pages 2110–2119, 2018.
158. Congcong Liu, Yuying Chen, Ming Liu, and Bertram E Shi. Avgcn: Trajectory prediction using graph convolutional networks guided by human attention. In *2021 IEEE International Conference on Robotics and Automation (ICRA)*, pages 14234–14240. IEEE, 2021c.
159. Irtiza Hasan, Francesco Setti, Theodore Tsesmelis, Alessio Del Bue, Marco Cristani, and Fabio Galasso. " seeing is believing": Pedestrian trajectory forecasting using visual frustum of attention. In *2018 IEEE Winter Conference on Applications of Computer Vision (WACV)*, pages 1178–1185. IEEE, 2018.
160. Renaud Bastien and Pawel Romanczuk. A model of collective behavior based purely on vision. *Science advances*, 6(6):eaay0792, 2020.
161. François A Lavergne, Hugo Wendehenne, Tobias Bäuerle, and Clemens Bechinger. Group formation and cohesion of active particles with visual perception–dependent motility. *Science*, 364(6435):70–74, 2019.
162. Faiza Gul, Wan Rahiman, and Syed Sahal Nazli Alhady. A comprehensive study for robot navigation techniques. *Cogent Engineering*, 6(1):1632046, 2019.
163. Thibault Kruse, Amit Kumar Pandey, Rachid Alami, and Alexandra Kirsch. Human-aware robot navigation: A survey. *Robotics and Autonomous Systems*, 61(12):1726–1743, 2013.
164. Angjoo Kanazawa, Michael J Black, David W Jacobs, and Jitendra Malik. End-to-end recovery of human shape and pose. In *proceedings IEEE/CVF International Conference on Computer Vision and Pattern Recognition*, pages 7122–7131, 2018.
165. Nikos Kolotouros, Georgios Pavlakos, Michael J Black, and Kostas Daniilidis. Learning to reconstruct 3d human pose and shape via model-fitting in the loop. In *proceedings IEEE/CVF International Conference on Computer Vision*, pages 2252–2261, 2019.
166. Hongwen Zhang, Yating Tian, Xinchi Zhou, Wanli Ouyang, Yebin Liu, Limin Wang, and Zhenan Sun. Pymaf: 3D human pose and shape regression with pyramidal mesh alignment feedback loop. In *proceedings IEEE/CVF International Conference on Computer Vision and Pattern Recognition*, pages 11446–11456, 2021a.
167. Hongsuk Choi, Gyeongsik Moon, JoonKyu Park, and Kyoung Mu Lee. Learning to estimate robust 3d human mesh from in-the-wild crowded scenes. In *proceedings IEEE/CVF International Conference on Computer Vision and Pattern Recognition*, 2022.
168. Wen Jiang, Nikos Kolotouros, Georgios Pavlakos, Xiaowei Zhou, and Kostas Daniilidis. Coherent reconstruction of multiple humans from a single image. In *proceedings IEEE/CVF International Conference on Computer Vision and Pattern Recognition*, pages 5579–5588, 2020.
169. Jianfeng Zhang, Dongdong Yu, Jun Hao Liew, Xuecheng Nie, and Jiashi Feng. Body meshes as points. In *proceedings IEEE/CVF International Conference on Computer Vision and Pattern Recognition*, pages 546–556, 2021b.
170. Yu Sun, Wu Liu, Qian Bao, Yili Fu, Tao Mei, and Michael J Black. Putting people in their place: Monocular regression of 3d people in depth. In *proceedings IEEE/CVF International Conference on Computer Vision and Pattern Recognition*, 2022.
171. Jianan Zhen, Qi Fang, Jiaming Sun, Wentao Liu, Wei Jiang, Hujun Bao, and Xiaowei Zhou. Smap: Single-shot multi-person absolute 3D pose estimation. In *proceedings European Conference on Computer Vision*, pages 550–566, 2020.
172. Gyeongsik Moon, Ju Yong Chang, and Kyoung Mu Lee. Camera distance-aware top-down approach for 3D multi-person pose estimation from a single RGB image. In *proceedings IEEE/CVF International Conference on Computer Vision*, pages 10133–10142, 2019.
173. Abdallah Benzine, Florian Chabot, Bertrand Luvison, Quoc Cuong Pham, and Catherine Achard. Pandanet: Anchor-based single-shot multi-person 3D pose estimation. In *proceedings IEEE/CVF International Conference on Computer Vision and Pattern Recognition*, pages 6856–6865, 2020.
174. Gregory Rogez, Philippe Weinzaepfel, and Cordelia Schmid. Lcr-net++: Multi-person 2D and 3D pose detection in natural images. *IEEE Transactions on Pattern Analysis and Machine Intelligence*, pages 1146–1161, 2019.

175. Can Wang, Jiefeng Li, Wentao Liu, Chen Qian, and Cewu Lu. Hmor: Hierarchical multi-person ordinal relations for monocular multi-person 3d pose estimation. In *proceedings European Conference on Computer Vision*, page 242–259, 2020b.
176. Dushyant Mehta, Oleksandr Sotnychenko, Franziska Mueller, Weipeng Xu, Mohamed Elgharib, Pascal Fua, Hans-Peter Seidel, Helge Rhodin, Gerard Pons-Moll, and Christian Theobalt. XNect: Real-time multi-person 3d motion capture with a single rgb camera. *ACM Transactions on Graphics*, 39(4):82–1, 2020.
177. Dushyant Mehta, Oleksandr Sotnychenko, Franziska Mueller, Weipeng Xu, Srinath Sridhar, Gerard Pons-Moll, and Christian Theobalt. Single-shot multi-person 3D pose estimation from monocular rgb. In *proceedings IEEE International Conference on 3D vision*, pages 120–130, 2018.
178. Matteo Fabbri, Fabio Lanzi, Simone Calderara, Stefano Alletto, and Rita Cucchiara. Compressed volumetric heatmaps for multi-person 3D pose estimation. In *proceedings IEEE/CVF International Conference on Computer Vision and Pattern Recognition*, pages 7204–7213, 2020.
179. Jogendra Nath Kundu, Ambareesh Revanur, Govind Vitthal Waghmare, Rahul Mysore Venkatesh, and R Venkatesh Babu. Unsupervised cross-modal alignment for multi-person 3D pose estimation. In *proceedings European Conference on Computer Vision*, pages 35–52, 2020.
180. Zhe Cao, Tomas Simon, Shih-En Wei, and Yaser Sheikh. Realtime multi-person 2D pose estimation using part affinity fields. In *proceedings IEEE/CVF International Conference on Computer Vision and Pattern Recognition*, pages 7291–7299, 2017.
181. Alex Krizhevsky, Ilya Sutskever, and Geoffrey E Hinton. Imagenet classification with deep convolutional neural networks. *Advances in neural information processing systems*, 25, 2012.
182. Matthew Loper, Naureen Mahmood, Javier Romero, Gerard Pons-Moll, and Michael J Black. Smpl: A skinned multi-person linear model. *ACM Transactions on Graphics*, 34(6):1–16, 2015.
183. Federica Bogo, Angjoo Kanazawa, Christoph Lassner, Peter Gehler, Javier Romero, and Michael J Black. Keep it SMPL: Automatic estimation of 3d human pose and shape from a single image. In *proceedings European Conference on Computer Vision*, pages 561–578, 2016.
184. Muhammed Kocabas, Nikos Athanasiou, and Michael J Black. VIBE: Video inference for human body pose and shape estimation. In *proceedings IEEE/CVF International Conference on Computer Vision and Pattern Recognition*, pages 5253–5263, 2020.
185. Muhammed Kocabas, Chun-Hao P Huang, Joachim Tesch, Lea Müller, Otmar Hilliges, and Michael J Black. SPEC: Seeing people in the wild with an estimated camera. In *proceedings IEEE/CVF International Conference on Computer Vision*, pages 11035–11045, 2021.
186. Gyeongsik Moon and Kyoung Mu Lee. Pose2pose: 3D positional pose-guided 3D rotational pose prediction for expressive 3D human pose and mesh estimation. *arXiv preprint*arXiv:2011.11534, 2020.
187. Georgios Pavlakos, Nikos Kolotouros, and Kostas Daniilidis. Texturepose: Supervising human mesh estimation with texture consistency. In *proceedings IEEE/CVF International Conference on Computer Vision*, pages 803–812, 2019.
188. Yu Sun, Yun Ye, Wu Liu, Wenpeng Gao, Yili Fu, and Tao Mei. Human mesh recovery from monocular images via a skeleton-disentangled representation. In *proceedings IEEE/CVF International Conference on Computer Vision*, pages 5349–5358, 2019b.
189. Wang Zeng, Wanli Ouyang, Ping Luo, Wentao Liu, and Xiaogang Wang. 3D human mesh regression with dense correspondence. In *proceedings IEEE/CVF International Conference on Computer Vision and Pattern Recognition*, pages 7054–7063, 2020.
190. Nicolas Ugrinovic, Adria Ruiz, Antonio Agudo, Alberto Sanfeliu, and Francesc Moreno-Noguer. Body size and depth disambiguation in multi-person reconstruction from single images. In *proceedings IEEE International Conference on 3D vision*, pages 53–63, 2021.
191. Yu Sun, Qian Bao, Wu Liu, Yili Fu, Michael J Black, and Tao Mei. Monocular, one-stage, regression of multiple 3D people. In *proceedings IEEE/CVF International Conference on Computer Vision*, pages 11179–11188, 2021.

192. Cristiane BR Ferreira, Helio Pedrini, Wanderley de Souza Alencar, William D Ferreira, Thyago Peres Carvalho, Naiane Sousa, and Fabrizzio Soares. Where's wally: A gigapixel image study for face recognition in crowds. In *International Symposium on Visual Computing*, pages 386–397. Springer, 2020.
193. Lingling Li, Xiaohui Guo, Yan Wang, Jingjing Ma, Licheng Jiao, Fang Liu, and Xu Liu. Region nms-based deep network for gigapixel level pedestrian detection with two-step cropping. *Neurocomputing*, 468:482–491, 2022.
194. Hao-Shu Fang, Shuqin Xie, Yu-Wing Tai, and Cewu Lu. RMPE: Regional multi-person pose estimation. In *proceedings IEEE/CVF International Conference on Computer Vision*, 2017.
195. Xiaohan Fei, Henry Wang, Lin Lee Cheong, Xiangyu Zeng, Meng Wang, and Joseph Tighe. Single view physical distance estimation using human pose. In *proceedings IEEE/CVF International Conference on Computer Vision*, pages 12406–12416, 2021.
196. Yi Zhou, Connelly Barnes, Jingwan Lu, Jimei Yang, and Hao Li. On the continuity of rotation representations in neural networks. In *proceedings IEEE/CVF International Conference on Computer Vision and Pattern Recognition*, pages 5745–5753, 2019b.
197. Mert Seker, Anssi Männistö, Alexandros Iosifidis, and Jenni Raitoharju. Automatic social distance estimation from images: Performance evaluation, test benchmark, and algorithm. *arXiv preprint*arXiv:2103.06759, 2021.
198. Martin A Fischler and Robert C Bolles. Random sample consensus: a paradigm for model fitting with applications to image analysis and automated cartography. *Communications of the ACM*, 24(6):381–395, 1981.
199. Vitor Albiero, Xingyu Chen, Xi Yin, Guan Pang, and Tal Hassner. img2pose: Face alignment and detection via 6dof, face pose estimation. In *proceedings IEEE/CVF International Conference on Computer Vision and Pattern Recognition*, pages 7617–7627, 2021.